Scheduling

HENRY LAURENCE GANTT
(1861–1919)

Henry Laurence Gantt was an industrial engineer and a disciple of Frederick W. Taylor. He developed his now famous charts during World War I to compare production schedules with their realizations. Gantt discussed the underlying principles in his paper "Efficiency and Democracy," which he presented at the annual meeting of the American Society of Mechanical Engineers in 1918. The Gantt charts currently in use are typically a simplification of the originals, both in purpose and in design.

Scheduling
Theory, Algorithms, and Systems
Second Edition

Michael Pinedo
New York University

Prentice Hall
Upper Saddle River, New Jersey 07458

Library of Congress Cataloging-in-Publication Data is on file.

Vice President and Editorial Director, ECS: *Marcia J. Horton*
Acquisitions Editor: *Laura Curless*
Editorial Assistant: *Erin Katchmar*
Vice President and Director of Production and Manufacturing, ESM: *David W. Riccardi*
Executive Managing Editor: *Vince O'Brien*
Managing Editor: *David A. George*
Production Editor: *Leslie Galen*
Director of Creative Services: *Paul Belfanti*
Creative Director: *Carole Anson*
Art Director: *Jayne Conte*
Art Editor: *Adam Velthaus*
Cover Designer: *Bruce Kenselaar*
Manufacturing Manager: *Trudy Pisciotti*
Marketing Manager: *Holly Stark*
Marketing Assistant: *Karen Moon*
Cover Image: *Courtesy of Framatome Technologies*

© 2002 by Prentice-Hall, Inc.
Upper Saddle River, New Jersey 07458

Printed in the United States of America

10 9 8 7 6 5 4 3 2 1

ISBN 0-13-028138-7

Prentice-Hall International (UK) Limited, *London*
Prentice-Hall of Australia Pty. Limited, *Sydney*
Prentice-Hall of Canada Inc., *Toronto*
Prentice-Hall Hispanoamericana, S.A., *Mexico*
Prentice-Hall of India Private Limited, *New Delhi*
Prentice-Hall of Japan, Inc., *Tokyo*
Pearson Education Asia Pte. Ltd., *Singapore*
Editora Prentice-Hall do Brasil, Ltda., *Rio de Janeiro*

Contents

Preface to the First Edition

Sequencing and scheduling is a form of decision-making that plays a crucial role in manufacturing and service industries. In the current competitive environment effective sequencing and scheduling has become a necessity for survival in the marketplace. Companies have to meet shipping dates that have been committed to customers, as failure to do so may result in a significant loss of goodwill. They also have to schedule activities in such a way as to use the resources available in an efficient manner.

Scheduling began to be taken seriously in manufacturing at the beginning of this century with the work of Henry Gantt and other pioneers. However, it took many years for the first scheduling publications to appear in the industrial engineering and operations research literature. Some of the first publications appeared in *Naval Research Logistics Quarterly* in the early 1950s and contained results by W.E. Smith, S.M. Johnson, and J.R. Jackson. During the 1960s a significant amount of work was done on dynamic programming and integer programming formulations of scheduling problems. After Richard Karp's famous paper on complexity theory, the research in the 1970s focused mainly on the complexity hierarchy of scheduling problems. In the 1980s several different directions were pursued in academia and industry with an increasing amount of attention paid to stochastic scheduling problems. Also, as personal computers started to permeate manufacturing facilities, scheduling systems were being developed for the generation of usable schedules in practice. This system design and development was, and is, being done by computer scientists, operations researchers and industrial engineers.

This book is the result of the development of courses in scheduling theory and applications at Columbia University. The book deals primarily with machine scheduling models. The first part covers deterministic models and the second part stochastic models. The third and final part deals with applications. In this last part scheduling problems in practice are discussed and the relevance of the theory to the real world is examined. From this examination it becomes clear that the advances in scheduling theory have had only a limited impact on scheduling problems in practice. Hopefully there will be, in a couple of years, a second edition in which the applications part will be expanded, showing a stronger connection with the more theoretical parts of the text.

This book has benefited from careful reading by numerous people. Reha Uzsoy and Alan Scheller Wolf went through the manuscript with a fine-tooth comb. Len Adler, Sid Browne, Xiuli Chao, Paul Glasserman, Chung-Yee Lee, Young-Hoon Lee, Joseph Leung, Elizabeth Leventhal, Rajesh Sah, Paul Shapiro, Jim Thompson, Barry Wolf, and the hundreds of students who had to take the (required) scheduling courses at Columbia provided many helpful comments that improved the manuscript.

The author is grateful to the National Science Foundation for its continued summer support, which made it possible to complete this project.

MICHAEL PINEDO
New York, 1994

Preface to the Second Edition

The book has been extended in a meaningful way. Five chapters have been added. In the deterministic section it is the treatment of the single machine, the job shop, and the open shop that have been expanded considerably. In the stochastic section a completely new chapter focuses on single machine scheduling with release dates. This chapter has been included because of multiple requests from instructors who wanted to see a connection between stochastic scheduling and priority queues. This chapter establishes such a link. Part III, the applications section, has been expanded the most. Instead of a single chapter on general-purpose procedures, there are now two chapters. The second chapter covers various techniques that are relatively new and that have started to receive a fair amount of attention over the last couple of years. There is also an additional chapter on the design and development of scheduling systems. This chapter focuses on rescheduling, learning mechanisms, and so on. The chapter with the examples of systems implementations is completely new. All systems described are of recent vintage. The last chapter contains a discussion on research topics that could become of interest in the next couple of years.

There is a companion website for this book:

http://www.stern.nyu.edu/~mpinedo

The intention is to keep the site as up-to-date as possible, including links to other sites that are potentially useful to instructors as well as students.

Many instructors who have used the book over the last couple of years have sent very useful comments and suggestions. Almost all of these comments have led to improvements in the manuscript.

Reha Uzsoy, as usual, went through the manuscript with a fine-tooth comb. Salah Elmaghraby, John Fowler, Celia Glass, Chung-Yee Lee, Sigrid Knust, Joseph Leung, Chris Potts, Steve Smith, Levent Tuncel, Amy Ward, Guochuan Zhang, Subhash Sarin, and Wilbert E. Wilhelm all made comments that led to substantial improvements.

A number of students, including Gabriel Adei, Yo Huh, Maher Lahmar, Sonia Leach, Michele Pfund, Edgar Possani, and Aysegul Toptal, have pointed out various errors in the original manuscript.

Without the help of a number of people from industry, it would not have been possible to produce a meaningful chapter on industrial implementations. Thanks are due to Heinrich Braun and Stephan Kreipl of SAP, Rama Akkiraju of IBM, Margie Bell of i2, Emanuela Rusconi and Fabio Tiozzo of Cybertec, and Paul Bender of SynQuest.

MICHAEL PINEDO
New York, 2001

1

Introduction

1.1 THE ROLE OF SCHEDULING

Scheduling deals with the allocation of scarce resources to tasks over time. It is a decision-making process with the goal of optimizing one or more objectives.

The resources and tasks in an organization can take many forms. The resources may be machines in a workshop, runways at an airport, crews at a construction site, processing units in a computing environment, and so on. The tasks may be operations in a production process, take-offs and landings at an airport, stages in a construction project, executions of computer programs, and so on. Each task may have a certain priority level, an earliest possible starting time, and a due date. The objectives can also take many forms. One objective may be the minimization of the completion time of the last task, and another may be the minimization of the number of tasks completed after their respective due dates.

1

Scheduling is a decision-making process that plays an important role in most manufacturing and production systems as well as in most information-processing environments. It also exists in transportation and distribution settings and in other types of service industries. The following three examples illustrate the role of the scheduling process in real-life situations.

Example 1.1.1 (A Paper Bag Factory)

Consider a factory that produces paper bags for cement, charcoal, dog food, and so on. The basic raw material for such an operation is rolls of paper. The production process consists of three stages: the printing of the logo, the gluing of the side of the bag, and the sewing of one end or both ends of the bag. At each stage, there are a number of machines that are not necessarily identical. The machines at a stage may differ slightly in the speed at which they can run, the number of colors they can print, or the size of bag they can handle. Each production order indicates a given quantity of a specific bag that has to be produced and shipped by a committed shipping date or due date. The processing times for the different operations are proportional to the size of the order (i.e., the number of bags ordered).

A late delivery implies a penalty in the form of loss of goodwill, and the magnitude of the penalty depends on the importance of the order or the client and the tardiness of the delivery. One of the objectives of the scheduling system is to minimize the sum of these penalties.

When a machine is switched over from one type of bag to another, a setup time is incurred. The length of the setup time on the machine depends on the similarities between the consecutive orders (the number of colors in common, the differences in bag size, etc.). Another objective of the scheduling system is to minimize the total time spent on setups.

Example 1.1.2 (Gate Assignments at an Airport)

Consider an airline terminal at a major airport. There are dozens of gates and hundreds of airplanes arriving and departing each day. The gates are not all identical and neither are the planes. Some of the gates are at locations with a lot of space where large planes (widebodies) can be accommodated easily. Other gates are in locations where it is difficult to bring in the planes. Certain planes may actually have to be towed to their gates.

Planes arrive and depart according to a certain schedule. However, the schedule is subject to a significant amount of randomness that may be weather related or due to events at other airports. During the time that a plane occupies a gate, the arriving passengers have to be deplaned, the plane has to be serviced, and the departing passengers have to be boarded. The scheduled departure time can be viewed as a due date, and the airline's performance is measured accordingly. However, if it is known in advance that the plane cannot land at the next airport because of anticipated congestion at the scheduled arrival time, then the plane does not take off (such a policy is followed to conserve fuel). If a plane is not allowed to take off, operating policies usually prescribe that passengers remain in the terminal rather than on the plane. If boarding is postponed, a plane may remain at a gate for an extended period of time, thus preventing other planes from using the gate.

The scheduler has to assign planes to gates in such a way that the assignment is physically feasible and optimizes a number of objectives. The scheduler has to assign planes to suitable gates that have to be available at the respective arrival times. The objectives include minimization of work for airline personnel and minimization of airplane delays.

In this scenario, the gates are the resources and the handling and servicing of the planes are the tasks. The arrival of a plane at a gate represents the starting time of a task, and the departure represents the completion time.

Example 1.1.3 (Scheduling Tasks in a Central Processing Unit [CPU])

One of the functions of a multitasking computer operating system is to schedule the time that the CPU devotes to the different programs that have to be executed. The exact processing times are usually not known in advance. However, the distribution of these random processing times may be known in advance, including their expected values and variances. In addition, each task usually has a certain priority level (the operating system typically allows operators and users to specify the priority level or weight of each task). In this case, the objective is to minimize the expected sum of the weighted completion times for all tasks.

To avoid the situation where relatively short tasks remain in the system for a long time waiting for much longer tasks with a higher priority, the operating system slices the tasks into little pieces. The operating system then rotates these slices on the CPU so that, in any given time interval, the CPU spends some amount of time on each task. This way, if by chance the processing time of one of the tasks is very short, the task will be able to leave the system relatively quickly.

An interruption of the processing of a task is often referred to as a *preemption*. It is clear that the optimal policy in such an environment makes heavy use of preemptions.

It may not be immediately clear what impact schedules have on given objectives. Does it make sense to invest time and effort searching for a good schedule rather than just choosing a schedule at random? Typically, the choice of schedule does have a major impact on system performance.

Scheduling can be difficult from a technical as well as an implementation point of view. The type of difficulties encountered on the technical side are similar to the difficulties encountered in other forms of combinatorial optimization and stochastic modeling. The difficulties on the implementation side are of a completely different kind. They may depend on the accuracy of the model used for the analysis of the actual scheduling problem and on the reliability of the input data required.

1.2 THE SCHEDULING FUNCTION IN AN ENTERPRISE

The scheduling function in a production system or service organization must interact with many other functions. These interactions are system-dependent and may differ substantially from one situation to another. They often take place within an enterprise-wide information system.

A modern factory or service organization often has an elaborate information system in place that includes a central computer and database. Local area networks of personal computers, workstations, and data entry terminals are connected to this central computer, which may be used either to retrieve data from the database or enter new data. The software controlling such an elaborate information system is typically referred to as an *enterprise resource planning* (ERP) system. A number of software companies specialize in the development of such systems, including SAP, J.D. Edwards, and PeopleSoft. Such an ERP system plays the role of an information highway that traverses the enterprise with, at all organizational levels, links to decision support systems.

Scheduling is often done interactively with a decision support system that is installed on a personal computer or workstation that is linked to the ERP system. Terminals at key locations linked to the ERP system can give departments throughout the enterprise access to all current scheduling information. These departments, in turn, can provide the scheduling system with up-to-date information concerning the statuses of jobs and machines.

There are, of course, also many other environments in which the communication between the scheduling function and other decision-making entities occur in meetings or through memos.

Scheduling in Manufacturing

Consider the following generic manufacturing environment and the role of its scheduling. Orders that are released in a manufacturing setting have to be translated into jobs with associated due dates. These jobs often have to be processed on the machines in a workcenter in a given order or sequence. The processing of jobs may sometimes be delayed if certain machines are busy, and preemptions may occur when high-priority jobs arrive at machines that are busy. Unexpected events on the shop floor, such as machine breakdowns or longer than expected processing times, also have to be taken into account because they may have a major impact on the schedules. Here, developing a detailed schedule of the tasks to be performed helps maintain efficiency and control of operations.

The shop floor is not the only part of the organization that impacts the scheduling process. It is also affected by the production planning process that handles medium- to long-term planning for the entire organization. This process attempts to optimize the firm's overall product mix and long-term resource allocation based on its inventory levels, demand forecasts, and resource requirements. Decisions made at this higher planning level may impact the scheduling process directly. Figure 1.1 depicts a diagram of the information flow in a manufacturing system.

In manufacturing, the scheduling function has to interact with other decision-making functions within the plant. One popular system that is widely used is the material requirements planning (MRP) system. After a schedule has been generated, it is necessary that all raw materials and resources are available at the specified times. The ready dates of all jobs have to be determined jointly by the production planning and scheduling system and the MRP system.

MRP systems are normally fairly elaborate. Each job has a bill of materials (BOM) itemizing the parts required for production. The MRP system keeps track of the inventory of each part. Furthermore, it determines the timing of the purchases of each one of the materials. In doing so, it uses techniques such as lot sizing

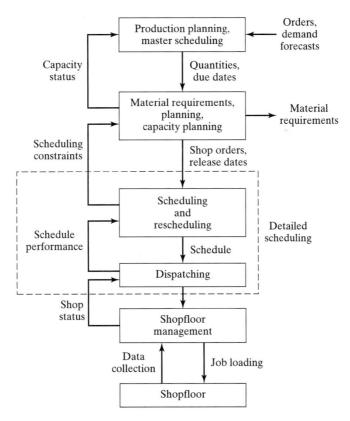

Figure 1.1   Information flow diagram in a manufacturing system.

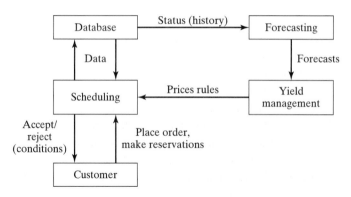

Figure 1.2   Information flow diagram in a service system.

and lot scheduling that are similar to those used in scheduling systems. There are many commercial MRP software packages available. As a result, there are many manufacturing facilities with MRP systems. In the cases where the facility does not have a scheduling system, the MRP system may be used for production planning purposes. However, in complex settings, it is not easy for an MRP system to do the detailed scheduling satisfactorily.

Scheduling in Services

To describe a generic service organization and a typical scheduling system is not easy. The scheduling function in a service organization may face a variety of different problems. It may have to deal with the reservation of resources (e.g., the assignment of planes to gates; see Example 1.1.2) or the reservation of meeting rooms or other facilities. The models used are at times somewhat different from those used in manufacturing settings. Scheduling in a service environment must be coordinated with other decision-making functions usually within elaborate information systems, much in the same way as the scheduling function in a manufacturing setting. These information systems usually rely on extensive databases that contain all the relevant information with regard to availability of resources and (potential) customers. The scheduling system often interacts with forecasting and yield management modules. Figure 1.2 depicts the information flow in a service organization such as a car rental agency. In contrast to manufacturing settings, there is usually no MRP system in a service environment.

1.3 OUTLINE OF THE BOOK

This book focuses on both the theory and applications of scheduling. The theoretical side deals with the detailed sequencing and scheduling of jobs. Given a collection of jobs to be processed in a certain machine environment, the problem is to sequence the jobs, subject to given constraints, in such a way that one or more performance criteria are optimized. The scheduler may have to deal with various forms of uncertainties, such as random job processing times, machines subject to breakdowns, rush orders, and so on.

Thousands of scheduling problems and models have been studied and analyzed in the past. Obviously, only a limited number are considered in this book; those discussed are selected based on the insight they provide, the methodology needed for their analysis, and their importance in applications.

Although the applications driving the models in this book come mainly from manufacturing and production environments, it is clear from the examples in Section 1.1 that scheduling plays a role in a wide variety of situations. The models and concepts dealt with in this book are also applicable in these other settings.

This book is divided into three parts. Part I (Chapters 2 to 8) deals with deterministic scheduling models. In these chapters, it is assumed that there are a finite number of jobs that have to be scheduled with one or more objectives to be minimized. Emphasis is placed on the analysis of relatively simple priority or dispatching rules. Chapter 2 discusses the notation and gives an overview of the models that are considered in the subsequent chapters. Chapters 3 to 8 consider the various machine environments. Chapters 3 and 4 deal with the single machine, Chapter 5 with machines in parallel, Chapter 6 with machines in series, and Chapter 7 with the more complicated job shop models. Chapter 8 focuses on open shops in which there are no restrictions on the routings of the jobs in the shop.

Part II (Chapters 9 to 13) deals with stochastic scheduling models. These chapters, in most cases, also assume that a given (finite) number of jobs have to be scheduled. The job data, such as processing times, release dates and due dates may not be exactly known in advance; only their distributions are known in advance. The actual processing times, release dates and due dates are known only after the *completion* of the processing or after the actual occurrence of the release or due date. In these models a single objective has to be minimized, usually in expectation. Again, emphasis is placed on the analysis of relatively simple priority or dispatching rules. Chapter 9 contains preliminary material. Chapter 10 covers the single machine environment. Chapter 11 also covers the single machine, but in this chapter it is assumed that the jobs are released at different points in time. This chapter establishes the relationship between stochastic scheduling and the theory of priority queues. Chapter 12 focuses on machines in parallel, and Chapter 13 describes the more complicated flow shop, job shop, and open shop models.

Part III (Chapters 14 to 20) deals with applications and implementations. Algorithms are described for a number of real-world scheduling problems. Design issues for scheduling systems are discussed and some examples of scheduling systems are given. Chapters 14 and 15 describe various general-purpose procedures that have proved to be useful in industrial scheduling systems. Chapter 16 describes a number of actual scheduling problems and how they have been solved in practice. Chapter 17 focuses on the basic issues concerning the design, development, and implementation of scheduling systems, and Chapter 18 discusses the more advanced concepts in the design and implementation of scheduling systems. Chapter 19 describes some examples of implementations. Chapter 20 ponders what lies ahead in scheduling.

Appendixes A, B, and C present short overviews of some of the basic methodologies: mathematical programming, dynamic programming and Markov decision processes, and complexity theory. Appendix D contains a complexity classification of the deterministic scheduling problems. Appendix E presents an overview of the stochastic scheduling problems. Appendix F lists a number of scheduling systems that have been developed in industry and academia.

This book is designed for either a senior or master's level course in Production Scheduling. When used for a senior level course, the topics most likely covered are from Parts I and III. Such a course can be given without getting into complexity theory: One can go through the chapters of Part I and skip all complexity proofs without loss of continuity. A master's level course may include topics from Part II. Even though all three parts are fairly self-contained, it is helpful to go through Chapter 2 before venturing into Part II.

Prerequisite knowledge for this book is an elementary course in Operations Research on the level of Hillier and Lieberman's *Introduction to Operations Research* and an elementary course in stochastic processes on the level of Ross's *Introduction to Probability Models*.

COMMENTS AND REFERENCES

Over the last four decades, many books have appeared that focus on sequencing and scheduling. These books range from the elementary to the more advanced.

A volume edited by Muth and Thompson (1963) contains a collection of papers focusing primarily on computational aspects of scheduling. One of the better known textbooks is by Conway, Maxwell, and Miller (1967), which, even though slightly out of date, is still very interesting; this book also deals with some of the stochastic aspects and with priority queues. A more recent text by Baker (1974) gives an excellent overview of the many aspects of deterministic scheduling. However, this book does not deal with computational complexity issues since it appeared just before research in computational complexity started to become popular. The book by Coffman (1976) is a compendium of papers on deterministic scheduling; it does cover computational complexity. An introductory textbook by French (1982) covers most of the techniques that are used in deterministic scheduling. The proceedings of a NATO workshop, edited by Dempster, Lenstra, and Rinnooy Kan (1982), contains a number of advanced papers on deterministic as well as stochastic scheduling. The relatively advanced book by Blazewicz, Cellary, Slowinski, and Weglarz (1986) focuses mainly on resource constraints and multi-objective deterministic scheduling. The book by Blazewicz, Ecker, Schmidt, and Weglarz (1993) is somewhat advanced and deals primarily with the computational aspects of deterministic scheduling models and their applications to manufacturing. The more applied text by Morton and Pentico (1993) presents a detailed analysis of a large number of scheduling heuristics that are useful for practitioners.

The monograph by Dauzère-Pérès and Lasserre (1994) focuses primarily on job shop scheduling. A collection of papers, edited by Zweben and Fox (1994), describes a number of scheduling systems and their actual implementation. Another collection of papers, edited by Brown and Scherer (1995), also describe various scheduling systems and their implementation. The proceedings of a workshop edited by Chrétienne, Coffman, Lenstra, and Liu (1995) contain a set of interesting papers concerning primarily deterministic scheduling. The textbook by Baker (1995) is very useful for an introductory course in sequencing and scheduling. Brucker (1995) presents a very detailed algorithmic analysis of the many deterministic scheduling models. Parker (1995) gives a similar overview and tends to focus on problems with precedence constraints or other graph-theoretic issues. Sule (1996) is a more applied text with a discussion of some interesting real-world problems. Blazewicz, Ecker, Pesch, Schmidt, and Weglarz (1996) is an extended edition of the earlier work by Blazewicz, Ecker, Schmidt, and Weglarz (1993). The monograph by Ovacik and Uzsoy (1997) is entirely dedicated to decomposition methods for complex job shops. The two volumes edited by Lee and Lei (1997) contain many interesting theoretical as well as applied papers. The book by Pinedo and Chao (1999) is more application oriented and describes a number of different scheduling models for problems arising in manufacturing as well as services.

Besides these books, a number of survey articles have appeared, each one with a large number of references. The articles by Graves (1981) and Rodammer and White (1988) review production scheduling. Atabakhsh (1991) presents a survey of constraint-based scheduling systems that use artificial intelligence techniques, and Noronha and Sarma (1991) review knowledge-based approaches for scheduling problems. Smith (1992) focuses in his survey on the development and implementation of scheduling systems. Lawler, Lenstra, Rinnooy Kan, and Shmoys (1993) give a detailed overview of deterministic sequencing and scheduling, and Righter (1994) does the same for stochastic scheduling. Queyranne and Schulz (1994) provide an in-depth analysis of polyhedral approaches to nonpreemptive machine scheduling problems. Chen, Potts, and Woeginger (1998) review computational complexity, algorithms, and approximability in deterministic scheduling. Sgall (1998) presents a survey of a relatively new area in deterministic scheduling called online scheduling. Even though online scheduling is considered part of deterministic scheduling, the theorems involved can often lead to interesting new insights into stochastic scheduling models.

PART 1
Deterministic Models

2

Deterministic Models: Preliminaries

In the past four decades, a considerable amount of theoretical research has been done in the field of deterministic scheduling. The number and variety of different models is astounding. During this time, a notation has evolved that succinctly captures the structure of many (but not all) deterministic models considered in the literature.

The first section in this chapter presents an adapted version of this notation. The second section contains a number of examples and describes some of the shortcomings of the framework and notation. The third section describes several classes of schedules. A class of schedules is typically characterized by the freedom the scheduler has in the decision-making process. The last section discusses the complexity of the scheduling problems introduced in the first section. This last section can be used, in conjunction with Appendixes C and D, to classify scheduling problems according to their complexity.

2.1 FRAMEWORK AND NOTATION

In all the scheduling problems considered, the number of jobs and machines are assumed to be finite. The number of jobs is denoted by n and the number of machines by m. Usually, the subscript j refers to a job, whereas the subscript i refers to a machine. If a job requires a number of processing steps or operations, then the pair (i, j) refers to the processing step or operation of job j on machine i. The following pieces of data are associated with job j:

Processing time (p_{ij}).   The p_{ij} represents the processing time of job j on machine i. The subscript i is omitted if the processing time of job j does not depend on the machine or if job j is only to be processed on one given machine.

Release date (r_j).   The release date r_j of job j may also be referred to as the *ready date*. It is the time the job arrives at the system (i.e., the earliest time at which job j can start its processing).

Due date (d_j).   The due date d_j of job j represents the committed shipping or completion date (the date the job is promised to the customer). Completion of a job after its due date *is* allowed, but then a penalty is incurred. When a due date *must* be met, it is referred to as a *deadline* and denoted by $\bar{d}_j$.

Weight (w_j).   The weight w_j of job j is basically a priority factor, denoting the importance of job j relative to the other jobs in the system. For example, this weight may represent the actual cost of keeping the job in the system. This cost could be a holding or inventory cost; it also could represent the amount of value already added to the job.

A scheduling problem is described by a triplet $\alpha \mid \beta \mid \gamma$. The α field describes the machine environment and contains a single entry. The β field provides details of processing characteristics and constraints and may contain no entry at all or multiple entries. The γ field describes the objective to be minimized and usually contains a single entry.

The possible machine environments specified in the α field are:

Single machine (1).   The case of a single machine is the simplest of all possible machine environments and is a special case of all other more complicated machine environments.

Identical machines in parallel (Pm).   There are m identical machines in parallel. Job j requires a single operation and may be processed on any one of the m machines or on any one that belongs to a given subset. If job j is not allowed to be processed on just any machine, but rather only on any one belonging to a specific subset M_j, then the entry M_j appears in the β field.

Machines in parallel with different speeds (Qm).   There are m machines in parallel with different speeds. The speed of machine i is denoted by v_i. The time

p_{ij} that job j spends on machine i is equal to p_j/v_i (assuming job j receives all its processing from machine i). This environment is referred to as *uniform* machines. If all machines have the same speed (i.e., $v_i = 1$ for all i and $p_{ij} = p_j$), then the environment is identical to the previous one.

Unrelated machines in parallel (Rm).   This environment is a generalization of the previous one. There are m different machines in parallel. Machine i can process job j at speed v_{ij}. The time p_{ij} that job j spends on machine i is equal to p_j/v_{ij} (again assuming job j receives all its processing from machine i). If the speeds of the machines are independent of the jobs (i.e., $v_{ij} = v_i$ for all i and j), then the environment is identical to the previous one.

Flow shop (Fm).   There are m machines in series. Each job has to be processed on each one of the m machines. All jobs have to follow the same route (i.e., they have to be processed first on machine 1, then machine 2, etc). After completion on one machine, a job joins the queue at the next machine. Usually, all queues are assumed to operate under the *First In First Out (FIFO)* discipline—that is, a job cannot "pass" another while waiting in a queue. If the *FIFO* discipline is in effect, the flow shop is referred to as a *permutation* flow shop and the β field includes the entry *prmu*.

Flexible flow shop (FFc).   A flexible flow shop is a generalization of the flow shop and the parallel machine environments. Instead of m machines in series, there are c stages in series with a number of identical machines in parallel at each stage. Each job has to be processed first at Stage 1, then Stage 2, and so on. A stage functions as a bank of parallel machines; at each stage, job j requires processing on only one machine and any machine can do. The queue at a stage may or may not operate according to the *First Come First Served (FCFS)* discipline.

Job shop (Jm).   In a job shop with m machines, each job has its own predetermined route to follow. A distinction is made between job shops in which each job visits each machine at most once and job shops in which a job may visit each machine more than once. In the latter case, the β-field contains the entry *recrc* for *recirculation*.

Flexible job shop (FJc).   A flexible job shop is a generalization of the job shop and the parallel machine environments. Instead of m machines, there are c work centers with a number of identical machines in parallel at each work center. Each job has its own route to follow through the shop; job j requires processing at each work center on only one machine and any machine can do. If a job on its route through the shop may visit a work center more than once, then the β-field contains the entry *recrc* for recirculation.

Open shop (*Om*).   There are m machines. Each job has to be processed again on each one of the m machines. However, some of these processing times may be zero. There are no restrictions with regard to the routing of each job through the machine environment. The scheduler is allowed to determine a route for each job, and different jobs may have different routes.

The processing restrictions and constraints specified in the β field may include multiple entries. Possible entries in the β field are:

Release dates (r_j).   If this symbol is present in the β field, job j may not start its processing before its release date r_j. If r_j does not appear in the β field, the processing of job j may start at any time. In contrast to the release dates, due dates are not specified in this field. The type of objective function gives sufficient indication whether there are due dates.

Sequence dependent setup times (s_{jk}).   The s_{jk} represents the sequence dependent setup time between jobs j and k; s_{0k} denotes the setup time for job k if job k is first in the sequence and s_{j0} the cleanup time after job j if job j is last in the sequence (of course, s_{0k} and s_{j0} may be zero). If the setup time between jobs j and k depends on the machine, then the subscript i is included (i.e., s_{ijk}). If no s_{jk} appears in the β field, all setup times are assumed to be 0 or sequence independent, in which case they are simply included in the processing times.

Preemptions (*prmp*).   Preemptions imply that it is not necessary to keep a job on a machine, once started, until completion. The scheduler is allowed to interrupt the processing of a job (preempt) at any point in time and put a different job on the machine instead. The amount of processing a preempted job already has received is not lost. When a preempted job is afterward put back on the machine (or on another machine in the case of parallel machines), it only needs the machine for its *remaining* processing time. When preemptions are allowed, *prmp* is included in the β field; when *prmp* is not included, preemptions are not allowed.

Precedence constraints (*prec*).   Precedence constraints may appear in a single machine or in a parallel machine environment, requiring that one or more jobs may have to be completed before another job is allowed to start its processing. There are several special forms of precedence constraints: If each job has at most one predecessor and at most one successor, the constraints are referred to as *chains*. If each job has at most one successor, the constraints are referred to as an *intree*. If each job has at most one predecessor, the constraints are referred to as an *outtree*. If no *prec* appears in the β field, the jobs are not subject to precedence constraints.

Breakdowns (*brkdwn*).   Machine breakdowns imply that machines are not continuously available. The periods that a machine is not available are, in this part of the book, assumed to be fixed (e.g., due to shifts or scheduled maintenance).

If there are a number of identical machines in parallel, the number of machines available at any point in time is a function of time (i.e., $m(t)$).

Machine eligibility restrictions (M_j).   The M_j symbol may appear in the β field when the machine environment is m machines in parallel (Pm). When the M_j is present, not all m machines are capable of processing job j. The set M_j denotes the set of machines that can process job j. If the β field does not contain M_j, job j may be processed on any one of the m machines.

Permutation *(prmu)*.   A constraint that may appear in the flow shop environment is that the queues in front of each machine operate according to the *First In First Out (FIFO)* discipline. This implies that the order (or *permutation*) in which the jobs go through the first machine is maintained throughout the system.

Blocking *(block)*.   Blocking is a phenomenon that may occur in flow shops. If a flow shop has a limited buffer in between two successive machines, then it may happen that when the buffer is full the upstream machine is not allowed to release a completed job. This phenomenon is known as *blocking*: The completed job has to remain on the upstream machine preventing or blocking that machine from working on another job. The most common occurrence of blocking considered in this book is the case with zero buffers in between any two successive machines. In this case, a job that has completed its processing on a given machine cannot leave the machine if the preceding job has not completed its processing yet on the next machine. Thus, the blocked job also prevents or blocks the next job from starting its processing on the given machine. In the models with blocking considered in the subsequent chapters, the assumption is made that the machines operate according to *FIFO*. That is, *block* implies *prmu*.

No-wait *(nwt)*.   The *no-wait* requirement is another phenomenon that may occur in flow shops. Jobs are not allowed to wait between two successive machines. This implies that the starting time of a job at the first machine has to be delayed to ensure that the job can go through the flow shop without having to wait for any machine. An example of such an operation is a steel rolling mill in which a slab of steel is not allowed to wait as it would cool off during a wait. It is clear that under *no-wait* the machines also operate under the *FIFO* discipline.

Recirculation *(recrc)*.   Recirculation may occur in a job shop or flexible job shop when a job may visit a machine or work center more than once.

Any other entry that may appear in the β field is self-explanatory. For example, $p_j = p$ implies that all processing times are equal, and $d_j = d$ implies that all due dates are equal. As stated before, due dates, in contrast to release dates, are usually not explicitly specified in this field; the type of objective function gives sufficient indication whether the jobs have due dates.

The objective to be minimized is always a function of the completion times of the jobs, which, of course, depend on the schedule. The completion time of the operation of job j on machine i is denoted by C_{ij}. The time job j exits the system (i.e., its completion time on the last machine on which it requires processing) is denoted by C_j. The objective may also be a function of the due dates. The *lateness* of job j is defined as

$$L_j = C_j - d_j,$$

which is positive when job j is completed late and negative when it is completed early. The *tardiness* of job j is defined as

$$T_j = \max(C_j - d_j, 0) = \max(L_j, 0).$$

The difference between the tardiness and the lateness lies in the fact that the tardiness never is negative. The *unit penalty* of job j is defined as

$$U_j = \begin{cases} 1 & \text{if } C_j > d_j \\ 0 & \text{otherwise} \end{cases}.$$

The lateness, tardiness, and unit penalty are the three basic due date-related penalty functions considered in this book. The shape of these functions are depicted in Figure 2.1.

Examples of possible objective functions to be minimized are:

Makespan $(C_{\max})$.   The makespan, defined as $\max(C_1, \dots, C_n)$, is equivalent to the completion time of the last job to leave the system. A minimum makespan usually implies a high utilization of the machine(s).

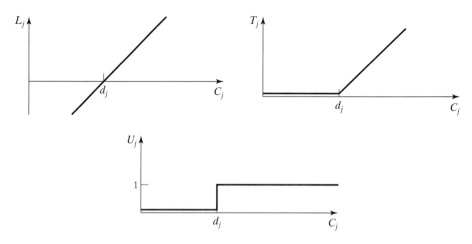

Figure 2.1   Due date-related penalty functions.

Maximum Lateness $(L_{\max})$.   The maximum lateness, $L_{\max}$, is defined as $\max(L_1, \ldots, L_n)$. It measures the worst violation of the due dates.

Total weighted completion time $(\sum w_j C_j)$.   The sum of the weighted completion times of the n jobs gives an indication of the total holding or inventory costs incurred by the schedule. The sum of the completion times is often referred to in the literature as the *flow time*. The total weighted completion time is then referred to as the *weighted flow time*.

Discounted total weighted completion time $(\sum w_j(1 - e^{-rC_j}))$.   This is a more general cost function than the previous one; the costs are now discounted at a rate of r, $0 < r < 1$, per unit time. That is, if job j is not completed by time t, an additional cost $w_j r e^{-rt} dt$ is incurred over the period $[t, t + dt]$. If job j is completed at time t, the total cost incurred over the period $[0, t]$ is $w_j(1 - e^{-rt})$. The value of r is usually close to 0, say 0.1 or 10%.

Total weighted tardiness $(\sum w_j T_j)$.   This is also a more general cost function than the total weighted completion time.

Weighted number of tardy jobs $(\sum w_j U_j)$.   The weighted number of tardy jobs is not only a measure of academic interest, but is often an objective in practice as it is a measure that can be recorded very easily.

All the objective functions listed are so-called *regular* performance measures. Regular performance measures are functions that are *nondecreasing* in $C_1, \ldots, C_n$. Recently, researchers have begun to study objective functions that are not regular. For example, when job j has a due date d_j, it may be subject to an earliness penalty, where the *earliness* of job j is defined as

$$E_j = \max(d_j - C_j, 0).$$

This earliness penalty is *nonincreasing* in C_j. An objective such as the total earliness plus the total tardiness—that is,

$$\sum_{j=1}^{n} E_j + \sum_{j=1}^{n} T_j,$$

is therefore not regular. A more general objective that is not regular is the total weighted earliness plus the total weighted tardiness—that is,

$$\sum_{j=1}^{n} w'_j E_j + \sum_{j=1}^{n} w''_j T_j.$$

The weight associated with the earliness of job j (w'_j) may be different from the weight associated with the tardiness of job j (w''_j).

2.2 EXAMPLES

The following examples illustrate the notation:

Example 2.2.1 (A Flexible Flow Shop)

$FFc \mid r_j \mid \sum w_j T_j$ denotes a flexible flow shop. The jobs have release dates and due dates, and the objective is the minimization of the total weighted tardiness. Example 1.1.1 in Section 1.1 (the paper bag factory) can be modeled as such. Actually, the problem described in Section 1.1 has some additional characteristics including sequence dependent setup times at each of the three stages. In addition, the processing time of job j on machine i has a special structure: It depends on the number of bags and the speed of the machine.

Example 2.2.2 (A Parallel Machine Environment)

$Pm \mid r_j, M_j \mid \sum w_j T_j$ denotes a system with m machines in parallel. Job j arrives at release date r_j and has to leave by the due date d_j. Job j may be processed only on one of the machines belonging to the subset M_j. If job j is not completed in time, a penalty $w_j T_j$ is incurred. This model can be used for the gate assignment problem in Example 1.1.2.

Example 2.2.3 (A Single Machine Environment)

$1 \mid r_j, prmp \mid \sum w_j C_j$ denotes a single machine system with job j entering the system at its release date r_j. Preemptions are allowed. The objective to be minimized is the sum of the weighted completion times. This model can be used to study the deterministic counterpart of the problem described in Example 1.1.3.

Example 2.2.4 (Sequence Dependent Setup Times)

$1 \mid s_{jk} \mid C_{\max}$ denotes a single machine system with n jobs subject to sequence dependent setup times, where the objective is to minimize the makespan. It is well known that this problem is equivalent to the so-called *Traveling Salesman Problem (TSP)*, where a salesman has to tour n cities in such a way that the total distance traveled is minimized (see Appendix C for a formal definition of the TSP).

Example 2.2.5 (A Project)

$P\infty \mid prec \mid C_{\max}$ denotes a scheduling problem with n jobs subject to precedence constraints and an unlimited number of machines (or resources) in parallel. The total time of the entire project has to be minimized. This type of problem is very common in project planning in the construction industry and has led to techniques such as the *Critical Path Method (CPM)* and the *Project Evaluation and Review Technique (PERT)*.

Example 2.2.6 (A Flow Shop)

$Fm \mid p_{ij} = p_j \mid \sum w_j C_j$ denotes a *proportionate* flow shop environment with m machines (i.e., m machines in series), with the processing times of job j on all m

machines identical and equal to p_j (hence the term *proportionate*). The objective is to find the order in which the n jobs go through the system so that the sum of the weighted completion times is minimized.

Example 2.2.7 (A Job Shop)

$Jm \parallel C_{max}$ denotes a job shop problem with m machines. There is no recirculation, so a job visits each machine at most once. The objective is to minimize the makespan. This problem is a famous problem in the scheduling literature and has received an enormous amount of attention.

Example 2.2.8 (A Flexible Job Shop)

$FJc \mid r_j, s_{ijk} \mid \sum w_j T_j$ refers to a flexible job shop with c work centers. The jobs have different release dates and are subject to sequence dependent setup times that are machine dependent. There is no recirculation, so a job visits each work center at most once. The objective is to minimize the total weighted tardiness. This particular problem has many applications in the real world, primarily in manufacturing.

Of course, there are many scheduling models that are not captured by this framework. One can define, for example, a more general flexible job shop in which each work center consists of a number of unrelated machines in parallel. When a job on its route through the system arrives at a bank of unrelated machines, it may be processed on any one of the machines, but its processing time now depends on the machine on which it is processed.

One can also define a model that is a mixture of a job shop and an open shop. The routes of some jobs are fixed, whereas the routes of other jobs are (partially) open.

The framework described in Section 2.1 has been designed primarily for models with a single objective. Most research in the past has concentrated on models with a single objective. Recently, researchers have begun studying models with multiple objectives, but a standard notation for such models has not yet been developed.

Various other scheduling features, which are not mentioned here, have been studied and analyzed in the literature. Such features include periodic or cyclic scheduling, personnel scheduling, resource-constrained scheduling, and batch processing.

2.3 CLASSES OF SCHEDULES

In scheduling terminology, a distinction is often made among a *sequence*, a *schedule*, and a *scheduling policy*. A sequence usually corresponds to a permutation of the n jobs or the order in which jobs are to be processed on a given machine. A schedule usually refers to an allocation of jobs within a more complicated setting of machines, allowing possibly for preemptions of jobs by other jobs that are released at later points in time. The concept of a scheduling policy is often used in stochastic settings: A policy prescribes an appropriate action for any one of the states the system may be in. In deterministic models, usually only sequences or schedules are of importance.

Assumptions have to be made with regard to what the scheduler is and is not allowed to do when he generates a schedule. For example, it may be that a schedule is not allowed to have any *unforced idleness* on any machine. This class of schedules is defined as follows.

Definition 2.3.1. (Nondelay Schedule) *A feasible schedule is called nondelay if no machine is kept idle while an operation is waiting for processing.*

Requiring a schedule to be nondelay is equivalent to prohibiting *unforced idleness*. For many models, including all those that allow preemptions and have regular objective functions, there are optimal schedules that are nondelay. For many models considered in this part of the book, the goal is to find an optimal nondelay schedule. However, there *are* models where it may be advantageous to have periods of unforced idleness.

A smaller class of schedules within the class of all nondelay schedules is the class of nonpreemptive, nondelay schedules. Nonpreemptive, nondelay schedules may lead to some interesting and unexpected anomalies.

Example 2.3.2 (A Scheduling Anomaly)

Consider an instance of $P2 \mid prec \mid C_{\max}$ with 10 jobs and the following processing times.

jobs	1	2	3	4	5	6	7	8	9	10
p_j	8	7	7	2	3	2	2	8	8	15

The jobs are subject to the precedence constraints depicted in Figure 2.2. The makespan of the nondelay schedule depicted in Figure 2.3a is 31 and the schedule is clearly optimal.

One would expect that, if each one of the 10 processing times is reduced by one time unit, the makespan would be less than 31. However, requiring the schedule to be nondelay results in the schedule depicted in Figure 2.3b, with a makespan of 32.

Suppose that now an additional machine is made available and that there are three machines instead of two. One would again expect that the makespan with the original processing times is less than 31. Again, the nondelay requirement has an unexpected effect. The makespan is 36.

Some heuristic procedures and algorithms for job shops are based on the construction of nonpreemptive schedules with certain special properties. Two classes of

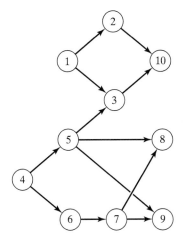

Figure 2.2   Precedence constraints graph for Example 2.3.2.

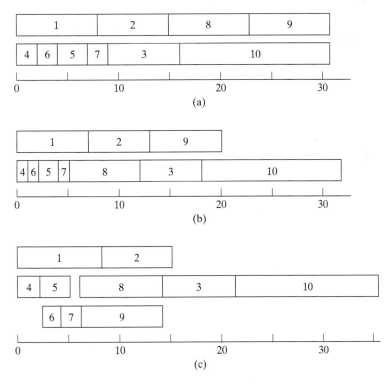

Figure 2.3   Gantt charts of nondelay schedules: (a) original schedules, (b) processing times one unit less, and (c) original processing times and three machines.

nonpreemptive schedules are of importance for certain algorithmic procedures for job shops.

Definition 2.3.3. (Active Schedule)   *A feasible schedule is called active if it is not possible to construct another schedule by changing the order of processing on the machines and having at least one operation finishing earlier and no operation finishing later.*

In other words, a schedule is active if no operation can be put into an empty hole earlier in the schedule while preserving feasibility. A nonpreemptive, nondelay schedule has to be active, but the reverse is not necessarily true. The following example describes a schedule that is active but not nondelay.

Example 2.3.4 (An Active Schedule)

Consider a job shop with three machines and two jobs. Job 1 needs one time unit on machine 1 and three time units on machine 2. Job 2 needs two time units on machine 3 and three time units on machine 2. Both jobs have to be processed last on machine 2. Consider the schedule that processes job 2 on machine 2 before job 1 (see Figure 2.4). It is clear that this schedule is active; reversing the sequence of the two jobs on machine 2 postpones the processing of job 2. However, the schedule is *not* nondelay. Machine 2 remains idle until time 2 while there is already a job available for processing at time 1.

It can be shown that there exists for $Jm \parallel \gamma$ an optimal schedule that is active provided the objective function γ is regular.

An even larger class of nonpreemptive schedules can be defined as follows.

Definition 2.3.5. (Semi-Active Schedule)   *A feasible schedule is called semi-active if no operation can be completed earlier without changing the order of processing on any one of the machines.*

It is clear that an active schedule has to be semi-active. However, the reverse is not necessarily true.

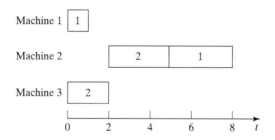

Figure 2.4   An active schedule that is not nondelay.

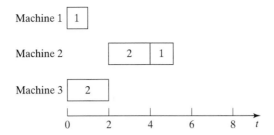

Figure 2.5   A semi-active schedule that is not active.

Example 2.3.6 (A Semi-Active Schedule)

Consider again a job shop with three machines and two jobs. The routing of the two jobs is the same as in the previous example. The processing times of job 1 on machines 1 and 2 are both equal to 1. The processing times of job 2 on machines 2 and 3 are both equal to 2. Consider the schedule under which job 2 is processed on machine 2 before job 1 (see Figure 2.5). This implies that job 2 starts its processing on machine 2 at time 2 and job 1 starts its processing on machine 2 at time 4. This schedule is semi-active. However, it is not active, as job 1 can be processed on machine 2 without delaying the processing of job 2 on machine 2.

An example of a schedule that is not even semi-active can be constructed easily. Postpone the start of the processing of job 1 on machine 2 for one time unit (i.e., machine 2 is kept idle for one unit of time between the processing of jobs 2 and 1). Clearly, this schedule is not even semi-active.

Figure 2.6 shows a Venn diagram of the three classes of nonpreemptive schedules: the nonpreemptive, nondelay schedules, the active schedules, and the semi-active schedules.

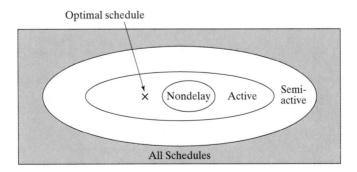

Figure 2.6   Venn diagram of classes of schedules for job shops.

2.4 COMPLEXITY HIERARCHY

Often an algorithm for one scheduling problem can be applied to another scheduling problem as well. For example, $1 \mid\mid \sum C_j$ is a special case of $1 \mid\mid \sum w_j C_j$, and a procedure for $1 \mid\mid \sum w_j C_j$ can, of course, also be used for $1 \mid\mid \sum C_j$. In complexity terminology, it is then said that $1 \mid\mid \sum C_j$ *reduces* to $1 \mid\mid \sum w_j C_j$. This is usually denoted by

$$1 \mid\mid \sum C_j \propto 1 \mid\mid \sum w_j C_j.$$

Based on this concept, a chain of reductions can be established. For example,

$$1 \mid\mid \sum C_j \propto 1 \mid\mid \sum w_j C_j \propto Pm \mid\mid \sum w_j C_j \propto Qm \mid prec \mid \sum w_j C_j.$$

Of course, there are also many problems that are not comparable with one another. For example, $Pm \mid\mid \sum w_j T_j$ is not comparable to $Jm \mid\mid C_{\max}$.

A considerable effort has been made to establish a problem hierarchy describing the relationships among the hundreds of scheduling problems. In the comparisons among the complexities of the different scheduling problems, it is of interest to know how a change in a single element in the classification of a problem affects its complexity. In Figure 2.7, a number of graphs are depicted that help determine the complexity hierarchy of deterministic scheduling problems. Most of the hierarchy depicted in these graphs is relatively straightforward. Two of the relationships may need an explanation—namely,

$$\alpha \mid \beta \mid L_{\max} \propto \alpha \mid \beta \mid \sum U_j$$

and

$$\alpha \mid \beta \mid L_{\max} \propto \alpha \mid \beta \mid \sum T_j.$$

However, it can be shown that a procedure for $\alpha \mid \beta \mid \sum U_j$ as well as a procedure for $\alpha \mid \beta \mid \sum T_j$ can be applied to $\alpha \mid \beta \mid L_{\max}$ with only minor modifications (see Exercise 2.23).

A significant amount of research in deterministic scheduling has been devoted to finding efficient, so-called polynomial time, algorithms for scheduling problems. However, many scheduling problems do not have a polynomial time algorithm; these problems are the so-called *NP-hard* problems. Verifying that a problem is NP-hard requires a formal mathematical proof (see Appendix C).

Research in the past has focused in particular on the borderline between polynomial time solvable problems and NP-hard problems. For example, in the string of problems described earlier, $1 \mid\mid \sum w_j C_j$ can be solved in polynomial time, whereas $Pm \mid\mid \sum w_j C_j$ is NP-hard, which implies that $Qm \mid prec \mid \sum w_j C_j$ is

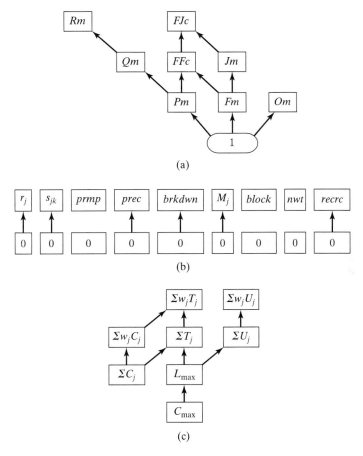

(a)

(b)

(c)

Figure 2.7   Complexity hierarchies of deterministic scheduling problems: (1) machine environments, (b) processing restrictions and constraints, and (c) objective functions.

also NP-hard. The following examples illustrate the borderlines between easy and hard problems within given sets of problems.

Example 2.4.1 (Complexity of Makespan Problems)

Consider the problems

    (i) $1 \mid\mid C_{\max}$,
    (ii) $P2 \mid\mid C_{\max}$,
   (iii) $F2 \mid\mid C_{\max}$,
    (iv) $Jm \mid\mid C_{\max}$,
    (v) $FFc \mid\mid C_{\max}$.

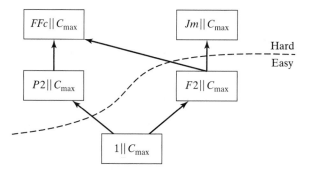

Figure 2.8   Complexity hierarchy of problems in Example 2.4.1.

The complexity hierarchy is depicted in Figure 2.8.

Example 2.4.2 (Complexity of Maximum Lateness Problems)

Consider the problems

    (i) $1 \| L_{\max}$,
   (ii) $1 \mid prmp \mid L_{\max}$,
  (iii) $1 \mid r_j \mid L_{\max}$,
  (iv) $1 \mid r_j, prmp \mid L_{\max}$,
   (v) $Pm \| L_{\max}$.

The complexity hierarchy is depicted in Figure 2.9.

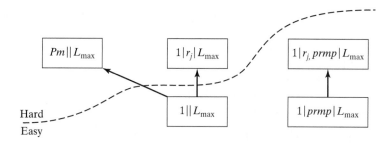

Figure 2.9   Complexity hierarchy of problems in Example 2.4.2.

EXERCISES (COMPUTATIONAL)

2.1. Consider the instance of $1 \| \sum w_j C_j$ with the following processing times and weights.

jobs	1	2	3	4
w_j	6	11	9	5
p_j	3	5	7	4

(a) Find the optimal sequence and compute the value of the objective.

(b) Give an argument for positioning jobs with larger weight more toward the beginning of the sequence and jobs with smaller weight more toward the end of the sequence.

(c) Give an argument for positioning jobs with smaller processing time more toward the beginning of the sequence and jobs with larger processing time more toward the end of the sequence.

(d) Determine which one of the following two generic rules is the most suitable for the problem:

    i. sequence the jobs in decreasing order of $w_j - p_j$;

    ii. sequence the jobs in decreasing order of w_j / p_j.

2.2. Consider the instance of $1 \parallel L_{\max}$ with the following processing times and due dates.

jobs	1	2	3	4
p_j	5	4	3	6
d_j	3	5	11	12

(a) Find the optimal sequence and compute the value of the objective.

(b) Give an argument for positioning jobs with earlier due dates more toward the beginning of the sequence and jobs with later due dates more toward the end of the sequence.

(c) Give an argument for positioning jobs with smaller processing time more toward the beginning of the sequence and jobs with larger processing time more toward the end of the sequence.

(d) Determine which one of the following four rules is the most suitable generic rule for the problem:

    i. sequence the jobs in increasing order of $d_j + p_j$;

    ii. sequence the jobs in increasing order of $d_j p_j$;

    iii. sequence the jobs in increasing order of d_j;

    iv. sequence the jobs in increasing order of p_j.

2.3. Consider the instance of $1 \parallel \sum U_j$ with the following processing times and due dates.

jobs	1	2	3	4
p_j	7	6	4	8
d_j	8	9	11	14

(a) Find all optimal sequences and compute the value of the objective.

(b) Formulate a generic rule based on the due dates and processing times that yields an optimal sequence for any instance.

2.4. Consider the instance of $1 \;||\; \sum T_j$ with the following processing times and due dates.

jobs	1	2	3	4
p_j	7	6	8	4
d_j	8	9	10	14

(a) Find all optimal sequences.

(b) Formulate a generic rule that is a function of the due dates and processing times that yields an optimal sequence for any instance.

2.5. Find the optimal sequence for $P5 \;||\; C_{\max}$ with the following 11 jobs.

jobs	1	2	3	4	5	6	7	8	9	10	11
p_j	9	9	8	8	7	7	6	6	5	5	5

2.6. Consider the instance of $F2 \;|\; prmu \;|\; C_{\max}$ with the following processing times.

jobs	1	2	3	4
p_{1j}	8	6	4	12
p_{2j}	4	9	10	6

Find all optimal sequences and determine the makespan under an optimal sequence.

2.7. Consider the instance of $F2 \;|\; block \;|\; C_{\max}$ with the same jobs and processing times as in Exercise 2.6. There is no (zero) buffer between the two machines. Find all optimal sequences, and compute the makespan under an optimal sequence.

2.8. Consider the instance of $F2 \;|\; nwt \;|\; C_{\max}$ with the same jobs and the same processing times as in Exercise 2.6. Find all optimal sequences, and compute the makespan under an optimal sequence.

2.9. Consider the instance of $O2 \;||\; C_{\max}$ with four jobs. The processing times of the four jobs on the two machines are again as in Exercise 2.6. Find all optimal schedules and compute the makespan under an optimal schedule.

2.10. Consider the instance of $J2 \;||\; C_{\max}$ with four jobs. The processing times of the four jobs on the two machines are again as in Exercise 2.6. Jobs 1 and 2 have to be processed first on machine 1 and then on machine 2 while jobs 3 and 4 have to be processed first on machine 2 and then on machine 1. Find all optimal schedules and determine the makespan under an optimal schedule.

EXERCISES (THEORY)

2.11. Explain why $\alpha \mid p_j = 1, r_j \mid \gamma$ is easier than $\alpha \mid prmp, r_j \mid \gamma$ when all processing times, release dates and due dates are integers.

2.12. Consider $1 \mid s_{jk} = a_k + b_j \mid C_{\max}$. That is, job j has two parameters associated with it—namely, a_j and b_j. If job j is followed by job k, there is a setup time $s_{jk} = a_k + b_j$ required before the start of job k's processing. The setup time of the first job in the sequence, s_{0k} is a_k, while the cleanup time at the completion of the last job in the sequence, s_{j0}, is b_j. Show that this problem is equivalent to $1 \mid\mid C_{\max}$ and that the makespan therefore does not depend on the sequence. Find an expression for the makespan.

2.13. Show that $1 \mid s_{jk} \mid C_{\max}$ is equivalent to the following Traveling Salesman Problem: A traveling salesman starts out from city 0, visits cities $1, 2, \ldots, n$ and returns to city 0 while minimizing the total distance traveled. The distance from city 0 to city k is s_{0k}; the distance from city j to city k is s_{jk}, and the distance from city j to city 0 is s_{j0}.

2.14. Show that $1 \mid brkdwn, prmp \mid \sum w_j C_j$ reduces to $1 \mid r_j, prmp \mid \sum w_j C_j$.

2.15. Show that $1 \mid p_j = 1 \mid \sum w_j T_j$ and $1 \mid p_j = 1 \mid L_{\max}$ are equivalent to the *assignment* problem (see Appendix A for a definition of the assignment problem).

2.16. Show that $Pm \mid p_j = 1 \mid \sum w_j T_j$ and $Pm \mid p_j = 1 \mid L_{\max}$ are equivalent to the *transportation* problem (see Appendix A for a definition of the transportation problem).

2.17. Consider $Pm \mid C_{\max}$. Show that for any nondelay schedule the following inequalities hold:

$$\frac{\sum p_j}{m} \le C_{\max} \le 2 \times \max\left(p_1, \ldots, p_n, \frac{\sum p_j}{m}\right).$$

2.18. Show that $Pm \mid M_j \mid \gamma$ reduces to $Rm \mid\mid \gamma$.

2.19. Show that $F2 \mid block \mid C_{\max}$ is equivalent to $F2 \mid nwt \mid C_{\max}$ and show that both problems are special cases of $1 \mid s_{jk} \mid C_{\max}$ and therefore special cases of the Traveling Salesman Problem.

2.20. Consider an instance of $Om \mid \beta \mid \gamma$ and an instance of $Fm \mid \beta \mid \gamma$. The two instances have the same number of machines and jobs, and the jobs have the same processing times on the m machines. The two instances are completely identical with the exception that one instance is an open shop and the other instance a flow shop. Show that the value of the objective under the optimal sequence in the flow shop is at least as large as the value of the objective under the optimal sequence in the open shop.

2.21. Consider $O2 \mid\mid C_{\max}$. Show that

$$C_{\max} \ge \max\left(\sum_{j=1}^{n} p_{1j}, \sum_{j=1}^{n} p_{2j}\right).$$

Find an instance of this problem where the optimal makespan is *strictly* larger than the RHS.

2.22. Describe the complexity relationships between the problems

   (a) $1 \mid\mid \sum w_j C_j$,
   (b) $1 \mid d_j = d \mid \sum w_j T_j$,
   (c) $1 \mid p_j = 1 \mid \sum w_j T_j$,
   (d) $1 \mid\mid \sum w_j T_j$,
   (e) $Pm \mid p_j = 1 \mid \sum w_j T_j$,
   (f) $Pm \mid\mid \sum w_j T_j$.

2.23. Show that $\alpha \mid \beta \mid L_{\max}$ reduces to $\alpha \mid \beta \mid \sum T_j$ as well as to $\alpha \mid \beta \mid \sum U_j$. (*Hint:* Note that if the minimum $L_{\max}$ is zero, the optimal solution with regard to $\sum U_j$ and $\sum T_j$ is zero as well. It suffices to show that a polynomial time procedure for $\alpha \mid \beta \mid \sum U_j$ can be adapted easily for application to $\alpha \mid \beta \mid L_{\max}$. This can be done through a parametric analysis on the d_j—i.e., solve $\alpha \mid \beta \mid \sum U_j$ with due dates $d_j + z$ and vary z.)

COMMENTS AND REFERENCES

One of the first classification schemes for scheduling problems appeared in Conway, Maxwell, and Miller (1967). In their survey paper, Lawler, Lenstra, and Rinnooy Kan (1982) modified and refined this scheme extensively. Herrmann, Lee, and Snowdon (1993) made another round of extensions. The framework presented here is another variation of the Lawler, Lenstra, and Rinnooy Kan (1982) notation with a slightly different emphasis.

For surveys on scheduling problems with nonregular objective functions, see Baker and Scudder (1990) and Raghavachari (1988).

The definitions of nondelay, active, and semi-active schedules have been around for a long time; see, for example, Giffler and Thompson (1960), and see French (1982) for a comprehensive overview of schedule classes. Example 2.3.2, which illustrates some of the anomalies of nondelay schedules, is from Graham (1966).

The complexity hierarchy of scheduling problems is motivated primarily by the the work of Rinnooy Kan (1976); Lenstra (1977); Lageweg, Lawler, Lenstra, and Rinnooy Kan (1981, 1982); and Lawler, Lenstra, Rinnooy Kan, and Shmoys (1993). For more on the reducibility among scheduling problems, see Timkovsky (2000).

3
Single Machine Models (Deterministic)

Single machine models are important for various reasons. The single machine environment is very simple and a special case of all other environments. Single machine models often have properties that neither machines in parallel nor machines in series have. The results that can be obtained for single machine models not only provide insights into the single machine environment, but they also provide a basis for heuristics that are applicable to more complicated machine environments. In practice, scheduling problems in more complicated machine environments are often decomposed into subproblems that deal with single machines. For example, a complicated machine environment with a single bottleneck may give rise to a single machine model.

In this chapter, various single machine models are analyzed in detail. The total weighted completion time objective is considered first, followed by several due date-related objectives, including maximum lateness, number of tardy jobs, total tardiness, and total weighted tardiness. All objective functions considered in this chapter are regular.

In most models considered in this chapter, there is no advantage in having preemptions. For these models, it can be shown that the optimal schedule in the class of preemptive schedules is nonpreemptive. However, if jobs are not all released at the same time, then preemptions may provide an advantage. If jobs are not all released at the same time in a nonpreemptive environment, then it may be advantageous to have unforced idleness (i.e., there may not be an optimal schedule that is nondelay).

3.1 THE TOTAL WEIGHTED COMPLETION TIME

The first objective to be considered is the total weighted completion time (i.e. $1 \mid\mid \sum w_j C_j$). The weight w_j of job j may be regarded as an importance factor; it may represent either a holding cost per unit time or the value already added to job j. This problem gives rise to one of the better known rules in scheduling theory, the *Weighted Shortest Processing Time first (WSPT)* rule. According to this rule, the jobs are ordered in decreasing order of w_j/p_j.

Theorem 3.1.1.   *The WSPT rule is optimal for $1 \mid\mid \sum w_j C_j$.*

Proof.   By contradiction. Suppose a schedule S that is not WSPT is optimal. In this schedule, there must be at least two adjacent jobs—say job j followed by job k, such that

$$\frac{w_j}{p_j} < \frac{w_k}{p_k}.$$

Assume job j starts its processing at time t. Perform a so-called *Adjacent Pairwise Interchange* on jobs j and k. Under the original schedule S, job j starts its processing at time t and is followed by job k, whereas under the new schedule, job k starts its processing at time t and is followed by job j. All other jobs remain in their original position. Call the new schedule S'. The total weighted completion time of the jobs processed before jobs j and k is not affected by the interchange. Neither is the total weighted completion time of the jobs processed after jobs j and k. Thus, the difference in the values of the objectives under schedules S and S' is due only to jobs j and k (see Figure 3.1). Under S the total weighted completion time of jobs j and k is

$$(t + p_j)w_j + (t + p_j + p_k)w_k,$$

Schedule S

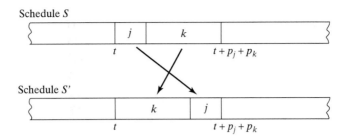

Schedule S'

Figure 3.1   A pairwise interchange of jobs j and k.

whereas under S' it is

$$(t + p_k)w_k + (t + p_k + p_j)w_j.$$

It is easily verified that if $w_j/p_j < w_k/p_k$, the sum of the two weighted completion times under S' is strictly less than under S. This contradicts the optimality of S and completes the proof of the theorem.   □

The computation time needed to order the jobs according to WSPT is the time required to sort the jobs according to the ratio of the two parameters. A simple sort can be done in $O(n \log(n))$ time; see Example C.1.1 in Appendix C.

How is the minimization of the total weighted completion time affected by precedence constraints? Consider the simplest form of precedence constraints (i.e., precedence constraints that take the form of parallel chains; see Figure 3.2). This problem can still be solved by a relatively simple and efficient (polynomial time) algorithm. This algorithm is based on some fundamental properties of scheduling with precedence constraints.

Consider two chains of jobs. One chain, say Chain I, consists of jobs $1, \dots, k$ and the other chain, say Chain II, consists of jobs $k + 1, \dots, n$. The precedence

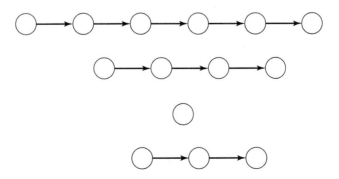

Figure 3.2   Precedence constraints in the form of chains.

constraints are as follows:

$$1 \to 2 \to \cdots \to k$$

and

$$k+1 \to k+2 \to \cdots \to n.$$

The next lemma is based on the assumption that if the scheduler decides to start processing jobs of one chain, he has to complete the *entire* chain before he is allowed to work on jobs of the other chain. The question is: If the scheduler wishes to minimize the total weighted completion time of the n jobs, which one of the two chains should he process first?

Lemma 3.1.2. *If*

$$\frac{\sum_{j=1}^{k} w_j}{\sum_{j=1}^{k} p_j} > (<) \frac{\sum_{j=k+1}^{n} w_j}{\sum_{j=k+1}^{n} p_j},$$

then it is optimal to process the chain of jobs $1, \ldots, k$ before (after) the chain of jobs $k+1, \ldots, n$.

Proof. By contradiction. Under sequence $1, \ldots, k, k+1, \ldots, n$ the total weighted completion time is

$$w_1 p_1 + \cdots + w_k \sum_{j=1}^{k} p_j + w_{k+1} \sum_{j=1}^{k+1} p_j + \cdots + w_n \sum_{j=1}^{n} p_j,$$

whereas under sequence $k+1, \ldots, n, 1, \ldots, k$ it is

$$w_{k+1} p_{k+1} + \cdots + w_n \sum_{j=k+1}^{n} p_j + w_1 \left(\sum_{j=k+1}^{n} p_j + p_1 \right) + \cdots + w_k \sum_{j=1}^{n} p_j.$$

The total weighted completion time of the first sequence is less than the total weighted completion time of the second sequence if

$$\frac{\sum_{j=1}^{k} w_j}{\sum_{j=1}^{k} p_j} > \frac{\sum_{j=k+1}^{n} w_j}{\sum_{j=k+1}^{n} p_j}.$$

The result follows.   □

An interchange between two adjacent chains of jobs is usually referred to as an *Adjacent Sequence Interchange*. Such an interchange is a generalization of an Adjacent Pairwise Interchange.

An important characteristic of chain

$$1 \rightarrow 2 \rightarrow \cdots \rightarrow k$$

is defined as follows: Let l^* satisfy

$$\frac{\sum_{j=1}^{l^*} w_j}{\sum_{j=1}^{l^*} p_j} = \max_{1 \le l \le k} \left(\frac{\sum_{j=1}^{l} w_j}{\sum_{j=1}^{l} p_j} \right).$$

The ratio on the left-hand side is called the ρ-factor of chain $1, \ldots, k$ and is denoted by $\rho(1, \ldots, k)$. Job l^* is referred to as the job that determines the ρ-factor of the chain.

Suppose now that the scheduler does not have to complete all the jobs in a chain before he is allowed to work on another chain. He may process some jobs of one chain (while adhering to the precedence constraints), switch over to another chain, and, at some later point in time, return to the first chain. If, in the case of multiple chains, the total weighted completion time is the objective function, then the following result holds.

Lemma 3.1.3.   *If job l^* determines $\rho(1, \ldots, k)$, then there exists an optimal sequence that processes jobs $1, \ldots, l^*$ one after another without any interruption by jobs from other chains.*

Proof. By contradiction. Suppose that under the optimal sequence the processing of the subsequence $1, \ldots, l^*$ is interrupted by a job, say job v, from another chain. The optimal sequence contains the subsequence $1, \ldots, u, v, u + 1, \ldots, l^*$, say subsequence S. It suffices to show that either with subsequence $v, 1, \ldots, l^*$, say S', or with subsequence $1, \ldots, l^*, v$, say S'', the total weighted completion time is less than with subsequence S. If it is not less with the first subsequence, then it has to be less with the second and vice versa. From Lemma 3.1.2, it follows that if the total weighted completion time with S is less than with S', then

$$\frac{w_v}{p_v} < \frac{w_1 + w_2 + \cdots + w_u}{p_1 + p_2 + \cdots + p_u}.$$

From Lemma 3.1.2, it also follows that if the total weighted completion time with S is less than with S'', then

$$\frac{w_v}{p_v} > \frac{w_{u+1} + w_{u+2} + \cdots + w_{l^*}}{p_{u+1} + p_{u+2} + \cdots + p_{l*}}.$$

If job l^* is the job that determines the ρ-factor of chain $1, \ldots, k$, then

$$\frac{w_{u+1} + w_{u+2} + \cdots + w_{l^*}}{p_{u+1} + p_{u+2} + \cdots + p_{l^*}} > \frac{w_1 + w_2 + \cdots + w_u}{p_1 + p_2 + \cdots + p_u}.$$

If $\mathcal{S}$ is better than $\mathcal{S}''$, then

$$\frac{w_v}{p_v} > \frac{w_{u+1} + w_{u+2} + \cdots + w_{l^*}}{p_{u+1} + p_{u+2} + \cdots + p_{l^*}} > \frac{w_1 + w_2 + \cdots + w_u}{p_1 + p_2 + \cdots + p_u}.$$

Therefore, $\mathcal{S}'$ is better than $\mathcal{S}$. The same argument goes through if the interruption of the chain is caused by more than one job. This completes the proof. □

The result in Lemma 3.1.3 is intuitive. The condition of the lemma implies that the ratios of the weight divided by the processing time of the jobs in the string $1, \ldots, l^*$ are increasing in some sense. If one had already decided to start processing a string of jobs, it makes sense to continue processing the string until job l^* is completed without processing any other job in between.

The two previous lemmas contain the basis for a simple algorithm that minimizes the total weighted completion time when the precedence constraints take the form of chains.

Algorithm 3.1.4. (Total Weighted Completion Time with Chains) *Whenever the machine is freed, select among the remaining chains the one with the highest ρ-factor. Process this chain without interruption up to and including the job that determines its ρ-factor.*

The following example illustrates the use of the algorithm.

Example 3.1.5 (Total Weighted Completion Time with Chains)

Consider the following two chains:

$$1 \rightarrow 2 \rightarrow 3 \rightarrow 4$$

and

$$5 \rightarrow 6 \rightarrow 7$$

The weights and processing times of the jobs are given in the following table.

jobs	1	2	3	4	5	6	7
w_j	6	18	12	8	8	17	18
p_j	3	6	6	5	4	8	10

The ρ-factor of the first chain is $(6 + 18)/(3 + 6)$ and is determined by job 2. The ρ-factor of the second chain is $(8 + 17)/(4 + 8)$ and is determined by job 6. As $24/9$

is larger than 25/12, jobs 1 and 2 are processed first. The ρ-factor of the remaining part of the first chain is 12/6 and determined by job 3. As 25/12 is larger than 12/6, jobs 5 and 6 follow jobs 1 and 2. The ρ-factor of the remaining part of the second chain is 18/10 and is determined by job 7; so job 3 follows job 6. As the w_j/p_j ratio of job 7 is higher than the ratio of job 4, job 7 follows job 3 and job 4 goes last.

Polynomial time algorithms have been obtained for $1 \mid prec \mid \sum w_j C_j$ with more general precedence constraints than the parallel chains just considered. However, with *arbitrary* precedence constraints, the problem is strongly NP-hard.

Up to now, all jobs were assumed to be available at time zero. Consider the problem where jobs are released at different points in time and the scheduler is allowed to preempt (i.e., $1 \mid r_j, prmp \mid \sum w_j C_j$). The first question that comes to mind is whether a preemptive version of the WSPT rule is optimal. A preemptive version of the WSPT rule can be formulated as follows: At any point in time, the available job with the highest ratio of weight to *remaining* processing time is selected for processing. The priority level of a job thus increases while being processed and a job can therefore not be preempted by another job that was already available at the start of its processing. However, a job may be preempted by a newly released job with a higher priority factor. Although this rule may appear a logical extension of the nonpreemptive WSPT rule, it does not necessarily lead to an optimal schedule since the problem is strongly NP-hard (see Appendix D).

If all the weights are equal, then the $1 \mid r_j, prmp \mid \sum C_j$ problem is easy (see Exercise 3.15). Yet the nonpreemptive version of this problem (i.e., $1 \mid r_j \mid \sum C_j$) is strongly NP-hard.

In Chapter 2, the total weighted discounted completion time $\sum w_j(1-e^{-rC_j})$, where r is the discount factor, is described as an objective that is, in a way, a generalization of the total weighted (undiscounted) completion time. The problem $1 \mid\mid \sum w_j(1 - e^{-rC_j})$ gives rise to a different priority rule—namely, the rule that schedules the jobs in decreasing order of

$$\frac{w_j e^{-rp_j}}{1 - e^{-rp_j}}.$$

In what follows, this rule is referred to as the *Weighted Discounted Shortest Processing Time first (WDSPT) rule.*

Theorem 3.1.6.   *For $1 \mid\mid \sum w_j(1 - e^{-rC_j})$, the WDSPT rule is optimal.*

Proof.   By contradiction. Again, assume that a different schedule, say schedule S, is optimal. Under this schedule, there have to be two jobs j and k, job j followed by job k, such that

$$\frac{w_j e^{-rp_j}}{1 - e^{-rp_j}} < \frac{w_k e^{-rp_k}}{1 - e^{-rp_k}}.$$

Assume job j starts its processing at time t. An Adjacent Pairwise Interchange between these two jobs results in a schedule $\mathcal{S}'$. It is clear that the only difference in the objective is due to jobs j and k. Under $\mathcal{S}$, the contribution of jobs j and k to the objective function equals

$$w_j \left(1 - e^{-r(t+p_j)}\right) + w_k \left(1 - e^{-r(t+p_j+p_k)}\right).$$

The contribution of jobs j and k to the objective under $\mathcal{S}'$ is obtained by interchanging the js and ks in this expression. Elementary algebra then shows that the value of objective function under $\mathcal{S}'$ is less than under $\mathcal{S}$. This leads to the contradiction that completes the proof.   □

As discussed in Chapter 2, the total undiscounted weighted completion time is basically a limiting case of the total discounted weighted completion time $\sum w_j(1 - e^{-rC_j})$. The WDSPT rule results in the same sequence as the WSPT rule if r is sufficiently close to zero (note that the WDSPT rule is not properly defined for $r = 0$).

Both $\sum w_j C_j$ and $\sum w_j(1 - e^{-rC_j})$ are special cases of the more general objective function $\sum w_j h(C_j)$. It has been shown that only the functions $h(C_j) = C_j$ and $h(C_j) = 1 - e^{-rC_j}$ lead to simple priority rules that order the jobs in decreasing order of some function $g(w_j, p_j)$. No such priority function g, which guarantees optimality, exists for any other cost function h. However, the objective $\sum h_j(C_j)$ can be dealt with via Dynamic Programming (see Appendix B).

In a similar way as Lemma 3.1.2 generalizes the Adjacent Pairwise Interchange argument for WSPT, there exists an Adjacent Sequence Interchange result that generalizes the Adjacent Pairwise Interchange argument used in the optimality proof for the WDSPT rule (see Exercise 3.21).

3.2 THE MAXIMUM LATENESS

The objectives considered in the next four sections are due date related. The first due date-related model is of a general nature—namely, the problem $1 \mid prec \mid h_{\max}$, where

$$h_{\max} = \max\left(h_1(C_1), \ldots, h_n(C_n)\right),$$

with $h_j, j = 1, \ldots, n$, being nondecreasing cost functions. This objective is clearly due date-related as the functions h_j may take any one of the forms depicted in Figure 2.1. This problem allows for an efficient *backward* dynamic programming algorithm even when the jobs are subject to arbitrary precedence constraints.

It is clear that the completion of the last job occurs at the makespan $C_{\max} = \sum p_j$, which is independent of the schedule. Let J denote the set of jobs already scheduled, which are processed during the time interval

$$\left[C_{\max} - \sum_{j \in J} p_j, C_{\max}\right].$$

The complement of set J, set J^c, denotes the set of jobs still to be scheduled and the subset J' of J^c denotes the set of jobs that can be scheduled immediately before set J (i.e., the set of jobs all of whose immediate successors are already in J). Set J' is referred to as the set of *schedulable* jobs. The following backward algorithm yields an optimal schedule.

Algorithm 3.2.1. (Minimizing Maximum Cost)

Step 1. *Set $J = \emptyset$, $J^c = \{1, \dots, n\}$ and J' the set of all jobs with no successors.*

Step 2. *Let j^* be such that*

$$h_{j^*}\left(\sum_{j \in J^c} p_j\right) = \min_{j \in J'}\left(h_j\left(\sum_{k \in J^c} p_k\right)\right)$$

Add j^* to J
Delete j^* from J^c
Modify J' to represent the new set of schedulable jobs.

Step 3. *If $J^c = \emptyset$ STOP, otherwise go to Step 2.*

Theorem 3.2.2.   *Algorithm 3.2.1 yields an optimal schedule for* $1 \mid prec \mid$ $h_{\max}$.

Proof.   By contradiction. Suppose at a given iteration job j^{**}, selected from J', does not have the minimum completion cost

$$h_{j^*}\left(\sum_{j \in J^c} p_j\right)$$

among the jobs in J'. Say job j^{**} is the job selected. The minimum cost job j^* must then be scheduled at a later iteration, implying that job j^* has to appear in the sequence before job j^{**}. A number of jobs may even appear between jobs j^* and j^{**} (see Figure 3.3).

To show that this sequence cannot be optimal, take job j^* and insert it in the schedule immediately following job j^{**}. All jobs in the original schedule between jobs j^* and j^{**}, including job j^{**}, are now completed earlier. The only job whose completion cost increases is job j^*. However, its completion cost now is, by definition, smaller than the completion cost of job j^{**} under the original schedule, so the

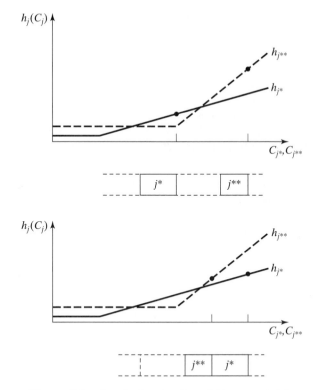

Figure 3.3   Proof of optimality of Theorem 3.2.2.

maximum completion cost decreases after the insertion of job j^*. This completes the proof.   □

The worst case computation time required by this algorithm can be established as follows. There are n steps needed to schedule the n jobs. At each step at most n jobs have to be considered. The overall running time of the algorithm is therefore bounded by $O(n^2)$.

The following example illustrates the application of this algorithm.

Example 3.2.3 (Minimizing Maximum Cost)

Consider the following three jobs.

jobs	1	2	3
p_j	2	3	5
$h_j(C_j)$	$1 + C_1$	$1.2\,C_2$	10

The makespan $C_{\max} = 10$ and $h_3(10) < h_1(10) < h_2(10)$ (as $10 < 11 < 12$). Job 3 is therefore scheduled last and has to start its processing at time 5. To determine which job is to be processed before job 3, $h_2(5)$ has to be compared with $h_1(5)$. Either job 1 or job 2 may be processed before job 3 in an optimal schedule as $h_1(5) = h_2(5) = 6$. So two schedules are optimal: 1, 2, 3 and 2, 1, 3.

The problem $1 \; || \; L_{\max}$ is the best known special case of $1 \mid prec \mid h_{\max}$. The function h_j is then defined as $C_j - d_j$, and the algorithm results in the schedule that orders the job in increasing order of their due dates (i.e., *Earliest Due Date [EDD] first*).

A generalization of $1 \; || \; L_{\max}$ is the problem $1 \mid r_j \mid L_{\max}$ with the jobs released at different points in time. This generalization, which does not allow pre-emption, is significantly harder than the problem with all jobs available at the same time. The optimal schedule is not necessarily a nondelay schedule. It may be advantageous to keep the machine idle just before the release of a new job.

Theorem 3.2.4. *The problem $1 \mid r_j \mid L_{\max}$ is strongly NP-hard.*

Proof. The proof is based on a reduction of 3-PARTITION to $1 \mid r_j \mid L_{\max}$. Given integers $a_1, \dots, a_{3t}, b$, such that $b/4 < a_j < b/2$ and $\sum_{j=1}^{3t} a_j = tb$, the following instance of $1 \mid r_j \mid L_{\max}$ can be constructed. The number of jobs, n, is equal to $4t - 1$ and

$$
\begin{aligned}
r_j &= jb + (j-1), & p_j &= 1, & d_j &= jb + j, & j &= 1, \dots, t-1, \\
r_j &= 0, & p_j &= a_{j-t+1}, & d_j &= tb + (t-1), & j &= t, \dots, 4t-1.
\end{aligned}
$$

Let $z = 0$. A schedule with $L_{\max} \le 0$ exists if and only if every job j, $j = 1, \dots, t-1$, can be processed between r_j and $d_j = r_j + p_j$. This can be done if and only if the remaining jobs can be partitioned over the t intervals of length b, which can be done if and only if 3-PARTITION has a solution (see Figure 3.4). $\square$

The $1 \mid r_j \mid L_{\max}$ problem is important because it appears frequently as a subproblem in heuristic procedures for flow shop and job shop problems. It has received considerable attention resulting in a number of reasonably effective enumerative branch and bound procedures. Branch and bound procedures are basically enumeration schemes where certain schedules or classes of schedules are discarded by

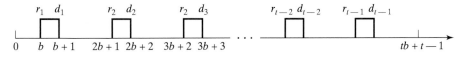

Figure 3.4 $1 \mid r_j \mid L_{\max}$ is strongly NP-hard.

showing that the values of the objective obtained with schedules from this class are larger than a provable lower bound; this lower bound is greater than or equal to the value of the objective of a schedule obtained earlier.

A branch and bound procedure for $1 \mid r_j \mid L_{\max}$ can be constructed as follows. The branching process may be based on the fact that schedules are developed starting from the beginning of the schedule. There is a single node at level 0 that is the top of the tree. At this node, none of the jobs has been put into any position in the sequence. There are n branches going down to n nodes at level 1. Each node at this level has a specific job put into the first position in the schedule. So at each one of these nodes, there are still $n - 1$ jobs whose position in the schedule has yet to be determined. There are $n - 1$ arcs emanating from each node at level 1 to level 2. There are therefore $(n - 1) \times (n - 2)$ nodes at level 2. At each node at level 2, the two jobs in the first two positions are specified; at level k, jobs in the first k positions are specified. Actually, it is not necessary to consider every remaining job as a candidate for the next position. If at a node at level $k - 1$ jobs $j_1, \ldots, j_{k-1}$ are scheduled as the first $k - 1$ jobs, then job j_k only has to be considered if

$$r_{j_k} < \min_{l \in J} (\max(t, r_l) + p_l),$$

where J denotes the set of jobs not yet scheduled and t denotes the time job j_k is supposed to start. The reason for this condition is clear: If job j_k does not satisfy this inequality, then selecting the job that minimizes the right-hand side instead of j_k does not increase the value of $L_{\max}$. The branching rule is thus fairly easy.

There are several ways in which bounds for nodes can be obtained. An easy lower bound for a node at level $k-1$ can be established by scheduling the remaining jobs J according to the *preemptive* EDD rule. The preemptive EDD rule is known to be optimal for $1 \mid r_j, prmp \mid L_{\max}$ (see Exercise 3.24) and thus provides a lower bound for the problem at hand. If a preemptive EDD rule results in a nonpreemptive schedule, then all nodes with a larger lower bound can be disregarded.

Example 3.2.5 (Branch and Bound Applied to Minimizing Maximum Lateness)

Consider the following four jobs.

jobs	1	2	3	4
p_j	4	2	6	5
r_j	0	1	3	5
d_j	8	12	11	10

At level 1 of the search tree, there are four nodes: $(1, *, *, *)$, $(2, *, *, *)$, $(3, *, *, *)$, and $(4, *, *, *)$. It is easy to see that nodes $(3, *, *, *)$ and $(4, *, *, *)$ may be disregarded immediately. Job 3 is released at time 3; if job 2 would start its

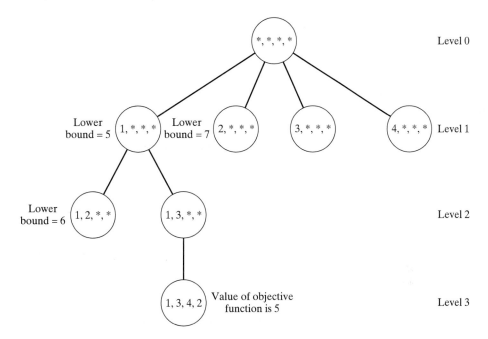

Figure 3.5   Branch and bound procedure for Example 3.2.5.

processing at time 1, job 3 still can start at time 3. Job 4 is released at time 5; if job 1 would start its processing at time 0, job 4 still can start at time 5 (see Figure 3.5).

Computing a lower bound for node $(1, *, *, *)$ according to the preemptive EDD rule results in a schedule where job 3 is processed during the time interval $[4, 5]$, job 4 during the time interval $[5, 10]$, job 3 (again) during interval $[10, 15]$, and job 2 during the time interval $[15, 17]$. The L_{max} of this schedule, which provides the lower bound for node $(1, *, *, *)$, is 5. In a similar way, a lower bound can be obtained for node $(2, *, *, *)$. The value of this lower bound is 7.

Consider node $(1, 2, *, *)$ at level 2. The lower bound for this node is 6 and is determined by the (nonpreemptive) schedule 1, 2, 4, 3. Proceed with node $(1, 3, *, *)$ at level 2. The lower bound is 5 and determined by the (nonpreemptive) schedule 1, 3, 4, 2. From the fact that the lower bound for node $(1, *, *, *)$ is 5 and the lower bound for node $(2, *, *, *)$ is larger than 5, it follows that schedule 1, 3, 4, 2 has to be optimal.

The problem $1 \mid r_j, prec \mid L_{max}$ can be handled in a similar way. This problem, from an enumeration point of view, is easier than the problem without precedence constraints, since the precedence constraints allow certain schedules to be ruled out immediately.

3.3 THE NUMBER OF TARDY JOBS

Another due date-related objective is $\sum U_j$. This objective may at first appear somewhat artificial and of no practical interest. However, in the real world, it is a performance measure that is often monitored and relative to which managers are measured. It is equivalent to the percentage of on-time shipments.

An optimal schedule for $1 \mid\mid \sum U_j$ takes the form of one set of jobs that will meet their due dates and that is scheduled first, followed by the set of remaining jobs that will not meet their due dates and that is scheduled last. It follows from the results in the previous section that the first set of jobs has to be scheduled according to EDD to make sure that $L_{\max}$ is negative; the order in which this second set of jobs is scheduled is immaterial. The problem $1 \mid\mid \sum U_j$ can be solved easily with a *forward* algorithm. Again, set J denotes the set of jobs already scheduled. Set J^d denotes the set of jobs already considered for scheduling, but that have been discarded as they will not meet their due dates in the optimal schedule. Set J^c denotes the set of jobs not yet considered for scheduling.

Algorithm 3.3.1. (Minimizing Number of Tardy Jobs)

Step 1. *Set $J = \emptyset$, $J^d = \emptyset$ and $J^c = \{1, \ldots, n\}$*

Step 2. *Let j^* denote the job which satisfies*

$$d_{j^*} = \min_{j \in J^c}(d_j)$$

> *Add j^* to J*
> *Delete j^* from J^c*
> *Go to Step 3.*

Step 3. *If*

$$\sum_{j \in J} p_j \leq d_{j^*}$$

> *Go to Step 4,*
> *otherwise*
> *let k^* denote the job which satisfies*

$$p_{k^*} = \max_{j \in J}(p_j)$$

> *Delete k^* from J*
> *Add k^* to J^d*

Step 4. *If $J^c = \emptyset$ STOP,*
> *otherwise go to Step 2.*

In words, the algorithm can be described as follows. Jobs are added to the set of on-time jobs in increasing order of due dates. If the inclusion of job j^* results in this job being completed late, the scheduled job with the longest processing time, say job k^*, is marked late and discarded. Since the algorithm basically orders the jobs according to their due dates, the worst case computation time is that of a simple sort—that is, $O(nlog(n))$.

Theorem 3.3.2.   *Algorithm 3.3.1 yields an optimal schedule for* $1 \parallel \sum U_j$.

Proof.   Without loss of generality, assume that $d_1 \leq d_2 \leq \cdots \leq d_n$. Let J_k denote a subset of jobs $\{1, \ldots , k\}$ that satisfies the following two conditions:

(i) It has the maximum number of jobs, say N_k, among jobs $\{1, \ldots , k\}$ completed by their due date.
(ii) Among all sets with N_k jobs among the first k jobs completed by their due date, the jobs in set J_k have the smallest total processing time.

Note that set J_n corresponds to an optimal schedule. The proof that the algorithm leads to J_n is by induction. It is true that the algorithm constructs J_1 according to the two conditions required for $k = 1$. The induction hypothesis is that the algorithm indeed constructs set J_k, which satisfies the two conditions. It remains to be shown that the manner in which set J_{k+1} is constructed starting out with set J_k also results in a set that satisfies the two conditions for $k + 1$. There are two cases to be considered.

Case 1.   Job $k + 1$ is added to set J_k and job $k + 1$ is completed by its due date. The two conditions clearly hold. It is impossible to have a set with more jobs among the first $k + 1$ completed in time. It is also clear that the last job has to be part of the set; the set therefore has a minimum total amount of processing among sets with that number of on-time completions.

Case 2.   Job $k + 1$ is added to set J_k and is not completed in time. From the fact that N_k is the maximum number of jobs to be completed in time among jobs $\{1, \ldots , k\}$ and that set J_k has the smallest total processing time among sets with N_k on-time completions, it follows that $N_{k+1} = N_k$. Adding job $k + 1$ to set J_k does not increase the number of jobs completed on time. But adding job $k + 1$ to the set and deleting the largest job among $J_k \cup \{k + 1\}$ keeps the same number of jobs completed on time and reduces the total time it takes to process these jobs. It can be shown easily that no other subset of $\{1, \ldots , k + 1\}$ can have N_k on-time completions and a smaller total processing time. The result follows. $\square$

Example 3.3.3 (Minimizing Number of Tardy Jobs)

Consider the following five jobs.

jobs	1	2	3	4	5
p_j	7	8	4	6	6
d_j	9	17	18	19	21

Jobs 1 and 2 can be positioned first and second in the sequence with both jobs being completed on time. Putting job 3 into the third position causes problems. Its completion time would be 19 while its due date is 18. Algorithm 3.3.1 prescribes the deletion of the job with the largest processing time among the first three jobs. Job 2 is therefore deleted, and jobs 1 and 3 remain in the first two positions. If job 4 now follows job 3, it is completed on time at 17; however, if job 5 follows job 4, it is completed late. The algorithm then prescribes to delete the job with the largest processing time among the ones already scheduled, which is job 1. So the optimal schedule is 3, 4, 5, 1, 2 with $\sum U_j = 2$.

The generalization of this problem with weights (i.e., $1 \mid\mid \sum w_j U_j$) is known to be NP-hard (see Appendix C). The special case with all due dates being equal is equivalent to the so-called *knapsack* problem. The due date is equivalent to the size of the knapsack, the processing times of the jobs are equivalent to the sizes of the items, and the weights are equivalent to the benefits obtained by putting the items into the knapsack. A popular heuristic for this problem is the WSPT rule, which sequences the jobs in decreasing order of w_j/p_j. A worst case analysis shows that this heuristic may perform arbitrarily badly, that is, that the ratio

$$\frac{\sum w_j U_j (WSPT)}{\sum w_j U_j (OPT)}$$

may be arbitrarily large.

Example 3.3.4 (A Knapsack and the WSPT Rule)

Consider the following three jobs.

jobs	1	2	3
p_j	11	9	90
w_j	12	9	89
d_j	100	100	100

Scheduling the jobs according to WSPT results in the schedule 1, 2, 3. The third job is completed late and $\sum w_j U_j (WSPT)$ is 89. Scheduling the jobs according to 2, 3, 1 results in $\sum w_j U_j (OPT)$ equal to 12.

3.4 THE TOTAL TARDINESS

The objective $\sum T_j$ is one that is important in practice as well. Minimizing the number of tardy jobs, $\sum U_j$, in practice cannot be the only objective to measure how due dates are being met. Some jobs may have to wait for an unacceptably long time if the number of late jobs is minimized. If instead the sum of the tardinesses is minimized, it is less likely that the wait of any given job will be unacceptably long.

The model $1 \parallel \sum T_j$ has received an enormous amount of attention in the literature. For many years, its computational complexity remained open until its NP-hardness was established by 1987. As $1 \parallel \sum T_j$ is NP-hard in the ordinary sense, it allows for a pseudopolynomial time algorithm based on dynamic programming (see Appendix C). The algorithm is based on two preliminary results.

Lemma 3.4.1.   *If $p_j \leq p_k$ and $d_j \leq d_k$, then there exists an optimal sequence in which job j is scheduled before job k.*

Proof. The proof of this result is left as an exercise.   □

This type of result is useful when an algorithm has to be developed for a problem that is NP-hard. Such a result, often referred to as a *Dominance Result* or *Elimination Criterion*, allows one to disregard a fairly large number of sequences. Such a dominance result may also be thought of as a set of precedence constraints on the jobs. The more precedence constraints created through such dominance results, the easier the problem becomes.

In the following lemma, the sensitivity of an optimal sequence to the due dates is considered. Two problem instances are considered, both of which have n jobs with processing times $p_1, \ldots, p_n$. The first instance has due dates $d_1, \ldots, d_n$. Let C_k' be the latest possible completion time of job k in any optimal sequence, say $\mathcal{S}'$, for this instance. The second instance has due dates $d_1, \ldots, d_{k-1}, \max(d_k, C_k')$, $d_{k+1}, \ldots, d_n$. Let $\mathcal{S}''$ denote the optimal sequence with respect to this second set of due dates and C_j'' the completion of job j under this second sequence.

Lemma 3.4.2.   *Any sequence that is optimal for the second instance is optimal for the first instance as well.*

Proof. Let $V'(\mathcal{S})$ denote the total tardiness under an arbitrary sequence $\mathcal{S}$ with respect to the first set of due dates and let $V''(\mathcal{S})$ denote the total tardiness under sequence $\mathcal{S}$ with respect to the second set of due dates. Now

$$V'(\mathcal{S}') = V''(\mathcal{S}') + A_k$$

and

$$V'(\mathcal{S}'') = V''(\mathcal{S}'') + B_k,$$

where, if $C'_k \leq d_k$, the two sets of due dates are the same and the sequence that is optimal for the second set is therefore also optimal for the first set. If $C'_k \geq d_k$, then

$$A_k = C'_k - d_k$$

and

$$B_k = \max(0, \min(C''_k, C'_k) - d_k).$$

It is clear that $A_k \geq B_k$. As $\mathcal{S}''$ is optimal for the second instance $V''(\mathcal{S}') \geq V''(\mathcal{S}'')$. Therefore, $V'(\mathcal{S}') \geq V'(\mathcal{S}'')$, which completes the proof. □

In the remainder of this section, it is assumed that $d_1 \leq \cdots \leq d_n$ and $p_k = \max(p_1, \ldots, p_n)$. That is, the job with the kth smallest due date has the largest processing time. From Lemma 3.4.1, it follows that there exists an optimal sequence in which jobs $\{1, \ldots, k - 1\}$ all appear, in some order, before job k. Of the remaining $n - k$ jobs (i.e., jobs $\{k + 1, \ldots, n\}$), some may appear before job k and some may appear after job k. The subsequent lemma focuses on these $n - k$ jobs.

Lemma 3.4.3.   *There exists an integer δ, $0 \leq \delta \leq n - k$, such that there is an optimal sequence S in which job k is preceded by all jobs j with $j \leq k + \delta$ and followed by all jobs j with $j > k + \delta$.*

Proof. Let C'_k denote the latest possible completion time of job k in any sequence that is optimal with respect to the given due dates $d_1, \ldots, d_n$. Let $\mathcal{S}''$ be a sequence that is optimal with respect to the due dates $d_1, \ldots, d_{k-1}, \max(C'_k, d_k)$, $d_{k+1}, \ldots, d_n$ and that satisfies the condition stated in Lemma 3.4.1. Let C''_k denote the completion time of job k under this sequence. By Lemma 3.4.2, sequence $\mathcal{S}''$ is also optimal with respect to the original due dates. This implies that $C''_k \leq \max(C'_k, d_k)$. One can assume that job k is not preceded in $\mathcal{S}''$ by a job with a due date later than $\max(C'_k, d_k)$ (if this would have been the case, this job would be on time, and repositioning this job by inserting it immediately after job k would not increase the objective function). Also, job k has to be preceded by all jobs with a due date earlier than $\max(C'_k, d_k)$ (otherwise Lemma 3.4.1 would be violated). So δ can be chosen to be the largest integer such that $d_{k+\delta} \leq \max(C'_k, d_k)$. This completes the proof. □

In the dynamic programming algorithm, a subroutine is required that generates an optimal schedule for the set of jobs $1, \ldots, l$ starting with the processing of this set at time t. Let k be the job with the largest processing time among these l jobs. From Lemma 3.4.3, it follows that for some δ ($0 \leq \delta \leq l - k$) there exists an optimal sequence starting at t that may be regarded as a concatenation of three subsets of jobs—namely,

  (i) jobs $1, 2, \ldots, k - 1, k + 1, \ldots, k + \delta$ in some order, followed by

  (ii) job k, followed by

  (iii) jobs $k + \delta + 1, k + \delta + 2, \ldots, l$ in some order.

The completion time of job k, $C_k(\delta)$, is given by

$$C_k(\delta) = \sum_{j \leq k+\delta} p_j.$$

It is clear that for the entire sequence to be optimal the first and third subsets must be optimally sequenced within themselves. This suggests a dynamic programming procedure that determines an optimal sequence for a larger set of jobs after having determined optimal sequences for proper subsets of the larger set. The subsets J used in this recursive procedure are of a very special type. A subset consists of all the jobs in a set $\{j, j + 1, \ldots, l - 1, l\}$ with processing times smaller than or equal to the processing time p_k of job k. However, job k is *not* part of this subset—not even when $j \leq k \leq l$. Such a subset is denoted by $J(j, l, k)$. Let $V(J(j, l, k), t)$ denote the total tardiness of this subset under an optimal sequence assuming that this subset starts at time t. The dynamic programming procedure can be stated as follows.

    Algorithm 3.4.4. (Minimizing Total Tardiness)

Initial Conditions:

$$V(\emptyset, t) = 0,$$

$$V(\{j\}, t) = \max(0, t + p_j - d_j).$$

Recursive Relation:

$$V(J(j, l, k), t) = \min_{\delta} \left(V(J(j, k' + \delta, k'), t) + \max(0, C_{k'}(\delta) - d_{k'}) \right.$$

$$\left. + V(J(k' + \delta + 1, l, k'), C_{k'}(\delta)) \right)$$

where k' is such that

$$p_{k'} = \max \left(p_{j'} \mid j' \in J(j, l, k) \right).$$

Optimal Value Function:

$$V(\{1, \ldots, n\}, 0).$$

The optimal $\sum T_j$ value is given by $V(\{1, \ldots, n\}, 0)$. The worst case computation time required by this algorithm can be established as follows. There are at most $O(n^3)$ subsets $J(j, l, k)$ and $\sum p_j$ points in time t. There are therefore at most $O(n^3 \sum p_j)$ recursive equations to be solved in the dynamic programming algorithm. As each recursive equation takes $O(n)$ time, the overall running time of the algorithm is bounded by $O(n^4 \sum p_j)$, which is clearly polynomial in n. However, because of the term $\sum p_j$, it only qualifies as a pseudopolynomial time algorithm.

Example 3.4.5 (Minimizing Total Tardiness)

Consider the following five jobs.

jobs	1	2	3	4	5
p_j	121	79	147	83	130
d_j	260	266	266	336	337

The job with the largest processing time is job 3. So $0 \leq \delta \leq 2$. The recursive equation yields:

$$V(\{1, 2, \ldots, 5\}, 0) = \min \begin{cases} V(J(1, 3, 3), 0) + 81 + V(J(4, 5, 3), 347) \\ V(J(1, 4, 3), 0) + 164 + V(J(5, 5, 3), 430) \\ V(J(1, 5, 3), 0) + 294 + V(\varnothing, 560) \end{cases}$$

The optimal sequences of the smaller sets can be determined easily. $V(J(1, 3, 3), 0)$ is clearly zero, and there are two sequences that yield zero: 1, 2 and 2, 1. The value of

$$V(J(4, 5, 3), 347) = 94 + 223 = 317$$

and this is achieved with sequence 4, 5. Also

$$V(J(1, 4, 3), 0) = 0.$$

This value is achieved with the sequences 1, 2, 4 and 2, 1, 4. The value of

$$V(J(5, 5, 3), 430)$$

is equal to 560 minus 337, which is 223. Finally,

$$V(J(1, 5, 3), 0) = 76.$$

This value is achieved with sequences 1, 2, 4, 5 and 2, 1, 4, 5.

$$V(\{1, 2, \ldots, 5\}, 0) = \min \begin{cases} 0 + 81 + 317 \\ 0 + 164 + 223 \\ 76 + 294 + 0 \end{cases} = 370.$$

Two optimal sequences are 1, 2, 4, 5, 3 and 2, 1, 4, 5, 3.

The $1 \mid\mid \sum T_j$ problem can also be solved with a branch and bound procedure. As this branch and bound procedure can also be applied to the more general problem with arbitrary weights, it is presented in the next section.

3.5 THE TOTAL WEIGHTED TARDINESS

The problem $1 \mid\mid \sum w_j T_j$ is an important generalization of the $1 \mid\mid \sum T_j$ problem discussed in the previous section. Dozens of researchers have worked on this problem and have experimented with many different approaches. The approaches range from very sophisticated computer-intensive techniques to fairly crude heuristics designed primarily for implementation purposes.

The dynamic programming algorithm for $1 \mid\mid \sum T_j$ described in the previous section can also deal with agreeable weights—that is, $p_j \geq p_k \implies w_j \leq w_k$. Lemma 3.4.1 can be generalized to this case as follows:

Lemma 3.5.1. *If there are two jobs j and k with $d_j \leq d_k$, $p_j \leq p_k$ and $w_j \geq w_k$, then there exists an optimal sequence in which job j appears before job k.*

Proof. The proof is based on a (not necessarily adjacent) pairwise interchange argument.   □

Unfortunately, no efficient algorithm can be obtained for $1 \mid\mid \sum w_j T_j$ with arbitrary weights.

Theorem 3.5.2. *The problem $1 \mid\mid \sum w_j T_j$ is strongly NP-hard.*

Proof.   The proof is done again by reducing 3-PARTITION to $1 \mid\mid \sum w_j T_j$. The reduction is based on the following transformation. Again, the number of jobs, n, is chosen to be equal to $4t - 1$ and

$$
\begin{aligned}
d_j &= 0, & p_j &= a_j, & w_j &= a_j, & j &= 1, \ldots, 3t, \\
d_j &= (j - 3t)(b + 1), & p_j &= 1, & w_j &= 2, & j &= 3t + 1, \ldots, 4t - 1.
\end{aligned}
$$

Let

$$
z = \sum_{1 \leq j \leq k \leq 3t} a_j a_k + \frac{1}{2}(t - 1)tb.
$$

It can be shown that there exists a schedule with an objective value z if and only if there exists a solution for the 3-PARTITION problem. The first $3t$ jobs have a w_j/p_j ratio equal to 1 and are due at time 0. There are $t - 1$ jobs with w_j/p_j ratio

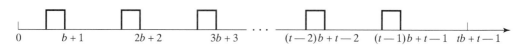

Figure 3.6   3-PARTITION reduces to $1 \mid\mid \sum \omega_j T_j$.

equal to 2 and their due dates are at $b + 1$, $2b + 2$, and so on. A solution with value z can be obtained if these $t - 1$ jobs can be processed exactly during the intervals

$$[b, b + 1], \ [2b + 1, 2b + 2], \ldots, \ [(t - 1)b + t - 2, (t - 1)b + t - 1]$$

(see Figure 3.6). To fit these $t - 1$ jobs in these $t - 1$ intervals, the first $3t$ jobs have to be partitioned into t subsets of three jobs each with the sum of the three processing times in each subset being equal to b. It can be verified that in this case the sum of the weighted tardinesses is equal to z.

     When such a partition is not possible, there is at least one subset of which the sum of the three processing times is larger than b and one other subset of which the sum of the three processing times is smaller than b. It can be verified that in this case the sum of the weighted tardinesses is larger than z.   □

Usually a branch and bound approach is used for $1 \mid\mid \sum w_j T_j$. More often, schedules are constructed starting from the end (i.e., backward in time). At the jth level of the search tree, jobs are put into the $(n - j + 1)$th position. So from each node at level $j - 1$, there are $n - j + 1$ branches going to level j. It may not be necessary to evaluate *all* possible nodes. Dominance results such as the one described in Lemma 3.5.1 may eliminate a number of nodes. The upper bound on the number of nodes at level j is $n!/(n - j)!$ The argument for constructing the sequence backward is that the larger terms in the objective function are likely to correspond to jobs positioned toward the end of the schedule. It appears to be advantageous to schedule these first.

     There are many different bounding techniques. One of the simpler bounding techniques is based on a *relaxation* of the problem to a transportation problem. In this procedure, each job j with (integer) processing time p_j is divided into p_j jobs, each with unit processing time. The decision variables x_{jk} is 1 if one unit of job j is processed during the time interval $[k - 1, k]$ and 0 otherwise. These decision variables x_{jk} must satisfy two sets of constraints:

$$\sum_{k=1}^{C_{\max}} x_{jk} = p_j, \qquad j = 1, \ldots, n$$

$$\sum_{j=1}^{n} x_{jk} = 1, \qquad k = 1, \ldots, C_{\max}.$$

Clearly, a solution satisfying these constraints does not guarantee a feasible schedule without preemptions. Define cost coefficients c_{jk} that satisfy

$$\sum_{k=l-p_j+1}^{l} c_{jk} \leq w_j \max(l - d_j, 0)$$

for $j = 1, \ldots, n; \ l = 1, \ldots, C_{\max}$. Then the minimum cost solution provides a lower bound since for any solution of the transportation problem with $x_{jk} = 1$ for $k = C_j - p_j + 1, \ldots, C_j$ the following holds

$$\sum_{k=1}^{C_{\max}} c_{jk} x_{jk} = \sum_{k=C_j-p_j+1}^{C_j} c_{jk} \leq w_j \max(C_j - d_j, 0).$$

It is fairly easy to find cost functions that satisfy this relationship. For example, set

$$c_{jk} = \begin{cases} 0, & \text{for } k \leq d_j \\ w_j, & \text{for } k > d_j. \end{cases}$$

The solution of the transportation problem provides a lower bound for $1 \ || \ \sum w_j T_j$. This bounding technique is applied to the set of unscheduled jobs at each node of the tree. If the lower bound is larger than the solution of any known schedule, the node may be eliminated.

Example 3.5.3 (Minimizing Total Weighted Tardiness)

Consider the following four jobs.

jobs	1	2	3	4
w_j	4	5	3	5
p_j	12	8	15	9
d_j	16	26	25	27

From Lemma 3.5.1, it immediately follows that in an optimal sequence job 4 follows job 2 and job 3 follows job 1. The branch and bound tree is constructed backward in time. Only two jobs have to be considered as candidates for the last position—namely, jobs 3 and 4. The nodes of the branch and bound tree that need to be investigated are depicted in Figure 3.7. To select a branch to search first, bounds are determined for both nodes at level 1.

A lower bound for an optimal sequence among the offspring of node $(*, *, *, 4)$ can be determined by considering the transportation problem described before applied to jobs 1, 2, and 3. The cost functions are chosen as follows

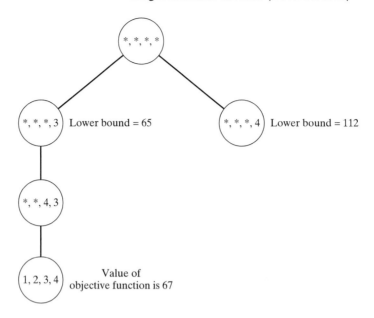

Figure 3.7   Branch and bound procedure for Example 3.5.3.

$$c_{1k} = 0, \qquad k = 1, \dots, 16$$
$$c_{1k} = 4, \qquad k = 17, \dots, 35$$
$$c_{2k} = 0, \qquad k = 1, \dots, 26$$
$$c_{2k} = 5, \qquad k = 27, \dots, 35$$
$$c_{3k} = 0, \qquad k = 1, \dots, 25$$
$$c_{3k} = 3, \qquad k = 26, \dots, 35$$

The optimal allocation of job segments to time slots puts job 1 in the first 12 slots, job 2 into slots 19 to 26, and job 3 into slots 13 to 18 and 27 to 35 (this optimal solution can be found by solving a transportation problem, but can also be found easily by trial and error). The cost of this allocation of the three jobs is 3×9 (the cost of allocating job 3 to slots 27 to 35). To obtain a lower bound for the node, the tardiness of job 4 has to be added; this results in the lower bound $27 + 85$, which equals 112.

In a similar fashion, a lower bound can be obtained for node $(*, *, *, 3)$. A lower bound for an optimal schedule for jobs 1, 2, and 4 yields 8, whereas the tardiness of job 3 is 57, resulting in a bound of 65.

As node $(*, *, *, 3)$ appears to be the more promising node, the offspring of this node is considered first. It turns out that the best schedule reachable from this node is 1, 2, 4, 3 with an objective value of 67.

From the fact that the lower bound for $(*, *, *, 4)$ is 112, it follows that 1, 2, 4, 3 is the best overall schedule.

There are many heuristic procedures for this problem. In Chapter 14, a composite dispatching rule, the so-called *Apparent Tardiness Cost (ATC)* rule, is described in detail.

3.6 DISCUSSION

All the models considered in this chapter have regular objective functions. This is one of the reasons that most of the models are relatively easy.

Some are solvable via simple priority (dispatching) rules (e.g., WSPT, EDD). Most of the other models that are not solvable via simple priority rules are still solvable either in polynomial or pseudopolynomial time. The models that are solvable in polynomial time are usually dealt with through dynamic programming (e.g., $1 \mid prec \mid h_{max}$, $1 \mid\mid \sum T_j$).

The only strongly NP-hard problem in this chapter is $1 \mid\mid \sum w_j T_j$. This problem has received an enormous amount of attention in the literature. There are two approaches for obtaining optimal solutions—namely branch and bound and dynamic programming. Section 3.5 presents a branch and bound approach. Appendix B describes a dynamic programming approach that is applied to the more general problem setting $1 \mid\mid \sum h_j(C_j)$.

The next chapter considers more general and harder single machine problems. It focuses on problems with nonregular objective functions as well as problems with multiple objective functions.

EXERCISES (COMPUTATIONAL)

3.1. Consider $1 \mid\mid \sum w_j C_j$ with the following weights and processing times.

jobs	1	2	3	4	5	6	7
w_j	0	18	12	8	8	17	16
p_j	3	6	6	5	4	8	9

    (a) Find *all* optimal sequences.
    (b) Determine the effect of a change in p_2 from 6 to 7 on the optimal sequence(s).
    (c) Determine the effect of the change under (b) on the value of the objective.

3.2. Consider $1 \mid chains \mid \sum w_j C_j$ with the same set of jobs as in Exercise 3.1.(a). The jobs are now subject to precedence constraints that take the form of chains:

$$1 \rightarrow 2$$
$$3 \rightarrow 4 \rightarrow 5$$
$$6 \rightarrow 7$$

Find all optimal sequences.

3.3. Consider $1 \ || \ \sum w_j (1 - e^{-rC_j})$ with the same set of jobs as in Exercise 3.1.
(a) Assume the discount rate r is 0.05. Find the optimal sequence. Is it unique?
(b) Assume the discount rate r is 0.5. Does the optimal sequence change?

3.4. Find all optimal sequences for the instance of $1 \ || \ h_{\max}$ with the following jobs.

jobs	1	2	3	4	5	6	7
p_j	4	8	12	7	6	9	9
$h_j(C_j)$	$3C_1$	77	C_3^2	$1.5C_4$	$70 + \sqrt{C_5}$	$1.6C_6$	$1.4C_7$

3.5. Consider $1 \ | \ prec \ | \ h_{\max}$ with the same set of jobs as in Exercise 3.4 and the following precedence constraints.

$$1 \rightarrow 7 \rightarrow 6$$
$$5 \rightarrow 7$$
$$5 \rightarrow 4$$

Find the optimal sequence.

3.6. Solve by branch and bound the following instance of the $1 \ | \ r_j \ | \ L_{\max}$ problem.

jobs	1	2	3	4	5	6	7
p_j	6	18	12	10	10	17	16
r_j	0	0	0	14	25	25	50
d_j	8	42	44	24	90	85	68

3.7. Consider the same problem as in the previous exercise. However, now the jobs are subject to the following precedence constraints:

$$2 \rightarrow 1 \rightarrow 4$$
$$6 \rightarrow 7$$

Find the optimal sequence.

3.8. Find the optimal sequence for the following instance of the $1 \ || \ \sum T_j$ problem.

jobs	1	2	3	4	5	6	7	8
p_j	6	18	12	10	10	11	5	7
d_j	8	42	44	24	26	26	70	75

Hint: Before applying the dynamic programming algorithm, consider first the elimination criterion in Lemma 3.4.1.

3.9. Find the optimal sequence for the following instance of the $1 \mid\mid \sum T_j$ problem.

jobs	1	2	3	4	5	6	7
p_j	6	12	10	18	10	17	16
d_j	8	42	44	24	89	80	63

3.10. Find the optimal sequence for the following instance of the $1 \mid\mid \sum w_j T_j$ problem.

jobs	1	2	3	4	5	6	7
p_j	6	18	12	10	10	17	16
w_j	1	5	2	4	1	4	2
d_j	8	42	44	24	90	85	68

EXERCISES (THEORY)

3.11. Consider $1 \mid\mid \sum w_j(1 - e^{-rC_j})$. Assume that $w_j/p_j \neq w_k/p_k$ for all j and k. Show that for r sufficiently close to zero the optimal sequence is WSPT.

3.12. Show that if all jobs have equal weights (i.e., $w_j = 1$ for all j), the WDSPT rule is equivalent to the *Shortest Processing Time first* (SPT) rule for any r, $0 < r < 1$.

3.13. Consider the problem $1 \mid prmp \mid \sum h_j(C_j)$. Show that if the functions h_j are *nondecreasing*, there exists an optimal schedule that is nonpreemptive. Does the result continue to hold for arbitrary functions h_j?

3.14. Consider the problem $1 \mid r_j \mid \sum C_j$.

(a) Show through a counterexample that the nonpreemptive rule, which schedules whenever a machine is freed the shortest job among those available for processing, does not always yield an optimal schedule. In parts (b) and (c), this rule is referred to as SPT*.

(b) Perform a worst case analysis of the SPT* rule; determine the maximum value that the ratio $\sum C_j(SPT^*)/\sum C_j(OPT)$ can take.

(c) Design a heuristic for $1 \mid r_j \mid C_j$ that performs better than SPT*.

3.15. Consider the problem $1 \mid r_j, prmp \mid \sum C_j$. Show that the preemptive *Shortest Remaining Processing Time first (SRPT)* rule is optimal.

3.16. Consider the problem $1 \mid prmp \mid \sum C_j$ with the additional restriction that job j *has to* be completed by a hard deadline $\bar{d}_j$. Assuming that there are feasible schedules, give an algorithm that minimizes the total completion time and prove that it leads to optimality.

3.17. Consider the following preemptive version of the WSPT rule: If $p_j(t)$ denotes the *remaining* processing time of job j at time t, then a preemptive version of the WSPT rule puts at every point in time the job with the highest $w_j/p_j(t)$ ratio on the machine. Show through a counterexample that this rule is not necessarily optimal for $1 \mid r_j, prmp \mid \sum w_j C_j$.

3.18. Give an algorithm for $1 \mid intree \mid \sum w_j C_j$ and prove that it leads to an optimal schedule (recall that in an intree each job has at most one successor).

3.19. Give an algorithm for $1 \mid outtree \mid \sum w_j C_j$ and show that it leads to an optimal schedule (recall that in an outtree each job has at most one predecessor).

3.20. Consider the problem $1 \mid\mid L_{\max}$. The *Minimum Slack first (MS)* rule selects at time t, when a machine is freed, among the remaining jobs the job with the minimum slack $\max(d_j - p_j - t, 0)$. Show through a counterexample that this rule is not necessarily optimal.

3.21. Perform an Adjacent Sequence Interchange for the weighted discounted flow time cost function. That is, state and prove a result analogous to Lemma 3.1.2.

3.22. Consider the problem $1 \mid chains \mid \sum w_j(1 - e^{-rC_j})$. Describe the algorithm that solves this problem and prove that it leads to the optimal sequence.

3.23. Consider the problem $1 \mid prec \mid \max(h_1(S_1), \dots, h_n(S_n))$, where S_j denotes the starting time of job j. The cost function h_j, $j = 1, \dots, n$ is *decreasing*. Unforced idleness of the machine is *not* allowed. Describe a dynamic programming type algorithm for this problem similar to the one in Section 3.2. Why does one have to use forward dynamic programming instead of backward dynamic programming?

3.24. Consider the problem $1 \mid r_j, prmp \mid L_{\max}$. Determine the optimal schedule and prove its optimality.

3.25. Show that
(a) SPT is optimal for $1 \mid brkdwn \mid \sum C_j$,
(b) Algorithm 3.3.1 is optimal for $1 \mid brkdwn \mid \sum U_j$,
(c) WSPT is not necessarily optimal for $1 \mid brkdwn \mid \sum w_j C_j$.

3.26. Consider $1 \mid\mid \sum w_j T_j$. Prove or disprove the following statement: If

$$w_j/p_j > w_k/p_k$$

$$p_j < p_k$$

and

$$d_j < d_k,$$

then there exists an optimal sequence in which job j appears before job k.

COMMENTS AND REFERENCES

The optimality of the WSPT rule for $1 \parallel \sum w_j C_j$ appears in the seminal paper by W.E. Smith (1956). Lawler (1978), Monma and Sidney (1979, 1987), Möhring and Radermacher (1985a), and Sidney and Steiner (1986) all present very elegant results for $1 \mid prec \mid \sum w_j C_j$; the classes of precedence constraints considered in these papers are fairly general and include chains as well as intrees and outtrees. The $1 \mid r_j, prmp \mid \sum C_j$ problem has been analyzed by Schrage (1968). The complexity proof for $1 \mid r_j, prmp \mid \sum w_j C_j$ is due to Labetoulle, Lawler, Lenstra, and Rinnooy Kan (1984). Rothkopf (1966a, 1966b) and Rothkopf and Smith (1984) analyze $1 \parallel \sum w_j (1 - e^{-rC_j})$.

The EDD rule is from Jackson (1955) and the algorithm for $1 \mid prec \mid h_{\max}$ is from Lawler (1973). The complexity proof for $1 \mid r_j \mid L_{\max}$ appears in Lenstra, Rinnooy Kan, and Brucker (1977). Many researchers have worked on branch and bound methods for $1 \mid r_j \mid L_{\max}$; see, for example, McMahon and Florian (1975), Carlier (1982), and Nowicki and Zdrzalka (1986). Potts (1980) analyzes a heuristic for $1 \mid r_j \mid L_{\max}$.

Algorithm 3.3.1, which minimizes the number of late jobs, is from Moore (1968). Kise, Ibaraki, and Mine (1978) consider the $1 \mid r_j \sum U_j$ problem. The NP-hardness of $1 \parallel \sum w_j U_j$ (i.e., the knapsack problem) is established in a classic paper by Karp (1972) on computational complexity. Sahni (1976) presents a pseudopolynomial time algorithm for this problem, and Gens and Levner (1981) and Ibarra and Kim (1978) give fast approximation algorithms. Potts and Van Wassenhove (1988) give a very efficient algorithm for a Linear Programming relaxation of the knapsack problem. A problem related to the knapsack problem is the so-called *due date assignment problem*. This problem has received a lot of attention as well; see Panwalkar, Smith and Seidmann (1982) and Cheng and Gupta (1989).

The dominance condition in Lemma 3.4.1 is due to Emmons (1969), and the pseudopolynomial time Algorithm 3.4.4 is from Lawler (1977). The NP-hardness of $1 \parallel \sum T_j$ is shown by Du and Leung (1990). For additional work on dynamic programming and other approaches for this problem, see Potts and van Wassenhove (1982, 1987).

The complexity of $1 \parallel \sum w_j T_j$ is established in Lawler (1977) as well as Lenstra, Rinnooy Kan, and Brucker (1977). Branch and bound methods using bounding techniques based on relaxations to transportation problems, are discussed in Gelders and Kleindorfer (1974, 1975). Many other approaches have been suggested for $1 \parallel \sum w_j T_j$; see, for example, Fisher (1976, 1981), Potts and van Wassenhove (1985), and Rachamadugu (1987).

4

More Advanced Single Machine Models (Deterministic)

This chapter focuses on a number of more advanced topics in single machine scheduling. Some topics are important because of the theoretical insights they provide, whereas others are important because of their practical implications.

The first section considers the same total tardiness problem discussed in Chapter 3. It describes a polynomial approximation scheme for the total tardiness problem that yields solutions close to optimal. The next section focuses on a generalization of the total tardiness problem; in addition to tardiness costs, there are also earliness costs. These scheduling problems have nonregular objective functions. The third section focuses on problems with primary and secondary objectives. The goal is to first determine the set of all schedules that are optimal with respect to the primary objective; within this set, a schedule has to be found that is optimal with respect to the secondary objective. The fourth section also focuses on problems with two objectives. However, in this section, the weights of the two objectives may be arbitrary. The last section considers the makespan when there are sequence-dependent setup times. There are two reasons for considering the makespan last. First, in most single machine environments, the makespan does not depend on the sequence and is therefore not that important. Second, in the case of sequence-dependent setup times, when the makespan does depend on the sequence, the algorithms tend to be complicated.

4.1 THE TOTAL TARDINESS: AN APPROXIMATION SCHEME

Since $1 \ || \ \sum T_j$ is NP-hard, neither branch and bound nor dynamic programming can yield an optimal solution in polynomial time. Therefore, it may be of interest to have an algorithm that finds, in polynomial time, a solution that is close to optimal.

An approximation scheme A is called fully polynomial if the value of the objective it achieves, say $\sum T_j(A)$, satisfies

$$\sum T_j(A) \leq (1 + \epsilon) \sum T_j(OPT),$$

where $\sum T_j(OPT)$ is the value of the objective under an optimal schedule. Moreover, for the approximation scheme to be fully polynomial, its worst case running time has to be bounded by a polynomial of a fixed degree in n and $1/\epsilon$. The remainder of this section discusses how the dynamic programming algorithm described in Section 3.4 can be used to construct a fully Polynomial Time Approximation Scheme (PTAS).

It can be shown that a given set of n jobs can be scheduled with zero total tardiness if and only if the EDD schedule results in a zero total tardiness. Let $\sum T_j(EDD)$ denote the total tardiness under the EDD sequence and $T_{\max}(EDD)$ the maximum tardiness—that is, $\max(T_1, \ldots, T_n)$, under the EDD sequence. Clearly,

$$T_{\max}(EDD) \leq \sum T_j(OPT) \leq \sum T_j(EDD) \leq n T_{\max}(EDD).$$

Let $V(J, t)$ denote the minimum total tardiness of the subset of jobs J, which starts processing at time t. For any given subset J, a time t^* can be computed such that $V(J, t) = 0$ for $t \leq t^*$ and $V(J, t) > 0$ for $t > t^*$. Moreover, it can easily be shown that

$$V(J, t^* + \delta) \geq \delta$$

for $\delta \geq 0$. So in executing the pseudopolynomial dynamic programming algorithm described before, one only has to compute $V(J, t)$ for

$$t^* \leq t \leq n T_{\max}(EDD).$$

Substituting $\sum p_j$ in the overall running time of the dynamic programming algorithm by $n T_{\max}(EDD)$ yields a new running time bound of $O(n^5 T_{\max}(EDD))$.

Now replace the given processing times p_j by the rescaled processing times

$$p'_j = \lfloor p_j/K \rfloor,$$

where K is a suitable chosen scaling factor. (This implies that p'_j is the largest integer that is smaller than or equal to p_j/K.) Replace the due dates d_j by new due dates

$$d'_j = d_j/K$$

(but without rounding). Consider an optimal sequence with respect to the rescaled processing times and the rescaled due dates and call this sequence $\mathcal{S}$. This sequence can be obtained within the time bound $O(n^5 T_{\max}(EDD)/K)$.

Let $\sum T_j^*(\mathcal{S})$ denote the total tardiness under sequence $\mathcal{S}$ with respect to the processing times $K p'_j$ and the original due dates and let $\sum T_j(\mathcal{S})$ denote the total tardiness with respect to the original processing times p_j (which may be slightly larger than $K p'_j$) and the original due dates. From the fact that

$$K p'_j \le p_j < K(p'_j + 1),$$

it follows that

$$\sum T_j^*(\mathcal{S}) \le \sum T_j(OPT) \le \sum T_j(\mathcal{S}) < \sum T_j^*(\mathcal{S}) + K\left(\frac{n(n+1)}{2}\right).$$

From this chain of inequalities, it follows that

$$\sum T_j(\mathcal{S}) - \sum T_j(OPT) < K\left(\frac{n(n+1)}{2}\right).$$

Recall that the goal is for $\mathcal{S}$ to satisfy

$$\sum T_j(\mathcal{S}) - \sum T_j(OPT) \le \epsilon \sum T_j(OPT).$$

If K is chosen such that

$$K = \left(\frac{2\epsilon}{n(n+1)}\right) T_{\max}(EDD),$$

then the stronger result

$$\sum T_j(\mathcal{S}) - \sum T_j(OPT) \le \epsilon T_{\max}(EDD)$$

is obtained. Moreover, for this choice of K, the time bound $O(n^5 T_{\max}(EDD)/K)$ becomes $O(n^7/\epsilon)$, making the approximation scheme fully polynomial.

This fully Polynomial Time Approximation Scheme can be summarized as follows.

Algorithm 4.1.1. (PTAS for Minimizing Total Tardiness)

Step 1. Apply EDD and determine T_{max}.
If $T_{max} = 0$, then $\sum T_j = 0$ and EDD is optimal; STOP.
Otherwise set

$$K = \left(\frac{2\epsilon}{n(n+1)} \right) T_{max}(EDD).$$

Step 2. Rescale processing times and due dates as follows:

$$p'_j = \lfloor p_j/K \rfloor,$$
$$d'_j = d_j/K.$$

Step 3. Apply Algorithm 3.4.4 to the rescaled data.

The sequence generated by this algorithm, say sequence $\mathcal{S}$, satisfies

$$\sum T_j(\mathcal{S}) \le (1+\epsilon) \sum T_j(OPT).$$

The following example illustrates the approximation scheme.

Example 4.1.2 (PTAS Minimizing Total Tardiness)

Consider a single machine and five jobs.

jobs	1	2	3	4	5
p_j	1210	790	1470	830	1300
d_j	1996	2000	2660	3360	3370

It can be verified (via dynamic programming) that the optimal sequence is 1, 2, 4, 5, 3, and that the total tardiness under this optimal sequence is 3700.

Applying EDD yields $T_{max}(EDD) = 2230$. If ϵ is chosen 0.02, then $K = 2.973$. The rescaled data are:

jobs	1	2	3	4	5
p_j	406	265	494	279	437
d_j	671.38	672.72	894.72	1130.17	1133.54

Solving this instance using the dynamic programming procedure described in Section 3.4 yields two optimal sequences: 1, 2, 4, 5, 3 and 2, 1, 4, 5, 3. If sequence 2, 1, 4, 5, 3 is applied to the original data set, then the total tardiness is 3704. Clearly,

$$\sum T_j(2, 1, 4, 5, 3) \le (1.02) \sum T_j(1, 2, 4, 5, 3).$$

4.2 THE TOTAL EARLINESS AND TARDINESS

All objective functions considered in Chapter 3 are regular performance measures (i.e., nondecreasing in C_j for all j). In practice, it may occur that if job j is completed before its due date d_j, an earliness penalty is incurred. The earliness of job j is defined as

$$E_j = \max(d_j - C_j, 0).$$

The objective function in this section is a generalization of the total tardiness objective. It is the sum of the total earliness and the total tardiness—that is,

$$\sum_{j=1}^{n} E_j + \sum_{j=1}^{n} T_j.$$

Since this problem is harder than the total tardiness problem, it makes sense to first analyze special cases that are tractable. Consider the special case with all jobs having the same due date (i.e., $d_j = d$ for all j).

An optimal schedule for this special case has a number of useful properties. For example, it can easily be shown that after the first job is started the n jobs have to be processed without interruption (i.e., there should be no unforced idleness in between the processing of any two consecutive jobs; see Exercise 4.11). However, it is possible that an optimal schedule does not start processing the jobs immediately at time 0; it may have to wait a while before it starts with the first job.

A second property concerns the actual sequence of the jobs. Under any given schedule, the jobs can be partitioned into two disjoint sets. One set contains the jobs that are completed early (i.e., $C_j \leq d$), and the other set contains the jobs that are completed late (i.e., $C_j > d$). The first set of jobs is referred to as set J_1 and the second set of jobs as set J_2.

Lemma 4.2.1. *In an optimal schedule, the early jobs, set J_1, are scheduled according to LPT, and the late jobs, set J_2, are scheduled according to SPT.*

Proof.   The proof is easy and left as an exercise (see Exercise 4.12).   □

Because of the property described in Lemma 4.2.1, it is often said that the optimal schedule has a V shape.

Consider an instance with the property that no optimal schedule starts processing its first job at $t = 0$ (i.e., the due date d is somewhat loose and the machine remains idle for some time before it starts processing its first job). If this is the case, then the following property holds.

Lemma 4.2.2. *There exists an optimal schedule in which one job is completed exactly at time d.*

Proof. The proof is by contradiction. Suppose there is no such schedule. Then there is always one job that starts its processing before d and completes its processing after d. Call this job j^*. Let $|J_1|$ denote the number of jobs that are early and $|J_2|$ the number of jobs that are late. If $|J_1| < |J_2|$, then shift the entire schedule to the left in such a way that job j^* completes its processing exactly at time d. This implies that the total tardiness decreases by $|J_2|$ times the length of the shift while the total earliness increases by $|J_1|$ times the shift. Clearly, the total earliness plus the total tardiness is reduced. The case $|J_1| > |J_2|$ can be treated in a similar way.

The case $|J_1| = |J_2|$ is somewhat special. In this case, there are many optimal schedules, of which only two satisfy the property stated in the lemma.   □

For an instance in which all optimal schedules start processing the first job some time after $t = 0$, the following algorithm yields the optimal allocations of jobs to sets J_1 and J_2. Assume $p_1 \geq p_2 \geq \cdots \geq p_n$.

Algorithm 4.2.3. (Minimizing Total Earliness and Tardiness with a Loose Due Date)

Step 1. *Assign job 1 to set J_1.*
Set $k = 2$.

Step 2. *Assign job k to set J_1 and job $k + 1$ to set J_2 or vice versa.*

Step 3. *If $k + 2 \leq n - 1$, set $k = k + 2$ and go to Step 2.*
If $k + 2 = n$, assign job n to either set J_1 or set J_2 and STOP.
If $k + 2 = n + 1$, all jobs have been assigned; STOP.

This algorithm is somewhat flexible in its assignment of jobs to sets J_1 and J_2. It can be implemented in such a way that in the optimal assignment the total processing time of the jobs assigned to J_1 is minimized. Given the total processing time of the jobs in J_1 and the due date d, it can be verified easily whether the machine indeed must remain idle before it starts processing its first job.

When the due date d is tight and it is necessary to start processing jobs immediately at time zero, the problem is NP-hard. However, the following heuristic, which assigns the n jobs to the n positions in the sequence, is very effective. Assume again $p_1 \geq p_2 \geq \cdots \geq p_n$.

Algorithm 4.2.4. (Minimizing Total Earliness and Tardiness with a Tight Due Date)

Step 1. *Set $\tau_1 = d$ and $\tau_2 = \sum p_j - d$.*
Set $k = 1$.

Step 2. *If $\tau_1 > \tau_2$, assign job k to the first unfilled position in the sequence and set $\tau_1 = \tau_1 - p_k$.*
If $\tau_1 < \tau_2$, assign job k to the last unfilled position in the sequence and set $\tau_2 = \tau_2 - p_k$.

Step 3. *If $k < n$, set $k = k + 1$ and go to Step 2.*
If $k = n$, STOP.

Example 4.2.5 (Minimizing Total Earliness and Tardiness with Tight Due Date)

Consider the following example with six jobs and $d = 180$.

jobs	1	2	3	4	5	6
p_j	106	100	96	22	20	2

Applying the heuristic yields the following results.

τ_1	τ_2	Assignment	Sequence
180	166	Job 1 Placed First	1xxxxx
74	166	Job 2 Placed Last	1xxxx2
74	66	Job 3 Placed First	13xxx2
−22	66	Job 4 Placed Last	13xx42
−22	44	Job 5 Placed Last	13x542
−22	12	Job 6 Placed Last	136542

Consider now the objective $\sum w'E_j + \sum w''T_j$ and assume again that all the due dates are the same (i.e., $d_j = d$), for all j. All jobs have exactly the same cost function, but the earliness penalty w' and the tardiness penalty w'' are not the same. All previous properties and algorithms can be generalized relatively easily to take the difference between w' and w'' into account (see Exercises 4.13 and 4.14).

Consider the even more general objective $\sum w'_j E_j + \sum w''_j T_j$ and $d_j = d$ for all j. So all jobs have the same due date, but the shapes of their cost functions are different (see Figure 4.1). The LPT-SPT sequence of Lemma 4.2.1 is not necessarily optimal in this case. The first part of the sequence must now be ordered in increasing order of w_j/p_j—that is, according to the Weighted Longest Processing Time first (WLPT) rule, and the second part of the sequence must be ordered according to the Weighted Shortest Processing Time first (WSPT) rule.

Consider the model with the objective function $\sum w'E_j + \sum w''T_j$ and with each job having a different due date (see Figure 4.2). It is clear that this problem is NP-hard since it is a more general model than the one considered in Section 3.4. This problem has an additional level of complexity. Because of the different due

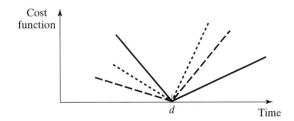

Figure 4.1   Cost functions with common due date and different shapes.

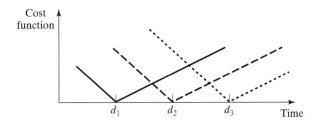

Figure 4.2   Cost functions with different due dates and similar shapes.

dates, it may not necessarily be optimal to process the jobs one after another without interruption; it may be necessary to have idle times between the processing of consecutive jobs. Therefore, this problem has two aspects: One aspect concerns the search for an optimal order in which to sequence the jobs, and the other aspect concerns the computation of the optimal starting times and completion times of the jobs. These two optimization problems are clearly not independent. Determining the optimal schedule is therefore a very hard problem. Approaches for dealing with this problem are typically based on dynamic programming or branch and bound. However, given any predetermined and fixed sequence, the timing of the processing of the jobs (and therefore also the idle times) can be determined via a relatively simple polynomial time algorithm. This polynomial time algorithm is also applicable in a more general setting that is described next.

The more general setting has as objective $\sum w'_j E_j + \sum w''_j T_j$, where the jobs have different due dates and different weights. This problem is clearly strongly NP-hard, since it is harder than the total weighted tardiness problem considered in Section 3.5. But, given a predetermined ordering of the jobs, the timings of the processings and the idle times can be computed in polynomial time. Some preliminary results are necessary to describe the algorithm that inserts the idle times in a given sequence. Assume that the job sequence $1, \ldots, n$ is fixed.

Lemma 4.2.6.   *If $d_{j+1} - d_j \leq p_{j+1}$, then there is no idle time between jobs j and j + 1.*

Proof. The proof is by contradiction. Consider three cases: Job j is early $(C_j < d_j)$, job j is completed exactly at its due date $(C_j = d_j)$, and job j is late $(C_j > d_j)$.

Case 1. If job j is completed early and there is an idle time between jobs j and $j + 1$, then the objective can be reduced by postponing the processing of job j and reducing the idle time. The schedule with the idle time can therefore not be optimal.

Case 2. If job j is completed at its due date and there is an idle time, then job $j + 1$ is completed late. Processing job $j + 1$ earlier and eliminating the idle time reduce the total objective. So the original schedule cannot be optimal.

Case 3. If job j is completed late and there is an idle time, then job $j + 1$ is also completed late. Processing job $j + 1$ earlier reduces the objective. □

Subsequence $u, \dots , v$ is called a job cluster if, for each pair of adjacent jobs j and $j + 1$, the inequality

$$d_{j+1} - d_j \le p_{j+1}$$

holds and if for $j = u - 1$ and $j = v$ the inequality does not hold. A cluster of jobs must therefore be processed without interruptions.

Lemma 4.2.7. *In each cluster in a schedule, the early jobs precede the tardy jobs. Moreover, if jobs j and $j + 1$ are in the same cluster and are both early, then $E_j \ge E_{j+1}$. If jobs j and $j + 1$ are both late, then $T_j \le T_{j+1}$.*

Proof. Assume jobs j and $j + 1$ belong to the same cluster. Let t denote the optimal start time of job j. Subtracting $t + p_j$ from both sides of

$$d_{j+1} - d_j \le p_{j+1}$$

and rearranging yields

$$d_{j+1} - t - p_j - p_{j+1} \le d_j - t - p_j.$$

This last inequality can be rewritten as

$$d_j - C_j \ge d_{j+1} - C_{j+1},$$

which implies the lemma. □

The given job sequence $1, \dots , n$ can be decomposed into a set of m clusters $\sigma_1, \sigma_2, \dots , \sigma_m$ with each cluster representing a subsequence. The algorithm that inserts the idle times starts out with the given sequence $1, \dots , n$ without idle times.

That is, the completion time of the kth job in the sequence is

$$C_k = \sum_{j=1}^{k} p_j.$$

Since the completion times in the original schedule are the earliest possible completion times, the algorithm has to determine how much to increase (or shift) the completion time of each job. In fact, it is sufficient to compute the optimal shift for each cluster since all its jobs are shifted by the same amount.

Consider a cluster σ_r that consists of jobs $k, k+1, \ldots, \ell$. Let

$$\Delta_j = \sum_{l=k}^{j} w_l' - \sum_{l=j+1}^{\ell} w_l'', \quad j = k, \ldots, \ell.$$

These numbers can be computed recursively by setting

$$\Delta_{k-1} = -\sum_{l=k}^{\ell} w_l''$$

and

$$\Delta_j = \Delta_{j-1} + w_j' + w_j'', \quad j = k, \ldots, \ell.$$

Define a block as a sequence of clusters that are processed without interruption. Consider block $\sigma_s, \sigma_{s+1}, \ldots, \sigma_m$. This block may have one or more clusters preceding it, but no clusters following it. Let j_r be the last job in cluster σ_r that is early (i.e., the job with the smallest earliness). Let

$$E(r) = E_{j_r} = d_{j_r} - C_{j_r}.$$

Clearly,

$$E(r) = \min_{k \leq j \leq j_r} (d_j - C_j).$$

Let

$$\Delta(r) = \Delta_{j_r} = \max_{k \leq j \leq j_r} \Delta_j.$$

If none of the jobs in cluster σ_r is early, then $E(r) = \infty$ and $\Delta(r) = -\sum_{l=k}^{\ell} w_l''$. If $d_{j_r} - C_{j_r} \geq 1$ for the last early job in every cluster $\sigma_r, r = s, s+1, \ldots, m$, then a shift of the entire block by one time unit to the right decreases the total cost by

$$\sum_{r=s}^{m} \Delta(r).$$

The idea behind the algorithm is to find the first (largest) block of clusters that cannot be shifted. This block stays in place for the time being. The procedure is repeated for the reduced set of clusters to find the next block that cannot be shifted. Otherwise, all remaining clusters are shifted by an amount that is equal to the smallest $E(r)$, and in one of these shifted clusters the last early job becomes an on-time job. All the completion times are updated and all the nonearly jobs are removed from the list of each cluster. The procedure is then repeated. The algorithm terminates once a block involving the last cluster cannot be shifted. The algorithm can be summarized as follows.

Algorithm 4.2.8. (Optimizing Timings Given a Predetermined Sequence)

Step 1. *Identify the clusters and compute $\Delta(r)$ for each cluster.*

Step 2. *Find the smallest s such that $\sum_{r=1}^{s} \Delta(r) \leq 0$.*
Set the original C_k for each job of the first s clusters.
If $s = m$, then STOP; otherwise go to Step 3.
If no such s exists, then go to Step 4.

Step 3. *Remove the first s clusters from the list.*
Goto Step 2 to consider the reduced set of clusters.

Step 4. *Find $\min(E(1), \dots , E(m))$.*
Increase all C_k by $\min(E(1), \dots , E(m))$.
Eliminate all early jobs that are no longer early.
Update $E(r)$ and $\Delta(r)$. Go to Step 2.

Example 4.2.9 (Optimizing Timings Given a Predetermined Sequence)

Consider seven jobs. The given sequence is $1, \dots , 7$.

jobs	1	2	3	4	5	6	7
p_j	3	2	7	3	6	2	8
d_j	12	4	26	18	16	25	30
w'_j	10	20	18	9	10	16	11
w''_j	12	25	38	12	12	18	15

The set of jobs can be decomposed into the three clusters $\sigma_1, \sigma_2, \sigma_3$, where $\sigma_1 = 1, 2, \sigma_2 = 3, 4, 5$, and $\sigma_3 = 6, 7$. The initial schedule has job completion times

$$3, 5, \quad 12, 15, 21, \quad 23, 31.$$

The sets of early jobs in clusters σ_1, σ_2, and σ_3 are, respectively, jobs (1), (3, 4), and (6). For each one of these jobs, the values of $d_j - C_j$ can be computed. Applying Step 1 of the algorithm results in the following table.

clusters	1	2	3
$E(r)$	9	3	2
$\Delta(r)$	-15	15	1

Step 2 of the Algorithm yields $s = 1$ and $\Delta(1) < 0$. So cluster σ_1 is not shifted and $C_1 = 3$ and $C_2 = 5$. Step 3 eliminates σ_1. Going back to Step 2 yields $\Delta(2) > 0$, $\Delta(2) + \Delta(3) > 0$, and no s exists. Going to Step 4 results in

$$\min(E(2), E(3)) = \min(3, 2) = 2.$$

Increase all completion times in the second and third cluster by two time units and eliminate job 6 from the list of early jobs. Update $d_k - C_k$. The new values of $E(r)$ and $\Delta(r)$ are presented in the following table.

clusters	2	3
$E(r)$	1	∞
$\Delta(r)$	15	-33

Returning to Step 2 yields $\Delta(2) > 0$ and $\Delta(2) + \Delta(3) < 0$. It follows that $s = 3 = m$. So the second and third cluster should not be shifted and the algorithm stops. The optimal completion times are

$$3, 5, 14, 17, 23, 25, 33.$$

As stated earlier, finding at the same time an optimal job sequence as well as an optimal timing of the jobs is strongly NP-hard.

A branch and bound procedure for this problem is more complicated than the one for the total weighted tardiness problem described in Section 3.5. The branching tree can be constructed in a manner that is similar to the one for the $1 \mid\mid \sum w_j T_j$ problem. However, finding good lower bounds for $1 \mid\mid \sum w'_j E_j + \sum w''_j T_j$ is considerably harder. One type of lower bound can be established by first setting $w'_j = 0$ for all j and then applying the lower bound described in Section 3.5 to the given instance by taking only the tardiness penalties into account. This lower bound is probably not that useful since it is based on two simplifications.

It is possible to establish certain dominance conditions. For example, if the due dates of two adjacent jobs both occur before the starting time of the first one of the two jobs, then the job with the higher w''_j / p_j ratio has to go first. Similarly, if the due dates of two adjacent jobs both occur after the completion time of the last one of the two jobs, then the job with the lower w'_j / p_j ratio has to go first.

Many heuristic procedures have been developed for this problem. These procedures are often based on a combination of decomposition and local search. The problem lends itself well to time-based decomposition procedures because it may be possible to tailor the decomposition process to the clusters and the blocks.

4.3 PRIMARY AND SECONDARY OBJECTIVES

In practice, a scheduler is often concerned with more than one objective. For example, he may want to minimize inventory costs and meet due dates. It would then be of interest to find, for example, a schedule that minimizes a combination of $\sum C_j$ and $L_{\max}$.

It is often the case that there is more than one schedule that is optimal with respect to a given objective. A decision maker may wish to consider the set of all schedules that are optimal with respect to a primary objective, and then search within this particular set of schedules for the schedule that is best with regard to a secondary objective. If the primary objective is denoted by γ_1 and the secondary by γ_2, then such a problem can be referred to as $\alpha \mid \beta \mid \gamma_1(opt), \gamma_2$.

Consider the following simple example. The primary objective is the total completion time $\sum C_j$ and the secondary objective is the maximum lateness $L_{\max}$—that is, $1 \mid\mid \sum C_j(opt), L_{\max}$. If there are no jobs with identical processing times, then there is exactly one schedule that minimizes the total completion time. Thus, there is no freedom remaining to minimize $L_{\max}$. If there are jobs with identical processing times, then there are multiple schedules that minimize the total completion time. A set of jobs with identical processing times is preceded by a job with a strictly shorter processing time and followed by a job with a strictly longer processing time. Jobs with identical processing times have to be processed one after another, but they may be done in any order. The decision maker now must find among all the schedules that minimize the total completion time the one that minimizes $L_{\max}$. So in an optimal schedule, a set of jobs with identical processing times has to be sequenced according to the EDD rule. The decision maker has to do so for each set of jobs with identical processing times. This rule may be referred to as SPT/EDD since the jobs are first scheduled according to SPT and ties are broken according to EDD (see Exercise 4.16 for a generalization of this rule).

Consider now the same two objectives with reversed priorities—that is, $1 \mid\mid L_{\max}(opt), \sum C_j$. In Chapter 3, it was shown that the EDD rule minimizes $L_{\max}$. Applying the EDD rule also yields the value of the minimum $L_{\max}$. Assume that the value of this minimum $L_{\max}$ is z. The original problem can be transformed into another problem that is equivalent. Create a new set of due dates $\bar{d}_j = d_j + z$. These new due dates are now deadlines. The problem is to find a schedule that minimizes $\sum C_j$ subject to the constraint that every job must be completed by its deadline (i.e, the maximum lateness with respect to the new due dates has to be zero or, equivalently, all the jobs have to be completed on time).

The algorithm for finding the optimal schedule is based on the result in the following lemma.

Lemma 4.3.1.   *For the single machine problem with n jobs subject to the constraint that all due dates have to be met, there exists a schedule that minimizes $\sum C_j$ in which job k is scheduled last, if and only if*

(i) $\bar{d}_k \geq \sum_{j=1}^{n} p_j,$

(ii) $p_k \geq p_\ell,$ *for all ℓ such that $\bar{d}_\ell \geq \sum_{j=1}^{n} p_j.$*

Proof.   By contradiction. Suppose that job k is not scheduled last. There is a set of jobs that is scheduled after job k and job ℓ is the one scheduled last. Condition (i) must hold for job ℓ otherwise job ℓ would not meet its due date. Assume that condition (ii) does not hold and that $p_\ell < p_k$. Perform a (nonadjacent) pairwise interchange between jobs k and ℓ. Clearly, the sum of the completion times of jobs k and ℓ decrease and the sum of the completion times of all jobs scheduled in between jobs k and ℓ go down as well. So the original schedule that positioned job ℓ last could not have minimized $\sum C_j$.   □

In the next algorithm, J^c denotes the set of jobs that remain to be scheduled.

Algorithm 4.3.2. (Minimizing Total Completion Time with Deadlines)

Step 1. *Set $k = n$, $\tau = \sum_{j=1}^{n} p_j$, $J^c = \{1, \ldots, n\}$.*

Step 2. *Find k^* in J^c such that*
  $\bar{d}_{k^*} \geq \tau$ *and*
  $p_{k^*} \geq p_\ell$, *for all jobs ℓ in J^c such that $\bar{d}_\ell \geq \tau$.*
  Put job k^* in position k of the sequence.

Step 3. *Decrease k by 1.*
  Decrease τ by p_{k^*}.
  Delete job k^* from J^c.

Step 4. *If $k \geq 1$ go to Step 2,*
  otherwise STOP.

This algorithm, similar to the algorithms in Sections 3.2 and 3.5, is a backward algorithm. The following example illustrates the use of this algorithm.

Example 4.3.3 (Minimizing Total Completion Time with Deadlines)

Consider the following instance with five jobs.

jobs	1	2	3	4	5
p_j	4	6	2	4	2
$\bar{d}_j$	10	12	14	18	18

Starting out, $\tau = 18$. Two jobs have a deadline larger than or equal to τ—namely, jobs 4 and 5. Job 4 has the longer processing time and should therefore go last. For the second iteration, the value of τ is reduced to 14. There are two jobs that have a deadline greater than or equal to 14—namely, 3 and 5. So either job can go in the second last position. For the third iteration, the value of τ is reduced further down to 12. Again, there are two jobs that have a deadline greater than or equal to 12—either jobs 2 and 3 or jobs 2 and 5. Clearly, job 2 (with a processing time of 6) should go in the third position. Proceeding in this manner yields two optimal schedules—namely, schedules 5, 1, 2, 3, 4 and 3, 1, 2, 5, 4.

It can be shown that, even when preemptions are allowed, the optimal schedules are nonpreemptive.

A fairly large number of problems of the type $1 \mid \beta \mid \gamma_1(opt), \gamma_2$ have been studied in the literature. Very few can be solved in polynomial time. However, problems of the type $1 \mid \beta \mid \sum w_j C_j(opt), \gamma_2$ tend to be easy (see Exercises 4.16 and 4.17).

4.4 MULTIPLE OBJECTIVES: A PARAMETRIC ANALYSIS

Suppose there are two objectives, γ_1 and γ_2. If the overall objective is $\theta_1\gamma_1 + \theta_2\gamma_2$, where θ_1 and θ_2 are the weights of the two objectives, then a scheduling problem can be denoted by $1 \mid \beta \mid \theta_1\gamma_1 + \theta_2\gamma_2$. Since multiplying both weights by the same constant does not change the problem, it is in what follows assumed that the weights add up to 1 (i.e., $\theta_1 + \theta_2 = 1$). The remaining part of this section focuses on a certain class of schedules.

Definition 4.4.1. (Pareto-Optimal Schedule)   *A schedule is called pareto-optimal if it is not possible to decrease the value of one objective without increasing the value of the other.*

All pareto-optimal solutions can be represented by a set of points in the (γ_1, γ_2) plane. This set of points illustrates the trade-offs between the two objectives. Consider the two objectives analyzed in the previous section (i.e., $\sum C_j$ and $L_{\max}$). The two cases considered in the previous section are the two extreme points of the trade-off curve. If $\theta_1 \to 0$ and $\theta_2 \to 1$, then

$$1 \mid \beta \mid \theta_1\gamma_1 + \theta_2\gamma_2 \ \to \ 1 \mid \beta \mid \gamma_2(opt), \gamma_1.$$

If $\theta_2 \to 0$ and $\theta_1 \to 1$, then

$$1 \mid \beta \mid \theta_1\gamma_1 + \theta_2\gamma_2 \ \to \ 1 \mid \beta \mid \gamma_1(opt), \gamma_2.$$

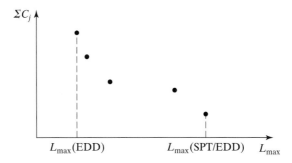

Figure 4.3   Trade-offs between total completion time and maximum lateness.

So the two extreme points of the trade-off curve in Figure 4.3 correspond to the problems discussed in the previous section. At one of the extreme points, the total completion time is minimized by the SPT rule with ties broken according to EDD; the $L_{\max}$ of this schedule can be computed easily and is denoted by $L_{\max}(SPT/EDD)$. At the other extreme point, the schedule is generated according to a more complicated backward procedure. The $L_{\max}$ is equal to $L_{\max}(EDD)$, and the total completion time of this schedule can be computed also. Clearly,

$$L_{\max}(EDD) \leq L_{\max}(SPT/EDD).$$

The algorithm that generates all pareto-optimal solutions in the trade-off curve contains two loops. One series of steps in the algorithm (the inner loop) is an adaptation of Algorithm 4.3.2. These steps determine, in addition to the optimal schedule with a maximum allowable $L_{\max}$, the minimum increment δ in the $L_{\max}$ that would allow for a decrease in the minimum $\sum C_j$. The second (outer) loop of the algorithm contains the structure that generates all the pareto-optimal points. The outer loop calls the inner loop at each pareto-optimal point to generate a schedule at that point and also to determine how to move to the next efficient point. The algorithm starts out with the EDD schedule that generates the first pareto-optimal point in the upper left part of the trade-off curve. It determines the minimum increment in the $L_{\max}$ needed to achieve a reduction in $\sum C_j$. Given this new value of $L_{\max}$, it uses the algorithm in the previous section to determine the schedule that minimizes $\sum C_j$ and proceeds to determine the next increment. This goes on until the algorithm reaches $L_{\max}(SPT/EDD)$.

Algorithm 4.4.2. (Determining Trade-Offs between Total Completion Time and Maximum Lateness)

Step 1. Set $r = 1$.
Set $L_{\max} = L_{\max}(EDD)$ and $\bar{d}_j = d_j + L_{\max}$.

Step 2. *Set $k = n$ and $J^c = \{1, \ldots, n\}$.*
Set $\tau = \sum_{j=1}^{n} p_j$ and $\delta = \tau$.

Step 3. *Find j^* in J^c such that*
$\bar{d}_{j^*} \geq \tau$, and
$p_{j^*} \geq p_l$ for all jobs l in J^c such that $\bar{d}_l \geq \tau$.
Put job j^* in position k of the sequence.

Step 4. *If there is no job ℓ such that $\bar{d}_\ell < \tau$ and $p_\ell > p_{j^*}$, go to Step 5.*
Otherwise find j^{**} such that

$$\tau - \bar{d}_{j^{**}} = \min_\ell (\tau - \bar{d}_\ell)$$

for all ℓ such that $\bar{d}_\ell < \tau$ and $p_\ell > p_{j^*}$.
Set $\delta^{**} = \tau - d_{j^{**}}$.
If $\delta^{**} < \delta$, then $\delta = \delta^{**}$.

Step 5. *Decrease k by 1.*
Decrease τ by p_{j^*}.
Delete job j^* from J^c.
If $k \geq 1$ go to Step 3,
otherwise go to Step 6.

Step 6. *Set $L_{\max} = L_{\max} + \delta$.*
If $L_{\max} > L_{\max}(SPT/EDD)$, then STOP.
Otherwise set $r = r + 1$, $\bar{d}_j = \bar{d}_j + \delta$, and go to Step 2.

The outer loop consists of Steps 1 and 6 while the inner loop consists of Steps 2, 3, 4, and 5. Steps 2, 3, and 5 represent an adaptation of Algorithm 4.3.2, and Step 4 computes the minimum increment in the $L_{\max}$ needed to achieve a reduction in $\sum C_j$.

It can be shown that the maximum number of pareto-optimal solutions is $n(n-1)/2$, which is $O(n^2)$ (see Exercise 4.19). Generating one pareto-optimal schedule can be done in $O(n\log(n))$. The total computation time of Algorithm 4.4.2 is therefore $O(n^3\log(n))$.

Example 4.4.3 (Determining Trade-Offs between Total Completion Time and Maximum Lateness)

Consider the following set of jobs.

jobs	1	2	3	4	5
p_j	1	3	6	7	9
d_j	30	27	20	15	12

The EDD sequence is 5, 4, 3, 2, 1 and $L_{\max}(EDD) = 2$. The SPT/EDD sequence is 1, 2, 3, 4, 5 and $L_{\max}(SPT/EDD) = 14$. Application of Algorithm 4.4.2 results in the following iterations.

Iteration r	$(\sum C_j, L_{\max})$	Pareto-optimal schedule	current $\tau + \delta$	δ
1	96, 2	5, 4, 3, 1, 2	32 29 22 17 14	1
2	77, 3	1, 5, 4, 3, 2	33 30 23 18 15	2
3	75, 5	1, 4, 5, 3, 2	35 32 25 20 17	1
4	64, 6	1, 2, 5, 4, 3	36 33 26 21 18	2
5	62, 8	1, 2, 4, 5, 3	38 35 28 23 20	3
6	60, 11	1, 2, 3, 5, 4	41 38 31 26 23	3
7	58, 14	1, 2, 3, 4, 5	44 41 34 29 26	Stop

However, when one would consider the objective $\theta_1 L_{\max} + \theta_2 \sum C_j$, then certain pareto-optimal schedules never may be optimal no matter what the weights are (see Exercise 4.9).

Consider the generalization $1 \mid\mid \theta_1 \sum w_j C_j + \theta_2 L_{\max}$. It is clear that the two extreme points of the trade-off curve can be determined in polynomial time (using WSPT/EDD and EDD). However, even though the two end-points of the trade-off curve can be analyzed in polynomial time, the problem with arbitrary weights θ_1 and θ_2 is NP-hard.

The trade-off curve that corresponds to the example in this section has the shape of a staircase. This shape is fairly common in a single machine environment with multiple objectives, especially when preemptions are not allowed. However, other machine environments (e.g., parallel machines), may exhibit smoother curves especially when preemptions are allowed (see Chapter 15).

4.5 THE MAKESPAN WITH SEQUENCE-DEPENDENT SETUP TIMES

For single machine scheduling problems with all $r_j = 0$ and no sequence-dependent setup times, the makespan is independent of the sequence and equal to the sum of the processing times. When there are sequence-dependent setup times, the makespan *does* depend on the schedule. In Appendix C, it is shown that $1 \mid s_{jk} \mid C_{\max}$ is strongly NP-hard.

However, the NP-hardness of $1 \mid s_{jk} \mid C_{\max}$ in the case of arbitrary setup times does not rule out the existence of efficient solution procedures when the setup times have a special form. In practice, setup times often do have a special structure.

Consider the following structure. Two parameters are associated with job j, say a_j and b_j, and $s_{jk} = \mid a_k - b_j \mid$. This setup time structure can be described as follows: After the completion of job j, the machine is left in state b_j; to be able to start job k, the machine has to be brought into state a_k. The total setup time necessary for bringing the machine from state b_j to state a_k is proportional to the absolute difference between the two states. This state variable could be, for example, temperature (in the case of an oven) or a measure of some other setting of the machine. In what follows, it is assumed that at time zero the state is b_0 and that after completing the last job the machine has to be left in state a_0 (this implies that an additional cleanup time is needed after the last job is completed).

This particular setup time structure *does* allow for a polynomial time algorithm. The description of the algorithm is actually easier in the context of the Traveling Salesman Problem (TSP). The algorithm is therefore presented here in the context of a TSP with $n + 1$ cities; the additional city being called city 0 with parameters a_0 and b_0. Without loss of generality, it may be assumed that $b_0 \leq b_1 \leq \cdots \leq b_n$. The traveling salesman leaving city j for city k (or, equivalently, job k following job j) is denoted by $k = \phi(j)$. The sequence of cities in a tour is denoted by Φ, which is a vector that maps each element of $\{0, 1, 2, \ldots, n\}$ onto a unique element of $\{0, 1, 2, \ldots, n\}$ by relations $k = \phi(j)$ indicating that the salesman visits city k after city j (or, equivalently, job k follows job j). Such mappings are called permutation mappings. Note that not all possible permutation mappings of $\{0, 1, 2, \ldots, n\}$ constitute feasible TSP tours. For example, $\{0, 1, 2, 3\}$ mapped onto $\{2, 3, 1, 0\}$ represents a feasible TSP. However, $\{0, 1, 2, 3\}$ mapped onto $\{2, 1, 3, 0\}$ does not represent a feasible tour since it represents two disjoint subtours—namely, subtour $0 \rightarrow 2 \rightarrow 3 \rightarrow 0$ and the subtour $1 \rightarrow 1$, which consists of a single city (see Figure 4.4). Define $\phi(k) = k$ to mean a redundant tour that starts and ends at k.

For the special cost structure of going from city j to k, it is clear that this cost is equal to the vertical height of the arrow connecting b_j with a_k in Figure 4.5. Define the cost of a redundant subtour (i.e., $\phi(k) = k$), as the vertical height of an arrow from b_k to a_k.

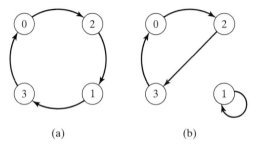

(a)   (b)

Figure 4.4 Permutation mappings. (a) $\{0, 1, 2, 3\} \rightarrow \{2, 3, 1, 0\}$ (b) $\{0, 1, 2, 3\} \rightarrow \{2, 1, 3, 0\}$.

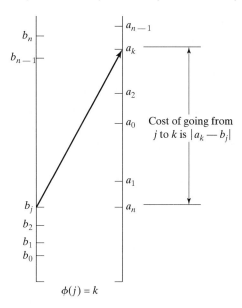

Figure 4.5   Cost of going from j to k.

Thus, any permutation mapping (which might possibly consist of subtours) can be represented as a set of arrows connecting b_j, $j = 0, \ldots, n$ to a_k, $k = 0, \ldots, n$, and the cost associated with such a mapping is simply the sum of the vertical heights of the $n + 1$ arrows.

Define now a *swap* $I(j, k)$ as that procedure which when applied to a permutation mapping Φ produces another permutation mapping Φ' by affecting only the assignments of j and k and leaving the others unchanged. More precisely, the new assignment $\Phi' = \Phi I(j, k)$ is defined as:

$$\phi'(k) = \phi(j),$$
$$\phi'(j) = \phi(k),$$

and

$$\phi'(l) = \phi(l)$$

for all l not equal to j or k. This transformation may also be denoted by $\phi'(j) = \phi(j)I(j, k)$. Note that this is *not* equivalent to an adjacent pairwise interchange within a sequence since a permutation mapping Φ does not always represent a sequence (a feasible TSP tour) to begin with. More intuitively, it only represents a swap of the arrows emanating from b_j and b_k leaving all other arrows unchanged. In particular, if these arrows crossed each other before, they will uncross now and vice versa. The implication of such a swap in terms of the actual tour and subtours is quite surprising, however. It can be easily verified that the swap $I(j, k)$ has the effect of creating two subtours out of one if j and k belong to the same subtour in Φ. Conversely, it combines two subtours to which j and k belong otherwise.

The following lemma quantifies the cost of the interchange $I(j, k)$ applied on the sequence Φ; the cost of this interchange is denoted by $c_\Phi I(j, k)$. In the lemma, the interval of the *unordered* pair $[a, b]$ refers to an interval on the real line and

$$|| [a, b] || = \begin{cases} 2(b - a) & \text{if } b \geq a \\ 2(a - b) & \text{if } b < a \end{cases}$$

Lemma 4.5.1.   *If the swap $I(j, k)$ causes two arrows that did not cross earlier to cross, then the cost of the tour increases and vice versa. The magnitude of this increase or decrease is given by*

$$c_\Phi I(j, k) = || [b_j, b_k] \cap [a_{\phi(j)}, a_{\phi(k)}] ||$$

So the change in cost is equal to twice the length of vertical overlap of the intervals $[b_j, b_k]$ and $[a_{\phi(j)}, a_{\phi(k)}]$.

Proof.   The proof can be divided into several cases and is fairly straightforward since the swap does not affect arrows other than the two considered. Hence it is left as an exercise (see Figure 4.6). ∎

The lemma is significant since it gives a visual cue to reducing costs by uncrossing the arrows that cross and helps quantify the cost savings in terms of amount of overlap of certain intervals. Such a visual interpretation immediately leads to the following result for optimal permutation mappings.

Lemma 4.5.2.   *An optimal permutation mapping Φ^* is obtained if*

$$b_j \leq b_k \implies a_{\phi^*(j)} \leq a_{\phi^*(k)}.$$

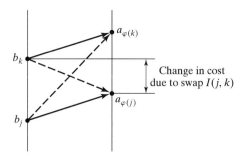

Figure 4.6   Change in cost due to swap $I(j, k)$.

Proof. The statement of the theorem is equivalent to no lines crossing in the diagram. Suppose two of the lines did cross. Performing a swap that uncrosses the lines leads to a solution as good or better than the previous.   □

As mentioned before, this is simply an optimal *permutation mapping* and not necessarily a feasible tour. It does, however, provide a lower bound for the optimal cost of any TSP. This optimal Φ^* may consist of p distinct subtours, say, $TR_1, \ldots, TR_p$. As seen before, performing a swap $I(j, k)$ such that j and k belong to distinct subtours will cause these subtours to coalesce into one and the cost will increase (since now two previously uncrossed lines do cross) by an amount $c_{\Phi^*}I(j, k)$. It is desirable to select j and k from different subtours in such a way that this cost of coalescing $c_{\Phi^*}I(j, k)$ is, in some way, minimized.

To determine these swaps, instead of considering the *directed* graph that represents the subtours of the traveling salesman, consider the undirected version of the same graph. The subtours represent distinct cycles and redundant subtours are simply independent nodes. To connect the disjoint elements (i.e., the cycles corresponding to the subtours) and construct a connected graph, additional arcs have to be inserted in this undirected graph. The costs of the arcs between cities belonging to different subtours in this undirected graph are chosen to be equal to the cost of performing the corresponding swaps in the tour of the traveling salesman in the directed graph. The cost of such a swap can be computed easily by Lemma 4.5.1. The arcs used to connect the disjoint subtours are selected according to the *Greedy Algorithm:* select the cheapest arc that connects two of the p subtours in the undirected graph; select among the remaining unused arcs the cheapest arc connecting two of the $p - 1$ remaining subtours; and so on. The arcs selected then satisfy the following property.

Lemma 4.5.3.   *The collection of arcs that connect the undirected graph with the least cost contain only arcs that connect city j with city $j + 1$.*

Proof.   The cost of the arcs $(c_{\Phi^*}I(j, k))$ needed to connect the distinct cycles of the undirected graph are computed from the optimal permutation mapping defined in Lemma 4.5.2 in which no two arrows cross. It is shown next that the cost of swapping two nonadjacent arrows is at least equal to the cost of swapping all arrows between them. This is easy to see if the cost is regarded as the intersection of two intervals given by Lemma 4.5.1. In particular, if $k > j + 1$,

$$c_{\Phi^*}I(j, k) = || [b_j, b_k] \cap [a_{\phi^*(j)}, a_{\phi^*(k)}] ||$$

$$\geq \sum_{i=j}^{k-1} || [b_i, b_{i+1}] \cap [a_{\phi^*(i)}, a_{\phi^*(i+1)}] ||$$

$$= \sum_{i=j}^{k-1} c_{\Phi^*} I(i, i+1).$$

So the arc (j, k) can be replaced by the sequence of arcs $(i, i+1)$, $i = j, \ldots, k-1$, to connect the two subtours to which j and k belong at as low or lower cost.   □

It is important to note that, in the construction of the undirected graph, the costs assigned to the arcs connecting the subtours were computed under the assumption that the swaps are performed on Φ^* in which no arrows cross. However, as swaps are performed to connect the subtours, this condition no longer remains valid. However, it can be shown that if the order in which the swaps are performed is determined with care, the costs of swaps are not affected by previous swaps. The following example shows that the sequence in which the swaps are performed can have an impact on the final cost.

Example 4.5.4 (Sequencing of Swaps)

Consider the situation depicted in Figure 4.7. The swap costs are $c_{\Phi} I(1, 2) = 1$ and $c_{\Phi} I(2, 3) = 1$. If the swap $I(2, 3)$ is performed followed by the swap $I(1, 2)$, the overlapping intervals that determine the costs of the interchange remain unchanged. However, if the sequence of swaps is reversed—that is, first swap $I(1, 2)$ is performed followed by swap $I(2, 3)$, then the costs do change: The cost of the first swap remains, of course, the same, but the cost of the second swap, $c_{\Phi} I(2, 3)$, now becomes 2 instead of 1.

The key point here is that the two swaps under consideration have an arrow in common—that is, $b_2 \rightarrow a_{\phi(2)}$. This arrow points up, and any swap that keeps it pointing up will not affect the cost of the swap below it as the overlap of intervals does not change.

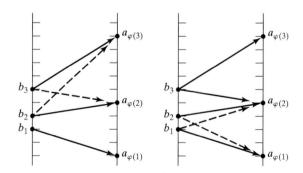

Figure 4.7   Situation in Example 4.5.4.

The example suggests that if a sequence of swaps needs to be performed, the swaps whose lower arcs point up can be performed starting from the top going down without changing the costs of swaps below them. A somewhat similar statement can be made with respect to swaps whose lower arrows point down.

To make this notion of up and down more rigorous, classify the nodes into two types. A node is said to be of Type I if $a_j \leq b_{\phi(j)}$ (i.e., the arrow points up), and it is of Type II if $a_j > b_{\phi(j)}$. A swap is of Type I if its lower node is of Type I and of Type II if its lower node is of Type II. From the previous arguments, it is easy to deduce that if the swaps $I(j, j+1)$ of Type I are performed in decreasing order of the node indexes, followed by swaps of Type II in increasing order of the node indexes, a single tour is obtained without changing any $c_{\Phi^*}I(j, j+1)$ involved in the swaps. The following algorithm sums up the entire procedure in detail.

Algorithm 4.5.5. (Finding Optimal Tour for TSP)

Step 1. Arrange the b_j in order of size and renumber the jobs so that

$$b_0 \leq b_1 \leq \cdots \leq b_n.$$

The permutation mapping Φ^* is defined by

$$\phi^*(j) = k,$$

k being such that a_k is the $(j+1)$th smallest of the a_j.

Step 2. Form an undirected graph with $n+1$ nodes and undirected arcs $A_{j,\phi^*(j)}$ connecting the jth and $\phi^*(j)$th nodes.
If the current graph has only one component, then STOP.
Otherwise go to Step 3.

Step 3. Compute the interchange costs $c_{\Phi^*}I(j, j+1)$ for $j = 0, \ldots, n-1$:

$$c_{\Phi^*}I(j, j+1) = 2\max\left(\min(b_{j+1}, a_{\phi^*(j+1)}) - \max(b_j, a_{\phi^*(j)}), 0\right)$$

Step 4. Select the smallest $c_{\Phi^*}I(j, j+1)$ such that j is in one component and $j+1$ in another (break ties arbitrarily).
Insert the undirected arc $A_{j,j+1}$ into the graph and repeat this step until all components in the undirected graph are connected.

Step 5. Divide the arcs selected in Step 4 into two groups.
Those $A_{j,j+1}$ for which $a_{\phi^*(j)} \geq b_j$ go in Group 1, the remaining go in Group 2.

Step 6. Find the largest index j_1 such that A_{j_1,j_1+1} is in Group 1.
Find the second largest j_2, and so on.

Find the smallest index k_1 such that A_{k_1, k_1+1} is in Group 2.
Find the second smallest k_2, and so on.

Step 7. *The optimal tour Φ^{**} is constructed by applying the following sequence of interchanges on the permutation Φ^*:*

$$\Phi^{**} = \Phi^* I(j_1, j_1 + 1) I(j_2, j_2 + 1)$$
$$\ldots I(j_l, j_l + 1) I(k_1, k_1 + 1) I(k_2, k_2 + 1)$$
$$\ldots I(k_m, k_m + 1).$$

The total cost of the resulting tour may be viewed as consisting of two components. One is the cost of the unrestricted permutation mapping Φ^* before the interchanges are performed. The other is the additional cost caused by the interchanges.

That this algorithm actually leads to the optimal tour can be shown in two steps. First, a lower bound is established for the total cost of an arbitrary permutation mapping. Second, it is shown that this lower bound, in case the permutation mapping represents an actual tour, is greater than or equal to the total cost of the tour constructed in the algorithm. These two steps then prove the optimality of the tour of the algorithm. As this proof is somewhat intricate, the reader is referred to the literature for its details.

A careful analysis of the algorithm establishes that the overall running time is bounded by $O(n^2)$.

Example 4.5.6 (Finding Optimal Tour for TSP)

Consider seven cities with the parameters given next.

cities	0	1	2	3	4	5	6
b_j	1	15	26	40	3	19	31
a_j	7	16	22	18	4	45	34

Step 1. Reordering the cities in such a way that $b_j \leq b_{j+1}$ results in the ordering and the $\phi^*(j)$ in the following table:

cities	0	1	2	3	4	5	6
b_j	1	3	15	19	26	31	40
$a_{\phi^*(j)}$	4	7	16	18	22	34	45
$\phi^*(j)$	1	0	2	6	4	5	3

Step 2. Form the undirected graph with j connected $\phi^*(j)$. Nodes 0 and 1 have to be connected with one another; nodes 3 and 6 have to be connected with one another; nodes 2, 4, and 5 are independent (each one of these three nodes is connected with itself).

Step 3. Computation of the interchange costs $c_{\phi^*} I (j, j+1)$ gives

$$c_{\phi^*} I (0, 1) = 0$$

$$c_{\phi^*} I (1, 2) = 2(15 - 7) = 16$$

$$c_{\phi^*} I (2, 3) = 2(18 - 16) = 4$$

$$c_{\phi^*} I (3, 4) = 2(22 - 19) = 6$$

$$c_{\phi^*} I (4, 5) = 2(31 - 26) = 10$$

$$c_{\phi^*} I (5, 6) = 2(40 - 34) = 12$$

Step 4. The undirected arcs $A_{1,2}$, $A_{2,3}$, $A_{3,4}$, and $A_{4,5}$ are inserted into the graph.

Step 5. The four arcs have to be partitioned into the two groups. To determine this, each b_j has to be compared with the corresponding $a_{\phi^*(j)}$.

arcs	b_j	$a_{\phi^*(j)}$	Group
$A_{1,2}$	$b_1 = 3$	$a_{\phi^*(1)} = a_0 = 7$	1
$A_{2,3}$	$b_2 = 15$	$a_{\phi^*(2)} = a_2 = 16$	1
$A_{3,4}$	$b_3 = 19$	$a_{\phi^*(3)} = a_6 = 18$	2
$A_{4,5}$	$b_4 = 26$	$a_{\phi^*(4)} = a_4 = 22$	2

Step 6. $j_1 = 2$, $j_2 = 1$, $k_1 = 3$, and $k_2 = 4$.

Step 7. The optimal tour is obtained after the following interchanges.

$$\Phi^{**} = \Phi^* I (2, 3) I (1, 2) I (3, 4) I (4, 5).$$

So the optimal tour is

$$0 \rightarrow 1 \rightarrow 6 \rightarrow 3 \rightarrow 4 \rightarrow 5 \rightarrow 2 \rightarrow 0.$$

The cost of this tour is

$$3 + 15 + 5 + 3 + 8 + 15 + 8 = 57.$$

In practice, when the setup times have an arbitrary structure, the myopic *Shortest Setup Time (SST) first* rule is often used. This rule implies that whenever a job is completed, the job with the smallest setup time is selected to go next. This SST rule is equivalent to the *Nearest Neighbour* rule for the TSP. Applying the SST rule on the instance in Example 4.5.6 results in the tour

$$0 \rightarrow 1 \rightarrow 2 \rightarrow 6 \rightarrow 3 \rightarrow 4 \rightarrow 5 \rightarrow 0.$$

The associated cost is

$$3 + 13 + 3 + 5 + 3 + 8 + 24 = 59.$$

This tour is not optimal.

Even though the SST rule usually leads to reasonable schedules, there are instances where the ratio

$$\frac{C_{\max}(SST) - \sum_{j=1}^{n} p_j}{C_{\max}(OPT) - \sum_{j=1}^{n} p_j}$$

is quite large. Nevertheless, the SST rule is often used as an integral component within more complicated dispatching rules (see Section 14.2).

4.6 DISCUSSION

Over the last decade, polynomial time approximation schemes (PTAS) have received an enormous amount of attention. Most of this attention has focused on NP-hard problems that lie close to the boundaries that separate NP-hard problems from polynomial time problems (e.g., $1 \mid r_j \mid \sum C_j$).

Problems with earliness and tardiness penalties have recently received a significant amount of attention as well. Even more general problems than those considered in this chapter have been studied. For example, some research has focused on problems with jobs that are subject to penalty functions such as the one presented in Figure 4.8.

Because of the importance of multiple objectives in practice, a considerable amount of research has been done on problems with multiple objectives. Of course, these problems are harder than the problems with just a single objective. So most

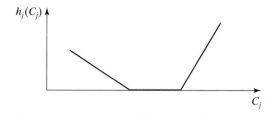

Figure 4.8   Cost function with due date range.

problems with two objectives are NP-hard. These types of problems may attract in the near future the attention of investigators who specialize in PTAS.

The makespan minimization problem when the jobs are subject to sequence-dependent setup times turns out to be equivalent to the Traveling Salesman Problem. Many combinatorial problems inspired by real-world settings are equivalent to Traveling Salesman Problems. Another scheduling problem discussed in Chapter 6 is also equivalent to the particular Traveling Salesman Problem described in Section 4.5.

EXERCISES (COMPUTATIONAL)

4.1. Consider a single machine and six jobs.

jobs	1	2	3	4	5	6
p_j	1190	810	1565	719	1290	482
d_j	1996	2000	2660	3360	3370	3375

Apply the PTAS described in Section 4.1 to this instance with $\epsilon = 0.02$. Are all sequences that are optimal for the rescaled data set also optimal for the original data set?

4.2. Consider the following instance with six jobs and $d = 156$.

jobs	1	2	3	4	5	6
p_j	4	18	25	93	102	114

Apply Algorithm 4.2.4 to find a sequence. Is the sequence generated by the heuristic optimal?

4.3. Consider the following instance with seven jobs. For each job, $w'_j = w''_j = w_j$. However, w_j is not necessarily equal to w_k.

jobs	1	2	3	4	5	6	7
p_j	4	7	5	9	12	2	6
w_j	4	7	5	9	12	2	6

All seven jobs have the same due date $d = 26$. Find all optimal sequences.

4.4. Consider again the instance of the previous exercise with seven jobs. Again, for each job, $w'_j = w''_j = w_j$. However, w_j is not necessarily equal to w_k. However, now the jobs have different due dates.

jobs	1	2	3	4	5	6	7
p_j	4	7	5	9	12	2	6
w_j	4	7	5	9	12	2	6
d_j	6	12	24	28	35	37	42

Find the optimal job sequence.

4.5. Give a numerical example of an instance with at most five jobs for which Algorithm 4.2.4 does not yield an optimal solution.

4.6. Consider the following instance of the $1 \ || \ \sum w_j C_j(opt), L_{max}$ problem.

jobs	1	2	3	4	5
w_j	4	6	2	4	20
p_j	4	6	2	4	10
d_j	14	18	18	22	0

Find all optimal schedules.

4.7. Consider the following instance of the $1 \ || \ L_{max}(opt), \sum w_j C_j$ problem.

jobs	1	2	3	4	5
w_j	4	6	2	4	20
p_j	4	6	2	4	10
d_j	14	18	18	22	0

Find all optimal schedules.

4.8. Apply Algorithm 4.4.2 to the following instance with five jobs and generate the entire trade-off curve.

jobs	1	2	3	4	5
p_j	4	6	2	4	2
d_j	2	4	6	10	10

4.9. Consider the instance of $1 \ || \ \theta_1 L_{max} + \theta_2 \sum C_j$ in Example 4.4.3. Find the ranges of θ_1 and θ_2 (assuming $\theta_1 + \theta_2 = 1$) for which each pareto-optimal schedule minimizes $\theta_1 L_{max} + \theta_2 \sum C_j$.

4.10. Consider an instance of the $1 \ | \ s_{jk} \ | \ C_{max}$ problem with the sequence-dependent setup times being of the form $s_{jk} = | \ a_k - b_j \ |$. The parameters a_j and b_k are in the following table. Find the optimal sequence.

cities	0	1	2	3	4	5	6
b_j	39	20	2	30	17	6	27
a_j	19	44	8	34	16	7	23

EXERCISES (THEORY)

4.11. Show that in an optimal schedule for an instance of $1 \mid d_j = d \mid \sum E_j + \sum T_j$ there is no unforced idleness in between two consecutive jobs.

4.12. Prove Lemma 4.2.1.

4.13. Consider the single machine scheduling problem with objective $\sum w' E_j + \sum w'' T_j$ and all jobs having the same due date (i.e., $d_j = d$). Note that the weight of the earliness penalty w' is different from the weight of the tardiness penalty w'', but the penalty structure is the same for each job. Consider an instance where the due date d is so far out that the machine will not start processing any job at time zero. Describe an algorithm that yields an optimal solution (i.e., a generalization of Algorithm 4.2.3).

4.14. Consider the same problem as described in the previous exercise. However, now the due date is not far out and the machine does have to start processing a job immediately at time zero. Describe a heuristic that would yield a good solution (i.e., a generalization of Algorithm 4.2.4).

4.15. Consider an instance where each job is subject to earliness and tardiness penalties and $w'_j = w''_j = w_j$ for all j. However, w_j is not necessarily equal to w_k. The jobs have different due dates. Prove or disprove that EDD minimizes the sum of the earliness and tardiness penalties.

4.16. Describe the optimal schedule for $1 \mid\mid \sum w_j C_j (opt), L_{\max}$.

4.17. Describe the optimal schedule for $1 \mid\mid \sum w_j C_j (opt), \sum U_j$.

4.18. Describe the algorithm for $1 \mid\mid L_{\max}(opt), \sum w_j C_j$ (i.e., generalize Lemma 4.3.1 and Algorithm 4.3.2).

4.19. Show that the maximum number of pareto-optimal solutions for $1 \mid\mid \theta_1 \sum C_j + \theta_2 L_{\max}$ is $n(n-1)/2$.

4.20. Describe the optimal schedule for $1 \mid\mid \theta_1 \sum U_j + \theta_2 L_{\max}$ under the agreeability conditions

$$d_1 \leq \cdots \leq d_n$$

and

$$p_1 \leq \cdots \leq p_n.$$

COMMENTS AND REFERENCES

An enormous amount of work has been done on Polynomial Time Approximation Schemes (PTAS). The algorithm described in Section 4.1 is one of the very first schemes developed for scheduling problems. This section is based on the paper by Lawler (1982). A significant amount of work has been done on approximation algorithms for $1 \mid r_j \mid \sum C_j$; see, for example, Chekuri, Motwani, Natarajan, and Stein (1997). However, there are many interesting examples of PTAS for other scheduling problems as well; see, for example, Hochbaum and Shmoys (1987), Sevastianov and Woeginger (1998), and Alon, Azar, Woeginger, and Yadid (1998). Schuurman and Woeginger (1999) present in their paper 10 open problems concerning Polynomial Time Approximation Schemes; their paper also contains an extensive reference list of PTAS papers. For a general overview of approximation techniques (covering more than just PTAS), see Chen, Potts, and Woeginger (1998).

The survey paper by Baker and Scudder (1990) focuses only on problems with earliness and tardiness penalties. The text by Baker (1995) has one chapter dedicated to problems with earliness and tardiness penalties. There are various papers on timing algorithms when the optimal order of the jobs is given. The optimal timing algorithm 4.2.8 is based on the paper by Szwarc and Mukhopadhyay (1995). An algorithm to find the optimal order of the jobs as well as their optimal start times and completion times, assuming $w'_j = w''_j = 1$ for all j, is presented by Kim and Yano (1994). For more results on models with earliness and tardiness penalties, see Sidney (1977), Hall and Posner (1991), and Hall, Kubiak, and Sethi (1991).

A fair amount of research has been done on single machine scheduling with multiple objectives. Some single machine problems with two objectives allow for polynomial time solutions; see, for example, Emmons (1975), Van Wassenhove and Gelders (1980), Nelson, Sarin, and Daniels (1986), Chen and Bulfin (1994), and Hoogeveen and Van de Velde (1995). Potts and Van Wassenhove (1983) as well as Posner (1985) consider the problem of minimizing the total weighted completion time with the jobs being subject to deadlines (this problem is strongly NP-hard). Chen and Bulfin (1993) present a detailed overview of the state of the art for multi-objective single machine scheduling.

The material in Section 4.5 dealing with the Traveling Salesman Problem is entirely based on the famous paper by Gilmore and Gomory (1964). For more results on scheduling with sequence-dependent setup times, see Bianco, Ricciardelli, Rinaldi, and Sassano (1988), Tang (1990), and Wittrock (1990).

5

Parallel Machine Models (Deterministic)

A bank of machines in parallel is a setting that is important from both a theoretical and a practical point of view. From a theoretical viewpoint, it is a generalization of the single machine and a special case of the flexible flow shop. From a practical point of view, it is important because the occurence of resources in parallel is common in the real world. Also, techniques for machines in parallel are often used in decomposition procedures for multistage systems.

In this chapter, several different objectives are considered. The three principal objectives are the minimization of the makespan, the total completion time, and the maximum lateness. With a single machine, the makespan objective is usually only of interest when there are sequence-dependent setup times; otherwise the makespan is equal to the sum of the processing times and is independent of the sequence. When dealing with machines in parallel, the makespan becomes an objective of considerable interest. In practice, one often has to deal with the problem of balancing the load on machines in parallel; by minimizing the makespan, the scheduler ensures a good balance.

One may consider scheduling parallel machines as a two-step process. First, one has to determine which jobs have to be allocated to which machines; second, one has to determine the sequence of the jobs allocated to each machine. With the makespan objective, only the allocation process is important.

With parallel machines, preemptions play a more important role than with a single machine. With a single machine, preemptions usually only play a role when jobs are released at different points in time. In contrast, with machines in parallel, preemptions are important even when all jobs are released at the same time.

For most models considered in this chapter, there are optimal schedules that are nondelay. However, when there are unrelated machines in parallel and the total completion time has to be minimized with no preemptions allowed, the optimal schedule is not necessarily nondelay.

The processing characteristics considered in this chapter include precedence constraints and the set functions M_j. It is assumed throughout that $p_1 \geq \cdots \geq p_n$.

5.1 THE MAKESPAN WITHOUT PREEMPTIONS

First, the problem $Pm \parallel C_{\max}$ is considered. This problem is of interest because minimizing the makespan has the effect of balancing the load over the various machines, which is an important objective in practice.

It is easy to see that $P2 \parallel C_{\max}$ is NP-hard in the ordinary sense as it is equivalent to PARTITION (see Appendix C). During the last couple of decades, many heuristics have been developed for $Pm \parallel C_{\max}$. One such heuristic is described here.

The Longest Processing Time first (LPT) rule assigns at $t = 0$ the m largest jobs to the m machines. After that, whenever a machine is freed, the longest job among those not yet processed is put on the machine. This heuristic tries to place the shorter jobs toward the end of the schedule, where they can be used for balancing the loads.

In the next theorem, an upper bound is presented for

$$\frac{C_{\max}(LPT)}{C_{\max}(OPT)},$$

where $C_{\max}(LPT)$ denotes the makespan of the LPT schedule and $C_{\max}(OPT)$ denotes the makespan of the (possibly unknown) optimal schedule. This type of worst case analysis is of interest as it gives an indication of how well the heuristic is guaranteed to perform as well as the type of instances for which the heuristic performs badly.

Theorem 5.1.1. *For Pm* $\|$ *C*$_{\max}$,

$$\frac{C_{\max}(LPT)}{C_{\max}(OPT)} \leq \frac{4}{3} - \frac{1}{3m}.$$

Proof. By contradiction. Assume that there exists one or more counterexamples where the ratio is *strictly* larger than $4/3 - 1/3m$. If more than one such counterexample exist, there must exist an example with the smallest number of jobs.

Consider this smallest counterexample and assume it has n jobs. This smallest counterexample has a useful property: Under LPT, the shortest job is the last job to start its processing and also the last job to finish its processing. That this is true can be seen as follows: First, under LPT by definition, the shortest job is the last to start its processing. Also, if this job is not the last to complete its processing, the deletion of this smallest job will result in a counterexample with fewer jobs (the $C_{\max}(LPT)$ remains the same while the $C_{\max}(OPT)$ may remain the same or decrease). So for the smallest counterexample, the starting time of the shortest job under LPT is $C_{\max}(LPT) - p_n$. Since at this point in time all other machines are still busy, it follows that

$$C_{\max}(LPT) - p_n \leq \frac{\sum_{j=1}^{n-1} p_j}{m}.$$

The right-hand side is an upper bound on the starting time of the shortest job. This upper bound is achieved when scheduling the first $n - 1$ jobs according to LPT results in each machine having exactly the same amount of processing to do. Now

$$C_{\max}(LPT) \leq p_n + \frac{\sum_{j=1}^{n-1} p_j}{m} = p_n \left(1 - \frac{1}{m}\right) + \frac{\sum_{j=1}^{n} p_j}{m}.$$

Since

$$C_{\max}(OPT) \geq \frac{\sum_{j=1}^{n} p_j}{m},$$

the following series of inequalities holds for the counterexample:

$$\frac{4}{3} - \frac{1}{3m} < \frac{C_{\max}(LPT)}{C_{\max}(OPT)} \leq \frac{p_n(1 - 1/m) + \sum_{j=1}^{n} p_j/m}{C_{\max}(OPT)}$$

$$= \frac{p_n(1 - 1/m)}{C_{\max}(OPT)} + \frac{\sum_{j=1}^{n} p_j/m}{C_{\max}(OPT)} \leq \frac{p_n(1 - 1/m)}{C_{\max}(OPT)} + 1.$$

Thus

$$\frac{4}{3} - \frac{1}{3m} < \frac{p_n(1 - 1/m)}{C_{max}(OPT)} + 1$$

and

$$C_{max}(OPT) < 3p_n.$$

Note that this last inequality is a *strict* inequality. This implies that for the smallest counterexample the optimal schedule can result in at most two jobs on each machine. It can be shown that if an optimal schedule is a schedule with at most two jobs on each machine, then the LPT schedule is optimal and the ratio of the two makespans is equal to one (see Exercise 5.11.b). This contradiction completes the proof of the theorem.   □

Example 5.1.2 (A Worst Case Example of LPT)

Consider four parallel machines and nine jobs whose processing times are given in the following table:

jobs	1	2	3	4	5	6	7	8	9
p_j	7	7	6	6	5	5	4	4	4

Scheduling the jobs according to LPT results in a makespan of 15. It can easily be shown that for this set of jobs a schedule can be found with a makespan of 12 (see Figure 5.1). This particular instance is thus a worst case when there are four machines in parallel.

What would the worst case be if instead of LPT an arbitrary priority rule is used? Consider the case where at time $t = 0$ the jobs are put in an arbitrary list. Whenever a machine is freed, the job (among the remaining jobs), that ranks highest on the list is put on the machine. It can be shown (see Exercise 5.28) that the worst case of this arbitrary list rule satisfies the inequality

$$\frac{C_{max}(LIST)}{C_{max}(OPT)} \leq 2 - \frac{1}{m}.$$

However, there are also several other heuristics for the $Pm \parallel C_{max}$ problem that are more sophisticated than LPT and that have tighter worst case bounds. These heuristics are beyond the scope of this book.

Consider now the same problem with the jobs subject to precedence constraints (i.e., $Pm \mid prec \mid C_{max}$). From the complexity point of view, this problem has to be at least as hard as the problem without precedence constraints. To obtain insights into the effects of precedence constraints, a number of special cases

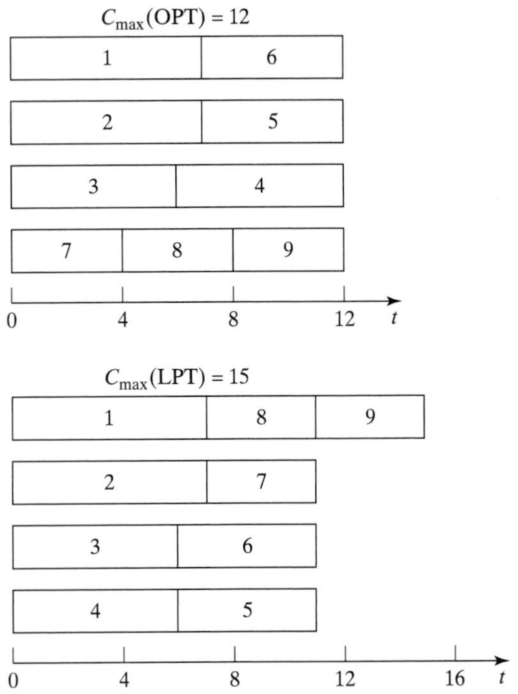

$C_{max}(OPT) = 12$

1	6
2	5
3	4
7	8	9

0   4   8   12 t

$C_{max}(LPT) = 15$

1	8	9
2	7
3	6
4	5

0   4   8   12   16 t

Figure 5.1   Worst case example of LPT.

may be considered. The special case with a single machine is clearly trivial. It is enough to keep the machine continuously busy and the makespan will be equal to the sum of the processing times. Consider the special case where there are an unlimited number of machines in parallel or where the number of machines is at least as large as the number of jobs (i.e., $m \geq n$). This problem may be denoted by $P\infty \mid prec \mid C_{max}$. This is a classical problem in the field of project planning and its study has led to the development of the well-known *Critical Path Method (CPM)* and *Project Evaluation and Review Technique (PERT)*. The optimal schedule and minimum makespan are determined through a very simple algorithm.

Algorithm 5.1.3. (Minimizing Makespan of a Project)   *Schedule the jobs one at a time starting at time zero. Whenever a job has been completed, start all jobs of which all predecessors have been completed (i.e., all schedulable jobs.)*

That this algorithm leads to an optimal schedule can be shown easily. The proof is left as an exercise. It turns out that in $P\infty \mid prec \mid C_{max}$ the start of the processing of some jobs usually can be postponed without increasing the makespan. These jobs are referred to as the *slack* jobs. The jobs that cannot be postponed are referred to as the *critical* jobs. The set of critical jobs is referred to as the *critical path(s)*. To determine the critical jobs, perform the same procedure applied in Algorithm 5.1.3 backward. Start at the makespan, which is now known, and work toward time zero while adhering to the precedence relationships. Doing this, all jobs are completed at the latest possible completion times and therefore started at their latest possible starting times as well. Those jobs whose earliest possible starting times are equal to their latest possible starting times are the critical jobs.

Example 5.1.4 (Minimizing Makespan of a Project)

Consider the nine jobs with the processing times given in the following table.

jobs	1	2	3	4	5	6	7	8	9
p_j	4	9	3	3	6	8	8	12	6

The precedence constraints are depicted in Figure 5.2.

The earliest completion time C'_j of job j can be computed easily.

jobs	1	2	3	4	5	6	7	8	9
C'_j	4	13	3	6	12	21	32	24	30

This implies that the makespan is 32. Assuming that the makespan is 32, the latest possible completion times C''_j can be computed.

jobs	1	2	3	4	5	6	7	8	9
C''_j	7	16	3	6	12	24	32	24	32

Those jobs of which the earliest possible completion times are equal to the latest possible completion times are said to be on the critical path. So the critical

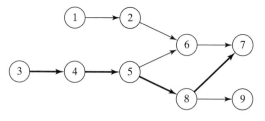

Figure 5.2 Precedence constraints graph with critical path in Example 5.1.4.

path is the chain

$$3 \rightarrow 4 \rightarrow 5 \rightarrow 8 \rightarrow 7.$$

The critical path in this case happens to be unique. The jobs that are not on the critical path are said to be slack. The amount of slack time for job j is the difference between its latest possible completion time and its earliest possible completion time.

In contrast to $1 \mid prec \mid C_{\max}$ and $P\infty \mid prec \mid C_{\max}$, the problem $Pm \mid prec \mid C_{\max}$ with $2 \leq m < n$ is strongly NP-hard. Even the special case with all processing times equal to 1 (i.e., $Pm \mid p_j = 1, prec \mid C_{\max}$), is not easy. However, constraining the problem further and assuming that the precedence graph takes the form of a tree (either an intree or an outtree) results in a problem (i.e., $Pm \mid p_j = 1, tree \mid C_{\max}$), that is easily solvable. This particular problem leads to a well-known scheduling rule, the *Critical Path (CP)* rule, which gives the highest priority to the job at the head of the longest string of jobs in the precedence graph (ties may be broken arbitrarily).

Before presenting the results concerning $Pm \mid p_j = 1, tree \mid C_{\max}$, it is necessary to introduce some notation. Consider an intree. The single job with no successors is called the *root* and is located at *level* 1. The jobs immediately preceding the root are at level 2; the jobs immediately preceding the jobs at level 2 are at level 3, and so on. In an outtree, all jobs with no successors are located at level 1. Jobs that have *only* jobs at level 1 as their *immediate* successors are said to be at level 2; jobs that have only jobs at levels 1 and 2 as their immediate successors are at level 3; and so on (see Figure 5.3). From this definition, it follows that the CP rule is equivalent to the *Highest Level first* rule. The number of jobs at level l is denoted by $N(l)$. The jobs with no predecessors may be referred to as *starting* jobs; the nodes in the graph corresponding to these jobs are often referred to in graph theory terminology as *leaves*. The highest level in the graph is denoted by $l_{\max}$. Let

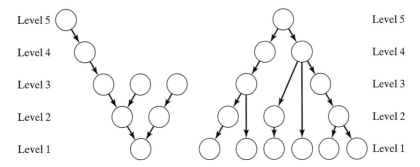

Figure 5.3   Intree and outtree.

$$H(l_{\max} + 1 - r) = \sum_{k=1}^{r} N(l_{\max} + 1 - k).$$

Clearly, $H(l_{\max} + 1 - r)$ denotes the total number of nodes at level $l_{\max} + 1 - r$ or higher—that is, at the highest r levels.

Theorem 5.1.5.   *The CP rule is optimal for $Pm \mid p_j = 1$, intree $\mid C_{\max}$ and for $Pm \mid p_j = 1$, outtree $\mid C_{\max}$.*

Proof.   The proof for intrees is slightly harder than the proof for outtrees. In what follows only the proof for intrees is given (the proof for outtrees is left as an exercise). In the proof for intrees, a distinction has to be made between two cases.

Case 1.   Assume the tree satisfies the following condition:

$$\max_{r} \left(\frac{\sum_{k=1}^{r} N(l_{\max} + 1 - k)}{r} \right) \leq m.$$

In this case, in every time interval, all the jobs available for processing can be processed and at most $l_{\max}$ time units are needed to complete all the jobs under the CP rule. But $l_{\max}$ is clearly a lower bound for $C_{\max}$. So the CP rule results in an optimal schedule.

Case 2.   Find for the tree the (smallest) integer $c \geq 1$ such that

$$\max_{r} \left(\frac{\sum_{k=1}^{r} N(l_{\max} + 1 - k)}{r + c} \right) \leq m < \max_{r} \left(\frac{\sum_{k=1}^{r} N(l_{\max} + 1 - k)}{r + c - 1} \right).$$

The c basically represents the smallest amount of time beyond time r needed to complete all jobs at the r highest levels. Let

$$\max_{r} \left(\frac{\sum_{k=1}^{r} N(l_{\max} + 1 - k)}{r + c - 1} \right) = \left(\frac{\sum_{k=1}^{r^*} N(l_{\max} + 1 - k)}{r^* + c - 1} \right) > m.$$

The number of jobs completed at time $(r^* + c - 1)$ is at most $m(r^* + c - 1)$. The number of jobs at levels higher than or equal to $l_{\max} + 1 - r^*$ is $\sum_{k=1}^{r^*} N(l_{\max} + 1 - k)$. As

$$\sum_{k=1}^{r^*} N(l_{\max} + 1 - k) > (r^* + c - 1)m,$$

there is at least one job at a level equal to or higher than $l_{\max} + 1 - r^*$ that is not processed by time $r^* + c - 1$. Starting with this job, there are at least $l_{\max} + 1 - r^*$ time units needed to complete all the jobs. A lower bound for the makespan under any type of scheduling rule is therefore

$$C_{\max} \geq (r^* + c - 1) + (l_{\max} + 1 - r^*) = l_{\max} + c.$$

To complete the proof, it suffices to show that the CP rule results in a makespan that is equal to this lower bound. This part of the proof is left as an exercise (see Exercise 5.14).   □

The question arises: How well does the CP rule perform under arbitrary precedence constraints when all jobs have equal processing times? It has been shown that for two machines in parallel

$$\frac{C_{\max}(CP)}{C_{\max}(OPT)} \leq \frac{4}{3}.$$

When there are more than two machines in parallel, the worst case ratio is larger. That the worst case bound for two machines can be attained is shown in the following example.

Example 5.1.6 (A Worst Case Example of CP)

Consider six jobs with unit processing times and two machines. The precedence relationships are depicted in Figure 5.4. Jobs 4, 5, and 6 are at level 1, while jobs 1, 2, and 3 are at level 2. As under the CP rule ties may be broken arbitrarily, a CP schedule may prescribe starting with jobs 1 and 2 at time zero. At their completion, only job 3 may be started. At time 2, jobs 4 and 5 are started. Job 6 goes last and is completed by time 4. Of course, an optimal schedule can be obtained by starting out at time zero with jobs 2 and 3. The makespan then equals 3.

Example 5.1.6 shows that processing the jobs with the largest number of successors first may result in a better schedule than processing the jobs at the highest level first. A priority rule often used when jobs are subject to arbitrary precedence

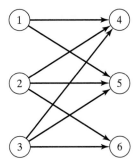

Figure 5.4 Worst case example of the CP rule for two machines (Example 5.1.6).

constraints is indeed the so-called *Largest Number of Successors first (LNS)* rule. Under this rule, the job with the largest total number of successors in the precedence constraints graph has the highest priority. Note that in the case of intrees, the CP rule and the LNS rule are equivalent; the LNS rule therefore results in an optimal schedule in the case of intrees. It can be shown fairly easily that the LNS rule is also optimal for $Pm \mid p_j = 1, outtree \mid C_{\max}$. The following example shows that the LNS rule may not yield an optimal schedule with arbitrary precedence constraints.

Example 5.1.7 (Application of LNS rule)

> Consider six jobs with unit processing times and two machines. The precedence constraints are depicted in Figure 5.5. The LNS rule may start at time 0 with jobs 4 and 6. At time 1, jobs 1 and 5 start. Job 2 then starts at time 2, and job 3 starts at time 3. The resulting makespan is 4. It is easy to see that the optimal makespan is 3 and that the CP rule actually achieves the optimal makespan.

Both the CP and LNS rules have more generalized versions that can be applied to problems with arbitrary job processing times. Instead of counting the *number* of jobs (as in the case with unit processing times), these more generalized versions prioritize based on the total amount of processing remaining to be done on the jobs in question. The CP rule then gives the highest priority to the job at the head of the string of jobs with the largest total amount of processing, where the processing time of the head is also included in the total. The generalization of the LNS rule gives the highest priority to that job which precedes the largest total amount of processing; again the processing time of the job is included in the total. The LNS name is clearly not appropriate for this generalization with arbitrary processing times as it refers to a number of jobs rather than to a total amount of processing.

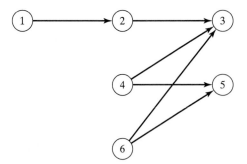

Figure 5.5   The LNS rule is not optimal with arbitrary precedence constraints (Example 5.1.7).

Another generalization of the $Pm \parallel C_{\max}$ problem that is of practical interest arises when job j is only allowed to be processed on subset M_j of the m parallel machines. Consider $Pm \mid p_j = 1, M_j \mid C_{\max}$ and assume that the sets M_j are *nested*—that is, one and only one of the following four conditions holds for jobs j and k.

    (i) M_j is equal to M_k $(M_j = M_k)$

    (ii) M_j is a subset of M_k $(M_j \subset M_k)$

    (iii) M_k is a subset of M_j $(M_k \subset M_j)$

    (iv) M_j and M_k do not overlap $(M_j \cap M_k = \emptyset)$

Under these conditions, a well-known dispatching rule, the *Least Flexible Job first (LFJ)* rule, is of importance. The LFJ rule selects, every time a machine is freed, among the available jobs the job that can be processed on the *smallest* number of machines (i.e., the least flexible job). Ties may be broken arbitrarily. This rule is rather crude as it does not specify, for example, which machine should be considered first when several machines become available at the same time.

    Theorem 5.1.8.   *The LFJ rule is optimal for $Pm \mid p_j = 1, M_j \mid C_{\max}$ when the M_j sets are nested.*

    Proof.   By contradiction. Suppose the LFJ rule does not yield an optimal schedule. Consider a schedule that is optimal and that differs from a schedule that could have been generated via the LFJ rule. Without loss of generality, the jobs on each machine can be ordered in increasing order of $\mid M_j \mid$. Now consider the earliest time slot that is filled by a job that could not have been placed there by the LFJ rule. Call this job j. There exists some job, say job j^*, that is scheduled to start later and could have been scheduled by LFJ in the position taken by job j. Note that job j^* is less flexible than job j (since $M_j \supset M_{j^*}$). Job j^* cannot start before job j since j is the earliest non-LFJ job. It cannot start at the same time as job j, for in that case job j is in a position where LFJ could have placed it. These two jobs, j and j^*, can be exchanged without altering $C_{\max}$ since $p_j = 1$ for all jobs. Note that the slot previously filled by j is now filled by j^*, which is an LFJ position. By repeating this process of swapping the earliest non-LFJ job, all jobs can be moved into positions where they could have been placed using LFJ without increasing $C_{\max}$. So it is possible to construct an LFJ schedule from an optimal schedule. This establishes the contradiction. $\square$

It can easily be shown that the LFJ rule is optimal for $P2 \mid p_j = 1, M_j \mid C_{\max}$ because with two machines, the M_j sets are always nested. However, with three or more machines the LFJ rule may not lead to an optimal solution for $Pm \mid p_j = 1, M_j \mid C_{\max}$ with arbitrary M_j, as illustrated in the following example.

Example 5.1.9 (Application of LFJ Rule)

Consider $P4 \mid p_j = 1, M_j \mid C_{\max}$ with eight jobs. The eight M_j sets are:

$$M_1 = \{1, 2\}$$
$$M_2 = M_3 = \{1, 3, 4\}$$
$$M_4 = \{2\}$$
$$M_5 = M_6 = M_7 = M_8 = \{3, 4\}$$

The M_j sets are clearly not nested. Under the LFJ rule, the machines can be considered in any order. Consider machine 1 first. The least flexible job that can be processed on machine 1 is job 1 as it can be processed on only two machines (jobs 2 and 3 can be processed on three machines). Consider machine 2 next. The least flexible job to be processed on machine 2 is clearly job 4. Least flexible jobs to be processed on machines 3 and 4 at time 0 could be jobs 5 and 6. At time 1, after jobs 1, 4, 5, and 6 have completed their processing on the four machines, the least flexible job to be processed on machine 1 is job 2. However, at this point, none of the remaining jobs can be processed on machine 2; so machine 2 remains idle. The least flexible jobs to go on machines 3 and 4 are jobs 7 and 8. This implies that job 3 only can be started at time 2, completing its processing at time 3. The makespan is therefore equal to 3.

A better schedule with a makespan equal to 2 can be obtained by assigning jobs 2 and 3 to machine 1; jobs 1 and 4 to machine 2; jobs 5 and 6 to machine 3; and jobs 7 and 8 to machine 4.

From Example 5.1.9, one may expect that, if a number of machines are free at the same point in time, it is advantageous to consider first the least flexible *machine*. The flexibility of a machine could be defined as the number of remaining jobs that can be processed (or the total amount of processing that can be done) on the machine. Assigning at each point in time first a job, any job, to the *Least Flexible Machine (LFM)*, however, does not guarantee an optimal schedule in the case of Example 5.1.9.

Heuristics can be designed that combine the LFJ rule with the LFM rule, giving priority to the least flexible jobs on the least flexible machines. That is, consider at each point in time first the Least Flexible Machine (LFM) (i.e., the machine that can process the smallest number of jobs) and assign to this machine the least flexible job that can be processed on it. Any ties may be broken arbitrarily. This heuristic may be referred to as the LFM-LFJ heuristic. However, in the case of Example 5.1.9, the LFM-LFJ does not lead to an optimal schedule either.

5.2 THE MAKESPAN WITH PREEMPTIONS

Consider the same problem as the one discussed in the beginning of the previous section, but now with preemptions allowed (i.e., $Pm \mid prmp \mid C_{max}$). Usually, but not always, allowing preemptions simplifies the analysis of a problem. This is indeed the case for this problem where it actually turns out that many schedules are optimal. First, consider the following linear programming (LP) formulation of the problem.

$$minimize \quad C_{max}$$

subject to

$$\sum_{i=1}^{m} x_{ij} = p_j, \qquad j = 1, \ldots, n$$

$$\sum_{i=1}^{m} x_{ij} \leq C_{max}, \qquad j = 1, \ldots, n$$

$$\sum_{j=1}^{n} x_{ij} \leq C_{max}, \qquad i = 1, \ldots, m$$

$$x_{ij} \geq 0, \qquad i = 1, \ldots, m, \quad j = 1, \ldots, n.$$

The variable x_{ij} represents the total time job j spends on machine i. The first set of constraints makes sure that each job receives the required amount of processing. The second set of constraints enforces that the total amount of processing each job receives is less than or equal to the makespan. The third set makes sure that the total amount of processing on each machine is less than the makespan. Since the C_{max} is basically a decision variable and not an element of the resource vector of the linear program, the second and third set of constraints may be rewritten as follows:

$$C_{max} - \sum_{i=1}^{m} x_{ij} \geq 0, \qquad j = 1, \ldots, n$$

$$C_{max} - \sum_{j=1}^{n} x_{ij} \geq 0, \qquad i = 1, \ldots, m$$

Example 5.2.1 (LP Formulation of Makespan Minimization with Preemptions)

Consider two machines and three jobs with $p_1 = 8$, $p_2 = 7$, and $p_3 = 5$. There are thus seven variables—namely, $x_{11}, x_{21}, x_{12}, x_{22}, x_{13}, x_{23}$, and C_{max} (see Appendix A). The **A** matrix is a matrix of 0s and 1s. The $\bar{c}$ vector contains six 0s and a single 1. The $\bar{b}$ vector contains the three processing times and five 0s.

This LP can be solved in polynomial time, but the solution of the LP does not prescribe an actual schedule; it merely specifies the amount of time job j should spend on machine i. However, with this information, a schedule can easily be constructed.

There are several other algorithms for $Pm \mid prmp \mid C_{max}$. One of these algorithms is based on the fact that it is easy to obtain an expression for the makespan under the optimal schedule. In the next lemma, a lower bound is established.

Lemma 5.2.2.   *Under the optimal schedule for $Pm \mid prmp \mid C_{max}$,*

$$C_{max} \geq \max\left(p_1, \sum_{j=1}^{n} p_j/m\right) = C_{max}^*.$$

Proof.   Recall that job 1 is the job with the largest processing time. The proof is easy and left as an exercise.   □

Having a lower bound allows for the construction of a very simple algorithm that minimizes the makespan. The fact that this algorithm actually produces a schedule with a makespan that is equal to the lower bound shows that the algorithm yields an optimal schedule.

Algorithm 5.2.3.   (Minimizing Makespan with Preemptions)

Step 1. *Take the n jobs and process them one after another on a single machine in any sequence. The makespan is then equal to the sum of the n processing times and is less than or equal to mC_{max}^*.*

Step 2. *Take this single machine schedule and cut it into m parts. The first part constitutes the interval $[0, C_{max}^*]$, the second part the interval $[C_{max}^*, 2C_{max}^*]$, the third part the interval $[2C_{max}^*, 3C_{max}^*]$, and so on.*

Step 3. *Take as the schedule for machine 1 in the bank of parallel machines the processing sequence of the first interval; take as the schedule for machine 2 the processing sequence of the second interval; and so on.*

It is obvious that the resulting schedule is feasible. Part of a job may appear at the end of the schedule for machine i while the remaining part may appear at the beginning of the schedule for machine $i + 1$. As preemptions are allowed and the processing time of each job is less than C_{max}^*, such a schedule is feasible. As this schedule has $C_{max} = C_{max}^*$, it is also optimal.

Another schedule that may appear appealing for $Pm \mid prmp \mid C_{max}$ is the *Longest Remaining Processing Time first (LRPT)* schedule. This schedule is the preemptive version of the (nonpreemptive) LPT schedule. It is a schedule that is structurally appealing, but mainly of academic interest. From a theoretical point of view, it is important because of similarities with optimal policies in stochastic scheduling (see Chapter 12). From a practical point of view, it has a serious drawback. The number of preemptions needed in the deterministc case is usually infinite.

Example 5.2.4 (Application of the LRPT Rule)

> Consider two jobs with unit processing times and a single machine. Under LRPT, the two jobs continuously have to rotate and wait for their next turn on the machine (i.e., a job stays on the machine for a time period ϵ, and after every time period ϵ the job waiting preempts the machine). The makespan is equal to 2 and is, of course, independent of the schedule. But note that the sum of the two completion times under LRPT is 4, whereas under the nonpreemptive schedule it is 3.

In the subsequent lemma and theorem, a proof technique based on a discrete time framework is used. All processing times are assumed to be integer, and the decision maker is allowed to preempt any machine only at integer times $1, 2, \ldots$ The proof that LRPT is optimal is based on a dynamic programming induction technique that requires some special notation. Assume that at some integer time t the remaining processing times of the n jobs are $p_1(t), p_2(t), \ldots, p_n(t)$. Let $\overline{p}(t)$ denote this vector of processing times. In the proof, two different vectors of remaining processing times at time t, say $\overline{p}(t)$ and $\overline{q}(t)$, are repeatedly compared with one another. The vector $\overline{p}(t)$ is said to *majorize* the vector $\overline{q}(t)$, $\overline{p}(t) \geq_m \overline{q}(t)$, if

$$\sum_{j=1}^{k} p_{(j)}(t) \geq \sum_{j=1}^{k} q_{(j)}(t),$$

for all $k = 1, \ldots, n$, where $p_{(j)}(t)$ denotes the jth largest element of vector $\overline{p}(t)$ and $q_{(j)}(t)$ denotes the jth largest element of vector $\overline{q}(t)$.

Example 5.2.5 (Vector Majorization)

> Consider the two vectors $\overline{p}(t) = (4, 8, 2, 4)$ and $\overline{q}(t) = (3, 0, 6, 6)$. Rearranging the elements within each vector and putting these in decreasing order results in vectors $(8, 4, 4, 2)$ and $(6, 6, 3, 0)$. It can be easily verified that $\overline{p}(t) \geq_m \overline{q}(t)$.

Lemma 5.2.6. *If $\overline{p}(t) >_m \overline{q}(t)$ then LRPT applied to $\overline{p}(t)$ results in a makespan that is larger than or equal to the makespan obtained by applying LRPT to $\overline{q}(t)$.*

Proof. The proof is by induction on the *total* amount of remaining processing. To show that the lemma holds for $\overline{p}(t)$ and $\overline{q}(t)$, with total remaining processing time $\sum_{j=1}^{n} p_j(t)$ and $\sum_{j=1}^{n} q_j(t)$, respectively, assume as induction hypothesis that the lemma holds for all pairs of vectors with total remaining processing less than or equal to $\sum_{j=1}^{n} p_j(t) - 1$ and $\sum_{j=1}^{n} q_j(t) - 1$, respectively. The induction base can be checked easily by considering the two vectors $1, 0, \ldots, 0$ and $1, 0, \ldots, 0$.

If LRPT is applied for one time unit on $\overline{p}(t)$ and $\overline{q}(t)$, respectively, then the vectors of remaining processing times at time $t+1$ are $\overline{p}(t+1)$ and $\overline{q}(t+1)$, respectively. Clearly,

$$\sum_{j=1}^{n} p_{(j)}(t+1) \le \sum_{j=1}^{n} p_{(j)}(t) - 1$$

and

$$\sum_{j=1}^{n} q_{(j)}(t+1) \le \sum_{j=1}^{n} q_{(j)}(t) - 1.$$

It can be shown that if $\overline{p}(t) \ge_m \overline{q}(t)$, then $\overline{p}(t+1) \ge_m \overline{q}(t+1)$. So if LRPT results in a larger makespan at time $t+1$ because of the induction hypothesis, it also results in a larger makespan at time t.

It is clear that if there are less than m jobs remaining to be processed, the lemma holds. □

Theorem 5.2.7. *LRPT yields an optimal schedule for $Pm \mid prmp \mid C_{\max}$ in discrete time.*

Proof. The proof is based on induction as well as contradiction arguments.

The first step of the induction is shown as follows. Suppose not more than m jobs have processing times remaining and that these jobs all have only one unit of processing time left. Then clearly LRPT is optimal.

Assume LRPT is optimal for any vector $\overline{p}(t)$ for which

$$\sum_{j=1}^{n} p_{(j)}(t) \le N - 1.$$

Consider now a vector $\overline{p}(t)$ for which

$$\sum_{j=1}^{n} p_{(j)}(t) = N.$$

The induction is based on the total amount of remaining processing, $N-1$, and *not* on the time t.

To show that LRPT is optimal for a vector of remaining processing times $\overline{p}(t)$ with a total amount of remaining processing $\sum_{j=1}^{n} p_j(t) = N$, assume that LRPT is optimal for all vectors with a smaller total amount of remaining processing. The proof of the induction step, showing that LRPT is optimal for $\overline{p}(t)$, is by contradiction. If LRPT is not optimal, another rule has to be optimal. This other rule does not act according to LRPT at time t, but from time $t+1$ onward it must act according to LRPT because of the induction hypothesis (LRPT is optimal from $t + 1$ on as the total amount of processing remaining at time $t + 1$ is strictly less than N). Call this supposedly optimal rule, which between t and $t + 1$ does not act according to LRPT, LRPT'. Now applying LRPT at time t on $\overline{p}(t)$ must be compared with applying LRPT' at time t on the same vector $\overline{p}(t)$. Let $\overline{p}(t + 1)$ and $\overline{p}'(t + 1)$ denote the vectors of remaining processing times at time $t + 1$ after applying LRPT and LRPT'. It is clear that $\overline{p}'(t+1) \geq_m \overline{p}(t+1)$. From Lemma 5.2.6 it follows that the makespan under LRPT' is larger than the makespan under LRPT. This completes the proof of the induction step and the proof of the theorem.   □

Example 5.2.8 (Application of LRPT in Discrete Time)

> Consider two machines and three jobs, say jobs 1, 2, and 3, with processing times 8, 7, and 6. The schedule under LRPT is depicted in Figure 5.6 and the makespan is 11.

That LRPT is also optimal in continuous time (resulting in an infinite number of preemptions) can be easily argued. Multiply all processing times by K, K being a very large integer. The problem intrinsically does not change as the relative lengths of the processing times remain the same. The optimal policy is, of course, again LRPT. But now there may be many more preemptions (recall preemptions must occur at integral time units). Basically, multiplying all processing times by K has the effect that the time slots become smaller relative to the processing times and the decision maker is allowed to preempt after shorter intervals. Letting K go to ∞ shows that LRPT is optimal in continuous time as well.

Example 5.2.9 (Application of LRPT in Continuous Time)

> Consider the same jobs as in the previous example. As preemptions may be done at any point in time, processor sharing takes place, see Figure 5.7. The makespan is now 10.5.

Consider the generalization to uniform machines—that is, m machines in parallel with machine i having speed v_i. Without loss of generality, it may be assumed that $v_1 \geq v_2 \geq \cdots \geq v_m$. Similar to Lemma 5.2.2, a lower bound can be established for the makespan here as well.

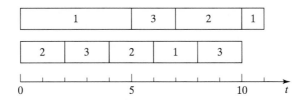

Figure 5.6   LRPT with three jobs on two machines with preemptions allowed at integer points in time (Example 5.2.8).

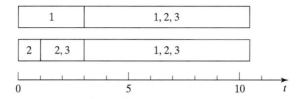

Figure 5.7 LRPT with three jobs on two machines with preemptions allowed any time (Example 5.2.9).

Lemma 5.2.10. *Under the optimal schedule for* $Qm \mid prmp \mid C_{max}$,

$$C_{max} \geq \max\left(\frac{p_1}{v_1}, \frac{p_1 + p_2}{v_1 + v_2}, \ldots, \frac{\sum_{j=1}^{m-1} p_j}{\sum_{i=1}^{m-1} v_i}, \frac{\sum_{j=1}^{n} p_j}{\sum_{i=1}^{m} v_i}\right).$$

Proof. The makespan has to be at least as large as the time it takes for the fastest machine to do the longest job. This time represents the first term within the "max" on the R.H.S. The makespan also has to be at least as large as the time needed for the fastest and second fastest machine to process the largest and second largest job while keeping the two machines occupied exactly the same amount of time. This amount of time represents the second term within the "max" expression. The remainder of the first $m - 1$ terms are determined in the same way. The last term is a bit different as it is the minimum time needed to process all n jobs on the m machines while keeping all the m machines occupied exactly the same amount of time. □

If the largest term in the lower bound is determined by the sum of the processing times of the k largest jobs divided by the sum of the speeds of the k fastest machines, then the $n - k$ smallest jobs under the optimal schedule do not receive any processing on the k fastest machines; these jobs only receive processing on the $m - k$ slowest machines.

Example 5.2.11 (Minimizing Makespan on Uniform Machines)

Consider three machines, 1, 2, and 3, with respective speeds 3, 2, and 1. There are three jobs, 1, 2, and 3, with respective processing times 36, 34, and 12. The optimal schedule assigns the two largest jobs to the two fastest machines. Job 1 is processed for 8 units of time on machine 1 and for 6 units of time on machine 2, whereas job 2 is processed for 8 units of time on machine 2 and for 6 units of time on machine 1. These two jobs are completed after 14 time units. Job 3 is processed only on machine 3 and is completed at time 12.

A generalization of the LRPT schedule described before is the so-called *Longest Remaining Processing Time on the Fastest Machine first (LRPT-FM)* rule. According to this rule at any point in time the job with the largest remaining processing time among the remaining jobs is assigned to the fastest machine; the job with the second largest remaining processing time is assigned to the second fastest machine, and so on.

This rule typically requires an infinite number of preemptions. If at time t two jobs have the same remaining processing time and this processing time is the longest among the jobs not yet completed at t, then one of the two jobs has to go on the fastest machine while the other has to go on the second fastest. At time $t + \epsilon$, ϵ being very small, the remaining processing time of the job on the second fastest machine is longer than the remaining processing time of the job on the fastest machine. So the job on the second fastest machine has to move to the fastest and vice versa. Thus, the LRPT-FM rule often results in so-called *processor sharing* where a number of machines, say m^*, process a number of jobs, say n^*, $n^* \geq m^*$, partitioning the total processing capability equally between the n^* jobs and keeping the remaining processing times of all n^* jobs equal at all times.

It can be shown that the LRPT-FM rule is indeed optimal for $Qm \mid prmp \mid C_{\max}$. It is again easier to construct first a proof of optimality in a discrete time framework and then extend the result to continuous time. Assume that the processing times as well as the speeds are integers. Replace the original machine i with speed v_i by v_i identical, so-called *unit-speed,* machines with speed 1. A job may be processed for one time unit on a subset of the unit-speed machines corresponding to machine i. It may not be processed at the same point in time by unit-speed machines that correspond to two different original machines.

Assume that preemptions are only allowed at integer times 1, 2, 3, ... If at time t job j has a remaining processing time p_j, its remaining processing time at time $t + 1$ is equal to $(p_j - l)$ after being processed for one time unit on l unit-speed machines corresponding to the original machine i. Lemma 5.2.6 can be shown for this framework as well. Based on this lemma, the following theorem can be shown.

Theorem 5.2.12.   *LRPT-FM yields an optimal schedule for $Qm \mid prmp \mid C_{\max}$ in discrete time.*

Example 5.2.13 (Application of the LRPT-FM Rule)

Consider again two machines and three jobs. The processing times of the three jobs are again 8, 7, and 6. The two machines have now different speeds: $v_1 = 2$ and $v_2 = 1$. The first machine can be replaced by two machines with speed 1. Note that a job may be processed simultaneously by machines 1 and 2, but not simultaneously by machines 1 and 3 or 2 and 3. These three machines may be scheduled as follows: Job 1 is assigned to the first two unit-speed machines for one time unit and job 2 is assigned to the third unit-speed machine for one time unit. At time 1, all three jobs have a remaining processing time equal to 6. So from time 1 on, each job occupies one unit-speed machine and the makespan is equal to 7.

To show that LRPT-FM is optimal in continuous time, two limits have to be taken. First, all the processing times, have to be multiplied by a large number K. In proportion to the new processing times a time unit is now very small. The possible number of preemptions goes up significantly this way, approaching continuous time preemptions. Second, the speeds of the machines have to be multiplied with a large number as well. Thus, the number of unit-speed machines for each original machine goes up dramatically. This implies that a finer partition of the processing capabilities of a given machine can be realized. Through such arguments, it can be shown that LRPT-FM is optimal in continuous time. In addition, it can be shown that the LRPT-FM rule among available jobs is also optimal when the jobs have different release dates (i.e., $Qm \mid r_j, prmp \mid C_{\max}$).

5.3 THE TOTAL COMPLETION TIME WITHOUT PREEMPTIONS

Consider m machines in parallel and n jobs. Recall that $p_1 \geq \cdots \geq p_n$. The objective to be minimized is the total unweighted completion time $\sum C_j$. From Theorem 3.1.1, it follows that for a single machine the *Shortest Processing Time first (SPT)* rule minimizes the total completion time. This single machine result can also be shown in a different way rather easily.

Let $p_{(j)}$ denote the processing time of the job in the jth position in the sequence. The total completion time can then be expressed as

$$\sum C_j = np_{(1)} + (n-1)p_{(2)} + \cdots + 2p_{(n-1)} + p_{(n)}.$$

This implies that there are n coefficients $n, n-1, \ldots, 1$ to be assigned to n different processing times. The processing times have to be assigned in such a way that the sum of the products is minimized. From the elementary Hardy, Littlewood, and Polya inequality as well as common sense, it follows that the highest coefficient, n, is assigned the smallest processing time, p_n; the second highest coefficient, $n-1$, is assigned the second smallest processing time, p_{n-1}; and so on. This implies that SPT is optimal.

This type of argument can be extended to the parallel machine setting as well.

Theorem 5.3.1. *The SPT rule is optimal for $Pm \mid\mid \sum C_j$.*

Proof. In the case of parallel machines, there are nm coefficients to which processing times can be assigned. These coefficients are m $n's$, m $(n-1)'s$, $\ldots$, m ones. The processing times have to be assigned to a subset of these coefficients to minimize the sum of the products. Assume that n/m is an integer. If it is not an integer, add a number of dummy jobs with zero processing times so that n/m is integer (adding jobs with zero processing times does not change the problem; these jobs would be instantaneously processed at time zero and would not contribute to the objective function). It is easy to see, in a similar manner as before that the set of m largest processing times have to be assigned to the m ones, the set of second m largest processing times have to be assigned to the m twos, and so on. This results in the m largest jobs each being processed on different machines and so on. That this class of schedules includes SPT can be shown as follows. According to the SPT schedule, the smallest job has to go on machine 1 at time zero, the second smallest one on machine 2, and so on; the $(m + 1)$th smallest job follows the smallest job on machine 1, the $(m + 2)$th smallest job follows the second smallest on machine 2, and so on. It is easy to verify that the SPT schedule corresponds to an optimal assignment of jobs to coefficients.   □

From the proof of the theorem, it is clear that the SPT schedule is not the only schedule that is optimal. Many more schedules also minimize the total completion time. The class of schedules that minimize the total completion time turns out to be fairly easy to characterize (see Exercise 5.21).

As pointed out in the previous chapter, the more general WSPT rule minimizes the total *weighted* completion time in the case of a single machine. Unfortunately, this result cannot be generalized to parallel machines, as shown in the following example.

Example 5.3.2 (Application of WSPT Rule)

Consider two machines and three jobs.

jobs	1	2	3
p_j	1	1	3
w_j	1	1	3

Scheduling jobs 1 and 2 at time zero and job 3 at time 1 results in a total weighted completion time of 14, whereas scheduling job 3 at zero and jobs 1 and 2 on the other machine results in a total weighted completion time of 12. Clearly, with this set of data, any schedule may be considered WSPT. However, making the weights of jobs 1 and 2 equal to $1 - \epsilon$ shows that WSPT does not necessarily yield an optimal schedule.

It has been shown in the literature that the WSPT heuristic is nevertheless a very good heuristic for the total weighted completion time on parallel machines. A worst case analysis of this heuristic yields the lower bound

$$\frac{\sum w_j C_j(WSPT)}{\sum w_j C_j(OPT)} < \frac{1}{2}(1 + \sqrt{2}).$$

What happens now in the case of precedence constraints? The problem $Pm \mid prec \mid$ $\sum C_j$ is known to be strongly NP-hard in the case of arbitrary precedence constraints. However, the special case with all processing times equal to 1 and precedence constraints that take the form of an outtree can be solved in polynomial time. In this special case, the Critical Path rule again minimizes the total completion time.

Theorem 5.3.3.   *The CP rule is optimal for $Pm \mid p_j = 1, outtree \mid \sum C_j$.*

Proof.   Up to some integer point in time, say t_1, the number of schedulable jobs is less than or equal to the number of machines. Under the optimal schedule, at each point in time before t_1, all schedulable jobs have to be put on the machines. Such actions are in accordance with the CP rule. Time t_1 is the first point in time when the number of schedulable jobs is strictly larger than m. There are at least $m + 1$ jobs available for processing, and each one of these jobs is at the head of a subtree that includes a string of a given length.

The proof that applying CP from t_1 is optimal is by contradiction. Suppose that after time t_1 another rule is optimal. This rule must, at least once, prescribe an action that is not according to CP. Consider the last point in time, say t_2, at which this rule prescribes an action not according to CP. So at t_2 there are m jobs, which are *not* heading the m longest strings, assigned to the m machines; from $t_2 + 1$, the CP rule is applied. Call the schedule from t_2 onward CP'. It suffices to show that applying CP from t_2 onward results in a schedule at least as good.

Consider under CP' the longest string headed by a job *not* assigned at t_2, say string 1, and the shortest string headed by a job that *is* assigned at t_2, say string 2. Let C_1' and C_2' denote the completion times of the last jobs of strings 1 and 2, respectively, under CP'. It is clear that under CP' all m machines have to be busy at least up to $C_2' - 1$. If $C_1' \geq C_2' + 1$ and there are machines idle before $C_1' - 1$, the application of CP at t_2 results in less idle time and a smaller total completion time. Under CP, the last job of string 1 is completed one time unit earlier, yielding one more completed job at or before $C_1' - 1$. In other cases, the total completion time under CP is equal to the total completion time under CP'. This implies that CP is optimal from t_2 on. As there is not then a last time for a deviation from CP, the CP rule is optimal.   □

In contrast to the makespan objective, the CP rule is, somewhat surprisingly, not necessarily optimal for intrees. Counterexamples can be found easily (see Exercise 5.24).

Consider the problem $Pm \mid p_j = 1, M_j \mid \sum C_j$. Again, if the M_j sets are nested, the Least Flexible Job first rule can be shown to be optimal.

Theorem 5.3.4. *The LFJ rule is optimal for $Pm \mid p_j = 1, M_j \mid \sum C_j$ when the M_j sets are nested.*

Proof.   The proof is similar to the proof of Theorem 5.1.8.   □

The previous model is a special case of $Rm \mid\mid \sum C_j$. As stated in Chapter 2, the machines in the Rm environment are entirely unrelated. That is, machine 1 may be able to process job 1 in a short time and may need a long time for job 2, whereas machine 2 may be able to process job 2 in a short time and may need a long time for job 1. That the Qm environment is a special case is clear. Identical machines in parallel with job j being restricted to machine set M_j is also a special case; the processing time of job j on a machine that is not part of M_j has to be considered very long, making it impossible to process the job on one of these machines.

The $Rm \mid\mid \sum C_j$ problem can be formulated as an integer program with a special structure that makes it possible to solve the problem in polynomial time. Recall that if job j is processed on machine i and there are $k - 1$ jobs following job j on this machine i, then job j contributes kp_{ij} to the value of the objective function. Let x_{ikj} denote $0-1$ integer variables, where $x_{ikj} = 1$ if job j is scheduled as the kth to last job on machine i and 0 otherwise. The integer program is then formulated as follows:

$$minimize \quad \sum_{i=1}^{m}\sum_{j=1}^{n}\sum_{k=1}^{n} kp_{ij}x_{ikj}$$

subject to

$$\sum_{i=1}^{m}\sum_{k=1}^{n} x_{ikj} = 1, \qquad j = 1, \ldots, n$$

$$\sum_{j=1}^{n} x_{ikj} \leq 1, \qquad i = 1, \ldots, m, k = 1, \ldots, n$$

$$x_{ikj} \in \{0, 1\}, \qquad i = 1, \ldots, m, k = 1, \ldots, n \quad j = 1, \ldots, n.$$

The constraints make sure that each job is scheduled exactly once and each position on each machine is taken by at most one job. Note that the processing times only appear in the objective function.

This is a so-called *weighted bipartite matching problem*, with the n jobs on one side and nm positions on the other side (each machine can process at most n jobs). If job j is matched with (assigned to) position ik, there is a cost kp_{ij}. The objective is to determine the matching in this so-called *bipartite graph* with a minimum total cost. It is known from the theory of network flows that the integrality constraints on the x_{ikj} may be replaced by non-negativity constraints without

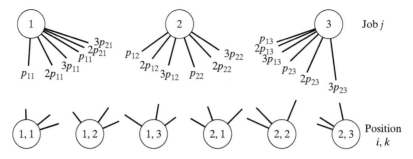

Figure 5.8   Bipartite graph for $Rm \parallel \sum C_j$ with three jobs.

changing the feasible set. This weighted bipartite matching problem then reduces to a regular linear program for which there exist polynomial time algorithms (see Appendix A).

Note that the optimal schedule may have unforced idleness.

Example 5.3.5 (Minimizing Total Completion Time with Unrelated Machines)

Consider two machines and three jobs. The processing times of the three jobs on the two machines are presented in the following table.

jobs	1	2	3
p_{1j}	4	5	3
p_{2j}	8	9	3

The bipartite graph associated with this problem is depicted in Figure 5.8. According to the optimal schedule, machine 1 processes job 1 first and job 2 second. Machine 2 processes job 3. This solution corresponds to $x_{121} = x_{112} = x_{213} = 1$ and all other x_{ikj} equal to zero. This optimal schedule is *not* nondelay (machine 2 is freed at time 3 and the waiting job is not put on the machine).

5.4 THE TOTAL COMPLETION TIME WITH PREEMPTIONS

In Theorem 5.3.1, it is shown that the nonpreemptive SPT rule minimizes $\sum C_j$ in a parallel machine environment. It turns out that the nonpreemptive SPT rule is also optimal when preemptions are allowed. This result is a special case of the more general result described next.

Consider m machines in parallel with different speeds (i.e., $Qm \mid prmp \mid \sum C_j$). This problem leads to the so-called *Shortest Remaining Processing Time on the Fastest Machine (SRPT-FM)* rule. According to this rule, at any point in time the job with the shortest remaining processing time is assigned to the fastest machine, the second shortest remaining processing time on the second fastest machine, and so on. Clearly, this rule requires preemptions. Every time the fastest machine completes a job, the job on the second fastest machine moves to the fastest machine, the job on the third fastest machine moves to the second fastest machine, and so on. So at the first job completion, there are $m - 1$ preemptions, at the second job completion there are $m - 1$ preemptions, and so on until the number of remaining jobs is less than the number of machines. From that point in time, the number of preemptions is equal to the number of remaining jobs.

The following lemma is needed for the proof.

Lemma 5.4.1.   *There exists an optimal schedule under which $C_j \leq C_k$ when $p_j \leq p_k$ for all j and k.*

Proof.   The proof is left as an exercise.   □

Without loss of generality, it may be assumed that there are as many machines as jobs. If the number of jobs is smaller than the number of machines, then the $m - n$ slowest machines are disregarded. If the number of jobs is larger than the number of machines, then $n - m$ machines are added with zero speeds.

Theorem 5.4.2.   *The SRPT-FM rule is optimal for $Qm \mid prmp \mid \sum C_j$.*

Proof.   Under SRPT-FM, $C_n \leq C_{n-1} \leq \cdots \leq C_1$. It is clear that under SRPT-FM the following equations have to be satisfied:

$$v_1 C_n = p_n$$
$$v_2 C_n + v_1(C_{n-1} - C_n) = p_{n-1}$$
$$v_3 C_n + v_2(C_{n-1} - C_n) + v_1(C_{n-2} - C_{n-1}) = p_{n-2}$$
$$\vdots$$
$$v_n C_n + v_{n-1}(C_{n-1} - C_n) + \cdots + v_1(C_1 - C_2) = p_1.$$

Adding these equations yields the following set of equations.

$$v_1 C_n = p_n$$
$$v_2 C_n + v_1 C_{n-1} = p_n + p_{n-1}$$
$$v_3 C_n + v_2 C_{n-1} + v_1 C_{n-2} = p_n + p_{n-1} + p_{n-2}$$
$$\vdots$$
$$v_n C_n + v_{n-1} C_{n-1} + \cdots + v_1 C_1 = p_n + p_{n-1} + \cdots + p_1.$$

Suppose schedule $\mathcal{S}'$ is optimal. From the previous lemma, it follows that $C'_n \leq C'_{n-1} \leq \cdots \leq C'_1$. The shortest job cannot be completed before p_n/v_1 (i.e., $C'_n \geq p_n/v_1$) or

$$v_1 C'_n \geq p_n.$$

Given that jobs n and $n-1$ are completed at C'_n and C'_{n-1},

$$(v_1 + v_2)C'_n + v_1(C'_{n-1} - C'_n)$$

is an upper bound on the amount of processing that can be done on these two jobs. This implies that

$$v_2 C'_n + v_1 C'_{n-1} \geq p_n + p_{n-1}.$$

Continuing in this manner, it is easy to show that

$$v_k C'_n + v_{k-1} C'_{n-1} + \cdots + v_1 C'_{n-k+1} \geq p_n + p_{n-1} + \cdots + p_{n-k+1}.$$

So

$$v_1 C'_n \geq v_1 C_n$$

$$v_2 C'_n + v_1 C'_{n-1} \geq v_2 C_n + v_1 C_{n-1}$$

$$\vdots$$

$$v_n C'_n + v_{n-1} C'_{n-1} + \cdots + v_1 C'_1 \geq v_n C_n + v_{n-1} C_{n-1} + \cdots + v_1 C_1.$$

If a collection of positive numbers α_j can be found, such that multiplying the jth inequality by α_j and adding all inequalities yields the inequality $\sum C'_j \geq \sum C_j$, then the proof is complete. It can be shown that these α_j must satisfy the equations

$$v_1 \alpha_1 + v_2 \alpha_2 + \cdots + v_n \alpha_n = 1$$

$$v_1 \alpha_2 + v_2 \alpha_3 + \cdots + v_{n-1} \alpha_n = 1$$

$$\vdots$$

$$v_1 \alpha_n = 1.$$

As $v_1 \geq v_2 \geq \cdots \geq v_n$, such a collection does exist.   □

Example 5.4.3 (Application of SRPT-FM Rule)

Consider 4 machines and 7 jobs. The machine data and job data are contained in the two tables below.

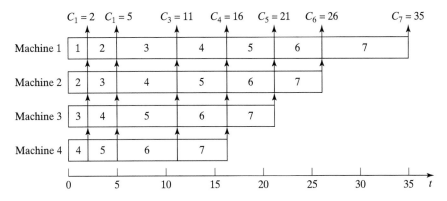

Figure 5.9 The SRPT-FM schedule (Example 5.4.3).

machines	1	2	3	4
v_i	4	2	2	1

jobs	1	2	3	4	5	6	7
p_j	8	16	34	40	45	46	61

Assuming that preemptions are only allowed at integer points in time, the SRPT-FM rule results in the schedule depicted in Figure 5.9. Under this optimal schedule, the total completion time is 116.

5.5 DUE DATE-RELATED OBJECTIVES

Single machine problems with due dates that can be solved in polynomial time usually have the minimization of the maximum lateness as objective (e.g., $1 \parallel L_{max}$, $1 \mid prmp \mid L_{max}$ and $1 \mid r_j, prmp \mid L_{max}$). Single machine problems with the total tardiness or the total weighted tardiness as objective tend to be very hard. It is easy to see that from a complexity point of view $Pm \parallel L_{max}$ is not as easy as $1 \parallel L_{max}$. Consider the special case where all jobs have due date 0. Finding a schedule with a minimum L_{max} is equivalent to $Pm \parallel C_{max}$ and is therefore NP-hard.

Consider $Qm \mid prmp \mid L_{max}$. This problem is one of the few parallel machine problems with a due date-related objective that can be solved in polynomial time. Suppose one has to verify whether there exists a feasible schedule with $L_{max} = z$. This implies that for job j the completion time C_j has to be less than or equal to $d_j + z$. Let $d_j + z$ be a hard deadline $\overline{d}_j$. Finding a feasible schedule with all jobs completing their processing before these deadlines is equivalent to solving the problem $Qm \mid r_j, prmp \mid C_{max}$. To see this, reverse the direction of time in the due date problem. Apply the LRPT-FM rule starting with the last deadline and work backward. The deadlines in the original problem play the role of the release dates in the reversed problem that is then equivalent to $Qm \mid r_j, prmp \mid C_{max}$. If applying the LRPT-FM rule backward results in a feasible schedule with all the jobs in the original problem starting at a time larger than or equal to zero, then there exists a schedule for $Qm \mid prmp \mid L_{max}$ with $L_{max} \leq z$. To find the minimum L_{max}, a simple search has to be done to determine the appropriate minimum value of z.

Example 5.5.1 (Minimizing Maximum Lateness with Preemptions)

Consider the following instance of $P2 \mid prmp \mid L_{max}$ with four jobs. The processing times and due dates are given in the following table. Preemptions are allowed at integer points in time.

jobs	1	2	3	4
d_j	4	5	8	9
p_j	3	3	3	8

First, it has to be checked whether there exists a feasible solution with $L_{max} = 0$. The data of the instance created through time reversal are determined as follows. The release dates are obtained by determining the maximum due date in the original problem, which is 9 and corresponds to job 4; the release date of job 4 in the new problem is then set equal to 0. The release dates of the remaining jobs are obtained by subtracting the original due dates from 9.

jobs	1	2	3	4
r_j	5	4	1	0
p_j	3	3	3	8

The question now is: In this new instance, can a schedule be created with a makespan less than 9? Applying LRPT immediately yields a feasible schedule.

Consider now $Qm \mid r_j, prmp \mid L_{max}$. Again a parametric study can be done. First an attempt is made to find a schedule with L_{max} equal to z. Due date d_j is replaced by a deadline $d_j + z$. Reversing this problem does not provide any additional insight as it results in a problem of the same type with release dates and due dates reversed. However, this problem can still be formulated as a network flow problem that can be solved in polynomial time.

5.6 DISCUSSION

This chapter focuses primarily on parallel machine problems that are either polynomial time solvable or have certain properties that are of interest. This chapter does not address the more complicated parallel machine problems that are strongly NP-hard and that have very little structure.

A significant amount of research has been done on parallel machine scheduling problems that are strongly NP-hard. A variety of integer programming formulations have appeared for $Rm \mid\mid \sum w_j C_j$ and $Rm \mid\mid \sum w_j U_j$. These integer programs can be solved using a special form of branch and bound, which is called *branch and price* and often also referred to as *column generation* (see Appendix A). However, there are many other parallel machine problems that are more complicated and that have not yet been tackled with exact methods.

An example of such a very hard problem is $Qm \mid s_{ijk} \mid \sum w_j T_j$. This problem is extremely hard to solve to optimality. It is already hard to find an optimal solution for instances with, say, five machines and 30 jobs. However, this problem is of considerable interest to industry, and many heuristics have been developed and experimented with. Part III of this book describes several heuristic methods that have been applied to this problem.

One category of parallel machine scheduling problems that has not been addressed in this chapter are the so-called *online scheduling problems*. In an online scheduling problem the jobs arrive one by one with processing times that only become known on arrival; they have to be scheduled immediately on one of the machines without knowledge of what jobs will come afterward or how many jobs are still to come. (In an offline scheduling problem, all information regarding all n jobs is known a priori. All problems considered in this chapter are offline problems.) An online version of $Pm \mid\mid C_{\max}$ can be described as follows. The jobs appear in a list (or sequence) and are presented to the decision maker according to this list one by one. When a job is presented to the decision maker, its processing time becomes known to the decision maker and the job has to be assigned to a machine and time slot before the next job on the list is seen. A job, when presented, can be assigned to any machine and time slot that is still available. (Because of this freedom in when and where to schedule a job, it may occur that one job gets to be processed before another job that had been scheduled earlier by the decision maker.) However, when a later job arrives on the stage, the schedule of the earlier jobs cannot be changed. A very basic example of an online algorithm is the list rule described in Section 5.1 for $Pm \mid\mid C_{\max}$. According to this model, the jobs are presented in a certain order (list); the LIST rule takes the job that ranks, among the remaining jobs, the highest on the list and schedules it on the machine with the smallest load as early as possible (see Exercises 5.28 and 5.29).

Online scheduling is important for several reasons. In practice, it often occurs that not all information is available before the first decision has to be made. From a theoretical point of view, online scheduling is important because it establishes a bridge between deterministic and stochastic scheduling. In stochastic scheduling, decisions typically have to be made with only a limited amount of information available.

EXERCISES (COMPUTATIONAL)

5.1. Consider $P6 \parallel C_{max}$ with 13 jobs.

jobs	1	2	3	4	5	6	7	8	9	10	11	12	13
p_j	6	6	6	7	7	8	8	9	9	10	10	11	11

(a) Compute the makespan under LPT.
(b) Find the optimal schedule.

5.2. Consider $P4 \mid prec \mid C_{max}$ with 12 jobs.

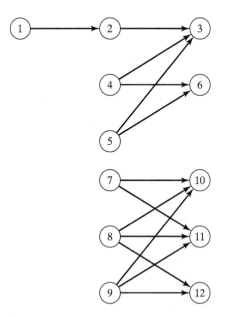

Figure 5.10   Precedence constraints graph (Exercise 5.2).

jobs	1	2	3	4	5	6	7	8	9	10	11	12
p_j	10	10	10	12	11	10	12	12	10	10	8	10

The jobs are subject to the precedence constraints depicted in Figure 5.10.

(a) Apply the generalized version of the CP rule: Every time a machine is freed, select the job at the head of the string with the largest total amount of processing.

(b) Apply the generalized version of the LNS rule: Every time a machine is freed, select the job that precedes the largest total amount of processing.

(c) Is either one of these two schedules optimal?

5.3. Consider $P3 \mid brkdwn, M_j \mid C_{max}$ with eight jobs.

jobs	1	2	3	4	5	6	7	8
p_j	10	10	7	7	7	7	7	7

Machines 1 and 2 are available continuously. Machine 3 is not available during the interval $[0, 1]$; after time 1, it is available throughout. The M_j sets are defined as follows:

$$M_1 = \{1, 3\}$$
$$M_2 = \{2, 3\}$$
$$M_3 = M_4 = M_5 = \{1\}$$
$$M_6 = M_7 = M_8 = \{2\}$$

(a) Apply the LPT rule (i.e., always give priority to the largest job that can be processed on the machine freed).

(b) Apply the LFJ rule (i.e., always give priority to the least flexible job while disregarding processing times).

(c) Compute the ratio $C_{max}(LPT)/C_{max}(LFJ)$.

5.4. Consider $P3 \mid prmp \mid \sum C_j$ with the additional constraint that the completion of job j has to be less than or equal to a given fixed deadline d_j. Preemptions may occur only at integer times $1, 2, 3, \ldots$

jobs	1	2	3	4	5	6	7	8	9	10	11
p_j	2	3	3	5	8	8	8	9	12	14	16
d_j	∞	∞	∞	∞	∞	∞	11	12	13	28	29

Find the optimal schedule and compute the total completion time.

5.5. Consider $P\infty \mid prec \mid C_{max}$.

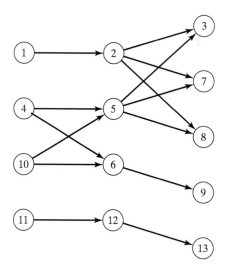

Figure 5.11   Precedence constraints graph ($P\infty \mid prec \mid C_{\max}$) for Exercise 5.5).

jobs	1	2	3	4	5	6	7	8	9	10	11	12	13
p_j	5	11	9	8	7	3	8	6	9	2	5	2	9

The precedence constraints are depicted in Figure 5.11. Determine the optimal makespan and which jobs are critical and which jobs are slack.

5.6. Consider $P5 \mid\mid \sum h(C_j)$ with 11 jobs.

jobs	1	2	3	4	5	6	7	8	9	10	11
p_j	5	5	5	6	6	7	7	8	8	9	9

The function $h(C_j)$ is defined as follows.

$$h(C_j) = \begin{cases} 0 & \text{if } C_j \leq 15 \\ C_j - 15 & \text{if } C_j > 15 \end{cases}$$

(a) Compute the value of the objective function under SPT.
(b) Compute the value of the objective function under the optimal schedule.

5.7. Consider again $P5 \mid\mid \sum h(C_j)$ with the 11 jobs of the previous exercise. The function $h(C_j)$ is now defined as follows:

$$h(C_j) = \begin{cases} C_j & \text{if } C_j \leq 15 \\ 15 & \text{if } C_j > 15. \end{cases}$$

(a) Compute the value of the objective function under SPT.

(b) Compute the value of the objective under the optimal schedule.

5.8. Consider $Q2 \mid prmp \mid C_{max}$ with the jobs

jobs	1	2	3	4
p_j	36	24	16	12

and machine speeds $v_1 = 2$ and $v_2 = 1$.

(a) Find the makespan under LRPT when preemptions can only be made at the time points 0, 4, 8, 12, and so on.

(b) Find the makespan under LRPT when preemptions can only be made at the time points 0, 2, 4, 6, 8, 10, 12, and so on.

(c) Find the makespan under LRPT when preemptions can be made at *any* time.

(d) Compare the makespans under (a), (b), and (c).

5.9. Consider the following example of $P3 \mid prmp, brkdwn \mid \sum C_j$ with six jobs. Three jobs have a processing time of 1 while the remaining three have a processing time of 2. There are three machines, but two machines are not available from time 2 onward. Determine the optimal schedule. Show that SRPT is not optimal.

5.10. Consider the following instance of $P2 \mid prmp \mid L_{max}$ with four jobs. Preemptions are allowed at integer points in time. Find an optimal schedule.

jobs	1	2	3	4
d_j	5	6	9	10
p_j	4	5	7	9

EXERCISES (THEORY)

5.11. Consider $Pm \mid\mid C_{max}$.

(a) Give an example showing that LPT is not necessarily optimal when the number of jobs is less than or equal to twice the number of machines ($n \leq 2m$).

(b) Show that if an optimal schedule results in at most two jobs on any machine, then LPT is optimal.

5.12. Consider $Pm \mid\mid C_{max}$. Describe the processing times of the instance that attains the worst case bound in Theorem 5.1.1 (as a function of m). (*Hint:* See Exercise 5.1.)

5.13. Show that the CP rule is optimal for $Pm \mid outtree, p_j = 1 \mid C_{max}$.

5.14. Complete the proof of Theorem 5.1.5. That is, show that the CP rule applied to $Pm \mid intree, p_j = 1 \mid C_{max}$ results in a makespan that is equal to $l_{max} + c$.

5.15. Consider $Pm \mid r_j, prmp \mid C_{max}$. Formulate the optimal policy and prove its optimality.

5.16. Consider $Pm \mid prmp, brkdwn \mid C_{\max}$ with the number of machines available being a function of time (i.e., $m(t)$). Show that for any function $m(t)$ LRPT minimizes the makespan.

5.17. Consider $Pm \mid brkdwn \mid \sum C_j$ with the number of machines available as a function of time (i.e., $m(t)$). Show that if $m(t)$ is increasing, the nonpreemptive SPT rule is optimal.

5.18. Consider $Pm \mid prmp, brkdwn \mid \sum C_j$ with the number of machines available being a function of time (i.e., $m(t)$). Show that the preemptive SRPT rule is optimal if $m(t) \geq m(s) - 1$ for all $s < t$.

5.19. Consider $Pm \mid prmp \mid \sum C_j$ with the added restriction that all jobs *must* be finished by some fixed deadline $\bar{d}$, where

$$\bar{d} \geq \max \left(\frac{\sum p_j}{m}, p_1, \ldots, p_n \right).$$

Find the rule that minimizes the total completion time and prove its optimality.

5.20. Consider $Pm \mid\mid \sum w_j C_j$. Show that in the worst case example of the WSPT rule, w_j has to be approximately equal to p_j for each j.

5.21. Give a characterization of the class of all schedules that are optimal for $Pm \mid\mid \sum C_j$. Determine the number of schedules in this class as a function of n and m.

5.22. Consider $P2 \mid\mid \sum C_j$. Develop a heuristic for minimizing the makespan subject to total completion time optimality. (*Hint:* Say a job is of *Rank j* if $j - 1$ jobs follow the job on its machine. With two machines in parallel, there are two jobs in each rank. Consider the difference in the processing times of the two jobs in the same rank. Base the heuristic on these differences.)

5.23. Consider $Pm \mid M_j \mid \gamma$. The sets M_j are given. Let J_i denote the set of jobs that are allowed to be processed on machine i. Show through a counterexample that the sets M_j being nested does not necessarily imply that sets J_i are nested. Give sufficiency conditions on the set structures under which the M_j and J_i sets are nested.

5.24. Show through a counterexample that the CP rule is not necessarily optimal for $Pm \mid intree, p_j = 1 \mid \sum C_j$.

5.25. Consider $Pm \mid r_j, prmp \mid L_{\max}$. Show through a counterexample that the preemptive EDD rule does *not* necessarily yield an optimal schedule.

5.26. Consider $Pm \mid intree, prmp \mid C_{\max}$ with the processing time of each job at level k equal to p_k. Show that a preemptive version of the generalized CP rule minimizes the makespan.

5.27. Consider $Q\infty \mid prec, prmp \mid C_{\max}$. There are an unlimited number of machines that operate at the same speed. There is *one* machine that is faster. Give an algorithm that minimizes the makespan and prove its optimality.

5.28. Consider the online version of $Pm \mid\mid C_{\max}$. The jobs are presented in an arbitrary list. Apply the LIST rule: As soon as a machine is freed, the job that ranks among the remaining jobs the highest is assigned to that machine.

(a) Show that for any list,

$$\frac{C_{\max}(LIST)}{C_{\max}(OPT)} \le 2 - \frac{1}{m}.$$

(b) Show that the bound is tight (i.e., create an instance of n jobs and a given list that achieves the bound).

5.29. Consider an online version of $Pm \mid r_j, prec \mid C_{\max}$. An online algorithm for this problem can be described as follows. The jobs are again presented in a list; whenever a machine is freed, the job that ranks highest among the remaining jobs that are ready for processing is assigned to that machine (i.e., it must be a job that already has been released and of which all predecessors already have been completed). Show that the bound presented in part (a) of Exercise 5.28 applies to this more general problem as well.

COMMENTS AND REFERENCES

The worst case analysis of the LPT rule for $Pm \parallel C_{\max}$ is from the classic paper by Graham (1969). This paper gives one of the first examples of worst case analyses of heuristics (see also Graham, 1966). It also provides a worst case analysis of an arbitrary list schedule for $Pm \parallel C_{\max}$. A more sophisticated heuristic for $Pm \parallel C_{\max}$, with a tighter worst case bound, is the so-called MULTIFIT heuristic; see Coffman, Garey, and Johnson (1978) and Friesen (1984a). Lee and Massey (1988) analyze a heuristic that is based on LPT as well as on MULTIFIT. For results on heuristics for the more general $Qm \parallel C_{\max}$, see Friesen and Langston (1983), Friesen (1984b), and Dobson (1984). Davis and Jaffe (1981) present an algorithm for $Rm \parallel C_{\max}$. The CPM and PERT procedures have been covered in many papers and textbooks; see for example, French (1982). The CP result in Theorem 5.1.5 is due to Hu (1961). See Lenstra and Rinnooy Kan (1978) with regard to $Pm \mid p_j = 1, prec \mid C_{\max}$, Du and Leung (1989) with regard to $P2 \mid tree \mid C_{\max}$, and Du, Leung, and Young (1991) with regard to $P2 \mid chains \mid C_{\max}$. Chen and Liu (1975) and Kunde (1976) analyze the worst case behavior of the CP rule for $Pm \mid p_j = 1, prec \mid C_{\max}$. Apparently, no worst case analysis has been done for the LFJ rule.

For $Pm \mid prmp \mid C_{\max}$, see McNaughton (1959). For $Qm \mid prmp \mid C_{\max}$, see Horvath, Lam, and Sethi (1977), Gonzalez and Sahni (1978a), and McCormick and Pinedo (1995).

Conway, Maxwell, and Miller (1967) discuss the SPT rule for $Pm \parallel \sum C_j$; they also give a characterization of the class of optimal schedules. For a discussion of $Qm \parallel \sum C_j$, see Horowitz and Sahni (1976). The worst case bound for the WSPT rule for $Pm \parallel \sum w_j C_j$ is from Kawaguchi and Kyan (1986). Elmaghraby and Park (1974) and Sarin, Ahn, and Bishop (1988) present branch and bound algorithms for this problem. Eck and Pinedo (1993) present a heuristic for minimizing the makespan and the total completion time simultaneously. The optimality of the CP rule for $Pm \mid p_j = 1, outtree \mid \sum C_j$ is due to Hu (1961). For complexity results with regard to $Pm \mid prec \mid \sum C_j$, see Sethi (1977) and Du, Leung, and Young (1990).

For an analysis of the $Qm \mid prmp \mid \sum C_j$ problem, see Lawler and Labetoulle (1978), Gonzalez and Sahni (1978), and McCormick and Pinedo (1995).

A significant amount of work has been done on $Qm \mid r_j, p_j = p, prmp \mid \gamma$; see Garey, Johnson, Simons, and Tarjan (1981), Federgruen and Groenevelt (1986), Lawler and Martel (1989), Martel (1982), and Simons (1983).

For results with regard to $Qm \mid prmp \mid L_{\max}$, see Bruno and Gonzalez (1976) and Labetoulle, Lawler, Lenstra, and Rinnooy Kan (1984). For other due date-related results, see Sahni and Cho (1979b).

Chen and Powell (1999) and Van den Akker, Hoogeveen, and Van de Velde (1999) applied branch and bound methods (including column generation) to $Rm \mid\mid \sum w_j C_j$ and $Rm \mid\mid \sum w_j U_j$.

The worst case analysis of an arbitrary list schedule for $Pm \mid\mid C_{\max}$ is regarded as one of the basic results in the area of online scheduling (see Exercises 5.28 and 5.29). For an overview of online scheduling on parallel machines with the makespan objective, see Fleischer and Wahl (2000). Research in online scheduling has focused on other objectives also; see, for example, Shmoys, Wein, and Williamson (1995). For a general survey of online scheduling, see Sgall (1998).

6

Flow Shops and Flexible Flow Shops (Deterministic)

In many manufacturing and assembly facilities, a number of operations have to be done on every job. Often these operations have to be done on all jobs in the same order implying that the jobs have to follow the same route. The machines are assumed to be set up in series, and the environment is referred to as a *flow shop*.

The storage or buffer capacities in between successive machines may sometimes be, for all practical purposes, virtually unlimited. This is often the case when the products being processed are physically small (e.g., printed circuit boards, integrated circuits), making it relatively easy to store large quantities between machines. When the products are physically large (e.g., TV sets, copiers), then the buffer space in between two successive machines may have a limited capacity, causing *blocking*. Blocking occurs when the buffer is full and the upstream machine is not allowed to release a job into the buffer after completing its processing. If this is the case, then the job has to remain at the upstream machine, preventing a job in queue at that machine from beginning its processing.

A somewhat more general machine environment consists of a number of stages in series with a number of machines in parallel at each stage. A job has to be processed at each stage on only one of the machines. This machine environment is often referred to as a flexible flow shop, compound flow shop, multiprocessor flow shop, or hybrid flow shop.

Most of the material in this chapter concerns the makespan objective. The makespan objective is of considerable practical interest as its minimization is to a certain extent equivalent to the maximization of the utilization of the machines. The models, however, tend to be of such complexity that makespan results are already relatively hard to obtain. Total completion time and due date-related objectives tend to be even harder.

6.1 FLOW SHOPS WITH UNLIMITED INTERMEDIATE STORAGE

When searching for an optimal schedule for $Fm \ || \ C_{\max}$, the question arises whether it suffices merely to determine a permutation in which the jobs traverse the entire system. Physically it may be possible for one job to pass another while waiting in queue for a machine that is busy. The machines may not operate according to the *First Come First Served* principle, and the sequence in which the jobs go through the machines may change from one machine to another. Changing the sequence of the jobs waiting in a queue between two machines may at times result in a smaller makespan. However, it can be shown that there always exists an optimal schedule without job sequence changes between the first two machines and between the last two machines (see Exercise 6.11). This implies that there are optimal schedules for $F2 \ || \ C_{\max}$ and $F3 \ || \ C_{\max}$ that do not require sequence changes between machines. One can find examples of flow shops with four machines in which the optimal schedule does require a job sequence change in between the second and third machines.

Finding an optimal schedule when sequence changes are allowed is significantly harder than finding an optimal schedule when they are not. Flow shops that do not allow sequence changes between machines are called *permutation* flow shops. In these flow shops, the same sequence or permutation of jobs is maintained throughout. The results in this chapter are mostly limited to permutation flow shops.

Given a permutation schedule $j_1, \dots, j_n$ for an m machine flow shop, the completion time of job j_k at machine i can be computed easily through a set of recursive equations:

$$C_{i,j_1} = \sum_{l=1}^{i} p_{l,j_1} \qquad\qquad i = 1, \dots, m$$

$$C_{1,j_k} = \sum_{l=1}^{k} p_{1,j_l} \qquad\qquad k = 1, \ldots, n$$

$$C_{i,j_k} = \max(C_{i-1,j_k}, C_{i,j_{k-1}}) + p_{i,j_k} \qquad i = 2, \ldots, m; \, k = 2, \ldots, n$$

The value of the makespan under a given permutation schedule can also be computed by determining the *critical path* in a directed graph corresponding to the schedule. For a given sequence $j_1, \ldots, j_n$, this directed graph is constructed as follows: For each operation, say the processing of job j_k on machine i, there is a node (i, j_k) with a weight that is equal to the processing time of job j_k on machine i. Node (i, j_k), $i = 1, \ldots, m-1$, and $k = 1, \ldots, n-1$, has arcs going to nodes $(i + 1, j_k)$ and (i, j_{k+1}). Nodes corresponding to machine m have only one outgoing arc, as do the nodes corresponding to job j_n. Node (m, j_n) has no outgoing arcs (see Figure 6.1). The total weight of the maximum weight path from node $(1, j_1)$ to node (m, j_n) corresponds to the makespan under the permutation schedule $j_1, \ldots, j_n$.

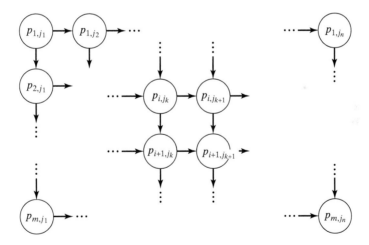

Figure 6.1   Directed graph for the computation of the makespan in $Fm \mid prmu \mid C_{\max}$ under sequence $j_1, j_2, \ldots j_n$.

Example 6.1.1 (Graph Representation of Flow Shop)

Consider five jobs on four machines with the processing times presented in the following table.

jobs	j_1	j_2	j_3	j_4	j_5
p_{1,j_k}	5	5	3	6	3
p_{2,j_k}	4	4	2	4	4
p_{3,j_k}	4	4	3	4	1
p_{4,j_k}	3	6	3	2	5

The corresponding graph and Gantt chart are depicted in Figure 6.2. From the directed graph, it follows that the makespan is 34. This makespan is determined by two critical paths.

An interesting result can be obtained by comparing two m machine permutation flow shops with n jobs. Let $p_{ij}^{(1)}$ and $p_{ij}^{(2)}$ denote the processing time of job j on machine i in the first and second flow shop, respectively. Assume

$$p_{ij}^{(1)} = p_{m+1-i,j}^{(2)}.$$

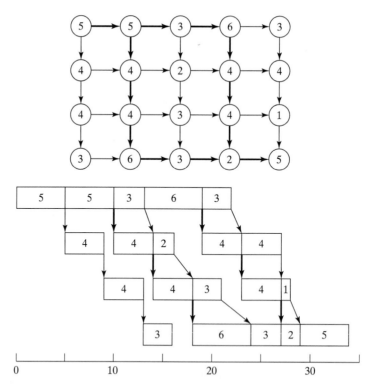

Figure 6.2   Directed graph, critical paths, and Gantt chart (the numerical entries represent the processing times of the jobs and not the job indexes).

This basically implies that the first machine in the second flow shop is identical to the last machine in the first flow shop; the second machine in the second flow shop is identical to the machine immediately before the last in the first flow shop; and so on. The following lemma applies to these two flow shops.

Lemma 6.1.2.   *Sequencing the jobs according to permutation $j_1, \ldots, j_n$ in the first flow shop results in the same makespan as sequencing the jobs according to permutation $j_n, \ldots, j_1$ in the second flow shop.*

Proof.   If the first flow shop under sequence $j_1, \ldots, j_n$ corresponds to the diagram in Figure 6.1, then the second flow shop under sequence $j_n, \ldots, j_1$ corresponds to the same diagram with all the arcs reversed. The weight of the maximum weight path from one corner node to the other corner node does not change if all the arcs are reversed.   □

Lemma 6.1.2 states the following *reversibility* result: The makespan does not change if the jobs traverse the flow shop in the opposite direction in reverse order.

Example 6.1.3 (Graph Representation and Reversibility)

Consider the instance of Example 6.1.1. The dual of this instance is given in the following table.

jobs	j_1	j_2	j_3	j_4	j_5
p_{1,j_k}	5	2	3	6	3
p_{2,j_k}	1	4	3	4	4
p_{3,j_k}	4	4	2	4	4
p_{4,j_k}	3	6	3	5	5

The corresponding digraph, its critical paths, and the Gantt charts are depicted in Figure 6.3. It is clear that the critical paths are determined by the same set of processing times and that the makespan, therefore, is 34 as well.   □

Consider now the $F2 \mid\mid C_{\max}$ problem: a flow shop with two machines in series with unlimited storage in between the two machines. There are n jobs and the processing time of job j on machine 1 is p_{1j} and the processing time on machine 2 is p_{2j}. This was one of the first problems to be analyzed in the early days of Operations Research and led to a classical paper in scheduling theory by S.M. Johnson. The rule that minimizes the makespan is commonly referred to as *Johnson's rule*.

An optimal sequence can be described as follows. Partition the jobs into two sets, with Set I containing all the jobs with $p_{1j} < p_{2j}$ and Set II all the jobs with $p_{1j} > p_{2j}$. The jobs with $p_{1j} = p_{2j}$ may be put in either set. The jobs in Set I go first and they go in in increasing order of p_{1j} (SPT); the jobs in Set II follow in decreasing order of p_{2j} (LPT). Ties may be broken arbitrarily. In what follows

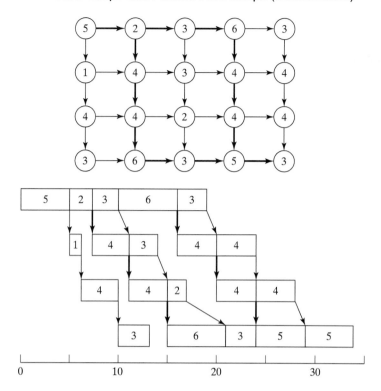

Figure 6.3   Directed graph, critical paths, and Gantt chart (the numerical entries represent the processing times of the jobs and not the job indexes).

such a schedule is referred to as an *SPT(1)-LPT(2)* schedule. Of course, there may be more than one schedule constructed this way.

Theorem 6.1.4.   *Any SPT(1)-LPT(2) schedule is optimal for F2 || C_{max}.*

Proof.   The proof is by contradiction. Suppose another type of schedule is optimal. In this optimal schedule, there must be a pair of adjacent jobs, say job j followed by job k, that satisfies one of the following three conditions:

  (i) job j belongs to Set II and job k to Set I;
 (ii) jobs j and k belong to Set I and $p_{1j} > p_{1k}$;
(iii) jobs j and k belong to Set II and $p_{2j} < p_{2k}$.

It suffices to show that under any of these three conditions the makespan is reduced after a pairwise interchange of jobs j and k. Assume that in the original schedule job l precedes job j and job m follows job k. Let C_{ij} denote the completion of job j on machine i under the original schedule and let C'_{ij} denote the completion time of job j on machine i after the pairwise interchange. Interchanging jobs j and k clearly does not affect the starting time of job m on machine 1 as its starting time on machine 1 equals $C_{1l} + p_{1j} + p_{1k}$. However, it is of interest to know at what time machine 2 becomes available for job m. Under the original schedule, this is the completion time of job k on machine 2 (i.e., C_{2k}), and after the interchange this is the completion time of job j on machine 2 (i.e., C'_{2j}). It suffices to show that $C'_{2j} \leq C_{2k}$ under any one of the three conditions described previously.

The completion time of job k on machine 2 under the original schedule is

$$C_{2k} = \max\left(\max(C_{2l}, C_{1l} + p_{1j}) + p_{2j}, C_{1l} + p_{1j} + p_{1k}\right) + p_{2k}$$

$$= \max\left(C_{2l} + p_{2k} + p_{2j}, C_{1l} + p_{1j} + p_{2j} + p_{2k}, C_{1l} + p_{1j} + p_{1k} + p_{2k}\right),$$

whereas the completion time of job j on machine 2 after the pairwise interchange is

$$C'_{2j} = \max\left(C_{2l} + p_{2k} + p_{2j}, C_{1l} + p_{1k} + p_{2k} + p_{2j}, C_{1l} + p_{1k} + p_{1j} + p_{2j}\right).$$

Under condition (i), $p_{1j} > p_{2j}$ and $p_{1k} < p_{2k}$. It is clear that the first terms within the max expressions of C_{2k} and C'_{2j} are identical. The second term in the last expression is smaller than the third in the first expression, and the third term in the last expression is smaller than the second in the first expression. So under condition (i), $C'_{2j} \leq C_{2k}$.

Under condition (ii), $p_{1j} < p_{2j}$, $p_{1k} < p_{2k}$, and $p_{1j} > p_{1k}$. Now the second as well as the third term in the last expression are smaller than the second term in the first expression. So under condition (ii), $C'_{2j} \leq C_{2k}$ as well.

Condition (iii) can be shown in a similar way as the second condition. Actually, condition (iii) follows immediately from the reversibility property of flow shops.   □

These SPT(1)-LPT(2) schedules are by no means the only schedules that are optimal for $F2 \mid\mid C_{\max}$. The class of optimal schedules appears to be hard to characterize and data dependent.

Example 6.1.5 (Multiple Schedules That Are Optimal)

Consider a set of jobs with one job that has a very small processing time on machine 1 and a very large processing time on machine 2, say K, with $K \geq \sum_{j=1}^{n} p_{1j}$. It is clear that under the optimal sequence this job should go first in the schedule. However, the order of the remaining jobs does not affect the makespan.

Unfortunately, the SPT(1)-LPT(2) schedule structure cannot be generalized to yield optimal schedules for flow shops with more than two machines. However, minimizing the makespan in a permutation flow shop with an arbitrary number of machines (i.e., $Fm \mid prmu \mid C_{\max}$), *can* be formulated as a Mixed Integer Program (MIP).

To formulate the problem as an MIP, a number of variables have to be defined: The decision variable x_{jk} equals 1 if job j is the kth job in the sequence and 0 otherwise. The auxiliary variable I_{ik} denotes the idle time on machine i between the processing of the jobs in the kth position and $(k + 1)$th position; the auxiliary variable W_{ik} denotes the waiting time of the job in the kth position in between machines i and $i + 1$. Of course, there exists a strong relationship between the variables W_{ik} and I_{ik}. For example, if $I_{ik} > 0$, then $W_{i-1,k+1}$ has to be zero. Formally, this relationship can be established by considering the difference between the time the job in the $(k + 1)$th position starts on machine $i + 1$ and the time the job in the kth position completes its processing on machine i. If Δ_{ik} denotes this difference and if $p_{i(k)}$ denotes the processing time on machine i of the job in the kth position in the sequence, then (see Figure 6.4)

$$\Delta_{ik} = I_{ik} + p_{i(k+1)} + W_{i,k+1} = W_{ik} + p_{i+1(k)} + I_{i+1,k}.$$

Note that minimizing the makespan is equivalent to minimizing the total idle time on the last machine, machine m. This idle time is equal to

$$\sum_{i=1}^{m-1} p_{i(1)} + \sum_{j=1}^{n-1} I_{mj},$$

which is the idle time that must occur before the job in the first position reaches the last machine and the sum of the idle times between the jobs on the last machine.

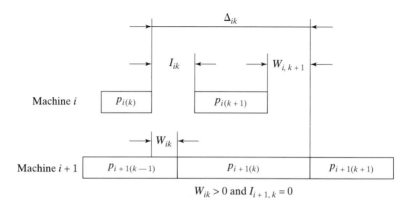

Figure 6.4 Constraints in the integer program formulation.

Using the identity

$$p_{i(k)} = \sum_{j=1}^{n} x_{jk} p_{ij},$$

the MIP can be formulated as follows.

$$\min \left(\sum_{i=1}^{m-1} \sum_{j=1}^{n} x_{j1} p_{ij} + \sum_{j=1}^{n-1} I_{mj} \right),$$

subject to

$$\sum_{j=1}^{n} x_{jk} = 1 \qquad k = 1, \ldots, n,$$

$$\sum_{k=1}^{n} x_{jk} = 1 \qquad j = 1, \ldots, n,$$

$$I_{ik} + \sum_{j=1}^{n} x_{j,k+1} p_{ij} + W_{i,k+1}$$

$$-W_{ik} - \sum_{j=1}^{n} x_{jk} p_{i+1,j} - I_{i+1,k} = 0 \qquad k = 1, \ldots, n-1; i = 1, \ldots, m-1,$$

$$W_{i1} = 0 \qquad i = 1, \ldots, m-1,$$

$$I_{1k} = 0 \qquad k = 1, \ldots, n-1.$$

The first set of constraints specifies that exactly one job has to be assigned to position k for any k. The second set of constraints specifies that job j has to be assigned to exactly one position. The third set of constraints relate the decision variables x_{jk} to the physical constraints. These physical constraints enforce the necessary relationships between the idle time and waiting time variables. Thus, the problem of minimizing the makespan in an m machine permutation flow shop is formulated as an MIP. The only integer variables are the binary (0-1) decision variables x_{jk}. The idle time and waiting time variables are non-negative continuous variables.

Example 6.1.6 (Mixed Integer Programming Formulation)

Consider again the instance in Example 6.1.1. Because the sequence is now not given, the subscript j_k in the table of Example 6.1.1 is replaced by the subscript j, and the headings j_1, j_2, j_3, j_4, and j_5 are replaced by 1, 2, 3, 4, and 5, respectively.

jobs	1	2	3	4	5
$p_{1,j}$	5	5	3	6	3
$p_{2,j}$	4	4	2	4	4
$p_{3,j}$	4	4	3	4	1
$p_{4,j}$	3	6	3	2	5

With these data, the objective of the MIP is

$$5x_{11} + 5x_{21} + 3x_{31} + 6x_{41} + 3x_{51} + 4x_{11} + 4x_{21} + 2x_{31} + 4x_{41} + 4x_{51}$$
$$+ 4x_{11} + 4x_{21} + 3x_{31} + 4x_{41} + x_{51} + I_{41} + I_{42} + I_{43} + I_{44}$$
$$= 13x_{11} + 13x_{21} + 8x_{31} + 14x_{41} + 8x_{51} + I_{41} + I_{42} + I_{43} + I_{44}.$$

The first and second set of constraints of the program contain five constraints each. The third set contains $(5 - 1)(4 - 1) = 12$ constraints. For example, the constraint corresponding to $k = 2$ and $i = 3$ is

$$I_{32} + 4x_{13} + 4x_{23} + 3x_{33} + 4x_{43} + x_{53} + W_{33}$$
$$- W_{32} - 3x_{12} - 6x_{22} - 3x_{32} - 2x_{42} - 5x_{52} - I_{42} = 0.$$

The fact that the problem can be formulated as an MIP does not immediately imply that the problem is NP-hard. It could be that the MIP contains a special structure that allows for a polynomial time algorithm (see, e.g., the integer programming formulation for $Rm \mid\mid \sum C_j$). In this case, however, it turns out that the problem *is* hard.

Theorem 6.1.7. $F3 \mid\mid C_{\max}$ *is strongly NP-hard.*

Proof.   By reduction from 3-PARTITION. Given integers $a_1, \ldots, a_{3t}, b$, under the usual assumptions, let the number of jobs n equal $4t + 1$ and let

$$
\begin{aligned}
p_{10} &= 0, & p_{20} &= b, & p_{30} &= 2b, \\
p_{1j} &= 2b, & p_{2j} &= b, & p_{3j} &= 2b, \quad j = 1, \ldots, t - 1, \\
p_{1t} &= 2b, & p_{2t} &= b, & p_{3t} &= 0, \\
p_{1,t+j} &= 0, & p_{2,t+j} &= a_j, & p_{3,t+j} &= 0, \quad j = 1, \ldots, 3t.
\end{aligned}
$$

Let $z = (2t + 1)b$. A makespan of value z can be obtained if the first $t + 1$ jobs are scheduled according to sequence $0, 1, \ldots, t$. These $t + 1$ jobs then form a framework, leaving t gaps on machine 2. Jobs $t + 1, \ldots, t + 3t$ have to be partitioned into t sets of three jobs each, and these t sets have to be scheduled in between the first $t + 1$ jobs. A makespan of value z can be obtained if and only if 3-PARTITION has a solution (see Figure 6.5).   □

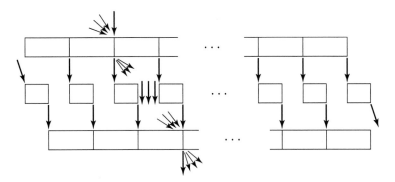

Figure 6.5   3-PARTITION reduces to $F3 \parallel C_{\max}$.

This complexity proof applies to permutation flow shops as well as to flow shops that allow sequence changes midstream (as said before, for three machine flow shops, it is known that a permutation schedule is optimal in the larger class of schedules).

Even though $Fm \mid prmu \mid C_{\max}$ is strongly NP-hard, it is of interest to study special cases that have nice structural properties. A number of special cases are important.

A special case of $Fm \mid prmu \mid C_{\max}$ that is of importance is the so-called *proportionate* permutation flow shop. In this flow shop, the processing times of job j on each of the m machines are equal to p_j (i.e., $p_{1j} = p_{2j} = \cdots = p_{mj} = p_j$). Minimizing the makespan in a proportionate permutation flow shop is denoted by $Fm \mid prmu, p_{ij} = p_j \mid C_{\max}$. For a proportionate flow shop model, an *SPT-LPT* sequence can be defined as follows. The jobs are partitioned into two sets; the jobs in one subset go first according to SPT, and the remaining jobs follow according to LPT. So a sequence $j_1, \ldots, j_n$ is SPT-LPT if and only if there is a job j_k such that

$$p_{j_1} \le p_{j_2} \le \cdots \le p_{j_k}$$

and

$$p_{j_k} \ge p_{j_{k+1}} \ge \cdots \ge p_{j_n}.$$

From Theorem 6.1.4, it follows that when $m = 2$, any SPT-LPT sequence must be optimal. As might be expected, these are not the only sequences that are optimal. This flow shop has a very special structure.

Theorem 6.1.8.   *For $Fm \mid prmu, p_{ij} = p_j \mid C_{\max}$, the makespan*

$$C_{\max} = \sum_{j=1}^{n} p_j + (m-1)\max(p_1, \ldots, p_n)$$

and is independent of the schedule.

Proof. The proof is left as an exercise. □

It can be shown that permutation schedules are also optimal in the larger class of schedules that allow jobs to pass one another while waiting for a machine (i.e., $Fm \mid p_{ij} = p_j \mid C_{\max}$; see Exercise 6.17). The result stated in the last theorem indicates that the proportionate flow shop is in one aspect similar to the single machine: The makespan does not depend on the sequence. Actually there are many more similarities between the proportionate flow shop and the single machine environment. The following examples illustrate this fact.

(i) The SPT rule is optimal for $1 \mid\mid \sum C_j$ as well as for $Fm \mid prmu, p_{ij} = p_j \mid \sum C_j$.

(ii) The algorithm that results in an optimal schedule for $1 \mid\mid \sum U_j$ also results in an optimal schedule for $Fm \mid prmu, p_{ij} = p_j \mid \sum U_j$.

(iii) The algorithm that results in an optimal schedule for $1 \mid\mid h_{\max}$ also results in an optimal schedule for $Fm \mid prmu, p_{ij} = p_j \mid h_{\max}$.

(iv) The pseudo-polynomial dynamic programming algorithm for $1 \mid\mid \sum T_j$ is also applicable to $Fm \mid prmu, p_{ij} = p_j \mid \sum T_j$.

(v) The elimination criteria that hold for $1 \mid\mid \sum w_j T_j$ also hold for $Fm \mid prmu, p_{ij} = p_j \mid \sum w_j T_j$.

Not all results that hold for the single machine hold for the proportionate flow shop. For example, WSPT does *not* necessarily minimize the total weighted completion time in proportionate flow shops. Counterexamples can be found easily.

The proportionate permutation flow shop model can be generalized to include machines with different speeds. If the speed of machine i is v_i, then the time job j spends on machine i is $p_{ij} = p_j/v_i$. The machine with the smallest v_i is called the *bottleneck* machine. However, the makespan is no longer schedule independent.

Theorem 6.1.9. *If in a proportionate permutation flow shop with different speeds the first (last) machine is the bottleneck, then LPT (SPT) minimizes the makespan.*

Proof. From the reversibility property, it immediately follows that it suffices to prove only one of the two results stated in the theorem. Only the case where the last machine is the bottleneck is shown here.

Consider first the special subcase with

$$v_m \leq v_1 \leq \min(v_2, \dots, v_{m-1});$$

that is, the last machine is the bottleneck and the first machine requires the second largest processing times for each one of the n jobs. It is easy to see that in such a flow shop the critical path only turns to the right at machine m and therefore turns down only once at some job j_k in the sequence $j_1, \dots, j_n$. So the critical path starts out on machine 1 going to the right, turns down at job j_k, and goes all the way down to machine m before turning to the right again. That SPT is optimal can be shown through a standard adjacent pairwise interchange argument. Consider a schedule that is not SPT. There are two adjacent jobs of which the first one is longer than the second. Interchanging these two jobs affects the makespan if and only if one of the two jobs is the job through which the critical path goes from machine 1 to m. It can be shown that an interchange then reduces the makespan and that SPT minimizes the makespan.

To complete the proof for the general case, call machine h an *intermediate bottleneck* if $v_h < \min(v_1, \dots, v_{h-1})$. There may be a number of intermediate bottlenecks in the proportionate flow shop. The arguments presented for the case with the only intermediate bottleneck being machine 1 extend to the general case with multiple intermediate bottlenecks. The critical path now only turns right at intermediate bottleneck machines. This structure can be exploited again with an adjacent pairwise interchange argument showing that SPT minimizes the makespan.

□

As $Fm \mid prmu \mid C_{\max}$ is one of the more basic scheduling problems, it has attracted a great deal of attention over the years. Many heuristics have been developed for dealing with this problem. One of the first heuristics developed for this problem is the *Slope* heuristic. According to this heuristic, a slope index is computed for each job. The slope index A_j for job j is defined as

$$A_j = -\sum_{i=1}^{m} (m - (2i - 1))\, p_{ij}.$$

The jobs are then sequenced in decreasing order of the slope index. The reasoning behind this heuristic is simple. From Theorem 6.1.4, it is already clear that jobs with small processing times on the first machine and large processing times on the second machine should be positioned more toward the beginning of the schedule, whereas jobs with large processing times on the first machine and small processing times on the second machine should be positioned more toward the end of the schedule. The slope index is large if the processing times on the downstream machines are large relative to the processing times on the upstream machines; the slope index is small if the processing times on the downstream machines are relatively small in comparison with the processing times on the upstream machines.

Example 6.1.10 (Application of Slope Heuristic)

Consider again the instance in Examples 6.1.1 and 6.1.6. Replace the j_k by j and the $j_1, \ldots, j_5$ by $1, \ldots, 5$. The slope indexes are:

$$A_1 = -(3 \times 5) - (1 \times 4) + (1 \times 4) + (3 \times 3) = -6$$

$$A_2 = -(3 \times 5) - (1 \times 4) + (1 \times 4) + (3 \times 6) = +3$$

$$A_3 = -(3 \times 3) - (1 \times 2) + (1 \times 3) + (3 \times 3) = +1$$

$$A_4 = -(3 \times 6) - (1 \times 4) + (1 \times 4) + (3 \times 2) = -12$$

$$A_5 = -(3 \times 3) - (1 \times 4) + (1 \times 1) + (3 \times 5) = +3.$$

The two sequences suggested by the heuristic are therefore 2, 5, 3, 1, 4 and 5, 2, 3, 1, 4, The makespan under both these sequences is 32. Complete enumeration verifies that both sequences are optimal.

In contrast to the makespan objective, results with regard to the total completion time objective are harder to obtain. It can be shown that $F2 \mid\mid \sum C_j$ is already strongly NP-hard. The proof of this result is somewhat involved and is therefore not presented here.

However, Theorem 6.1.8 facilitates the analysis of the flow time $Fm \mid p_{ij} = p_j \mid \sum C_j$ problem considerably, and it can be shown fairly easily that SPT minimizes the total completion time in a proportionate flow shop.

6.2 FLOW SHOPS WITH LIMITED INTERMEDIATE STORAGE

Consider m machines in series with *zero* intermediate storage between successive machines. If a given machine finishes the processing of any given job, the job cannot proceed to the next machine while that machine is busy, but must remain on the machine, which therefore remains idle. As stated before, this phenomenon is referred to as *blocking*.

In what follows, only flow shops with zero intermediate storages are considered since any flow shop with positive (but finite) intermediate storages between machines can be modeled as a flow shop with zero intermediate storages. This follows from the fact that a storage space capable of containing one job may be regarded as a *machine* on which the processing times of all jobs are equal to zero.

The problem of minimizing the makespan in a flow shop with zero intermediate storages is referred to in what follows as $Fm \mid block \mid C_{\max}$.

Let D_{ij} denote the time that job j actually departs machine i. Clearly, $D_{ij} \geq C_{ij}$. Equality holds when job j is not blocked. The time job j starts its processing

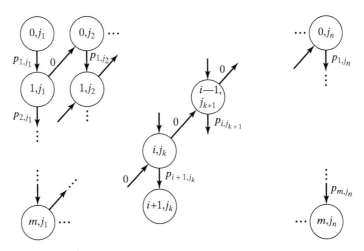

Figure 6.6   Directed graph for the computation of the makespan.

at the first machine is denoted by D_{0j}. The following recursive relationships hold under sequence $j_1, \ldots, j_n$.

$$D_{i,j_1} = \sum_{l=1}^{i} p_{l,j_1}$$

$$D_{i,j_k} = \max(D_{i-1,j_k} + p_{i,j_k}, D_{i+1,j_{k-1}})$$

$$D_{m,j_k} = D_{m-1,j_k} + p_{m,j_k}.$$

For this model, the makespan under a given permutation schedule can also be computed by determining the critical path in a directed graph. In this directed graph, node (i, j_k) denotes the departure time of job j_k from machine i. In contrast with the graph in Section 6.1 for flow shops with unlimited intermediate storages, in this graph the arcs, rather than the nodes, have weights. Node (i, j_k), $i = 1, \ldots, m-1; k = 1, \ldots, n-1$, has two outgoing arcs; one arc goes to node $(i+1, j_k)$ and has a weight or distance p_{i+1,j_k}, the other arc goes to node $(i-1, j_{k+1})$ and has weight zero. Node (m, j_k) has only one outgoing arc to node $(m-1, j_{k+1})$ with zero weight. Node (i, j_n) has only one outgoing arc to node $(i+1, j_n)$ with weight p_{i+1,j_n}. Node (m, j_n) has no outgoing arcs (see Figure 6.6). The $C_{\max}$ under sequence $j_1, \ldots, j_n$ is equal to the length of the maximum weight path from node $(0, j_1)$ to node (m, j_n).

Example 6.2.1 (Graph Representation of a Flow Shop with Blocking)

Consider the instance of Example 6.1.1. Assume that the same five jobs with the same processing times have to traverse the same four machines. The only difference

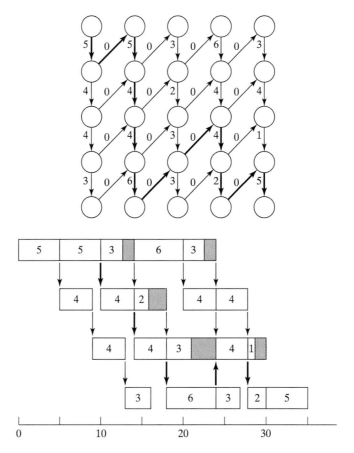

Figure 6.7 Directed graph, critical path, and Gantt chart (the numerical entries represent the processing times of the jobs and not the job indexes).

is that now there is zero intermediate storage between the machines. The directed graph and the Gantt chart corresponding to this situation are depicted in Figure 6.7. There is now only one critical path that determines the makespan of 35.

The following lemma shows that the reversibility property extends to flow shops with zero intermediate storage. Consider two m machine flow shops with blocking and let $p_{ij}^{(1)}$ and $p_{ij}^{(2)}$ denote the processing times of job j on machine i in the first and second flow shops, respectively.

Lemma 6.2.2. *If*

$$p_{ij}^{(1)} = p_{m+1-i,j}^{(2)},$$

then sequence $j_1, \ldots, j_n$ in the first flow shop results in the same makespan as sequence $j_n, \ldots, j_1$ in the second flow shop.

Proof. It can be shown that there is a one-to-one correspondence between paths of equal weight in the two directed graphs corresponding to the two flow shops. This implies that the paths with maximal weights in the two directed graphs must have the same total weight.   □

This reversibility result is similar to the result in Lemma 6.1.2. Actually, one can argue that the result in Lemma 6.1.2 is a special case of the result in Lemma 6.2.2. The unlimited intermediate storages can be regarded as sets of machines on which all processing is equal to zero.

Consider the $F2 \mid block \mid C_{\max}$ problem with two machines in series and zero intermediate storage in between. Note that in this flow shop, whenever a job starts its processing on the first machine, the preceding job starts its processing on the second machine. The time job j_k spends on machine 1, in process or blocked, is therefore $\max(p_{1,j_k}, p_{2,j_{k-1}})$. The first job in the sequence spends only p_{1,j_k} on machine 1. This makespan minimization problem is equivalent to a Traveling Salesman Problem with $n + 1$ cities. Let the distance from city j to city k be equal to

$$d_{0k} = p_{1k}$$
$$d_{j0} = p_{2j}$$
$$d_{jk} = \max(p_{2j}, p_{1k}).$$

The total distance traveled is then equal to the makespan of the flow shop. Actually, the distance matrix can be simplified somewhat. Instead of minimizing the makespan, one can minimize the total time between 0 and $C_{\max}$ that one of the two machines is either idle or blocked. The two objectives are equivalent since twice the makespan is equal to the sum of the $2n$ processing times plus the sum of all idle times. Minimizing the sum of all idle times is equivalent to the following Traveling Salesman Problem with $n + 1$ cities:

$$d_{0k} = p_{1k}$$
$$d_{j0} = p_{2j}$$
$$d_{jk} = \| p_{2j} - p_{1k} \| .$$

The idle time on one of the machines when job j starts on machine 2 and job k starts on machine 1 is the difference between the processing times p_{2j} and p_{1k}. If p_{2j} is larger than p_{1k}, job k will be blocked for the time difference on machine 1; otherwise machine 2 will remain idle for the time difference. The distance matrix of this Traveling Salesman Problem is identical to the one discussed in Section 4.5. The values for b_0 and a_0 in Section 4.5 have to be chosen equal to 0.

The algorithm for the $1 \mid s_{jk} \mid C_{max}$ problem with $s_{jk} = \parallel a_k - b_j \parallel$ can now be used for $F2 \mid block \mid C_{max}$ as well, which implies that there exists an $O(n^2)$ algorithm for $F2 \mid block \mid C_{max}$.

Example 6.2.3 (A Two Machine Flow Shop with Blocking and the TSP)

Consider a four job instance with processing times

jobs	1	2	3	4
$p_{1,j}$	2	3	3	9
$p_{2,j}$	8	4	6	2

This translates into a TSP with five cities. In the notation of Section 4.5, this instance of the TSP is specified by following a_j and b_j values.

cities	0	1	2	3	4
b_j	0	2	3	3	9
a_j	0	8	4	6	2

Applying Algorithm 4.5.5 on this instance results in the tour $0 \to 1 \to 4 \to 2 \to 3 \to 0$. There are actually two schedules optimal for the flow shop with blocking—namely, schedule $1, 4, 2, 3$ and schedule $1, 4, 3, 2$. These are different from the SPT(1)-LPT(2) schedules that are optimal in the case of unlimited buffers. With the same four jobs and unlimited buffers the following three schedules are optimal: $1, 3, 4, 2$; $1, 2, 3, 4$; and $1, 3, 2, 4$.

The three machine version of this problem cannot be described as a Traveling Salesman Problem and is known to be strongly NP-hard. The proof, however, is rather complicated and therefore omitted.

There are special cases of $Fm \mid block \mid C_{max}$ that are tractable. Consider the proportionate case where $p_{1j} = \cdots = p_{mj} = p_j$, for $j = 1, \ldots, m$. That is, consider $Fm \mid block, p_{ij} = p_j \mid C_{max}$.

Theorem 6.2.4. *A schedule is optimal for $Fm \mid block, p_{ij} = p_j \mid C_{max}$ if and only if it is an SPT-LPT schedule.*

Proof. The makespan has to satisfy the inequality

$$C_{max} \geq \sum_{j=1}^{n} p_j + (m-1) \max(p_1, \ldots, p_n),$$

as the R.H.S. is the optimal makespan when there are unlimited buffers between any two successive machines. Clearly, the makespan with limited or no buffers has to be at least as large. It suffices to show that the makespan under any SPT-LPT schedule is equal to the lower bound, whereas the makespan under any schedule that is not SPT-LPT is strictly larger than the lower bound.

That the makespan under any SPT-LPT schedule is equal to the lower bound can be shown easily. Under the SPT part of the schedule the jobs are never blocked. In other words, each job in this first part of the schedule, once started on the first machine, proceeds through the system without stopping. If job j_k is the job in the SPT-LPT schedule $j_1, \ldots, j_k$ with the largest processing time, then job j_k departs the system at

$$C_{j_k} = \sum_{l=1}^{k-1} p_{j_l} + m p_{j_k}.$$

The jobs in the LPT part of the sequence, of course, do experience blocking as shorter jobs follow longer jobs. However, it is clear that now, in this part of the schedule, a machine never has to wait for a job. Every time a machine has completed processing a job from the LPT part of the schedule, the next job is ready to start (as shorter jobs follow longer jobs). So after job j_k has completed its processing on machine m, machine m remains continuously busy until it has completed all the remaining jobs. The makespan under an SPT-LPT schedule is therefore equal to the makespan under an SPT-LPT schedule in the case of unlimited buffers. SPT-LPT schedules therefore have to be optimal.

That SPT-LPT schedules are the only optimal schedules can be shown by contradiction. Suppose that another schedule that is not SPT-LPT is also optimal. Again, the job with the longest processing time, job j_k, contributes m times its processing to the makespan. However, there must be some job, say job j_h (not the longest job), that is positioned in between two jobs that are both longer. If job j_h appears in the schedule before job j_k, it remains on machine 1 for an amount of time that is larger than its processing time since it is blocked by the preceding job on machine 2. So its contribution to the makespan is strictly larger than its processing time, causing the makespan to be strictly larger than the lower bound. If job j_h appears in the schedule after job j_k, then the jobs following job j_k on machine m are not processed one after another without any idle times in between. After job j_h, there is an idle time on machine m as the next job has a processing time that is strictly larger than the processing time of job j_h.   □

There exists some similarity between this model and the proportionate flow shop model with unlimited intermediate storage. For the case with blocking, it can also be shown that the SPT rule minimizes $\sum C_j$.

As with flow shops with unlimited intermediate storage, a fair amount of research has been done in the development of heuristics for the minimization of the makespan in flow shops with limited intermediate storage and blocking. One popular heuristic for $Fm \mid block \mid C_{\max}$ is the *Profile Fitting (PF)* heuristic, which works as follows: One job is selected to go first, possibly according to some scheme (e.g., the job with the smallest sum of processing times). This job, say job j_1, does not encounter any blocking and proceeds smoothly from one machine to the next, generating a profile. The profile is determined by its departure from machine i. If job j_1 corresponds to job k, then

$$D_{i,j_1} = \sum_{h=1}^{i} p_{h,j_1} = \sum_{h=1}^{i} p_{hk}.$$

To determine which job should go second, every remaining unscheduled job is tried out. For each candidate job, a computation is carried out to determine the amount of time machines are idle and the amount of time the job is blocked at a machine. The departure epochs of a candidate for the second position, say job j_2, can be computed recursively:

$$D_{1,j_2} = \max(D_{1,j_1} + p_{1,j_2}, D_{2,j_1})$$

$$D_{i,j_2} = \max(D_{i-1,j_2} + p_{i,j_2}, D_{i+1,j_1}), \quad i = 2, \ldots, m-1.$$

$$D_{m,j_2} = \max(D_{m-1,j_2}, D_{m,j_1}) + p_{m,j_2}$$

The time wasted at machine i—that is, the time the machine is either idle or blocked—is $D_{i,j_2} - D_{i,j_1} - p_{i,j_2}$. The sum of these idle and blocked times over all m machines is then computed. The candidate with the smallest total is selected as the second job.

After selecting the job that fits best as the job for second position, the new profile (i.e., the departure times of this second job from all machines), is computed and the procedure repeats. From the remaining jobs in the set of unscheduled jobs, the best fit is again selected and so on.

In this description, the goodness of fit of a particular job was measured by the total time wasted on all m machines. Each machine was considered equally important. It is intuitive that lost time on a bottleneck machine is worse than lost time on a machine that does not have much processing to do. When measuring the total amount of lost time, it may be appropriate to multiply each of these inactive time periods at a given machine with a factor proportional to the degree of congestion at the machine. The higher the degree of congestion at a particular machine, the larger the weight. One measure for the degree of congestion of a machine that is easy to calculate is simply the total amount of processing to be done on all the jobs at the machine in question. Experiments have shown that such a weighted version of the PF heuristic works quite well.

Example 6.2.5 (Application of PF Heuristic)

Consider again five jobs and four machines. The processing times of the five jobs are the ones in the table in Examples 6.1.1, 6.1.6, and 6.1.10. Assume that there is zero storage in between the successive machines.

Take as the first job the job with the smallest total processing time (i.e., job 3). Apply the unweighted PF heuristic. Each one of the four remaining jobs has to be tried out. If either job 1 or job 2 would go second, the total idle time of the four machines would be 11; if job 4 would go second, the total idle time of the four machines would be 15; and if job 5 would be second, the total idle time would be 3. It is clear that job 5 is the best fit. Continuing in this manner, the PF heuristic results in the sequence 3, 5, 1, 2, 4 with a makespan equal to 32. From the fact that this makespan is equal to the minimum makespan in the case of unlimited intermediate storage, it follows that sequence 3, 5, 1, 2, 4 is also optimal in the case of zero intermediate storage.

To study the effect of the selection of the first job, consider the application of the PF heuristic after selecting the job with the largest total processing time as the initial job. So job 2 goes first. Application of the unweighted PF heuristic leads to the sequence 2, 1, 3, 5, 4 with makespan 35. This sequence is clearly not optimal.

Consider now a flow shop with zero intermediate storage subject to different operating procedures. A job, when it goes through the system, is not allowed to wait at any machine. That is, whenever it has completed its processing on one machine, the next machine has to be idle so that the job does not have to wait. In contrast to the blocking case, where jobs are *pushed* down the line by machines upstream that have completed their processing, in this case the jobs are actually *pulled* down the line by machines that have become idle. This constraint is referred to as the *no-wait* constraint, and minimizing the makespan in such a flow shop is referred to as the $Fm \mid nwt \mid C_{\max}$ problem. It is easy to see that $F2 \mid block \mid C_{\max}$ is equivalent to $F2 \mid nwt \mid C_{\max}$. However, when there are more than two machines in series, the two problems are different. The $Fm \mid nwt \mid C_{\max}$ problem, in contrast to the $Fm \mid block \mid C_{\max}$ problem, can still be formulated as a Traveling Salesman Problem. The intercity distances are

$$
d_{jk} = \max_{1 \le i \le m} \left(\sum_{h=1}^{i} p_{hj} - \sum_{h=1}^{i-1} p_{hk} \right)
$$

for $j, k = 0, \ldots, n$. When there are more than two machines in series, this Traveling Salesman Problem is known to be strongly NP-hard.

6.3 FLEXIBLE FLOW SHOPS WITH UNLIMITED INTERMEDIATE STORAGE

The flexible flow shop is a machine environment with c stages in series; at stage $l, l = 1, \ldots, c$, there are m_l identical machines in parallel. There is unlimited intermediate storage between any two successive stages. The machine environment of the first example in Section 1.1 constitutes a flexible flow shop. Job $j, j = 1, \ldots, n$ has to be processed at each stage on one machine—any one will do. The processing times of job j at the various stages are $p_{1j}, p_{2j}, \ldots, p_{cj}$. Minimizing the makespan and total completion time are referred to as $FFc \parallel C_{\max}$ and $FFc \parallel \sum C_j$, respectively. The parallel machine environment as well as the flow shop with unlimited intermediate storages are special cases of this machine environment. As this environment is rather complex, only the special case with proportionate processing times (i.e., $p_{1j} = p_{2j} = \cdots = p_{cj} = p_j$), is considered here.

Consider $FFc \mid p_{ij} = p_j \mid C_{\max}$. One would expect the LPT heuristic to perform well in the nonpreemptive case and the LRPT heuristic to perform well in the preemptive case. Of course, the LPT rule cannot guarantee an optimal schedule; a single stage (a parallel machine environment) is already NP-hard. The worst case behavior of the LPT rule when applied to multiple stages in series may actually be worse than when applied to a single stage.

Example 6.3.1 (Minimizing Makespan in a Flexible Flow Shop)

Consider two stages with two machines in parallel at the first stage and a single machine at the second stage. There are two jobs with $p_1 = p_2 = 100$ and a hundred jobs with $p_3 = p_4 = \cdots = p_{102} = 1$. It is clear that, to minimize the makespan, one long job should go at time zero on machine 1 and the 100 short jobs should be processed on machine 2 between time 0 and time 100. Under this schedule, the makespan is 301. Under the LPT schedule, the makespan is 400.

In a preemptive setting, the LRPT rule is optimal for a single stage. When there are multiple stages, this is not true anymore. The LRPT schedule has the disadvantage that at the first stage all jobs are finished very late, leaving the machines at the second stage idle for a very long time.

Consider now the proportionate flexible flow shop problem $FFc \mid p_{ij} = p_j \mid \sum C_j$. The SPT rule is known to be optimal for a single stage and for any number of stages with a single machine at each stage. Consider now the additional constraint where each stage has at least as many machines as the previous stage (the flow shop is said to be diverging).

Theorem 6.3.2.   *The SPT rule is optimal for $FFc \mid p_{ij} = p_j \mid \sum C_j$ if each stage has at least as many machines as the preceding stage.*

Proof. From Theorem 5.3.1, it is known that SPT minimizes the total completion time when the flexible flow shop consists of a single stage. It is clear that SPT not only minimizes the total completion time in this case, but also the sum of the starting times (the only difference between the sum of the completion times and the sum of the starting times is the sum of the processing times, which is independent of the schedule).

In a proportionate flexible flow shop with c stages, the completion time of job j at the last stage occurs at the earliest cp_j time units after its starting time at the first stage.

Consider now a flexible flow shop with the same number of machines at each stage, say m. It is clear that under SPT each job when completed at one stage does not have to wait for processing at the next stage. Immediately after completion at one stage it can start its processing at the following stage (as all preceding jobs have smaller processing times than the current job). So under SPT, the sum of the completion times is equal to the sum of the starting times at the first stage plus $\sum_{j=1}^{n} cp_j$. As SPT minimizes the sum of the starting times at the first stage and job j must remain at least cp_j time units in the system, SPT has to be optimal. $\square$

It is easy to verify that the SPT rule does not always lead to an optimal schedule for arbitrary proportionate flexible flow shops. A counterexample with only two stages can be found easily.

EXERCISES (COMPUTATIONAL)

6.1. Consider $F4 \mid prmu \mid C_{\max}$ with the following five jobs under the given sequence $j_1, \ldots, j_5$.

jobs	j_1	j_2	j_3	j_4	j_5
p_{1,j_k}	5	3	6	4	9
p_{2,j_k}	4	8	2	9	13
p_{3,j_k}	7	8	7	6	5
p_{4,j_k}	8	4	2	9	1

Find the critical path and compute the makespan under the given sequence.

6.2. Write the integer programming formulation of $F4 \mid prmu \mid C_{\max}$ with the set of jobs in Exercise 6.1.

6.3. Apply the Slope heuristic to the set of jobs in Exercise 6.1. Is (Are) the resulting sequence(s) actually optimal?

6.4. Consider $F4 \mid block \mid C_{\max}$ with five jobs and the same set of processing times as in Exercise 6.1. Assume there is zero buffer in between any two successive machines. Apply the Profile Fitting heuristic to determine a sequence for this problem. Take job j_1 as the first job. If there are ties, consider all the possibilities. Is (any one of) the resulting sequence(s) optimal?

6.5. Consider $F4 \mid prmu \mid C_{max}$ with the following jobs.

jobs	1	2	3	4	5
$p_{1,j}$	18	16	21	16	22
$p_{2,j}$	6	5	6	6	5
$p_{3,j}$	5	4	5	5	4
$p_{4,j}$	4	2	1	3	4

(a) Can this problem be reduced to a similar problem with a smaller number of machines and the same optimal sequence?

(b) Determine whether Theorem 6.1.4 can be applied to the reduced problem.

(c) Find the optimal sequence.

6.6. Apply Algorithm 3.3.1 to find an optimal schedule for the proportionate flow shop $F3 \mid p_{ij} = p_j \mid \sum U_j$ with the following jobs.

jobs	1	2	3	4	5	6
p_j	5	3	4	4	9	3
d_j	17	19	21	22	24	24

6.7. Find the optimal schedule for the instance of the proportionate flow shop $F2 \mid p_{ij} = p_j \mid h_{max}$ with the following jobs.

jobs	1	2	3	4	5
p_j	5	3	6	4	9
$h_j(C_j)$	$12\sqrt{C_1}$	72	$2C_3$	$54 + .5C_4$	$66 + \sqrt{C_5}$

6.8. Apply a variant of Algorithm 3.4.4 to find an optimal schedule for the instance of the proportionate flow shop $F2 \mid p_{ij} = p_j \mid \sum T_j$ with the following five jobs.

jobs	1	2	3	4	5
p_j	5	3	6	4	9
d_j	4	11	2	9	13

6.9. Consider $F2 \mid block \mid C_{max}$ with zero intermediate storage and four jobs.

jobs	1	2	3	4
$p_{1,j}$	2	5	5	11
$p_{2,j}$	10	6	6	4

(a) Apply Algorithm 4.5.5 to find the optimal sequence.

(b) Find the optimal sequence when there is an unlimited intermediate storage.

6.10. Find the optimal schedule for a proportionate flexible flow shop $FF2 \mid p_{ij} = p_j \mid \sum C_j$ with three machines at the first stage and one machine at the second stage. There are five jobs. Determine whether SPT is optimal.

jobs	1	2	3	4	5
p_j	2	2	2	2	5

EXERCISES (THEORY)

6.11. Consider the problem $Fm \mid\mid C_{\max}$. Assume that the schedule does allow one job to pass another while they are waiting for processing on a machine.

(a) Show that there always exists an optimal schedule that does not require sequence changes between machines 1 and 2 and between machines $m - 1$ and m. (*Hint:* By contradiction. Suppose the optimal schedule requires a sequence change between machines 1 and 2. Modify the schedule in such a way that there is no sequence change and the makespan remains the same.)

(b) Find an instance of $F4 \mid\mid C_{\max}$ where a sequence change between machines 2 and 3 results in a smaller makespan than in the case where sequence changes are not allowed.

6.12. Consider $Fm \mid prmu \mid C_{\max}$. Let

$$p_{i1} = p_{i2} = \cdots = p_{in} = p_i$$

for $i = 2, \ldots, m - 1$. Furthermore, let

$$p_{11} \leq p_{12} \leq \cdots \leq p_{1n}$$

and

$$p_{m1} \geq p_{m2} \geq \cdots \geq p_{mn}.$$

Show that sequence $1, 2, \ldots, n$ (i.e., SPT(1)-LPT(m)), is optimal).

6.13. Consider $Fm \mid prmu \mid C_{\max}$ where $p_{ij} = a_i + b_j$ (i.e., the processing time of job j on machine i consists of a component that is job dependent as well as a component that is machine dependent). Find the optimal sequence when $a_1 \leq a_2 \leq \cdots \leq a_m$ and prove your result.

6.14. Consider $Fm \mid prmu \mid C_{\max}$. Let $p_{ij} = a_j + i b_j$ with $b_j > -a_j/m$.

(a) Find the optimal sequence.

(b) Does the Slope heuristic lead to an optimal schedule?

6.15. Consider $F2 \mid\mid C_{\max}$.

  (a) Show that the Slope heuristic for two machines reduces to sequencing the jobs in decreasing order of $p_{2j} - p_{1j}$.

  (b) Show that the Slope heuristic is not necessarily optimal for two machines.

  (c) Show that sequencing the jobs in decreasing order of p_{2j}/p_{1j} is not necessarily optimal either.

6.16. Consider $F3 \mid\mid C_{\max}$. Assume

$$\max_{j \in \{1,\dots,n\}} p_{2j} \leq \min_{j \in \{1,\dots,n\}} p_{1j}$$

and

$$\max_{j \in \{1,\dots,n\}} p_{2j} \leq \min_{j \in \{1,\dots,n\}} p_{3j}.$$

Show that the optimal sequence is the same as the optimal sequence for $F2 \mid\mid C_{\max}$ with processing times p'_{ij} where $p'_{1j} = p_{1j} + p_{2j}$ and $p'_{2j} = p_{2j} + p_{3j}$.

6.17. Show that a permutation sequence is optimal in a proportionate flow shop $Fm \mid p_{ij} = p_j \mid C_{\max}$ in the class of schedules that does allow sequence changes midstream.

6.18. Show that if in a sequence for $F2 \mid\mid C_{\max}$ any two adjacent jobs j and k satisfy the condition

$$\min(p_{1j}, p_{2k}) \leq \min(p_{1k}, p_{2j})$$

then the sequence minimizes the makespan. (Note that this is a sufficiency condition but not a necessary condition for optimality.)

6.19. Show that for $Fm \mid prmu \mid C_{\max}$ the makespan under an arbitrary permutation sequence cannot be larger than m times the makespan under the optimal sequence. Show how this worst case bound actually can be attained.

6.20. Consider a proportionate flow shop with two objectives—namely, the flow time and the maximum lateness (i.e., $Fm \mid p_{ij} = p_j \mid \sum C_j + L_{\max}$). Develop a polynomial time algorithm for this problem. (*Hint:* Parametrize on the maximum lateness. Assume the maximum lateness to be z; then consider new due dates $d_j + z$ which basically are hard deadlines. Start out with the SPT rule and modify when necessary.)

6.21. Consider a proportionate flow shop with n jobs. Assume that there are no two jobs with equal processing times. Determine the number of different SPT-LPT schedules.

6.22. Consider $Fm \mid prmu, p_{ij} = p_j \mid \sum w_j C_j$. Show that if $w_j/p_j > w_k/p_k$ and $p_j < p_k$, then there exists an optimal sequence in which job j precedes job k.

6.23. Consider the following hybrid between $Fm \mid prmu \mid C_{\max}$ and $Fm \mid block \mid C_{\max}$. Between some machines there is zero intermediate storage and between other machines there is an infinite intermediate storage. Suppose a job sequence is given. Give a description of the graph through which the length of the makespan can be computed.

COMMENTS AND REFERENCES

The solution for the $F2 \parallel C_{max}$ problem is presented in the famous paper by S.M. Johnson (1954). The integer programming formulation of $Fm \parallel C_{max}$ is from Wagner (1959) and the NP-hardness proof for $F3 \parallel C_{max}$ is from Garey, Johnson, and Sethi (1976). A definition of SPT-LPT schedules appears in Pinedo (1982). Theorem 6.1.9 is from Eck and Pinedo (1988). For results regarding proportionate flow shops, see Ow (1985), Pinedo (1985), and Shakhlevich, Hoogeveen, and Pinedo (1998). For an overview of $Fm \parallel C_{max}$ models with special structures that can be solved easily, see Monma and Rinnooy Kan (1983); their framework includes the results obtained earlier by Smith, Panwalkar, and Dudek (1975, 1976) and Szwarc (1971, 1973, 1978). The slope heuristic for permutation flow shops is from Palmer (1965). Many other heuristics have been developed for $Fm \parallel C_{max}$; see, for example, Campbell, Dudek, and Smith (1970), Gupta (1972), Baker (1975), Dannenbring (1977), Widmer and Hertz (1989), and Taillard (1990). For complexity results with regard to various objective functions, see Gonzalez and Sahni (1978b) and Du and Leung (1993a, 1993b).

The flow shop with limited intermediate storage $Fm \mid block \mid C_{max}$ is studied in detail by Levner (1969), Reddy and Ramamoorthy (1972), and Pinedo (1982). The reversibility result in Lemma 6.2.1 is from Muth (1979). The Profile Fitting heuristic is from McCormick, Pinedo, Shenker, and Wolf (1989). Wismer (1972) establishes the link between $Fm \mid nwt \mid C_{max}$ and the Traveling Salesman Problem. Sahni and Cho (1979a), Papadimitriou and Kannelakis (1980), and Röck (1984) obtain complexity results for $Fm \mid nwt \mid C_{max}$. Goyal and Sriskandrajah (1988) present a review of complexity results and approximation algorithms for $Fm \mid nwt \mid \gamma$. For an overview concerning all models in the classes $Fm \parallel \gamma$, $Fm \mid block \mid \gamma$ and $Fm \mid nwt \mid \gamma$, see Hall and Sriskandarajah (1996).

Theorem 6.3.2 is from Eck and Pinedo (1988). For makespan results with regard to the flexible flow shops, see Sriskandarajah and Sethi (1989). Yang, Kreipl, and Pinedo (2000) present heuristics for the flexible flow shop with the total weighted tardiness as objective. For more applied issues concerning flexible flow shops, see Hodgson and McDonald (1981a, 1981b, 1981c).

7

Job Shops
(Deterministic)

This chapter deals with multi-operation models that are different from the flow shop models discussed in the previous chapter. In a flow shop model, all jobs follow the same route. When the routes are fixed, but not necessarily the same for each job, the model is called a *job shop*. If a job in a job shop has to visit certain machines more than once, the job is said to recirculate. Recirculation is a common phenomenon in the real world. For example, in semiconductor manufacturing, jobs have to recirculate many times before they complete all their processing.

The first section focuses on representations and formulations of the classical job shop problem with the makespan objective and no recirculation. It also describes a branch and bound procedure that is designed to find the optimal solution. The second section describes a popular heuristic for job shops with the makespan objective and no recirculation. This heuristic is typically referred to as the *Shifting Bottleneck* heuristic. The third section focuses on a more elaborate version of the shifting bottleneck heuristic that is designed for the total weighted tardiness objective. The last section discusses possible extensions.

7.1 DISJUNCTIVE PROGRAMMING AND BRANCH AND BOUND

Consider $J2 \mid\mid C_{\max}$. There are two machines and n jobs. Some jobs have to be processed first on machine 1 and then on machine 2 while the remaining jobs have to be processed first on machine 2 and then on machine 1. The processing time of job j on machine 1 (2) is p_{1j} (p_{2j}). The objective is to minimize the makespan.

This problem can be reduced to $F2 \mid\mid C_{\max}$ as follows. Let $J_{1,2}$ denote the set of jobs that have to be processed first on machine 1 and $J_{2,1}$ the set of jobs that have to be processed first on machine 2. Observe that when a job from $J_{1,2}$ has completed its processing on machine 1, postponing its processing on machine 2 does not affect the makespan as long as machine 2 is kept busy. The same can be said about a job from $J_{2,1}$; if such a job has completed its processing on machine 2, postponing its processing on machine 1 (as long as machine 1 is kept busy) does not affect the makespan. Hence, a job from $J_{1,2}$ has on machine 1 a higher priority than any job from $J_{2,1}$, whereas a job from $J_{2,1}$ has on machine 2 a higher priority than any job from $J_{1,2}$. It remains to be determined in what sequence jobs in $J_{1,2}$ go through machine 1 and jobs in $J_{2,1}$ go through machine 2. One of these two sequences can be determined by considering $J_{1,2}$ as an $F2 \mid\mid C_{\max}$ problem with machine 1 set up first and machine 2 set up second; the other sequence can be determined by considering $J_{2,1}$ as another $F2 \mid\mid C_{\max}$ problem with machine 2 set up first and machine 1 second. This leads to SPT(1)-LPT(2) sequences for each of the two sets, with priorities between sets as specified earlier.

This two machine problem is one of the few job shop scheduling problems for which a polynomial time algorithm can be found. The few other job shop scheduling problems for which polynomial time algorithms can be obtained usually require all processing times to be either 0 or 1.

The remainder of this section is dedicated to the $Jm \mid\mid C_{\max}$ problem with arbitrary processing times and no recirculation.

Minimizing the makespan in a job shop without recirculation, $Jm \mid\mid C_{\max}$, can be represented in a very nice way by a disjunctive graph. Consider a directed graph G with a set of nodes N and two sets of arcs A and B. The nodes N correspond to all the operations (i, j) that must be performed on the n jobs. The so-called *conjunctive* (solid) arcs A represent the routes of the jobs. If arc $(i, j) \rightarrow (k, j)$ is part of A, then job j has to be processed on machine i before it is processed on machine k—that is, operation (i, j) precedes operation (k, j). Two operations that belong to two different jobs and that have to be processed on the same machine are connected to one another by two so-called *disjunctive* (broken) arcs that go in opposite directions. The disjunctive arcs B form m cliques of double arcs, one clique for each machine. (A *clique* is a term in graph theory that refers to a graph in which any two nodes are connected to one another; in this case, each connection within a clique consists of a pair of disjunctive arcs.) All operations (nodes) in the same clique have to be done on the same machine. All arcs emanating from a node, conjunctive as well as disjunctive, have as length the processing time of the operation that is represented by that node. In addition, there is a source U and a sink V, which are dummy nodes. The source node U has n conjunctive arcs emanating to the first operations of the n jobs, and the sink node V has n conjunctive arcs coming from all the last operations. The arcs emanating from the source have length zero (see Figure 7.1). This graph is denoted by $G = (N, A, B)$.

A feasible schedule corresponds to a *selection* of one disjunctive arc from each pair such that the resulting directed graph is acyclic. This implies that a selection of disjunctive arcs from a clique has to be acyclic. Such a selection determines the sequence in which the operations are to be performed on that machine. That a selection from a clique has to be acyclic can be argued as follows: If there were a cycle within a clique, a feasible sequence of the operations on the corresponding machine would not have been possible. It may not be immediately obvious why there should not be any cycle formed by conjunctive arcs and disjunctive arcs from different cliques. However, such a cycle would also correspond to a situation that is infeasible. For example, let (h, j) and (i, j) denote two consecutive operations that belong to job j, and let (i, k) and (h, k) denote two consecutive operations that belong to job k. If under a given schedule operation (i, j) precedes operation (i, k) on machine i and operation (h, k) precedes operation (h, j) on machine h, then the graph contains a cycle with four arcs—two conjunctive arcs and two disjunctive arcs from different cliques. Such a schedule is physically impossible. Summarizing, if D denotes the subset of the selected disjunctive arcs and the graph $G(D)$ is defined by the set of conjunctive arcs and the subset D, then D corresponds to a feasible schedule if and only if $G(D)$ contains no directed cycles.

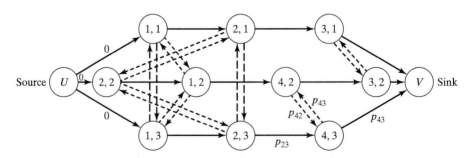

Figure 7.1   Directed graph for job shop with makespan as objective.

The makespan of a feasible schedule is determined by the longest path in $G(D)$ from the source U to the sink V. This longest path consists of a set of operations of which the first starts at time 0 and the last finishes at the time of the makespan. Each operation on this path is immediately followed by either the next operation on the same machine or the next operation of the same job on another machine. The problem of minimizing the makespan is reduced to finding a selection of disjunctive arcs that minimizes the length of the longest path (i.e., the *critical path*).

There are several mathematical programming formulations for the job shop without recirculation, including a number of integer programming formulations. However, the formulation most often used is the so-called disjunctive programming formulation (see Appendix A). This disjunctive programming formulation is closely related to the disjunctive graph representation of the job shop.

To present the disjunctive programming formulation, let the variable y_{ij} denote the starting time of operation (i, j). Recall that set N denotes the set of all operations (i, j) and set A the set of all routing constraints $(i, j) \rightarrow (k, j)$ that require job j to be processed on machine i before it is processed on machine k. The following mathematical program minimizes the makespan.

$$\text{minimize } C_{\max}$$

subject to

$$y_{kj} - y_{ij} \geq p_{ij} \qquad\qquad \text{for all } (i, j) \rightarrow (k, j) \in A$$

$$C_{\max} - y_{ij} \geq p_{ij} \qquad\qquad \text{for all } (i, j) \in N$$

$$y_{ij} - y_{il} \geq p_{il} \text{ or } y_{il} - y_{ij} \geq p_{ij} \qquad \text{for all } (i, l) \text{ and } (i, j), \ i = 1, \dots, m$$

$$y_{ij} \geq 0 \qquad\qquad \text{for all } (i, j) \in N.$$

In this formulation, the first set of constraints ensure that operation (k, j) cannot start before operation (i, j) is completed. The third set of constraints are called the *disjunctive constraints*; they ensure that some ordering exists among operations of different jobs that have to be processed on the same machine. Because of these constraints, this formulation is referred to as a disjunctive programming formulation.

Example 7.1.1 (Disjunctive Programming Formulation)

Consider the following example with four machines and three jobs. The route (i.e., the machine sequence) as well as the processing times are given in the the following table.

jobs	machine sequence	processing times
1	1, 2, 3	$p_{11} = 10,\ p_{21} = 8,\ p_{31} = 4$
2	2, 1, 4, 3	$p_{22} = 8,\ p_{12} = 3,\ p_{42} = 5,\ p_{32} = 6$
3	1, 2, 4	$p_{13} = 4,\ p_{23} = 7,\ p_{43} = 3$

The objective consists of the single variable $C_{\max}$. The first set of constraints consists of seven constraints: two for job 1, three for job 2, and two for job 3. For example, one of these is

$$y_{21} - y_{11} \geq 10\ (= p_{11}).$$

The second set consists of 10 constraints, one for each operation. An example is

$$C_{\max} - y_{11} \geq 10\ (= p_{11}).$$

The set of disjunctive constraints contains eight constraints: three each for machines 1 and 2 and one each for machines 3 and 4 (there are three operations to be performed on machines 1 and 2 and two operations on machines 3 and 4). An example of a disjunctive constraint is

$$y_{11} - y_{12} \geq 3(= p_{12})\ \text{or}\ y_{12} - y_{11} \geq 10\ (= p_{11}).$$

The last set includes 10 non-negativity constraints, one for each starting time.

That a scheduling problem can be formulated as a disjunctive program does not imply that there is a standard solution procedure available that will work satisfactorily. Minimizing the makespan in a job shop is a very hard problem, and solution procedures are based on either enumeration or heuristics.

To obtain optimal solutions, branch and bound methods are required. The branching and bounding procedures that can be applied to this problem are typically of a special design. To describe one of the branching procedures, a specific class of schedules is considered.

Definition 7.1.2. (Active Schedule)   *A feasible schedule is called active if it cannot be altered in any way such that some operation is completed earlier and no other operation is completed later.*

A schedule being active implies that when a job arrives at a machine, this job is processed in the prescribed sequence as early as possible. An active schedule cannot have any idle period in which the operation of a waiting job could fit.

From the definition, it follows that an active schedule has the property that it is impossible to reduce the makespan without increasing the starting time of some operation. Of course, there are many different active schedules. It can be shown that there exists among all possible schedules an active schedule that minimizes the makespan.

A popular branching scheme is based on the generation of all active schedules. All such active schedules can be generated by a simple algorithm. In this algorithm, Ω denotes the set of all operations of which all predecessors already have been scheduled (i.e., the set of all schedulable operations) and r_{ij} the earliest possible starting time of operation (i, j) in Ω. The set Ω' is a subset of set Ω.

Algorithm 7.1.3. (Generation of all Active Schedules)

Step 1. *(Initial Conditions) Let Ω contain the first operation of each job;*
Let $r_{ij} = 0$, for all $(i, j) \in \Omega$.

Step 2. *(Machine Selection) Compute for the current partial schedule*

$$t(\Omega) = \min_{(i,j)\in\Omega} \{r_{ij} + p_{ij}\}$$

and let i^* denote the machine on which the minimum is achieved.

Step 3. *(Branching) Let Ω' denote the set of all operations (i^*, j) on machine i^* such that*

$$r_{i^*j} < t(\Omega).$$

For each operation in Ω', consider an (extended) partial schedule with that operation as the next one on machine i^*.
For each such (extended) partial schedule, delete the operation from Ω, include its immediate follower in Ω, and return to Step 2.

Algorithm 7.1.3 is the basis for the branching process. Step 3 performs the branching from the node that is characterized by the given partial schedule; the number of branches is equal to the number of operations in Ω'. With this algorithm, one can generate the entire tree, and the nodes at the very bottom of the tree correspond to all the active schedules.

So a node $\mathcal{V}$ in the tree corresponds to a partial schedule and the partial schedule is characterized by a selection of disjunctive arcs that corresponds to the order in which all the predecessors of a given set Ω have been scheduled. A branch out of node $\mathcal{V}$ corresponds to the selection of an operation $(i^*, j) \in \Omega'$ as the next to go on machine i^*. The disjunctive arcs $(i^*, j) \to (i^*, k)$ then have to be added to machine i^* for all operations (i^*, k) still to be scheduled on machine i^*. This implies that the newly created node at the lower level, say node $\mathcal{V}'$, which corresponds to a partial schedule with only one more operation in place, contains various additional disjunctive arcs that are selected (see Figure 7.2). Let D' denote the set of disjunctive arcs selected at the newly created node. Refer to the graph, which includes all

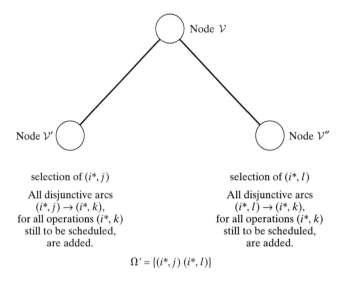

Figure 7.2   Branching tree for branch and bound approach.

the conjunctive arcs and set D' as graph $G(D')$. The number of branches sprouting from node V is equal to the number of operations in Ω'.

To find a lower bound for the makespan at node V', consider graph $G(D')$. The length of the critical path in this graph already results in a lower bound for the makespan at node V'. Call this lower bound $LB(V')$. Better (higher) lower bounds for this node can be obtained as follows.

Consider machine i and assume that all *other* machines are allowed to process, at any point in time, multiple operations simultaneously (since not all disjunctive arcs are selected yet in $G(D')$, it may be the case that, at some points in time, multiple operations require processing on a machine simultaneously). However, machine i must process its operations one after another. First, compute the earliest possible starting times r_{ij} of all the operations (i, j) on machine i—that is, determine in graph $G(D')$ the length of the longest path from the source to node (i, j). Second, for each operation (i, j) on machine i, compute the minimum amount of time needed between the completion of operation (i, j) and the lower bound $LB(V')$ by determining the longest path from node (i, j) to the sink in $G(D')$. This amount of time, together with the lower bound on the makespan, translates into a due date d_{ij} for operation (i, j)—that is, d_{ij} is equal to $LB(V')$ minus the length of the longest path from node (i, j) to the sink plus p_{ij}. Consider now the problem of sequencing the operations on machine i as a single machine problem with jobs arriving at different release dates, no preemptions allowed, and the maximum lateness as the objective to be minimized—that is, $1 \mid r_j \mid L_{max}$ (see Section 3.2). Even though this problem is strongly NP-hard, there are relatively effective algorithms that generate good solutions. The optimal sequence obtained for this problem implies a selection of disjunctive arcs that can be added (temporarily) to D'. This then may lead to a longer overall critical path in the graph, a larger makespan, and a better (higher) lower bound for node V'. At node V', this can be done for each of the m machines separately. The largest makespan obtained this way can be used as a lower bound at node V'. Of course, the temporary disjunctive arcs inserted to obtain the lower bound are deleted as soon as the best lower bound is determined.

Although it appears somewhat of a burden to have to solve m strongly NP-hard scheduling problems to obtain one lower bound for another strongly NP-hard problem, this type of bounding procedure has performed reasonably well in computational experiments.

Example 7.1.4 (Application of Branch and Bound)

Consider the instance described in Example 7.1.1. The initial graph contains only conjunctive arcs and is depicted in Figure 7.3.a. The makespan corresponding to this graph is 22. Applying the branch and bound procedure to this instance results in the following branch and bound tree.

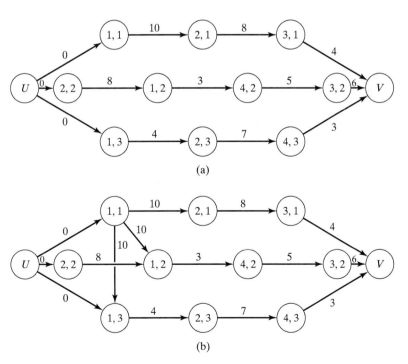

Figure 7.3 Precedence graphs at level 1 of Example 7.1.4.

Level 1: Applying Algorithm 7.1.3 yields

$$\Omega = \{(1, 1), (2, 2), (1, 3)\},$$

$$t(\Omega) = \min(0 + 10, 0 + 8, 0 + 4) = 4,$$

$$i^* = 1,$$

$$\Omega' = \{(1, 1), (1, 3)\}.$$

So there are two nodes of interest at level 1, one corresponding to operation $(1, 1)$ being processed first on machine 1 and the other to operation $(1, 3)$ being processed first on machine 1.

If operation $(1, 1)$ is scheduled first, then the two disjunctive arcs depicted in Figure 7.3.b are added to the graph. The node is characterized by the two disjunctive arcs

$$(1, 1) \rightarrow (1, 2),$$

$$(1, 1) \rightarrow (1, 3).$$

The addition of these two disjunctive arcs immediately increases the lower bound on the makespan to 24. To improve this lower bound, one can generate for machine 1 an instance of $1 \mid r_j \mid L_{\max}$. The release date of job j in this single machine problem is determined by the longest path from the source to node $(1, j)$ in Figure 7.3.b. The due date of job j is computed by finding the longest path from node $(1, j)$ to the sink, subtracting p_{1j} from the length of this longest path, and subtracting the remaining value from 24. These computations lead to the following single machine problem for machine 1.

jobs	1	2	3
p_{1j}	10	3	4
r_{1j}	0	10	10
d_{1j}	10	13	14

The sequence that minimizes $L_{\max}$ is 1, 2, 3 with $L_{\max} = 3$. This implies that a lower bound for the makespan at the corresponding node is $24 + 3 = 27$. An instance of $1 \mid r_j \mid L_{\max}$ corresponding to machine 2 can be generated in the same way. The release dates and due dates also follow from Figure 7.3.b (assuming a makespan of 24) and are as follows.

jobs	1	2	3
p_{2j}	8	8	7
r_{2j}	10	0	14
d_{2j}	20	10	21

The optimal sequence is 2, 1, 3 with $L_{max} = 4$. This yields a better lower bound for the makespan at the node that corresponds to operation (1, 1) being scheduled first (i.e., $24 + 4 = 28$). Analyzing machines 3 and 4 in the same way does not yield a better lower bound.

The second node at Level 1 corresponds to operation (1, 3) being scheduled first. If (1, 3) is scheduled to go first, two different disjunctive arcs are added to the original graph, yielding a lower bound of 26. The associated instance of the maximum lateness problem for machine 1 has an optimal sequence 3, 1, 2 with $L_{max} = 2$. This implies that the lower bound for the makespan at this node, corresponding to operation (1, 3) scheduled first, is also equal to 28. Analyzing machines 2, 3, and 4 does not result in a better lower bound.

The next step is to branch from node (1, 1) at Level 1 and generate the nodes at the next level.

Level 2:　Applying Algorithm 7.1.3 now yields

$$\Omega = \{(2, 2), (2, 1), (1, 3)\},$$

$$t(\Omega) = \min(0 + 8, 10 + 8, 10 + 4) = 8,$$

$$i^* = 2,$$

$$\Omega' = \{(2, 2)\}.$$

There is one node of interest at this part of Level 2—the node corresponding to operation (2, 2) being processed first on machine 2 (see Figure 7.4). Two disjunctive arcs are added to the graph—namely, $(2, 2) \rightarrow (2, 1)$ and $(2, 2) \rightarrow (2, 3)$. So this node is characterized by a total of four disjunctive arcs:

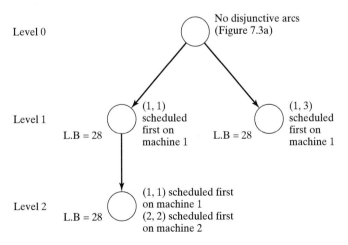

Figure 7.4　Branching tree in Example 7.1.4.

$$(1, 1) \to (1, 2),$$
$$(1, 1) \to (1, 3),$$
$$(2, 2) \to (2, 1),$$
$$(2, 2) \to (2, 3).$$

This leads to an instance of $1 \mid r_j \mid L_{max}$ for machine 1 with the following release dates and due dates (assuming a makespan of 28).

jobs	1	2	3
p_{1j}	10	3	4
r_{1j}	0	10	10
d_{1j}	14	17	18

The optimal job sequence is 1, 3, 2 and $L_{max} = 0$. This implies that the lower bound for the makespan at the corresponding node is $28 + 0 = 28$. Analyzing machines 2, 3, and 4 in the same way does not increase the lower bound.

Continuing the branch and bound procedure results in the following job sequences for the four machines.

machine	job sequence
1	1, 3, 2 (or 1, 2, 3)
2	2, 1, 3
3	1, 2
4	2, 3

The makespan under this optimal schedule is 28 (see Figure 7.5).

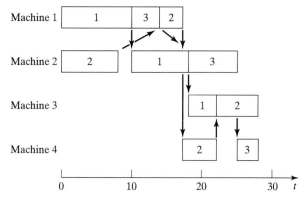

Figure 7.5   Gantt chart for $J4 \mid\mid C_{max}$ (Example 7.1.4).

The approach described previously is based on complete enumeration and is guaranteed to lead to an optimal schedule. However, with a large number of machines and jobs the computation time is prohibitive. Already with 20 machines and 20 jobs, it is hard to find an optimal schedule.

It is therefore necessary to develop heuristics that lead to reasonably good schedules in a reasonably short time. In the next section, we describe a well-known heuristic that has a very good track record in extensive computational experiments.

7.2 THE SHIFTING BOTTLENECK HEURISTIC AND THE MAKESPAN

One of the most successful heuristic procedures developed for $Jm \mid\mid C_{max}$ is the *Shifting Bottleneck* heuristic.

In the following overview of the Shifting Bottleneck heuristic, M denotes the set of all m machines. In the description of an iteration of the heuristic, it is assumed that in previous iterations a selection of disjunctive arcs has already been fixed for a subset M_0 of machines. So for each one of the machines in M_0, a sequence of operations has already been determined.

An iteration determines which machine in $M - M_0$ has to be included next in set M_0. The sequence in which the operations on this machine have to be processed is also generated in this iteration. To determine which machine should be included next in M_0, an attempt is made to determine which unscheduled machine causes in one sense or another the severest disruption. To determine this, the original directed graph is modified by deleting *all* disjunctive arcs of the machines yet to be scheduled (i.e., the machines in set $M - M_0$) and keeping only the relevant disjunctive arcs of the machines in set M_0 (one from every pair). Call this graph G'. Deleting all disjunctive arcs of a specific machine implies that all operations on this machine, which originally were supposed to be done on this machine one after another, now may be done in parallel (as if the machine has infinite capacity, or equivalently, each one of these operations has the machine for itself). The graph G' has one or more critical paths that determine the corresponding makespan. Call this makespan $C_{max}(M_0)$.

Suppose that operation (i, j), $i \in \{M - M_0\}$, has to be processed in a time window of which the release date and due date are determined by the critical (longest) paths in G'—that is, the release date is equal to the longest path in G' from the source to node (i, j) and the due date is equal to $C_{max}(M_0)$, minus the longest path from node (i, j) to the sink, plus p_{ij}. Consider each of the machines in $M - M_0$ as a separate $1 \mid r_j \mid L_{max}$ problem. As stated in the previous section, this problem is strongly NP-hard, but procedures have been developed that perform reasonably well. The minimum L_{max} of the single machine problem corresponding to machine i is denoted by $L_{max}(i)$ and is a measure of the criticality of machine i.

After solving all these single machine problems, the machine with the *largest* maximum lateness is chosen. Among the remaining machines, this machine is in a sense the most critical, or the "bottleneck," and therefore the one to be included next in M_0. Label this machine k, call its maximum lateness $L_{\max}(k)$, and schedule it according to the optimal solution obtained for the single machine problem associated with this machine. If the disjunctive arcs that specify the sequence of operations on machine k are inserted in graph G', then the makespan of the current partial schedule increases by at least $L_{\max}(k)$—that is,

$$C_{\max}(M_0 \cup k) \geq C_{\max}(M_0) + L_{\max}(k).$$

Before starting the next iteration and determining the next machine to be scheduled, one additional step has to be done within the current iteration. In this additional step, all the machines in the original set M_0 are resequenced to see if the makespan can be reduced. That is, a machine, say machine l, is taken out of set M_0 and a graph G'' is constructed by modifying graph G' through the inclusion of the disjunctive arcs that specify the sequence of operations on machine k and the exclusion of the disjunctive arcs associated with machine l. Machine l is resequenced by solving the corresponding $1 \mid r_j \mid L_{\max}$ problem with the release and due dates determined by the critical paths in graph G''. Resequencing each of the machines in the original set M_0 completes the iteration.

In the next iteration, the entire procedure is repeated and another machine is added to the current set $M_0 \cup k$.

The shifting bottleneck heuristic can be summarized as follows.

Algorithm 7.2.1. (Shifting Bottleneck Heuristic)

Step 1. *(Initial conditions)*
Set $M_0 = \emptyset$.
Graph G contains all the conjunctive arcs and no disjunctive arcs.
Set $C_{\max}(M_0)$ equal to the longest path in graph G.

Step 2. *(Analysis of machines still to be scheduled)*
Do for each machine i in set $M - M_0$ the following:
generate an instance of $1 \mid r_j \mid L_{\max}$
(with the release date of operation (i, j) determined by the longest path in graph G from the source to node (i, j);
and the due date of operation (i, j) determined by the longest path in graph G from node (i, j) to the sink minus p_{ij}).
Minimize the $L_{\max}$ in each one of these single machine subproblems.

Let $L_{max}(i)$ denote the minimum L_{max} in the subproblem corresponding to machine i.

Step 3. *(Bottleneck selection and sequencing)*
Let

$$L_{max}(k) = \max_{i \in \{M - M_0\}} (L_{max}(i))$$

Sequence machine k according to the sequence obtained in Step 2 for that machine.
Insert all the corresponding disjunctive arcs in graph G.
Insert machine k in M_0.

Step 4. *(Resequencing of all machines scheduled earlier)*
Do for each machine $i \in \{M_0 - k\}$ the following:
delete the corresponding disjunctive arcs from G;
formulate a single machine subproblem for machine i with release dates and due dates of the operations determined by longest path calculations in G.
Find the sequence that minimizes $L_{max}(i)$ and insert the corresponding disjunctive arcs in graph G.

Step 5. *(Stopping criterion)*
If $M_0 = M$ then STOP, otherwise go to Step 2.

The structure of the shifting bottleneck heuristic shows the relationship between the bottleneck concept and more combinatorial concepts such as critical (longest) path and maximum lateness. A critical path indicates the location and timing of a bottleneck. The maximum lateness gives an indication of the amount by which the makespan increases if a machine is added to the set of machines already scheduled.

The remainder of this section contains two examples that illustrate the use of the shifting bottleneck heuristic.

Example 7.2.2 (Application of Shifting Bottleneck Heuristic)

Consider the instance with four machines and three jobs described in Examples 7.1.1 and 7.1.4. The routes of the jobs (i.e., the machine sequences) and the processing times are given in the following table.

jobs	machine sequence	processing times
1	1, 2, 3	$p_{11} = 10, p_{21} = 8, p_{31} = 4$
2	2, 1, 4, 3	$p_{22} = 8, p_{12} = 3, p_{42} = 5, p_{32} = 6$
3	1, 2, 4	$p_{13} = 4, p_{23} = 7, p_{43} = 3$

Iteration 1:   Initially, set M_0 is empty and graph G' contains only conjunctive arcs and no disjunctive arcs. The critical path and makespan $C_{max}(\emptyset)$ can be determined easily: This makespan is equal to the maximum total processing time required for any job. The maximum of 22 is achieved in this case by both jobs 1 and 2. To determine which machine to schedule first, each machine is considered as a $1 \mid r_j \mid L_{max}$ problem with the release dates and due dates determined by the longest paths in G' (assuming a makespan of 22).

The data for the $1 \mid r_j \mid L_{max}$ problem corresponding to machine 1 are presented in the following table.

jobs	1	2	3
p_{1j}	10	3	4
r_{1j}	0	8	0
d_{1j}	10	11	12

The optimal sequence turns out to be 1, 2, 3 with $L_{max}(1) = 5$.
The data for the subproblem regarding machine 2 are:

jobs	1	2	3
p_{2j}	8	8	7
r_{2j}	10	0	4
d_{2j}	18	8	19

The optimal sequence for this problem is 2, 3, 1 with $L_{max}(2) = 5$. Similarly, it can be shown that

$$L_{max}(3) = 4$$

and

$$L_{max}(4) = 0.$$

From this it follows that either machine 1 or machine 2 may be considered a bottleneck. Breaking the tie arbitrarily, machine 1 is selected to be included in M_0. The graph G'' is obtained by fixing the disjunctive arcs corresponding to the sequence of the jobs on machine 1 (see Figure 7.6). It is clear that

$$C_{max}(\{1\}) = C_{max}(\emptyset) + L_{max}(1) = 22 + 5 = 27.$$

Iteration 2:   Given that the makespan corresponding to G'' is 27, the critical paths in the graph can be determined. The three remaining machines have to be analyzed separately as $1 \mid r_j \mid L_{max}$ problems. The data for the instance concerning machine 2 are:

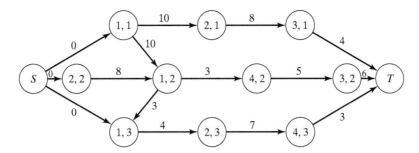

Figure 7.6 Interation 1 of shifting bottleneck heuristic (Example 7.2.2).

jobs	1	2	3
p_{2j}	8	8	7
r_{2j}	10	0	17
d_{2j}	23	10	24

The optimal schedule is 2, 1, 3 and the resulting $L_{max}(2) = 1$. The data for the instance corresponding to machine 3 are:

jobs	1	2
p_{3j}	4	6
r_{3j}	18	18
d_{3j}	27	27

Both sequences are optimal and $L_{max}(3) = 1$. Machine 4 can be analyzed in the same way and the resulting $L_{max}(4) = 0$. Again, there is a tie, and machine 2 is selected to be included in M_0. So $M_0 = \{1, 2\}$ and

$$C_{max}(\{1, 2\}) = C_{max}(\{1\}) + L_{max}(2) = 27 + 1 = 28.$$

The disjunctive arcs corresponding to the job sequence on machine 2 are added to G'' and graph G''' is obtained. At this point, still as a part of iteration 2, an attempt may be made to decrease $C_{max}(\{1, 2\})$ by resequencing machine 1. It can be checked that resequencing machine 1 does not give any improvement.

Iteration 3:   The critical path in G''' can be determined, and machines 3 and 4 remain to be analyzed. These two problems turn out to be very simple with both having a zero maximum lateness. Neither of the machines constitutes a bottleneck in any way.

The final schedule is determined by the following job sequences on the four machines: job sequence 1, 2, 3 on machine 1; job sequence 2, 1, 3 on machine 2;

job sequence 2, 1 on machine 3; and job sequence 2, 3 on machine 4. The makespan is 28.

The implementation of the shifting bottleneck technique in practice often tends to be more complicated than the heuristic described earlier. The solution procedure for the single machine subproblem must deal with some additional complications.

The single machine maximum lateness problem that has to be solved repeatedly within each iteration of the heuristic may at times be slightly different and more complicated than the $1 \mid r_j \mid L_{\max}$ problem described in Chapter 3 (which is also the problem used for determining the lower bounds in the previous section). In the single machine problem that has to be solved in the shifting bottleneck heuristic, the operations on a given machine may have to be subject to a special type of precedence constraint. It may be that an operation that has to be processed on a particular machine can only be processed on that machine after certain other operations have completed their processing on that machine. Such a precedence constraint may be imposed by the sequences of the operations on the machines that have already been scheduled in earlier iterations.

It may even be that two operations that are subject to such a constraint not only have to be processed in the given order, but they may also have to be processed a given amount of time apart from one another. That is, in between the processing of two operations that are subject to such a precedence constraint, a certain minimum amount of time (i.e., a delay) has to elapse.

The lengths of the delays are also determined by the sequences of the operations on the machines already scheduled. These precedence constraints are therefore referred to as *delayed* precedence constraints.

The next example illustrates the potential need for delayed precedence constraints in the single machine subproblem. Without these constraints, the shifting bottleneck heuristic may end up in a situation where there is a cycle in the disjunctive graph with the corresponding schedule being infeasible. The following example illustrates how sequences on machines already scheduled (machines in M_0) impose constraints on machines still to be scheduled (machines in $M - M_0$).

Example 7.2.3 (Delayed Precedence Constraints)

Consider the following instance.

jobs	machine sequence	processing times
1	1, 2	$p_{11} = 1, p_{21} = 1$
2	2, 1	$p_{22} = 1, p_{12} = 1$
3	3	$p_{33} = 4$
4	3	$p_{34} = 4$

Applying the shifting bottleneck heuristic results in the following three iterations.

Iteration 1:   The first iteration consists of the optimization of three subproblems. The data for the three subproblems associated with machines 1, 2, and 3 are tabulated next.

jobs	1	2	jobs	1	2	jobs	1	2
p_{1j}	1	1	p_{2j}	1	1	p_{3j}	4	4
r_{1j}	0	1	r_{2j}	1	0	r_{3j}	0	0
d_{1j}	3	4	d_{2j}	4	3	d_{3j}	4	4

The optimal solutions for machines 1 and 2 have $L_{\max} \leq 0$, whereas that for machine 3 has $L_{\max} = 4$. So machine 3 is scheduled first and arc $(3, 4) \rightarrow (3, 3)$ is inserted.

Iteration 2:   The new set of subproblems are associated with machines 1 and 2.

jobs	1	2	jobs	1	2
p_{1j}	1	1	p_{2j}	1	1
r_{1j}	0	1	r_{2j}	1	0
d_{1j}	7	8	d_{2j}	8	7

The optimal solutions for machines 1 and 2 both have $L_{\max} = -6$, so we arbitrarily select machine 1 to be scheduled next. Arc $(1, 2) \rightarrow (1, 1)$ is inserted (see Figure 7.7.a).

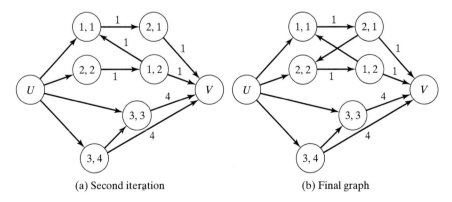

(a) Second iteration   (b) Final graph

Figure 7.7   Application of the shifting bottleneck heuristic in Example 7.2.3.

Iteration 3:   One subproblem remains and is associated with machine 2.

jobs	1	2
p_{2j}	1	1
r_{2j}	3	0
d_{2j}	8	5

Any schedule for machine 2 yields an $L_{max} \leq 0$. If a schedule would be selected arbitrarily and arc $(2, 1) \rightarrow (2, 2)$ would be inserted, then a cycle is created in the graph, and the overall schedule is infeasible (see Figure 7.7.b).

This situation could have been prevented by imposing delayed precedence constraints. After scheduling machine 1 (in iteration 2), there is a path from $(2, 2)$ to $(2, 1)$ with length 3. After iteration 2 has been completed, a delayed precedence constraint can be generated for subsequent iterations. Operation $(2, 2)$ must precede operation $(2, 1)$. Furthermore, there must be a delay of two time units in between the completion of operation $(2, 2)$ and the start of operation $(2, 1)$. With this constraint, iteration 3 generates a sequence for machine 2 that results in a feasible schedule.

Extensive numerical research has shown that the Shifting Bottleneck heuristic is extremely effective. When applied to a standard test problem with 10 machines and 10 jobs that had remained unsolved for more than 20 years, the heuristic obtained a good solution very fast. This solution turned out to be optimal after a branch and bound procedure yielded the same result and verified its optimality. The branch and bound approach, in contrast to the heuristic, needed many hours of CPU time. The disadvantage of the heuristic is, of course, that there is no guarantee that the solution it reaches is optimal.

The Shifting Bottleneck heuristic can be adapted to be applied to more general models than the job shop model considered here (e.g., flexible job shops with recirculation).

7.3 THE SHIFTING BOTTLENECK HEURISTIC AND THE TOTAL WEIGHTED TARDINESS

This section describes an approach for $Jm \ || \ \sum w_j T_j$ that combines a variant of the Shifting Bottleneck heuristic discussed in the previous section with a priority rule called the Apparent Tardiness Cost first (ATC) rule.

The disjunctive graph representation for $Jm \ || \ \sum w_j T_j$ is different from that for $Jm \ || \ C_{max}$. In the makespan problem, only the completion time of the last job that leaves the system is of importance. Therefore, there is a single sink in the disjunctive graph. In the total weighted tardiness problem, the completion times of all n jobs are of importance. Instead of a single sink, there are now n sinks—that

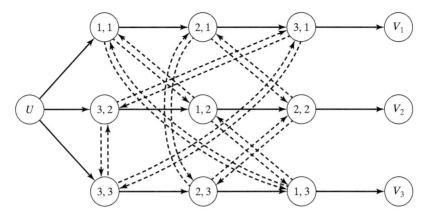

Figure 7.8   Directed graph for job shop with total weighted tardiness as the objective.

is, $V_1, \ldots, V_n$ (see Figure 7.8). The length of the longest path from the source U to the sink V_k represents the completion time of job k.

The approach can be described as follows. Machines are again scheduled one at a time. At the start of a given iteration, all machines in set M_0 have already been scheduled (i.e., the disjunctive arcs for these machines have been selected), and in this iteration it has to be decided which machine should be scheduled next and how it should be scheduled. Each of the remaining machines has to be analyzed separately, and for each of these machines a measure of criticality has to be computed. The steps done within an iteration can be described as follows.

In the disjunctive graph representation, all disjunctive arcs belonging to the machines still to be scheduled are deleted and all disjunctive arcs selected for the machines already scheduled (set M_0) are kept in place. Given this directed graph, the completion times of all n jobs can be computed easily. Let C'_k denote the completion time of job k. Now consider a machine i that still has to be scheduled (machine i is an element of set $M - M_0$). To avoid an increase in the completion time C'_k, operation (i, j), $j = 1, \ldots, n$, must be completed on machine i by some local due date d^k_{ij}. This local due date can be computed by considering the longest path from operation (i, j) to the sink corresponding to job k (i.e., V_k). If there is no path from node (i, j) to sink V_k, then the local due date d^k_{ij} is infinity. So, because of job k, there may be a local due date d^k_{ij} for operation (i, j). That is, if operation (i, j) is completed after d^k_{ij}, then job k's overall completion time is postponed, resulting in a penalty. If the completion of job k, C'_k, is already past the due date d_k of job k, any increase in the completion time increases the penalty at a rate w_k. Because operation (i, j) may cause a delay in the completion of any one of the n jobs, one

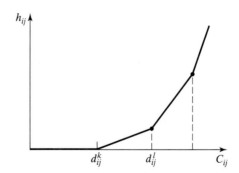

Figure 7.9   Cost function h_{ij} of operation (i, j) in single-machine sub-problem.

can assume that operation (i, j) is subject to n local due dates. This implies that operation (i, j) is subject to a piece-wise linear cost function h_{ij} (see Figure 7.9).

Thus, a measure of criticality of machine i can be obtained by solving a single machine problem with each operation subject to a piece-wise linear cost function— that is, $1 \mid\mid \sum h_j(C_j)$, where h_j is a piece-wise linear cost function corresponding to job j. As in the previous section, the operations may be subject to delayed precedence constraints to ensure feasibility.

This single machine subproblem is a generalization of the $1 \mid\mid \sum w_j T_j$ problem (see Chapter 3). A well-known priority rule for $1 \mid\mid \sum w_j T_j$ is the so-called *Apparent Tardiness Cost (ATC)* rule. This ATC heuristic is a composite dispatching rule that combines the WSPT rule and the so-called *Minimum Slack first (MS)* rule (under the MS rule, the slack of job j at time t, $\max(d_j - p_j - t, 0)$, is computed and the job with the minimum slack is scheduled). Under the ATC rule, jobs are scheduled one at a time; that is, every time the machine becomes free, a ranking index is computed for each remaining job. The job with the highest ranking index is then selected to be processed next. This ranking index is a function of the time t at which the machine became free as well as of the p_j, the w_j, and the d_j of the remaining jobs. The index is defined as

$$I_j(t) = \frac{w_j}{p_j} \exp\left(-\frac{\max(d_j - p_j - t, 0)}{K\bar{p}}\right),$$

where K is a scaling parameter that can be determined empirically, and $\bar{p}$ is the average of the processing times of the remaining jobs. The ATC rule is discussed in detail in Chapter 14.

The piece-wise linear and convex function h_j in the subproblem $1 \mid\mid \sum h_j(C_j)$ may be regarded as a sum of linear penalty functions, for each of which an ATC priority index can be computed. One can think of several composite priority index functions for this more complicated cost function. A reasonably effective one assigns to operation (i, j) the priority value

$$I_{ij}(t) = \sum_{k=1}^{n} \frac{w_k}{p_{ij}} \exp\left(-\frac{(d_{ij}^k - p_{ij} + (r_{ij} - t))^+}{K \bar{p}}\right),$$

where t is the earliest time at which machine i can be used, K is a scaling parameter, and $\bar{p}$ is the integer part of the average length of the operations to be processed on machine i. This composite dispatching rule yields a reasonably good schedule for machine i.

A measure for the criticality of machine i can now be computed in a number of ways. For example, consider the solutions of all the single machine subproblems and set the measure for the criticality of a machine equal to the corresponding value of the objective function. However, there are more involved and effective methods for measuring machine criticality. For example, by selecting the disjunctive arcs implied by the schedule for machine i, one can easily compute in the new disjunctive graph the new (overall) completion times of all n jobs, say C_k''. Clearly, $C_k'' \geq C_k'$. The contribution of job k to the measure of criticality of machine i is computed as follows. If $C_k' > d_k$, then the contribution of job k to the measure of criticality of machine i is $w_k(C_k'' - C_k')$. However, if $C_k' < d_k$, then the penalty due to an increase of the completion of job k is more difficult to estimate. This penalty would then be a function of C_k', C_k'', and d_k. Several functions have been experimented with and appear to be promising. One such function is

$$w_k(C_k'' - C_k') \exp\left(-\frac{(d_k - C_k'')^+}{K}\right),$$

where K is a scaling parameter. Summing over all jobs, that is,

$$\sum_{k=1}^{n} w_k(C_k'' - C_k') \exp\left(-\frac{(d_k - C_k'')^+}{K}\right),$$

provides a measure of criticality for machine i. This last expression plays a role that is similar to the one of $L_{\max}(i)$ in Step 2 of Algorithm 7.2.1. After the criticality measures of all the machines in $M - M_0$ have been computed, the machine with the highest measure is selected as the next one to be included in set M_0.

However, this process does not yet complete an iteration. The original shifting bottleneck approach, as described in Algorithm 7.2.1, suggests that rescheduling all the machines in the original set M_0 is advantageous. This rescheduling may result in different and better schedules. After this step has been completed, the entire process repeats itself and the next iteration is started.

Example 7.3.1 (Shifting Bottleneck and Total Weighted Tardiness)

Consider the instance with three machines and three jobs depicted in Figure 7.8.

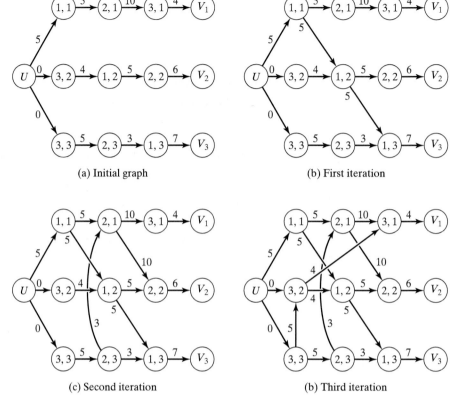

Figure 7.10 Directed graphs in Example 7.3.1.

job	w_j	r_j	d_j	machine sequence	processing times
1	1	5	24	1, 2, 3	$p_{11} = 5, p_{21} = 10, p_{31} = 4$
2	2	0	18	3, 1, 2	$p_{32} = 4, p_{12} = 5, p_{22} = 6$
3	2	0	16	3, 2, 1	$p_{33} = 5, p_{23} = 3, p_{13} = 7$

The initial graph is depicted in Figure 7.10.a.

   Iteration 1.   The first iteration requires the optimization of three subproblems—one for each machine in the job shop. The data for these three subproblems, corresponding to machines 1, 2, and 3, are tabulated next.

jobs	1	2	3
p_{1j}	5	5	7
r_{1j}	5	4	8
$d_{1j}^1, d_{1j}^2, d_{1j}^3$	10, −, −	−, 12, −	−, −, 16

jobs	1	2	3
p_{2j}	10	6	3
r_{2j}	10	9	5
$d_{2j}^1, d_{2j}^2, d_{2j}^3$	20, –, –	–, 18, –	–, –, 9

jobs	1	2	3
p_{3j}	4	4	5
r_{3j}	20	0	0
$d_{3j}^1, d_{3j}^2, d_{3j}^3$	24, –, –	–, 7, –	–, –, 6

The entry "–" indicates that the corresponding due date d_{ij}^k is infinite (i.e., there is no path from operation (i, j) to the sink corresponding to job k). The subproblems are solved using a dispatching rule that is based on the priority index $I_{ij}(t)$ for operation (i, j), where t is the earliest time at which machine i can be used. Set the scaling parameter K equal to 0.1.

Since none of the operations has been scheduled yet, the priority index for the operations assigned to machine 1 yields $I_{11}(4) = 1.23 \times 10^{-6}$, $I_{12}(4) = 3.3 \times 10^{-7}$, and $I_{13}(4) = 1.46 \times 10^{-12}$ since $t = 4$ and $\bar{p} = 5$. Put the operation with the highest priority—namely, operation $(1, 1)$—in the first position and recalculate the remaining indexes to select the operation that has to be scheduled next. The solutions obtained for these three subproblems are:

machine i	sequence	value
1	(1, 1), (1, 2), (1, 3)	18
2	(2, 3), (2, 1), (2, 2)	16
3	(3, 3), (3, 2), (3, 1)	4

Since the solution of subproblem 1 has the highest value, schedule machine 1 by inserting the disjunctive arcs $(1, 1) \rightarrow (1, 2)$ and $(1, 2) \rightarrow (1, 3)$ as shown in Figure 7.10.b.

Iteration 2.   The data for the new subproblems, corresponding to machines 2 and 3, are tabulated next.

jobs	1	2	3
p_{2j}	10	6	3
r_{2j}	10	15	5
$d_{2j}^1, d_{2j}^2, d_{2j}^3$	20, –, –	–, 21, –	–, –, 15

jobs	1	2	3
p_{3j}	4	4	5
r_{3j}	20	0	0
$d_{3j}^1, d_{3j}^2, d_{3j}^3$	24, −, −	−, 10, 10	−, −, 12

In this iteration, operation (3, 2) has two due dates because there is a (directed) path from node (3, 2) to V_2 and V_3. This makes its index equal to

$$I_{32}(0) = 1.53 \times 10^{-7} + 1.53 \times 10^{-7} = 3.06 \times 10^{-7}$$

since $t = 0$ and $\bar{p} = 4$. The solutions obtained for the two subproblems are:

machine i	sequence	value
2	(2, 3), (2, 1), (2, 2)	10
3	(3, 2), (3, 3), (3, 1)	0

The solution for subproblem 2 has the highest value (10). Schedule machine 2 by inserting the disjunctive arcs (2, 3) → (2, 1) and (2, 1) → (2, 2) as shown in Figure 7.10.c.

Iteration 3.   The only subproblem that remains is the one for machine 3.

jobs	1	2	3
p_{3j}	4	4	5
r_{3j}	20	0	0
$d_{3j}^1, d_{3j}^2, d_{3j}^3$	24, −, −	−, 15, 10	7, 7, 12

Its optimal solution is (3, 3), (3, 2), (3, 1) with value equal to zero, so insert the arcs (3, 3) → (3, 2) and (3, 2) → (3, 1), as shown in Figure 7.10.d. The final solution is depicted in Figure 7.11, with objective function equal to

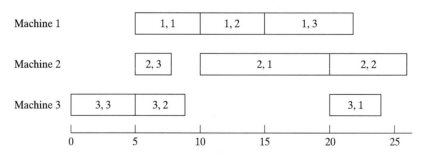

Figure 7.11   Final schedule in Example 7.3.1.

$$\sum_{j=1}^{3} w_j T_j = 1 \times (24 - 24)^+ + 2 \times (26 - 18)^+ + 2 \times (22 - 16)^+ = 28.$$

It happens that, in this case, the heuristic does not yield an optimal solution. The optimal solution with a value of 18 can be obtained with more elaborate versions of this heuristic. These versions make use of backtracking techniques as well as machine reoptimization (similar to Step 4 in Algorithm 7.2.1).

7.4 DISCUSSION

The disjunctive graph formulation for $Jm \mid\mid C_{\max}$ extends to $Jm \mid recrc \mid C_{\max}$. The set of disjunctive arcs for a machine may now not be a clique. If two operations of the same job have to be performed on the same machine, a precedence relationship is given. These two operations are not connected by a pair of disjunctive arcs since they are already connected by conjunctive arcs. The branch and bound approach described in Section 7.1 still applies. However, the bounding mechanism is now not based on the solution of a $1 \mid r_j \mid L_{\max}$ problem, but rather on the solution of a $1 \mid r_j, prec \mid L_{\max}$ problem. The precedence constraints are the routing constraints on the different operations of the same job to be processed on the machine.

In the same way that a flow shop can be generalized into a flexible flow shop, a job shop can be generalized into a flexible job shop. The fact that the flexible flow shop allows for few structural results already gives an indication how hard it is to obtain results for the flexible job shop. Even the proportionate case (i.e., $p_{ij} = p_j$ for all i) is hard to analyze.

The Shifting Bottleneck heuristic can be adapted in such a way that it can be applied to models that are more general than $Jm \mid\mid C_{\max}$. These more general models include recirculation as well as multiple machines at every stage.

One variation of the Shifting Bottleneck heuristic is based on decomposition principles. This variation is especially suitable for the scheduling of flexible job shops. The following five-phase approach can be utilized for the flexible job shop:

Phase 1. The shop is divided into a number of workcenters that have to be scheduled. A workcenter may consist of a single machine or a bank of machines in parallel.

Phase 2. The entire job shop is represented through a disjunctive graph.

Phase 3. A performance measure is computed in order to rank the workcenters in order of criticality. The schedule of the most critical workcenter, among the workcenters of which the sequences still have to be determined, is fixed.

Phase 4. The disjunctive graph representation is used to capture the interactions between the workcenters already scheduled and those not yet scheduled.

Phase 5. Those workcenters that already have been sequenced are rescheduled using the new information obtained in Phase 4. If all workcenters have been scheduled, stop. Otherwise go to Phase 3.

The subproblem now becomes a nonpreemptive parallel machine scheduling problem with the jobs subject to different release dates and the maximum lateness as the objective to be minimized. A significant amount of computational research has been done on this parallel machine problem.

EXERCISES (COMPUTATIONAL)

7.1. Consider the following heuristic for $Jm \mid\mid C_{max}$. Each time a machine is freed, select the job (among the ones immediately available for processing on the machine) with the largest *total* remaining processing (including the processing on the machine freed). If at any point in time more than one machine is freed, consider first the machine with the largest remaining workload. Apply this heuristic to the instance in Example 7.1.1.

7.2. Consider the following instance of $Jm \mid\mid C_{max}$.

jobs	machine sequence	processing times
1	1, 2, 3	$p_{11} = 9, p_{21} = 8, p_{31} = 4$
2	1, 2, 4	$p_{12} = 5, p_{22} = 6, p_{42} = 3$
3	3, 1, 2	$p_{33} = 10, p_{13} = 4, p_{23} = 9$

Give the disjunctive programming formulation of this instance.

7.3. Consider the following instance of the $Jm \mid\mid C_{max}$ problem.

jobs	machine sequence	processing times
1	1, 2, 3, 4	$p_{11} = 9, p_{21} = 8, p_{31} = 4, p_{41} = 4$
2	1, 2, 4, 3	$p_{12} = 5, p_{22} = 6, p_{42} = 3, p_{32} = 6$
3	3, 1, 2, 4	$p_{33} = 10, p_{13} = 4, p_{23} = 9, p_{43} = 2$

Give the disjunctive programming formulation of this instance.

7.4. Apply the heuristic described in Exercise 7.1 to the instance in Exercise 7.3.

7.5. Consider the instance in Exercise 7.2.

(a) Apply the Shifting Bottleneck heuristic to this instance (doing the computation by hand).

(b) Compare your result with the result of the shifting bottleneck routine in the LEKIN system (see Chapter 19).

7.6. Consider again the instance in Exercise 7.2.

(a) Apply the branch and bound algorithm to this instance of job shop problem.

(b) Compare your result with the result of the local search routine in the LEKIN system (see Chapter 19).

7.7. Consider the following instance of the two machine flow shop with the makespan as objective (i.e., an instance of $F2 \parallel C_{\max}$, which is a special case of $J2 \parallel C_{\max}$).

jobs	1	2	3	4	5	6	7	8	9	10	11
p_{1j}	3	6	4	3	4	2	7	5	5	6	12
p_{2j}	4	5	5	2	3	3	6	6	4	7	2

(a) Apply the heuristic described in Exercise 7.1 to this two machine flow shop.

(b) Apply the Shifting Bottleneck heuristic to this two machine flow shop.

(c) Construct a schedule using Johnson's rule (see Chapter 6).

(d) Compare the schedules found under (a), (b), and (c).

7.8. Consider the instance of the job shop with the total weighted tardiness objective described in Example 7.3.1. Apply the Shifting Bottleneck heuristic again, but now use as scaling parameter $K = 5$. Compare the resulting schedule with the schedule obtained in Example 7.3.1.

7.9. Consider the following instance of $Jm \parallel \sum w_j T_j$.

job	w_j	r_j	d_j	machine sequence	processing times
1	1	3	23	1, 2, 3	$p_{11} = 4, p_{21} = 9, p_{31} = 5$
2	2	2	17	3, 1, 2	$p_{32} = 4, p_{12} = 5, p_{22} = 5$
3	2	0	15	3, 2, 1	$p_{33} = 6, p_{23} = 4, p_{13} = 6$

(a) Apply the Shifting Bottleneck heuristic for the total weighted tardiness.

(b) Compare your result with the result of the shifting bottleneck routine in the LEKIN system (see Chapter 19).

(c) Compare your result with the result of the local search routine in the LEKIN system.

7.10. Consider the following instance of $F2 \parallel \sum w_j T_j$.

jobs	1	2	3	4	5
p_{1j}	12	4	6	8	2
p_{2j}	10	5	4	6	3
d_j	12	32	21	14	28
w_j	3	2	4	3	2

Apply the Shifting Bottleneck heuristic to minimize the total weighted tardiness.

EXERCISES (THEORETICAL)

7.11. Design a branching scheme for a branch and bound approach that is based on the insertion of disjunctive arcs. The root node of the tree corresponds to a disjunctive graph without any disjunctive arcs. Each node in the branching tree corresponds to a particular selection of a subset of the disjunctive arcs. That is, for any particular node in the tree, a subset of the disjunctive arcs has been fixed in certain directions while the remaining set of disjunctive arcs has not been fixed yet. From every node, there are two arcs emanating to two nodes at the next level. One of the two nodes at the next level corresponds to an additional disjunctive arc being fixed in one direction while the other node corresponds to the selection of the reverse arc.

Develop an algorithm that generates the nodes of such a branching tree and show that your algorithm generates every possible schedule.

7.12. Determine an upper and a lower bound for the makespan in an m machine job shop when preemptions are not allowed. The processing time of job j on machine i is p_{ij} (i.e., no restrictions on the processing times).

7.13. Show that when preemptions are allowed there always exists an optimal schedule for the job shop that is nondelay.

7.14. Consider $J2 \mid recrc, p_{ij} = 1 \mid C_{\max}$. Each job has to be processed a number of times on each one of the two machines. A job always has to alternate between the two machines (i.e., after a job has completed one operation on one of the machines, it has to go to the other machine for the next operation). The processing time of each operation is 1. Determine the schedule that minimizes the makespan and prove its optimality.

COMMENTS AND REFERENCES

Job shop scheduling has received an enormous amount of attention in the research literature as well as in books.

An algorithm for the minimization of the makespan in a two machine job shop without recirculation is from Jackson (1956), and the disjunctive programming formulation described in Section 7.1 is from Roy and Sussmann (1964).

Branch and bound techniques have often been applied to minimize the makespan in job shops; see, for example, Lomnicki (1965), Brown and Lomnicki (1966), McMahon and Florian (1975), Barker and McMahon (1985), Carlier and Pinson (1989), Applegate and Cook (1991), Hoitomt, Luh, and Pattipati (1993), Brucker, Jurisch, and Sievers (1994), and Brucker, Jurisch, and Krämer (1994). For an overview of branch and bound techniques applied to the job shop problem, see Pinson (1995). Some of the branching schemes of these branch and bound approaches are based on the generation of active schedules (the concept of an active schedule was first introduced by Giffler and Thompson, 1960), whereas other branching schemes are based on the directions of the disjunctive arcs to be selected.

The famous Shifting Bottleneck heuristic is from Adams, Balas, and Zawack (1988). Their algorithm makes use of a single machine scheduling algorithm developed by Carlier (1982). Earlier work on this particular single machine subproblem was done by McMahon and Florian (1975). Nowicki and Zdrzalka (1986), Dauzère-Pérès and Lasserre (1993, 1994), and Balas, Lenstra, and Vazacopoulos (1995) all developed more refined versions of the Carlier algorithm. The monograph by Ovacik and Uzsoy (1997) presents an excellent treatise of the application of decomposition methods and Shifting Bottleneck techniques to large-scale job shops with several objectives (e.g., the makespan and the maximum lateness). This monograph is based on a number of papers by the authors; see, for example, Uzsoy (1993) for the application of decomposition methods to flexible job shops.

Job shops with the total weighted tardiness as objective have been the focus of a number of studies. Vepsalainen and Morton (1987) developed heuristics based on priority rules. Singer and Pinedo (1998) developed a branch and bound approach, and Pinedo and Singer (1999) developed the shifting bottleneck approach described in Section 7.3.

In addition to the procedures discussed in this chapter, job shop problems have also been tackled with local search procedures; see, for example, Matsuo, Suh, and Sullivan (1988), Dell'Amico and Trubian (1991), Della Croce, Tadei, and Volta (1992), Storer, Wu, and Vaccari (1992), Nowicki and Smutnicki (1996), and Kreipl (2000). Examples of such local search procedures are presented in Chapter 14.

For a broader view of the job shop scheduling problem, see Wein and Chevelier (1992). For an interesting special case of the job shop (i.e., a flow shop with reentry) see Graves, Meal, Stefek, and Zeghmi (1983).

8

Open Shops
(Deterministic)

This chapter deals with multi-operation models that are different from the job shop models considered in the previous chapter. In a job shop, each job has a fixed route that is predetermined. In practice, it often occurs that the route of the job is immaterial and up to the scheduler to decide. When the routes of the jobs are open, the model is referred to as an *open* shop.

The first section covers nonpreemptive open shop models with the makespan as objective. The second section deals with preemptive open shop models with the makespan as objective. The third and fourth sections focus on nonpreemptive and preemptive models with the maximum lateness as objective. The fifth section considers nonpreemptive models with the number of tardy jobs as objective.

186

8.1 THE MAKESPAN WITHOUT PREEMPTIONS

Consider $O2 \mid\mid C_{max}$—that is, there are two machines and n jobs. Job j may be processed first on machine 1 and then on machine 2 or vice versa; the decision maker may determine the routes. The makespan has to be minimized. It is clear that

$$C_{max} \geq \max \left(\sum_{j=1}^{n} p_{1j}, \sum_{j=1}^{n} p_{2j} \right)$$

since the makespan cannot be less than the workload on either machine. One would typically expect the makespan to be *equal* to the RHS of the inequality; one would expect only in very special cases the makespan to be larger than the RHS. It is worthwhile to investigate the special cases where the makespan is strictly greater than the maximum of the two workloads.

This section considers only nondelay schedules. That is, if there is a job waiting for processing when a machine is free, then that machine is not allowed to remain idle. It immediately follows that an idle period can occur on a machine if and only if one job remains to be processed on that machine and, when that machine is available, this last job is just then being processed on the other machine. It can be shown that at most one such idle period can occur on at most one of the two machines (see Figure 8.1). Such an idle period *may* cause an unnecessary increase in the makespan; if this last job turns out to be the very last job to complete all its processing, then the idle period does cause an increase in the makespan (see Figure 8.1.a). If this last job, after having completed its processing on the machine

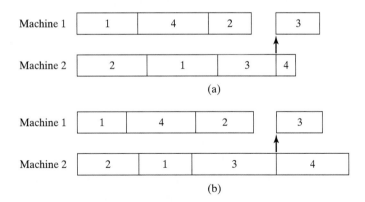

Figure 8.1   Idle periods in two machine open shops: (a) idle period causes unnecessary increase in makespan, (b) idle period does not cause an unnecessary increase in makespan.

that was idle, is *not* the very last job to leave the system, then the makespan is still equal to the maximum of the two workloads (see Figure 8.1.b).

Consider the following rule: Whenever a machine is freed, start processing among the jobs that have not yet received processing on either machine the one with the longest processing time on the *other* machine. This rule is in what follows referred to as the *Longest Alternate Processing Time first (LAPT)* rule. At time zero, when both machines are idle, it may occur that the same job qualifies to be first on both machines. In this case, it does not matter on which machine this job is processed first. According to this LAPT rule, whenever a machine is freed, jobs that already have completed their processing on the other machine have the lowest—that is, zero—priority on the machine just freed. There is therefore no distinction between the priorities of two jobs that both have already been processed on the other machine.

Theorem 8.1.1. *The LAPT rule yields an optimal schedule for $O2 \parallel C_{\max}$ with makespan*

$$C_{\max} = \max \left(\max_{j \in \{1, \ldots, n\}} (p_{1j} + p_{2j}), \sum_{j=1}^{n} p_{1j}, \sum_{j=1}^{n} p_{2j} \right).$$

Proof. Actually, a more general (and less restrictive) scheduling rule already guarantees a minimum makespan. This more general rule may result in many different schedules that are all optimal. This class of optimal schedules includes the LAPT schedule. This general rule also assumes that unforced idleness is not allowed.

Assume, without loss of generality, that the longest processing time among the $2n$ processing times belongs to operation $(1, k)$–that is,

$$p_{ij} \le p_{1k}, \qquad i = 1, 2, j = 1, \ldots, n.$$

The more general rule can be described as follows. If operation $(1, k)$ is the longest operation, then job k must be started at time 0 on machine 2. The processing of operation $(1, k)$ has to be postponed as long as possible (i.e., its priority is at all times lower than the priority of any other operation available for processing on machine 1 and it can only be processed on machine 1 if no other job is available for processing on machine 1; this can only happen either if it is the last operation to be done on machine 1 or if it is the second to last operation and the last operation is not available because it is just then being processed on machine 2). The $2(n-1)$ operations of the remaining $n-1$ jobs can be processed on the two machines in any order; however, unforced idleness is not allowed.

That this rule results in a schedule with a minimum makespan can be shown as follows. If the resulting schedule has no idle period on either machine, then, of course, it is optimal. However, an idle period may occur either on machine 1 or machine 2. So two cases have to be considered.

Case 1.   Suppose an idle period occurs on machine 2. If this is the case, then only one more operation needs processing on machine 2, but this operation still has to complete its processing on machine 1. Assume this operation is of job l. When job l starts on machine 2, job k starts on machine 1 and $p_{1k} > p_{2l}$. So the makespan is determined by the completion of job k on machine 1, and no idle period has occurred on machine 1. So the schedule is optimal.

Case 2.   Suppose an idle period occurs on machine 1. An idle period on machine 1 can occur only when machine 1 is freed after completing all its operations with the exception of operation $(1, k)$ and operation $(2, k)$ of job k is at that point still being processed on machine 2. In this case, the makespan is equal to $p_{2k} + p_{1k}$ and the schedule is optimal. $\Box$

Another rule that may seem appealing at first sight is the rule that gives the highest priority to the job with the largest *total* remaining processing time on *both* machines. It turns out that there are instances, even with two machines, when this rule results in a schedule that is not optimal (see Exercise 8.12). The fact that the priority of a job on one machine is determined only by the amount of processing to be done afterward on the *other* machine is important.

The LAPT rule described earlier may be regarded as a special case of a more general rule that can be applied to open shops with more than two machines. This more general rule may be referred to as the *Longest Total Remaining Processing on Other Machines first* rule. According to this rule, again, the processing required on the machine currently available does not affect the priority level of a job. However, this rule does not always result in an optimal schedule as $Om \parallel C_{max}$ is NP-hard when $m \geq 3$.

Theorem 8.1.2.   *The problem $O3 \parallel C_{max}$ is NP-hard.*

Proof.   The proof is based on a reduction of PARTITION to $O3 \parallel C_{max}$. The PARTITION problem is formulated as follows. Given positive integers $a_1, \ldots, a_t$ and

$$b = \frac{1}{2} \sum_{j=1}^{t} a_j,$$

do there exist two disjoint subsets, S_1 and S_2, such that

$$\sum_{j \in S_1} a_j = \sum_{j \in S_2} a_j = b?$$

The reduction is based on the following transformation. Consider $3t + 1$ jobs. Of these $3t + 1$ jobs, there are $3t$ jobs that have only one nonzero operation and one job that has to be processed on each one of the three machines.

$$p_{1j} = a_j, \, p_{2j} = p_{3j} = 0, \qquad \text{for } 1 \leq j \leq t,$$

$$p_{2j} = a_j, \, p_{1j} = p_{3j} = 0, \qquad \text{for } t + 1 \leq j \leq 2t,$$

$$p_{3j} = a_j, \, p_{1j} = p_{2j} = 0, \qquad \text{for } 2t + 1 \leq j \leq 3t,$$

$$p_{1,3t+1} = p_{2,3t+1} = p_{3,3t+1} = b,$$

where

$$\sum_{j=1}^{t} a_j = \sum_{j=t+1}^{2t} a_j = \sum_{j=2t+1}^{3t} a_j = 2b$$

and $z = 3b$. The open shop problem now has a schedule with a makespan equal to z if and only if there exists a partition. It is clear that to have a makespan equal to $3b$ job $3t + 1$ has to be processed on the three machines without interruption. Consider the machine on which job $3t + 1$ is processed second—that is, during the interval $(b, 2b)$. Without loss of generality, it may be assumed that this is machine 1. Jobs $1, \ldots, t$ have to be processed only on machine 1. If there exists a partition of these t jobs in such a way that one set can be processed during the interval $(0, b)$ and the other set can be processed during the interval $(2b, 3b)$, then the makespan is $3b$ (see Figure 8.2). If there does not exist such a partition, then the makespan has to be larger than $3b$.   □

The LAPT rule for $O2 \parallel C_{\max}$ is one of the few polynomial time algorithms for nonpreemptive open shop problems. Most of the more general open shop models within the framework of Chapter 2 are NP-hard (e.g., $O2 \mid r_j \mid C_{\max}$). However, the

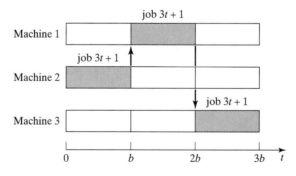

Figure 8.2   Reduction of PARTITION to $O3 \parallel C_{\max}$.

problem $Om \mid r_j, p_{ij} = 1 \mid C_{\max}$ can be solved in polynomial time. This problem is discussed in a more general setting in Section 8.3.

8.2 THE MAKESPAN WITH PREEMPTIONS

Preemptive open shop problems tend to be somewhat easier. In contrast to $Om \parallel C_{\max}$, the $Om \mid prmp \mid C_{\max}$ problem is solvable in polynomial time.

From the fact that with two machines the value of the makespan under LAPT is a lower bound for the makespan even when preemptions are allowed, it follows that the nonpreemptive LAPT rule is also optimal for $O2 \mid prmp \mid C_{\max}$.

It is easy to establish a lower bound for the makespan with m machines when preemptions are allowed:

$$C_{\max} \geq \max\left(\max_{j\in\{1,\dots,n\}} \sum_{i=1}^{m} p_{ij}, \max_{i\in\{1,\dots,m\}} \sum_{j=1}^{n} p_{ij}\right).$$

That is, the makespan is at least as large as the maximum workload of each of the m machines and at least as large as the total amount of processing to be done on each of the n jobs. It turns out that it is rather easy to obtain a schedule with a makespan that is equal to this lower bound.

To see how the algorithm works, consider the $m \times n$ matrix $\mathbf{P}$ of the processing times p_{ij}. Row i or column j is called *tight* if its sum is equal to the lower bound and *slack* otherwise. Suppose it is possible to find in this matrix a subset of nonzero entries with exactly one entry in each tight row and one entry in each tight column and at most one entry in each slack row and slack column. Such a subset would be called a *decrementing* set. This subset is used to construct a partial schedule of length Δ for some appropriately chosen Δ. In this partial schedule, machine i works on job j for an amount of time that is equal to $\min(p_{ij}, \Delta)$ for each element p_{ij} in the decrementing set. In the original matrix $\mathbf{P}$, the entries corresponding to the decrementing set are then reduced to $\max(0, p_{ij} - \Delta)$ and the resulting matrix is then called $\mathbf{P}'$. If Δ is chosen appropriately, the makespan $C'_{\max}$ that corresponds to the new matrix $\mathbf{P}'$ is equal to $C_{\max} - \Delta$. This value for Δ has to be chosen carefully. First, it is clear that the Δ has to be smaller than every p_{ij} in the decrementing set that is in a tight row or column, otherwise there will be a row or column in $\mathbf{P}'$ that is strictly larger than $C'_{\max}$. For the same reason, if p_{ij} is an element in the decrementing set in a slack row, say row i, it is necessary that

$$\Delta \leq p_{ij} + C_{\max} - \sum_k p_{ik},$$

where $C_{\max} - \sum p_{ik}$ is the amount of slack time in row i. Similarly, if p_{ij} is an entry in the slack column j, then

$$\Delta \leq p_{ij} + C_{\max} - \sum_k p_{kj},$$

where $C_{\max} - \sum p_{kj}$ is the amount of slack time in column j. If row i or column j does not contain an element in the decrementing set, then

$$\Delta \leq C_{\max} - \sum_j p_{ij}$$

or

$$\Delta \leq C_{\max} - \sum_i p_{ij}.$$

If Δ is chosen to be as large as possible subject to these conditions, then either $\mathbf{P}'$ will contain at least one less strictly positive element than $\mathbf{P}$ or $\mathbf{P}'$ will contain at least one more tight row or column than $\mathbf{P}$. It is then clear that there cannot be more than $r + m + n$ iterations where r is the number of strictly positive elements in the original matrix.

It turns out that it is always possible to find a decrementing set for a non-negative matrix $\mathbf{P}$. This property is the result of a basic theorem due to Birkhoff and von Neumann regarding stochastic matrices and permutation matrices. However, the proof of this theorem is beyond the scope of this book.

Example 8.2.1 (Minimizing Makespan with Preemptions)

Consider three machines and four jobs with the processing times specified by the matrix

$$\mathbf{P} = \begin{bmatrix} 3 & 4 & 0 & 4 \\ 4 & 0 & 6 & 0 \\ 4 & 0 & 0 & 6 \end{bmatrix}.$$

It is easily verified that $C_{\max} = 11$ and that the first row and first column are tight. A possible decrementing set comprises the processing times $p_{12} = 4$, $p_{21} = 4$, and $p_{34} = 6$. If Δ is set equal to 4, then $C'_{\max} = 7$. A partial schedule is constructed by scheduling job 2 on machine 1 for four time units, job 1 on machine 2 for four time units, and job 4 on machine 3 for four time units. The matrix is now

$$\mathbf{P}' = \begin{bmatrix} 3 & 0 & 0 & 4 \\ 0 & 0 & 6 & 0 \\ 4 & 0 & 0 & 2 \end{bmatrix}.$$

Again, the first row and the first column are tight. A decrementing set is obtained with the processing times $p_{11} = 3$, $p_{23} = 6$, and $p_{34} = 2$. Choosing $\Delta = 3$, the

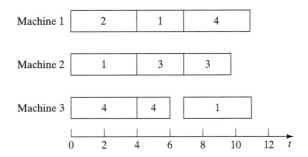

Figure 8.3 Optimal schedule for $O3 \mid prmp \mid C_{max}$ with four jobs (Example 8.2.1).

partial schedule can be augmented by assigning job 1 to machine 1 for three time units, job 3 to machine 2 for three time units, and job 4 again to machine 3 but only for two time units. The matrix is

$$\mathbf{P}'' = \begin{bmatrix} 0 & 0 & 0 & 4 \\ 0 & 0 & 3 & 0 \\ 4 & 0 & 0 & 0 \end{bmatrix}.$$

The last decrementing set is obtained with the remaining three positive processing times. The final schedule is obtained by augmenting the partial schedule by assigning job 4 on machine 1 for four time units, job 3 to machine 2 for three time units, and job 1 to machine 3 for four time units (see Figure 8.3).

8.3 THE MAXIMUM LATENESS WITHOUT PREEMPTIONS

The $Om \mid\mid L_{max}$ problem is a generalization of the $Om \mid\mid C_{max}$ problem and is therefore at least as hard.

Theorem 8.3.1.   *The problem $O2 \mid\mid L_{max}$ is strongly NP-hard.*

Proof.   The proof is done by reducing 3-PARTITION to $O2 \mid\mid L_{max}$. The 3-PARTITION problem is formulated as follows. Given positive integers $a_1, \ldots, a_{3t}$, b, such that

$$\frac{b}{4} < a_j < \frac{b}{2}$$

and

$$\sum_{j=1}^{3t} a_j = tb,$$

do there exist t pairwise disjoint three element subsets $S_i \subset \{1, \ldots, 3t\}$ such that

$$\sum_{j \in S_i} a_j = b$$

for $i = 1, \ldots, t$?

The following instance of $O2 \parallel L_{\max}$ can be constructed. The number of jobs, n, is equal to $4t$ and

$$
\begin{array}{llll}
p_{1j} = 0 & p_{2j} = a_j & d_j = 3tb & j = 1, \ldots, 3t \\
p_{1j} = 0 & p_{2j} = 2b & d_j = 2b & j = 3t + 1 \\
p_{1j} = 3b & p_{2j} = 2b & d_j = (3(j - 3t) - 1)b & j = 3t + 2, \ldots, 4t.
\end{array}
$$

There exists a schedule with $L_{\max} \le 0$ if and only if jobs $1, \ldots, 3t$ can be divided into t groups each containing three jobs and requiring b units of processing time on machine 2 (i.e., if and only if 3-PARTITION has a solution).   □

It can be shown that $O2 \parallel L_{\max}$ is equivalent to $O2 \mid r_j \mid C_{\max}$. Consider the $O2 \parallel L_{\max}$ problem with deadlines $\bar{d}_j$ rather than due dates d_j. Let

$$\bar{d}_{\max} = \max(\bar{d}_1, \ldots, \bar{d}_n).$$

Apply a time reversal to $O2 \parallel L_{\max}$. Finding a feasible schedule with $L_{\max} = 0$ is now equivalent to finding a schedule for the $O2 \mid r_j \mid C_{\max}$ with

$$r_j = \bar{d}_{\max} - \bar{d}_j$$

and a makespan that is less than $\bar{d}_{\max}$. So the $O2 \mid r_j \mid C_{\max}$ problem is therefore also strongly NP-hard.

Consider now the special case $Om \mid r_j, p_{ij} = 1 \mid L_{\max}$. The fact that all processing times are equal to 1 makes the problem considerably easier. The polynomial time solution procedure consists of three phases—namely,

Phase 1. Parametrizing and a binary search.
Phase 2. Solving a network flow problem.
Phase 3. Coloring a bipartite graph.

The first phase of the procedure involves a parametrization. Let L be a free parameter and assume that each job has a deadline $d_j + L$. The objective is to find a schedule in which each job is completed before or at its deadline ensuring that $L_{\max} \le L$. Let

$$t_{\max} = \max(d_1, \ldots, d_n) + L,$$

that is, no job should receive any processing after time $t_{\max}$.

The second phase focuses on the following network flow problem: There is a source node U that has n arcs emanating to nodes $1, \ldots, n$. Node j corresponds to job j. The arc from the source node U to node j has capacity m (equal to the number of machines and number of operations of each job). There is a second set of $t_{\max}$ nodes, each node corresponding to one time unit. Node $t, t = 1, \ldots, t_{\max}$, corresponds to the time slot $[t - 1, t]$. Node j has arcs emanating to nodes $r_j + 1, r_j + 2, \ldots, d_j + L$. Each one of these arcs has unit capacity. Each node of the set of $t_{\max}$ nodes has an arc with capacity m going to sink V (see Figure 8.4). The capacity limit on each one of these arcs is necessary to ensure that no more than m operations are processed in any given time period. The solution of this network flow problem indicates in which time slots the m operations of job j have to be processed.

However, the network flow solution cannot be translated immediately into a feasible schedule for the open shop because in the network flow formulation no distinction is made between the different machines (i.e., in this solution, it may be possible that two different operations of the same job are processed in two different time slots on the same machine). However, it turns out that the assignment of operations to time slots prescribed by the network flow solution can be transformed into a feasible schedule in such a way that each operation of job j is processed on a different machine.

The third phase of the algorithm generates a feasible schedule. Consider a graph coloring problem with a bipartite graph that consists of two sets of nodes

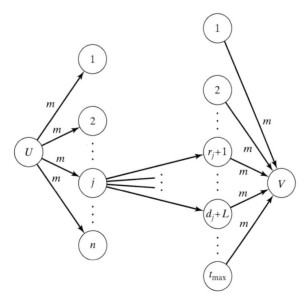

Figure 8.4   Network flow problem of phase 2.

N_1 and N_2 and a set of undirected arcs. Set N_1 has n nodes and set N_2 has t_{max} nodes. Each node in N_1 is connected to m nodes in N_2; a node in N_1 is connected to those m nodes in N_2 that correspond to the time slots in which its operations are supposed to be processed (according to the solution of the network flow problem in the second phase). So each one of the nodes in N_1 is connected to exactly m nodes in N_2 while each node in N_2 is connected to at most m nodes in N_1. A result in graph theory states that if each node in a bipartite graph has at most m arcs, then the arcs can be colored with m different colors in such a way that no node has two arcs of the same color. Each color then corresponds to a given machine.

The coloring algorithm that achieves this can be described as follows. Let g_j, $j = 1, \ldots, n$ denote the degree of a node from set N_1, and let h_t, $t = 1, \ldots, t_{max}$ denote the degree of a node from set N_2. Let

$$\Delta = \max(g_1, \ldots, g_n, h_1, \ldots, h_{t_{max}}).$$

To describe the algorithm that yields a coloring with Δ colors, let $a_{jt} = 1$ if node j from N_1 is connected to node t from N_2, and let $a_{jt} = 0$ otherwise. The a_{jt} are elements of a matrix with n rows and t_{max} columns. Clearly,

$$\sum_{j=1}^{n} a_{jt} \leq \Delta \qquad t = 1, \ldots, t_{max}$$

and

$$\sum_{t=1}^{t_{max}} a_{jt} \leq \Delta \qquad j = 1, \ldots, n.$$

The entries (j, t) in the matrix with $a_{jt} = 1$ are referred to as *occupied cells*. Each occupied cell in the matrix has to be assigned one of the Δ colors in such a way that the same color is not assigned twice in any row or column.

The assignment of colors to occupied cells is done by visiting the occupied cells of the matrix row by row from left to right. When visiting occupied cell (j, t), a color c, not yet assigned in column t, is selected. If c is assigned to another cell in row j, say (j, t^*), then there exists a color c' not yet assigned in row j that can be used to replace the assignment of c to (j, t^*). If another cell (j^*, t^*) in column t^* already has assignment c', this assignment is replaced with c. This conflict resolution process stops when there is no remaining conflict. If the partial assignment before coloring (j, t) was feasible, then the conflict resolution procedure yields a feasible coloring in at most n steps.

Example 8.3.2 (Minimizing Maximum Lateness without Preemptions)

Consider the following instance of $O3 \mid r_j, p_{ij} = 1 \mid L_{max}$ with three machines and seven jobs.

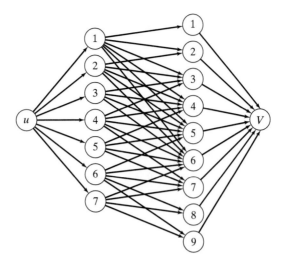

Figure 8.5   Network flow problem in phase 2 of Example 8.3.2.

j	1	2	3	4	5	6	7
r_j	0	1	2	2	3	4	5
d_j	5	5	5	6	6	8	8

Assume that $L = 1$. Each job has a deadline $\bar{d}_j = d_j + 1$. So $t_{\max} = 9$. Phase 2 results in the network flow problem described in Figure 8.5. On the left there are seven nodes that correspond to the seven jobs and on the right there are nine nodes that correspond to the nine time units.

The result of the network flow problem is that the jobs are processed during the time units given in the following table.

job	1	2	3	4	5	6	7
time units	1, 2, 3	2, 3, 4	4, 5, 6	4, 5, 6	5, 6, 7	7, 8, 9	7, 8, 9

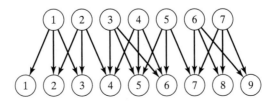

Figure 8.6   Bipartite graph coloring in phase 3 of Example 8.3.2.

It can be easily verified that at no point in time more than three jobs are being processed simultaneously.

Phase 3 leads to the graph coloring problem. The graph is depicted in Figure 8.6 and the matrix with the appropriate coloring is

$$
\begin{Vmatrix}
1 & 1 & 1 & 0 & 0 & 0 & 0 & 0 & 0 \\
0 & 1 & 1 & 1 & 0 & 0 & 0 & 0 & 0 \\
0 & 0 & 0 & 1 & 1 & 1 & 0 & 0 & 0 \\
0 & 0 & 0 & 1 & 1 & 1 & 0 & 0 & 0 \\
0 & 0 & 0 & 0 & 1 & 1 & 1 & 0 & 0 \\
0 & 0 & 0 & 0 & 0 & 0 & 1 & 1 & 1 \\
0 & 0 & 0 & 0 & 0 & 0 & 1 & 1 & 1
\end{Vmatrix}.
$$

It is easy to find a red (r), blue (b), and white (w) coloring that corresponds to a feasible schedule.

$$
\begin{Vmatrix}
r & b & w & - & - & - & - & - & - \\
- & r & b & w & - & - & - & - & - \\
- & - & - & r & b & w & - & - & - \\
- & - & - & b & w & r & - & - & - \\
- & - & - & - & r & b & w & - & - \\
- & - & - & - & - & - & r & b & w \\
- & - & - & - & - & - & b & w & r
\end{Vmatrix}.
$$

Since there is a feasible schedule for $L = 1$, it has to be verified at this point whether there is a feasible schedule for $L = 0$. It can be easily shown that there does not exist a schedule in which every job is completed on time.

8.4 THE MAXIMUM LATENESS WITH PREEMPTIONS

It is often the case that the preemptive version of a scheduling problem is easier than its nonpreemptive counterpart. That is also the case with $Om \mid prmp \mid L_{max}$ and $Om \mid\mid L_{max}$. Consider $O2 \mid prmp \mid L_{max}$ and assume that $d_1 \leq \cdots \leq d_n$. Let

$$
A_k = \sum_{j=1}^{k} p_{1j}
$$

and

$$
B_k = \sum_{j=1}^{k} p_{2j}.
$$

The procedure to minimize the maximum lateness first considers the due dates as absolute deadlines and then tries to generate a feasible solution. The jobs are scheduled in increasing order of their deadlines (i.e., first job 1, then job 2, etc.). Suppose that jobs $1, \ldots, j - 1$ have been scheduled successfully and job j has to be scheduled next. Let x_j (y_j) denote the total amount of time prior to d_j that machine 1 (2) is idle while machine 2 (1) is busy. Let z_j denote the total amount of time prior to d_j that machines 1 and 2 are idle simultaneously. Note that x_j, y_j, and z_j are not independent since

$$x_j + z_j = d_j - A_{j-1}$$

and

$$y_j + z_j = d_j - B_{j-1}.$$

The minimum amount of processing that must be done on operation $(1, j)$ while both machines are available is $\max(0, p_{1j} - x_j)$, and the minimum amount of processing on operation $(2, j)$ while both machines are available is $\max(0, p_{2j} - y_j)$. It follows that job j can be scheduled successfully if and only if

$$\max(0, p_{1j} - x_j) + \max(0, p_{2j} - y_j) \le z_j.$$

This inequality is equivalent to the following three inequalities:

$$p_{1j} - x_j \le z_j$$
$$p_{2j} - y_j \le z_j$$
$$p_{1j} - x_j + p_{2j} - y_j \le z_j.$$

So job j can be scheduled successfully if and only if each one of the following three feasibility conditions holds:

$$A_j \le d_j$$
$$B_j \le d_j$$
$$A_j + B_j \le 2d_j - z_j.$$

These inequalities indicate that, to obtain a feasible schedule, an attempt has to be made in each iteration to minimize the value of z_j. The smallest possible values of $z_1, \ldots, z_n$ are defined recursively by

$$z_1 = d_1$$
$$z_j = d_j - d_{j-1} + \max(0, z_{j-1} - p_{1,j-1} - p_{2,j-1}), \qquad j = 2, \ldots, n.$$

To verify the existence of a feasible schedule, the values of $z_1, \ldots, z_n$ have to be computed recursively, and for each z_j it has to be checked whether it satisfies the third one of the feasibility conditions. There exists a feasible schedule if all the z_j satisfy the conditions.

To minimize $L_{\max}$, a parametrized version of the preceding computation has to be done. Replace each d_j by $d_j + L$, where L is a free parameter. The smallest value of L for which there exists a feasible schedule is equal to the minimum value of $L_{\max}$ that can be achieved with the original due dates d_j.

It turns out that there exists also a polynomial time algorithm for the more general open shop with m machines even when the jobs have different release dates—that is, $Om \mid r_j, prmp \mid L_{\max}$. Again, as in the case with two machines, the due dates d_j are considered deadlines $\bar{d}_j$, and an attempt is made to find a feasible schedule where each job is completed before or at its due date. Let

$$a_1 < a_2 < \cdots < a_{p+1}$$

denote the ordered collection of all distinct release dates r_j and deadlines $\bar{d}_j$. So there are p intervals $[a_k, a_{k+1}]$. Let I_k denote the length of interval k—that is,

$$I_k = a_{k+1} - a_k.$$

Let the decision variable x_{ijk} denote the amount of time that operation (i, j) is processed during interval k. Consider the following linear program:

$$\max \sum_{k=1}^{p} \sum_{i=1}^{m} \sum_{j=1}^{n} x_{ijk}$$

subject to

$$\sum_{i=1}^{m} x_{ijk} \leq I_k \qquad \text{for all } 1 \leq j \leq n, 1 \leq k \leq p$$

$$\sum_{j=1}^{n} x_{ijk} \leq I_k \qquad \text{for all } 1 \leq i \leq m, 1 \leq k \leq p$$

$$\sum_{k=1}^{p} x_{ijk} \leq p_{ij} \qquad \text{for all } 1 \leq j \leq n, 1 \leq i \leq m$$

$$x_{ijk} \geq 0 \qquad \text{if } r_j \leq a_k \text{ and } d_j \geq a_{k+1}$$

$$x_{ijk} = 0 \qquad \text{if } r_j \geq a_{k+1} \text{ or } d_j \leq a_k.$$

The first inequality requires that no job is scheduled for more than I_k units of time in interval k. The second inequality requires that the amount of processing assigned to any machine is not more than the length of the interval. The third inequality requires that each job is not processed longer than necessary. The constraints on x_{ijk} ensure that no job is assigned to a machine either before its release date or after its due date. An initial feasible solution for this problem is clearly $x_{ijk} = 0$. However, since the objective is to maximize the sum of the x_{ijk}, the third inequality is tight under the optimal solution assuming there exists a feasible solution for the scheduling problem.

If there exists a feasible solution for the linear program, then there exists a schedule with all jobs completed on time. However, the solution of the linear program only gives the optimal values for the decision variables x_{ijk}. It does not specify how the operations should be scheduled within the interval $[a_k, a_{k+1}]$. This scheduling problem within each interval can be solved as follows: Consider interval k as an independent open shop problem with the processing time of operation (i, j) being the value x_{ijk} that came out of the linear program. The objective for the open shop scheduling problem for interval k is equivalent to the minimization of the makespan (i.e., $Om \mid prmp \mid C_{\max}$). The polynomial algorithm described in section 8.2 can then be applied to each interval separately.

If the outcome of the linear program indicates that no feasible solution exists, then (similar to the $m = 2$ case) a parametrized version of the entire procedure has to be carried out. Replace each $\bar{d}_j$ by $\bar{d}_j + L$, where L is a free parameter. The smallest value of L for which there exists a feasible schedule is equal to the minimum value of $L_{\max}$ that can be achieved with the original due dates d_j.

Example 8.4.1 (Minimizing Maximum Lateness with Preemptions)

Consider the following instance of $O3 \mid r_j, prmp \mid L_{\max}$ with three machines and five jobs.

j	1	2	3	4	5
p_{1j}	1	2	2	2	3
p_{2j}	3	1	2	2	1
p_{3j}	2	1	1	2	1
r_j	1	1	3	3	3
d_j	9	7	6	7	9

There are four intervals that are determined by $a_1 = 1$, $a_2 = 3$, $a_3 = 6$, $a_4 = 7$, and $a_5 = 9$. The lengths of the four intervals are $I_1 = 2$, $I_2 = 3$, $I_3 = 1$, and $I_4 = 2$. There are $4 \times 3 \times 5 = 60$ decision variables x_{ijk}.

The first set of constraints of the linear program has 20 constraints. The first one of this set (i.e., $j = 1, k = 1$) is

$$x_{111} + x_{211} + x_{311} = 2.$$

The second set of constraints has 12 constraints. The first one of this set (i.e., $i = 1, k = 1$) is

$$x_{111} + x_{121} + x_{131} + x_{141} + x_{151} = 2.$$

The third set of constraints has 15 constraints. The first one of this set (i.e., $i = 1, j = 1$) is

$$x_{111} + x_{112} + x_{113} + x_{114} + x_{115} = 1.$$

It turns out that this linear program has no feasible solution. Replacing d_j by $d_j + 1$ yields another linear program that also does not have a feasible solution. Replacing the original d_j by $d_j + 2$ results in the following data set:

j	1	2	3	4	5
p_{1j}	1	2	2	2	3
p_{2j}	3	1	2	2	1
p_{3j}	2	1	1	2	1
r_j	1	1	3	3	3
d_j	11	9	8	9	11

There are four intervals that are determined by $a_1 = 1, a_2 = 3, a_3 = 8, a_4 = 9$, and $a_5 = 11$. The lengths of the four intervals are $I_1 = 2, I_2 = 5, I_3 = 1$, and $I_4 = 2$. The resulting linear program has feasible solutions and the optimal solution is the following:

$x_{111} = 1$	$x_{211} = 0$	$x_{311} = 1$
$x_{121} = 1$	$x_{221} = 1$	$x_{321} = 0$
$x_{131} = 0$	$x_{231} = 0$	$x_{331} = 0$
$x_{141} = 0$	$x_{241} = 0$	$x_{341} = 0$
$x_{151} = 0$	$x_{251} = 0$	$x_{351} = 0$
$x_{112} = 0$	$x_{212} = 0$	$x_{312} = 1$
$x_{122} = 1$	$x_{222} = 0$	$x_{322} = 0$
$x_{132} = 2$	$x_{232} = 2$	$x_{332} = 1$
$x_{142} = 1$	$x_{242} = 2$	$x_{342} = 2$
$x_{152} = 1$	$x_{252} = 1$	$x_{352} = 1$
$x_{113} = 0$	$x_{213} = 1$	$x_{313} = 0$
$x_{123} = 0$	$x_{223} = 0$	$x_{323} = 1$
$x_{133} = 0$	$x_{233} = 0$	$x_{333} = 0$
$x_{143} = 1$	$x_{243} = 0$	$x_{343} = 0$
$x_{153} = 0$	$x_{253} = 0$	$x_{353} = 0$

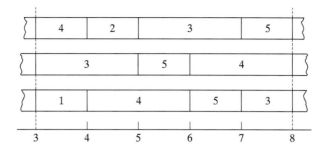

Figure 8.7   Schedule for interval [3, 8] in Example 8.4.1.

$x_{114} = 0$	$x_{214} = 2$	$x_{314} = 0$
$x_{124} = 0$	$x_{224} = 0$	$x_{324} = 0$
$x_{134} = 0$	$x_{234} = 0$	$x_{334} = 0$
$x_{144} = 0$	$x_{244} = 0$	$x_{344} = 0$
$x_{154} = 2$	$x_{254} = 0$	$x_{354} = 0$

Each one of the four intervals has to be analyzed now as a separate $O3 \mid prmp \mid C_{\max}$ problem. Consider, for example, the second interval [3, 8] (i.e., x_{ij2}). The $O3 \mid prmp \mid C_{\max}$ problem for this interval contains the following data.

j	1	2	3	4	5
p_{1j}	0	1	2	1	1
p_{2j}	0	0	2	2	1
p_{3j}	1	0	1	2	1

Applying the algorithm described in Section 8.2 results in the schedule presented in Figure 8.7 (which turns out to be nonpreemptive). The schedules in the other three intervals can be determined very easily.

8.5 THE NUMBER OF TARDY JOBS

The $Om \mid p_{ij} = 1 \mid \sum U_j$ problem is strongly related to the problems discussed in the previous sections. In this problem, again each job consists of m operations and each operation requires one time unit. Assume, without loss of generality, that $d_1 \leq d_2 \leq \cdots \leq d_n$.

It can be easily shown that the set of jobs that are completed on time in an optimal schedule belong to a set $k^*, k^* + 1, \ldots, n$. So the search for an optimal schedule has two aspects. First, it has to be determined what the optimal value of

k^* is; second, given k^*, a schedule has to be constructed in which each job of this set finishes on time.

The value of k^* can be determined via binary search. Given a specific set of jobs that have to be completed on time, a schedule can be generated as follows: Consider the problem $Om \mid r_j, p_{ij} = 1 \mid C_{\max}$, which is a special case of the $Om \mid r_j, p_{ij} = 1 \mid L_{\max}$ problem that is solvable by the polynomial time algorithm described in Section 8.3. Set r_j in this corresponding problem equal to $d_{\max} - d_j$ in the original problem. In essence, the $Om \mid r_j, p_{ij} = 1 \mid C_{\max}$ problem is a time reversed version of the original $Om \mid p_{ij} = 1 \mid \sum U_j$ problem. If for the makespan minimization problem a schedule can be found with a makespan less than $d_{\max}$, then the reverse schedule is applicable to the $Om \mid p_{ij} = 1 \mid \sum U_j$ problem with all jobs completing their processing on time.

8.6 DISCUSSION

This chapter, as many other chapters in this book, focuses mainly on models that are polynomial time solvable. Most open shop models tend to be NP-hard.

For example, little can be said about the total completion time objective. The $Om \mid\mid \sum C_j$ problem is strongly NP-hard when $m \geq 2$. The $Om \mid prmp \mid \sum C_j$ problem is known to be strongly NP-hard when $m \geq 3$. When $m = 2$, the $Om \mid prmp \mid \sum C_j$ problem is NP-hard in the ordinary sense.

In the same way that a flow shop can be generalized to a flexible flow shop, an open shop can be generalized to a flexible open shop. The fact that the flexible flow shop allows for few structural results already gives an indication that it may be hard to obtain results for the flexible open shop. Even the proportionate cases (i.e., $p_{ij} = p_j$ for all i or $p_{ij} = p_i$ for all j) are hard to analyze.

EXERCISES (COMPUTATIONAL)

8.1. Consider the following instance of $O2 \mid\mid C_{\max}$. Determine the number of optimal schedules that are nondelay.

jobs	1	2	3	4
p_{1j}	9	7	5	13
p_{2j}	5	10	11	7

8.2. Consider the following instance of $O5 \mid\mid C_{\max}$ with six jobs and all processing times either 0 or 1. Find an optimal schedule.

jobs	1	2	3	4	5	6
p_{1j}	1	0	0	1	1	1
p_{2j}	1	1	1	0	1	0
p_{3j}	0	1	0	1	1	1
p_{4j}	1	1	1	0	0	1
p_{5j}	1	1	1	1	0	0

8.3. Consider the proportionate open shop $O4 \mid p_{ij} = p_j \mid C_{max}$ with six jobs. Compute the makespan under the optimal schedule.

jobs	1	2	3	4	5	6
p_j	3	5	6	6	8	9

8.4. Consider the problem $O4 \parallel C_{max}$ and consider the *Longest Total Remaining Processing on Other Machines (LTRPOM)* rule. Every time a machine is freed, the job with the longest total remaining processing time on all *other* machines, among available jobs, is selected for processing. Unforced idleness is not allowed. Consider the following processing times.

jobs	1	2	3	4
p_{1j}	5	5	13	0
p_{2j}	5	7	3	8
p_{3j}	12	5	7	0
p_{4j}	0	5	0	15

(a) Apply the LTRPOM rule. Consider at time 0 first machine 1, then machine 2, followed by machines 3 and 4. Compute the makespan.
(b) Apply the LTRPOM rule. Consider at time 0 first machine 4, then machine 3, followed by machines 2 and 1. Compute the makespan.
(c) Find the optimal schedule and the minimum makespan.

8.5. Find an optimal schedule for the instance of $O4 \mid prmp \mid C_{max}$ with four jobs and with the same processing times as in Exercise 8.4.

8.6. Consider the following instance of $O4 \mid r_j, p_{ij} = 1 \mid L_{max}$ with four machines and seven jobs.

job j	1	2	3	4	5	6	7
r_j	0	1	2	2	3	4	5
d_j	6	6	6	7	7	9	9

    (a) Find the optimal schedule and the minimum $L_{\max}$.

    (b) Compare your result with the result in Example 8.3.2.

8.7. Solve the following instance of the $O2 \mid prmp \mid L_{\max}$ problem.

job j	1	2	3	4	5	6	7
p_{1j}	7	3	2	5	3	2	3
p_{2j}	3	4	2	4	3	4	5
d_j	5	6	6	11	14	17	20

8.8. Solve the following instance of the proportionate $O2 \mid prmp \mid L_{\max}$ problem.

job j	1	2	3	4	5	6	7
p_{1j}	7	3	2	5	3	2	3
p_{2j}	7	3	2	5	3	2	3
d_j	5	6	6	11	14	17	20

Can the algorithm described in Section 8.4 be simplified when the processing times are proportionate?

8.9. Consider the Linear Programming formulation of the instance in Exercise 8.7. Write out the objective function. How many constraints are there?

8.10. Consider the following instance of $Om \mid p_{ij} = 1 \mid \sum U_j$ with three machines and eight jobs.

job j	1	2	3	4	5	6	7	8
d_j	3	3	4	4	4	4	5	5

Find the optimal schedule and the maximum number of jobs completed on time.

EXERCISES (THEORETICAL)

8.11. Show that nondelay schedules for $Om \mid\mid C_{\max}$ have at most $m - 1$ idle times on one machine. Show also that if there are $m - 1$ idle times on one machine, there can be at most $m - 2$ idle times on any other machine.

8.12. Consider the following rule for $O2 \mid\mid C_{\max}$. Whenever a machine is freed, start processing the job with the largest sum of remaining processing times on the two machines. Show through a counterexample that this rule does not necessarily minimize the makespan.

8.13. Give an example of $Om \mid\mid C_{\max}$ where the optimal schedule is not nondelay.

8.14. Consider $O2 \mid\mid \sum C_j$. Show that the rule that always gives priority to the job with the smallest total remaining processing is not necessarily optimal.

8.15. Consider $O2 \mid prmp \mid \sum C_j$. Show that the rule that always gives preemptive priority to the job with the smallest total remaining processing time is not necessarily optimal.

8.16. Consider $Om \mid\mid C_{\max}$. The processing time of job j on machine i is either 0 or 1. Consider the following rule: At each point in time, select from the machines that have not been assigned a job yet the machine that still has the largest number of jobs to do. Assign to that machine the job that still has to undergo processing on the largest number of machines (ties may be broken arbitrarily). Show through a counterexample that this rule does not necessarily minimize the makespan.

8.17. Consider a flexible open shop with two workcenters. Workcenter 1 consists of a single machine and workcenter 2 consists of two identical machines. Determine whether LAPT minimizes the makespan.

8.18. Consider the proportionate open shop $Om \mid p_{ij} = p_j \mid C_{\max}$. Find the optimal schedule and prove its optimality.

8.19. Consider the proportionate open shop $Om \mid prmp, p_{ij} = p_j \mid \sum C_j$. Find the optimal schedule and prove its optimality.

8.20. Consider the following two machine hybrid of an open shop and a job shop. Job j has processing time p_{1j} on machine 1 and p_{2j} on machine 2. Some jobs have to be processed first on machine 1 and then on machine 2. Other jobs have to be processed first on machine 2 and then on machine 1. The routing of the remaining jobs may be determined by the scheduler. Describe a schedule that minimizes the makespan.

8.21. Find an upper and a lower bound for the makespan in an m machine open shop when preemptions are not allowed. The processing time of job j on machine i is p_{ij} (i.e., no restrictions on the processing times).

8.22. Compare $Om \mid p_j = 1 \mid \gamma$ with $Pm \mid p_j = 1, chains \mid \gamma$ in which there are n chains consisting of m jobs each. Let Z_1 denote the value of the objective function in the open shop problem and let Z_2 denote the value of the objective function in the parallel machine problem. Find conditions under which $Z_1 = Z_2$ and give examples where $Z_1 > Z_2$.

COMMENTS AND REFERENCES

The LAPT rule for $O2 \mid\mid C_{\max}$ appears to be new. Different algorithms have been introduced before for $O2 \mid\mid C_{\max}$; see, for example, Gonzalez and Sahni (1976). Gonzalez and Sahni (1976) provide an NP-hardness proof for $O3 \mid\mid C_{\max}$, and Sevastianov and Woeginger (1998) present a Polynomial Time Approximation Scheme (PTAS) for $Om \mid\mid C_{\max}$. The NP-hardness proof for $O2 \mid\mid \sum C_j$ is due to Achugbue and Chin (1982).

The polynomial time algorithm for $Om \mid prmp \mid C_{\max}$ is from Lawler and Labetoulle (1978). This algorithm is based on a property of stochastic matrixes that is due to Birkhoff and Von Neumann; for more on this property, see Marshall and Olkin (1979). For additional work on $Om \mid prmp \mid C_{max}$, see Gonzalez (1979).

Lawler, Lenstra, and Rinnooy Kan (1981) provide a polynomial time algorithm for $O2 \mid prmp \mid L_{max}$ and show that $O2 \mid\mid L_{max}$ is NP-Hard in the strong sense. Cho and Sahni (1981) analyze preemptive open shops with more than two machines and present the linear programming formulation for $Om \mid prmp \mid L_{max}$.

For results on the minimization of the (weighted) number of late jobs in open shops with unit processing times, see Brucker, Jurisch, and Jurisch (1993), Galambos and Woeginger (1995), and Kravchenko (2000).

Achugbue and Chin (1982) present an NP-hardness proof for $O2 \mid\mid \sum C_j$, and Liu and Bulfin (1985) provide an NP-hardness proof for $O3 \mid prmp \mid \sum C_j$. Tautenhahn and Woeginger (1997) analyze the total completion time when all the processing times are equal to 1.

Vairaktarakis and Sahni (1995) obtain results for flexible open shop models.

PART 2
Stochastic Models

9

Stochastic Models: Preliminaries

Production environments in the real world are subject to many sources of uncertainty or randomness. Sources of uncertainty with a major impact include machine breakdowns and unexpected releases of high-priority jobs. In the processing times, which are often not precisely known in advance, lies another source of uncertainty. A good model for a scheduling problem should address these forms of uncertainty.

There are several ways in which such forms of randomness can be modeled. To take an example, one may model the possibility of machine breakdown as an integral part of the processing times. This is done by modifying the distribution of the processing times to take the possibility of breakdowns into account. Alternatively, one may model breakdowns as a separate stochastic process that determines when a machine is available and when it is not.

The first section of this chapter describes the framework and notation. The second section deals with distributions and classes of distributions. The third section goes over various forms of stochastic dominance. The fourth section discusses the effect of randomness on the expected value of the objective function given that a schedule is fixed. In the fifth section, several classes of scheduling policies are specified.

9.1 FRAMEWORK AND NOTATION

In what follows, it is assumed that the *distributions* of the processing times, the release dates, and the due dates are all known in advance—that is, at time zero. The actual *outcome* or *realization* of a random processing time only becomes known at the completion of the processing; the realization of a release date or due date becomes known only at that point in time when it actually occurs.

In this part of the book, the following notation is used. Random variables are capitalized, whereas the actual realized values are in lowercase. Job j has the following quantities of interest associated with it.

X_{ij} = the random processing time of job j on machine i; if job j
is only to be processed on one machine, or if it has the same
processing times on each one of the machines it may visit,
the subscript i is omitted.

$1/\lambda_{ij}$ = the mean or expected value of the random variable X_{ij}.

R_j = the random release date of job j.

D_j = the random due date of job j.

w_j = the weight (or importance factor) of job j.

This notation is not completely analogous to the notation used for the deterministic models. The reason that X_{ij} is used as the processing time in stochastic scheduling is because P usually refers to a probability. The weight w_j, similar to that in the deterministic models, is basically equivalent to the cost of keeping job j in the system for one unit of time. In the queueing theory literature, which is closely related to stochastic scheduling, c_j is often used for the weight or cost of job j. The c_j and the w_j are equivalent.

9.2 DISTRIBUTIONS AND CLASSES OF DISTRIBUTIONS

Distributions and density functions may take many forms. In what follows, for obvious reasons, only distributions of non-negative random variables are considered.

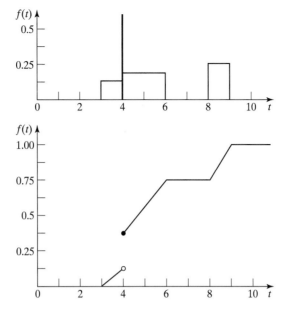

Figure 9.1   Example of density function and distribution function.

A density function may be continuous over given intervals and may have mass concentrated at given discrete points. This implies that the distribution function may not be differentiable everywhere (see Figure 9.1). In what follows, a distinction is made between continuous time distributions and discrete time distributions.

A random variable from a continuous time distribution may assume any real non-negative value within one or more intervals. The distribution function of a continuous time distribution is usually denoted by $F(t)$ and its density function by $f(t)$—that is,

$$F(t) = P(X \leq t) = \int_0^t f(t)\, dt,$$

where

$$f(t) = \frac{dF(t)}{dt}$$

provided the derivative exists. Furthermore,

$$\bar{F}(t) = 1 - F(t) = P(X \geq t).$$

An important example of a continuous time distribution is the *exponential* distribution. The density function of an exponentially distributed random variable

X is

$$f(t) = \lambda e^{-\lambda t},$$

and the corresponding distribution function is

$$F(t) = 1 - e^{-\lambda t},$$

which is equal to the probability that X is smaller than t (see Figure 9.2). The mean or expected value of X is

$$E(X) = \int_0^\infty t f(t) \, dt = \int_0^\infty t \, dF(t) = \frac{1}{\lambda}.$$

The parameter λ is called the *rate* of the exponential distribution.

A random variable from a discrete time distribution may assume only values on the non-negative integers—that is, $P(X = t) \geq 0$ for $t = 0, 1, 2, \ldots$ and $P(X = t) = 0$ otherwise. An important discrete time distribution is the *deterministic* distribution. A deterministic random variable assumes a given value with probability one.

Another important example of a discrete time distribution is the *geometric* distribution. The probability that a geometrically distributed random variable X assumes the value $t, t = 0, 1, 2, \ldots$, is

$$P(X = t) = (1 - q)q^t.$$

Its distribution function is

$$P(X \leq t) = \sum_{s=0}^{t}(1 - q)q^s = 1 - \sum_{s=t+1}^{\infty}(1 - q)q^s = 1 - q^{t+1},$$

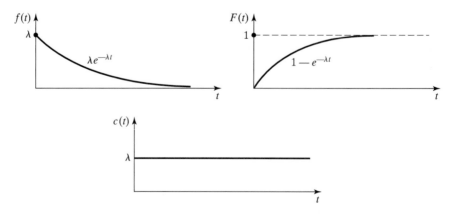

Figure 9.2   The exponential distribution.

and its mean is

$$E(X) = \frac{q}{1-q}.$$

The *completion rate* $c(t)$ of a continuous time random variable X with density function $f(t)$ and distribution function $F(t)$ is defined as follows:

$$c(t) = \frac{f(t)}{1 - F(t)}.$$

This completion rate is equivalent to the failure rate or hazard rate in reliability theory. For an exponentially distributed random variable, $c(t) = \lambda$ for all t. That the completion rate is independent of t is one of the reasons that the exponential distribution plays an important role in stochastic scheduling. This property is closely related to the *memoryless* property of the exponential distribution, which implies that the distribution of the *remaining* processing time of a job that already has been processed for an amount of time t is exponentially distributed with rate λ and therefore identical to its processing time distribution at the very start of its processing.

The completion rate of a discrete time random variable is defined as

$$c(t) = \frac{P(X = t)}{P(X \geq t)}.$$

The discrete time completion rate of the geometric distribution is

$$c(t) = \frac{P(X = t)}{P(X \geq t)} = \frac{(1-q)q^t}{q^t} = 1 - q, \quad t = 0, 1, 2, \ldots,$$

which is a constant independent of t. This implies that the probability a job is completed at t, given it has not been completed before t, is $1 - q$. So the geometric distribution has the memoryless property as well. The geometric distribution is, in effect, the discrete time counterpart of the exponential distribution.

Distributions, either discrete time or continuous time, can be classified based on the completion rate. An *Increasing Completion Rate (ICR)* distribution is defined as a distribution whose completion rate $c(t)$ is increasing in t, whereas a *Decreasing Completion Rate (DCR)* distribution is defined as a distribution whose completion rate is decreasing in t.

A subclass of the class of continuous time ICR distributions is the class of *Erlang(k, λ)* distributions. The *Erlang(k, λ)* distribution is defined as

$$F(t) = 1 - \sum_{r=0}^{k-1} \frac{(\lambda t)^r e^{-\lambda t}}{r!}.$$

The *Erlang*(k, λ) is a k-fold convolution of the same exponential distribution with rate λ. The mean of this *Erlang*(k, λ) distribution is k/λ. If k equals one, then the distribution is the exponential. If both k and λ go to ∞ while $k/\lambda = 1$, then the *Erlang*(k, λ) approaches the deterministic distribution with mean 1.

A subclass of the class of continuous time DCR distributions is the class of *mixtures of exponentials*. A random variable X is distributed according to a mixture of exponentials if it is exponentially distributed with rate λ_j with probability p_j, $j = 1, \ldots, n$, and

$$\sum_{j=1}^{n} p_j = 1.$$

The exponential as well as the deterministic distribution are special cases of ICR distributions. The exponential distribution is DCR as well as ICR. The class of DCR distributions contains other special distributions. For example, let X with probability p be exponentially distributed with mean $1/p$ and with probability $1-p$ be zero. The mean and variance of this distribution are $E(X) = 1$ and Var$(X) = 2/p - 1$. When p is very close to zero, this distribution is in what follows referred to as an *Extreme Mixture of Exponentials (EME)* distribution. Of course, similar distributions can be constructed for the discrete time case as well.

One way of measuring the variability of a distribution is through its coefficient of variation $C_v(X)$, which is defined as the square root of the variance (i.e., the standard deviation) divided by the mean—that is,

$$C_v(X) = \frac{\sqrt{\text{Var}(X)}}{E(X)} = \frac{\sqrt{E(X^2) - (E(X))^2}}{E(X)}.$$

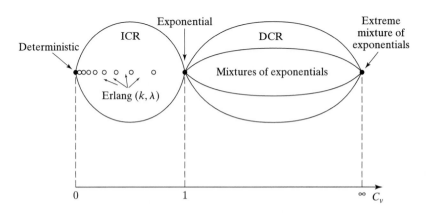

Figure 9.3   Class of distributions.

It can be verified easily that the $C_v(X)$ of the deterministic distribution is zero and the $C_v(X)$ of the exponential distribution is one (see Figure 9.3). The $C_v(X)$ of an extreme mixture of exponentials may be arbitrarily large (it goes to ∞ when p goes to 0). One may expect that the $C_v(X)$ of the geometric is 1, since the geometric is a discrete time counterpart of the exponential distribution. However, the $C_v(X)$ of the geometric, as defined earlier, is $1/\sqrt{q}$ (see Exercise 9.16).

9.3 STOCHASTIC DOMINANCE

It occurs often in stochastic scheduling that two random variables have to be compared with one another. There are many ways in which comparisons between random variables can be made. Comparisons are based on properties referred to as *stochastic dominance* (i.e., a random variable *dominates* another with respect to some stochastic property). All forms of stochastic dominance presented in this section apply to continuous random variables as well as discrete random variables. Only in a few cases are the discrete time and the continous time definitions presented separately. Most forms of stochastic dominance can also be applied in comparisons between a continuous random variable and a discrete random variable.

Definition 9.3.1. (Stochastic Dominance Based on Expectation)

(i) *The random variable X_1 is said to be larger in expectation than the random variable X_2 if $E(X_1) \geq E(X_2)$.*

(ii) *The random variable X_1 is said to be stochastically larger than the random variable X_2 if*

$$P(X_1 > t) \geq P(X_2 > t)$$

or

$$1 - F_1(t) \geq 1 - F_2(t)$$

for all t. This ordering is usually referred to as stochastic ordering and is denoted by $X_1 \geq_{st} X_2$.

(iii) *The continuous time random variable X_1 is larger than the continuous time random variable X_2 in the likelihood ratio sense if $f_1(t)/f_2(t)$ is nondecreasing in $t, t \geq 0$. The discrete time random variable X_1 is larger than the discrete time random variable X_2 in the likelihood ratio sense if $P(X_1 = t)/P(X_2 = t)$ is nondecreasing in $t, t = 0, 1, 2, \ldots$ This form of stochastic dominance is denoted by $X_1 \geq_{lr} X_2$.*

(iv) The random variable X_1 is almost surely larger than or equal to the random variable X_2 if $P(X_1 \geq X_2) = 1$. This ordering implies that the density functions f_1 and f_2 may overlap at most on one point and is denoted by $X_1 \geq_{a.s.} X_2$.

Ordering in expectation is the crudest form of stochastic dominance. Stochastic ordering implies ordering in expectation since

$$E(X_1) = \int_0^\infty t f_1(t)\, dt = \int_0^\infty (1 - F_1(t))\, dt = \int_0^\infty \bar{F}_1(t)\, dt$$

(see Exercise 9.11). It can easily be shown that likelihood ratio ordering implies stochastic ordering and that the reverse does not hold.

Example 9.3.2 (Stochastically Ordered Random Variables)

Consider two discrete time random variables X_1 and X_2. Both take values on 1, 2, and 3:

$$P(X_1 = 1) = \tfrac{1}{8}, \quad P(X_1 = 2) = \tfrac{3}{8}, \quad P(X_1 = 3) = \tfrac{1}{2};$$
$$P(X_2 = 1) = \tfrac{1}{8}, \quad P(X_2 = 2) = \tfrac{5}{8}, \quad P(X_2 = 3) = \tfrac{1}{4}.$$

Note that X_1 and X_2 are stochastically ordered but not likelihood ratio ordered as the ratio $P(X_1 = t)/P(X_2 = t)$ is not monotone.

It is also easy to find an example of a pair of random variables that are monotone likelihood ratio ordered, but not almost surely ordered.

Example 9.3.3 (Likelihood Ratio Ordered Random Variables)

Consider two exponential distributions with rates λ_1 and λ_2. These two distributions are likelihood ratio ordered as

$$\frac{f_1(t)}{f_2(t)} = \frac{\lambda_1 e^{-\lambda_1 t}}{\lambda_2 e^{-\lambda_2 t}} = \frac{\lambda_1}{\lambda_2} e^{-(\lambda_1 - \lambda_2)t},$$

which is monotone in t. Of course, the two exponentials are not almost surely ordered as their density functions overlap everywhere.

The four forms of stochastic dominance described previously all imply that the random variables being compared, in general, have different means. They lead to the following chain of implications.

almost surely larger $\Longrightarrow$ larger in likelihood ratio sense $\Longrightarrow$

stochastically larger $\Longrightarrow$ larger in expectation

There are several other important forms of stochastic dominance that are based on the *variability* of the random variables under the assumption that the *means are equal*. In the subsequent definitions, three such forms are presented. One of these is defined for density functions that are symmetric around the mean—that is,

$$f(E(X) + t) = f(E(X) - t)$$

for all $0 \leq t \leq E(X)$. Such a density function then has an upper bound of $2E(X)$. A Normal (Gaussian) density function with mean μ that is cut off at 0 and 2μ is a symmetric density function.

Definition 9.3.4. (Stochastic Dominance Based on Variance)

(i) *The random variable X_1 is said to be larger than the random variable X_2 in the variance sense if the variance of X_1 is larger than the variance of X_2.*

(ii) *The continuous random variable X_1 is said to be more variable than the continuous random variable X_2 if*

$$\int_0^\infty h(t)dF_1(t) \geq \int_0^\infty h(t)dF_2(t)$$

for all convex functions h. The discrete random variable X_1 is said to be more variable than the discrete random variable X_2 if

$$\sum_{t=0}^\infty h(t)P(X_1 = t) \geq \sum_{t=0}^\infty h(t)P(X_2 = t)$$

for all convex functions h. This ordering is denoted by $X_1 \geq_{cx} X_2$.

(iii) *The random variable X_1 is said to be symmetrically more variable than the random variable X_2 if the density functions $f_1(t)$ and $f_2(t)$ are symmetric around the same mean $1/\lambda$ and $F_1(t) \geq F_2(t)$ for $0 \leq t \leq 1/\lambda$ and $F_1(t) \leq F_2(t)$ for $1/\lambda \leq t \leq 2/\lambda$.*

Again, the first form of stochastic dominance is somewhat crude. However, any two random variables with equal means can be compared with one another in this way.

From the fact that the functions $h(t) = t$ and $h(t) = -t$ are convex, it follows that if X_1 is more variable than X_2, then $E(X_1) \geq E(X_2)$ and $E(X_1) \leq E(X_2)$. So $E(X_1)$ has to be equal to $E(X_2)$. From the fact that $h(t) = t^2$ is convex, it follows that $\text{Var}(X_1)$ is larger than $\text{Var}(X_2)$. Variability ordering is a partial ordering (i.e., not every pair of random variables with equal means can be ordered in this way). At times, variability ordering is also referred to as ordering *in the convex sense*.

It can be shown easily that symmetrically more variable implies more variable in the convex sense but not vice versa.

Example 9.3.5 (Random Variables That Are Variability Ordered)

Consider a deterministic random variable X_1 that always assumes the value $1/\lambda$ and an exponentially distributed random variable X_2 with mean $1/\lambda$. It can be easily verified that X_2 is more variable, but not symmetrically more variable than X_1.

Example 9.3.6 (Symmetric Random Variables That Are Variability Ordered)

Consider X_1 with a uniform distribution over the interval $[1, 3]$—that is, $f_1(t) = 0.5$ for $1 \leq t \leq 3$, and X_2 with a uniform distribution over the interval $[0, 4]$—that is, $f_2(t) = 0.25$ for $0 \leq t \leq 4$. It is easily verified that X_2 is symmetrically more variable than X_1.

The forms of stochastic dominance described in Definition 9.3.4 lead to the following chain of implications:

> symmetrically more variable $\Longrightarrow$ more variable $\Longrightarrow$ larger in variance

9.4 IMPACT OF RANDOMNESS ON FIXED SCHEDULES

The stochastic ordering ($\geq_{st}$) as well as the variability ordering ($\geq_{cx}$) described in the previous section are restricted versions of another form of dominance known as *increasing convex* ordering.

> **Definition 9.4.1. (Increasing Convex Ordering)** *A continuous time random variable X_1 is said to be larger than a continuous time random variable X_2 in the increasing convex sense if*
>
> $$\int_0^\infty h(t)dF_1(t) \geq \int_0^\infty h(t)dF_2(t)$$
>
> *for all increasing convex functions h. The discrete time random variable X_1 is said to be larger than the discrete time random variable X_2 in the increasing convex sense if*
>
> $$\sum_{t=0}^\infty h(t)P(X_1 = t) \geq \sum_{t=0}^\infty h(t)P(X_2 = t)$$
>
> *for all increasing convex functions h. This ordering is denoted by $X_1 \geq_{icx} X_2$.*

Clearly, $E(X_1)$ is larger than, but not necessarily equal to, $E(X_2)$, and Var(X_1) is not necessarily larger than Var(X_2). However, if $E(X_1) = E(X_2)$, then indeed

$\text{Var}(X_1) \geq \text{Var}(X_2)$. From Definition 9.4.1, it immediately follows that

> stochastically larger $\Longrightarrow$ larger in the increasing convex sense
>
> more variable $\Longrightarrow$ larger in the increasing convex sense

To see the importance of this form of stochastic dominance, consider two vectors of independent random variables—namely, $X_1^{(1)}, \ldots, X_n^{(1)}$ and $X_1^{(2)}, \ldots, X_n^{(2)}$. All $2n$ random variables are independent. Let

$$Z_1 = g(X_1^{(1)}, \ldots, X_n^{(1)})$$

and

$$Z_2 = g(X_1^{(2)}, \ldots, X_n^{(2)}),$$

where the function g is increasing convex in each one of the n arguments.

Lemma 9.4.2. *If $X_j^{(1)} \geq_{icx} X_j^{(2)}$, $j = 1, \ldots, n$, then $Z_1 \geq_{icx} Z_2$.*

Proof.   The proof is by induction on n. When $n = 1$ it has to be shown that

$$E\left(h(g(X_1^{(1)}))\right) \geq E\left(h(g(X_1^{(2)}))\right),$$

with both g and h increasing convex and $X_1^{(1)} \geq_{icx} X_1^{(2)}$. This follows from the definition of variability ordering as the function $h(g(t))$ is increasing and convex in t since

$$\frac{d}{dt}h(g(t)) = h'(g(t))g'(t) \geq 0$$

and

$$\frac{d^2}{dt^2}h(g(t)) = h''(g(t))(g'(t))^2 + h'(g(t))g''(t) \geq 0.$$

Assume as induction hypothesis that the lemma holds for vectors of size $n-1$. Now

$$E\left(h(g(X_1^{(1)}, X_2^{(1)}, \ldots, X_n^{(1)})) \mid X_1^{(1)} = t\right)$$

$$= E\left(h(g(t, X_2^{(1)}, \ldots, X_n^{(1)})) \mid X_1^{(1)} = t\right)$$

$$= E\left(h(g(t, X_2^{(1)}, \ldots, X_n^{(1)}))\right)$$

$$\geq E\left(h(g(t, X_2^{(2)}, \ldots, X_n^{(2)}))\right)$$

$$= E\left(h(g(t, X_2^{(2)}, \ldots, X_n^{(2)})) \mid X_1^{(1)} = t\right).$$

Taking expectations yields

$$E\left(h(g(X_1^{(1)}, X_2^{(1)}, \ldots, X_n^{(1)}))\right) \geq E\left(h(g(X_1^{(1)}, X_2^{(2)}, \ldots, X_n^{(2)}))\right).$$

Conditioning on $X_2^{(2)}, \ldots, X_n^{(2)}$ and using the result for $n = 1$ shows that

$$E\left(h(g(X_1^{(1)}, X_2^{(2)}, \ldots, X_n^{(2)}))\right) \geq E\left(h(g(X_1^{(2)}, X_2^{(2)}, \ldots, X_n^{(2)}))\right),$$

which completes the proof.   □

The following two examples illustrate the significance of the previous lemma.

Example 9.4.3 (Stochastic Comparison of Makespans)

Consider two scenarios, each with two machines in parallel and two jobs. The makespan in the first scenario is

$$C_{\max}^{(1)} = \max\left(X_1^{(1)}, X_2^{(1)}\right),$$

and the makespan in the second scenario is

$$C_{\max}^{(2)} = \max\left(X_1^{(2)}, X_2^{(2)}\right).$$

The "max" function is increasing convex in both arguments. From Lemma 9.4.2, it immediately follows that if $X_j^{(1)} \geq_{cx} X_j^{(2)}$, then $X_j^{(1)} \geq_{icx} X_j^{(2)}$ and therefore $C_{\max}^{(1)} \geq_{icx} C_{\max}^{(2)}$. This implies $E(C_{\max}^{(1)}) \geq E(C_{\max}^{(2)})$.

Example 9.4.4 (Stochastic Comparison of Total Completion Times)

Consider the problem $1 \mid\mid \sum h(C_j)$ with the function h increasing convex. Consider two scenarios, each one with a single machine and n jobs. The processing time of job j in the first (second) scenario is $X_j^{(1)}$ ($X_j^{(2)}$). Assume that in both cases the jobs are scheduled in the sequence $1, 2, \ldots, n$. The objective function in scenario $i, i = 1, 2$, is therefore

$$\sum_{j=1}^{n} h\left(C_j^{(i)}\right) = h\left(X_1^{(i)}\right) + h\left(X_1^{(i)} + X_2^{(i)}\right) + \cdots + h\left(X_1^{(i)} + X_2^{(i)} + \cdots + X_n^{(i)}\right).$$

The objective function is increasing convex in each one of the n arguments. From Lemma 9.4.2, it follows that if $X_j^{(1)} \geq_{cx} X_j^{(2)}$, then $X_j^{(1)} \geq_{icx} X_j^{(2)}$ and therefore $E(\sum_{j=1}^{n} h(C_j^{(1)})) \geq E(\sum_{j=1}^{n} h(C_j^{(2)}))$. Note that if the function h is linear, the values of the two objectives are equal (since $E(X_j^{(1)}) = E(X_j^{(2)})$).

Lemma 9.4.2 turns out to be very useful for determining bounds on performance measures of given schedules when the processing time distributions satisfy certain properties. In the next lemma, four distributions—F_1, F_2, F_3, and F_4—are considered, all with mean 1.

Lemma 9.4.5.   *If F_1 is deterministic, F_2 is ICR, F_3 is exponential and F_4 is DCR, then*

$$F_1 \leq_{cx} F_2 \leq_{cx} F_3 \leq_{cx} F_4.$$

Proof.   For a proof of this result, the reader is referred to Barlow and Proschan (1975). The result shown by Barlow and Proschan is actually more general than the result stated here: The distributions F_2 and F_4 do not necessarily have to be ICR and DCR, respectively. These distributions may belong to larger classes of distributions.   □

The result in Lemma 9.4.5 can also be extended in another direction. It can be shown that for any DCR distribution with mean 1, there exists an Extreme Mixture of Exponentials (EME) distribution with mean 1 that is more variable.

These orderings make it easy to obtain upper and lower bounds on performance measures when processing times are either all ICR or all DCR.

Example 9.4.6 (Bounds on the Expected Makespan)

Consider the scenario of two machines in parallel and two jobs (see Example 9.4.3). Suppose X_1 and X_2 are independent and identically distributed (i.i.d.) according to F with mean 1. If F is deterministic, then the makespan is 1. If F is exponential, then the makespan is $3/2$. If F is an EME distribution as defined in Section 9.1, then the makespan is $2 - (p/2)$ (i.e., if p goes to 0, the makespan goes to 2). Combining the conclusion of Example 9.4.3 with Lemma 9.4.5 results, when F is ICR, in the inequalities

$$1 \leq E(C_{\max}) \leq \frac{3}{2},$$

and, when F is DCR, in the inequalities

$$\frac{3}{2} \leq E(C_{\max}) < 2.$$

It is easy to see that the makespan never can be larger than 2. If both jobs are processed on the same machine one after another, the expected makespan is equal to 2.

9.5 CLASSES OF POLICIES

In stochastic scheduling, certain conventions have to be made that are not needed in deterministic scheduling. During the evolution of a stochastic process, new information becomes available continuously. Job completions and occurrences of random release dates and due dates represent additional information that the decision maker may wish to take into account when scheduling the remaining part of the process. The amount of freedom the decision maker has in using this additional information is the basis for the various classes of decision-making policies. In this section, four classes of policies are defined.

The first class of policies is, in what follows, only used in scenarios where all the jobs are available for processing at time zero; the machine environments considered are the single machine, parallel machines, and permutation flow shops.

Definition 9.5.1. (Nonpreemptive Static List Policy) *Under a nonpreemptive static list policy, the decision maker orders the jobs at time zero according to a priority list. This priority list does not change during the evolution of the process, and every time a machine is freed the next job on the list is selected for processing.*

Under this class of policies, the decision maker puts at time zero the n jobs in a list (permutation) and the list does not change during the evolution of the process. In the case of machines in parallel, every time a machine is freed, the job at the top of the list is selected as the next one for processing. In the case of a permutation flow shop, the jobs are also put in a list in front of the first machine at time zero; every time the first machine is freed the next job on the list is scheduled for processing. This class of nonpreemptive static list policies is also referred to as the class of *permutation* policies. This class of policies is in a sense similar to the static priority rules usually considered in deterministic models.

Example 9.5.2 (Application of a Nonpreemptive Static List Policy)

Consider a single machine and three jobs. All three jobs are available at time zero. All three jobs have the same processing time distributions, which is 2 with probability .5 and 8 with probability .5. The due date distributions are the same, too. The due date is 1 with probability .5 and 5 with probability .5. If a job is completed at the same time as its due date, it is considered to be on time. It would be of interest to know the expected number of jobs completed in time under a permutation policy.

Under a permutation policy, the first job is completed in time with probability .25 (its processing time has to be 2 and its due date has to be 5); the second job is completed in time with probability .125 (the processing times of the first and second job have to be 2 and the due date of the second job has to be 5); the third job will never be completed in time. The expected number of on-time completions is therefore .375, and the expected number of tardy jobs is $3 - 0.375 = 2.625$.

The second class of policies is a preemptive version of the first class and is only used in scenarios where jobs are released at *different* points in time.

Definition 9.5.3. (Preemptive Static List Policy)   *Under a preemptive static list policy, the decision maker orders the jobs at time zero according to a priority list. This list includes jobs with nonzero release dates (i.e., jobs that are to be released later). This priority list does not change during the evolution of the process, and at any point in time the job at the top of the list of available jobs is the one to be processed on the machine.*

Under this class of policies, the following may occur. When there is a job release at some point in time and the job released is higher on the static list than the job currently being processed, then the job being processed is preempted and the job released is put on the machine.

Under the third and fourth class of policies, the decision maker is allowed to make his decisions during the evolution of the process. That is, every time he makes a decision, he may take all information available at that time into account. The third class of policies does not allow preemptions.

Definition 9.5.4. (Nonpreemptive Dynamic Policy)   *Under a nonpreemptive dynamic policy, every time a machine is freed, the decision maker is allowed to determine which job goes next. His decision at such a time point may depend on all the information available (e.g., the current time, the jobs waiting for processing, the jobs currently being processed on other machines, and the amount of processing these jobs have already received on these machines). However, the decision maker is not allowed to preempt; once a job begins processing, it has to be completed without interruption.*

Example 9.5.5 (Application of a Nonpreemptive Dynamic Policy)

Consider the same problem as in Example 9.5.2. It is of interest to know the expected number of jobs completed in time under a nonpreemptive dynamic policy. Under a nonpreemptive dynamic policy, the probability the first job is completed in time is again .25. With probability .5, the first job is completed at time 2. With probability .25, the due dates of both remaining jobs already occurred at time 1 and there will be no more on-time completions. With probability .75, at least one of the remaining two jobs has a due date at time 5. The probability that the second job put on the machine is completed in time is 3/16 (the probability that the first job has completion time 2 times the probability at least one of the two remaining jobs has due date 5 times the probability that the second job has processing time 2). The expected number of on-time completions is therefore .4375 and the expected number of tardy jobs is 2.5625.

The last class of policies is a preemptive version of the third class.

Definition 9.5.6. (Preemptive Dynamic Policy)) *Under a preemptive dynamic policy, the decision maker may decide at any point in time which jobs should be processed on the machines. His decision at any given point may take into account all information available at that point and may require preemptions.*

Example 9.5.7 (Application of a Preemptive Dynamic Policy)

Consider again the problem of Example 9.5.2. It is of interest to know the expected number of jobs completed in time under a preemptive dynamic policy. Under a preemptive dynamic policy, the probability that the first job is completed in time is again .25. This first job is either taken off the machine at time 1 (with probability .5) or at time 2 (with probability .5). The probability the second job put on the machine is completed in time is $3/8$ since the second job enters the machine either at time 1 or at time 2 and the probability of being completed on time is 0.75 times the probability it has processing time 2, which equals $3/8$ (regardless of when the first job was taken off the machine). However, unlike under the nonpreeemptive dynamic policy, the second job put on the machine is taken off with probability .5 at time 3 and with probability 0.5 at time 4. So there is actually a chance that the third job that goes on the machine will be completed in time. The probability the third job is completed in time is $1/16$ (the probability that the due date of the first job is 1 ($= .5$) times the probability that the due dates of both remaining jobs are 5 ($= .25$) times the probability that the processing time of the third job is 2 ($= .5$)). The total expected number of on-time completions is therefore $11/16 = 0.6875$ and the expected number of tardy jobs is 2.3125.

It is clear that the optimal preemptive dynamic policy leads to the best possible value of the objective as in this class of policies the decision maker has the most information available and the largest amount of freedom. No general statement can be made with regard to a comparison between the optimal preemptive static list policy and the optimal nonpreemptive dynamic policy. The static list policy has the advantage that preemptions are allowed, whereas the nonpreemptive dynamic policy has the advantage that all current information can be taken into account during the process. However, if all jobs are present at time zero and the environment is either a bank of machines in parallel or a permutation flow shop, then the optimal nonpreemptive dynamic policy is at least as good as the optimal nonpreemptive static list policy (see Examples 9.5.2 and 9.5.5).

There are several forms of minimization in stochastic scheduling. Whenever an objective function has to be minimized, it should be specified in what *sense* the objective has to be minimized. The crudest form of optimization is in the *expectation* sense (e.g., one wishes to minimize the *expected* makespan—that is, $E(C_{\max})$—and find a policy under which the expected makespan is smaller than the expected makespan under any other policy). A stronger form of optimization is optimization in the *stochastic* sense. If a schedule or policy minimizes $C_{\max}$

stochastically, the makespan under the optimal schedule or policy is *stochastically* less than the makespan under any other schedule or policy. Stochastic minimization, of course, implies minimization in expectation. In the subsequent chapters, the objective is usually minimized in expectation. Frequently, however, the policies that minimize the objective in expectation minimize the objective stochastically as well.

EXERCISES (COMPUTATIONAL)

9.1. Determine the completion rate of the discrete Uniform distribution with $P(X = i) = 0.1$, for $i = 1, \dots, 10$.

9.2. Determine the completion rate of the continuous Uniform distribution, with density function $f(t) = 0.1$, for $0 \leq t \leq 10$, and 0 otherwise.

9.3. Consider two discrete time random variables X_1 and X_2. Both take values on 1, 2, and 3:

$$P(X_1 = 1) = .2, \quad P(X_1 = 2) = .3, \quad P(X_1 = 3) = .5;$$

$$P(X_2 = 1) = .2, \quad P(X_2 = 2) = .6, \quad P(X_2 = 3) = .2.$$

(a) Are X_1 and X_2 likelihood ratio ordered?
(b) Are X_1 and X_2 stochastically ordered?

9.4. Consider two discrete time random variables X_1 and X_2. Both take values on 1, 2, 3, and 4:

$$P(X_1 = 1) = .125, \quad P(X_1 = 2) = .375, \quad P(X_1 = 3) = .375, \quad P(X_1 = 4) = .125;$$

$$P(X_2 = 1) = .150, \quad P(X_2 = 2) = .400, \quad P(X_2 = 3) = .250, \quad P(X_2 = 4) = .200.$$

(a) Are X_1 and X_2 symmetric variability ordered?
(b) Are X_1 and X_2 variability ordered?

9.5. Consider three jobs on two machines. The processing times of the three jobs are independent and identically distributed (i.i.d.) according to the discrete distribution $P(X_j = 0.5) = P(X_j = 1.5) = 0.5$.
(a) Compute the expected makespan.
(b) Compute the total expected completon time (*Hint:* Note that the sum of the expected completion times is equal to the sum of the expected starting times plus the sum of the expected processing times.)
(c) Compare the results of (a) and (b) with the results in case all three jobs have deterministic processing times equal to 1.

9.6. Do the same as in the previous exercise, but now with two machines and four jobs.

9.7. Consider a flow shop of two machines with unlimited intermediate storage. There are two jobs and the four processing times are i.i.d. according to the discrete distribution

$$P(X_j = 0.5) = P(X_j = 1.5) = 0.5.$$

Compute the expected makespan and the sum of the expected completion times.

9.8. Consider a single machine and three jobs. The three jobs have i.i.d. processing times. The distribution is exponential with mean 1. The due dates of the three jobs are also i.i.d. exponential with mean 1. The objective is to minimize the expected number of tardy jobs—that is, $E(\sum_{j=1}^{n} U_j)$. Consider a nonpreemptive static list policy.
  (a) Compute the probability of the first job being completed on time.
  (b) Compute the probability of the second job being completed on time.
  (c) Compute the probability of the third job being completed on time.

9.9. Do the same as in the previous exercise but now for a nonpreemptive dynamic policy.

9.10. Do the same as in Exercise 9.8 but now for a preemptive dynamic policy.

EXERCISES (THEORETICAL)

9.11. Show that

$$E(X) = \int_0^\infty \bar{F}(t)\,dt$$

(recall that $\bar{F}(t) = 1 - F(t)$).

9.12. Show that

$$E(\min(X_1, X_2)) = \int_0^\infty \bar{F}_1(t)\bar{F}_2(t)\,dt,$$

when X_1 and X_2 are independent.

9.13. Show that

$$E(\min(X_1, X_2)) = \frac{1}{\lambda + \mu},$$

when X_1 is exponentially distributed with rate λ and X_2 exponentially distributed with rate μ.

9.14. Show that

$$E(\max(X_1, X_2)) = E(X_1) + E(X_2) - E(\min(X_1, X_2)),$$

with X_1 and X_2 arbitrarily distributed.

9.15. Compute the coefficient of variation of the *Erlang(k, λ)* distribution (recall that the *Erlang(k, λ)* distribution is a convolution of k i.i.d. exponential random variables with rate λ).

9.16. Consider the following variant of the geometric distribution:

$$P(X = t + a) = (1 - q)q^t, \quad t = 0, 1, 2, \ldots,$$

where $a = (\sqrt{q} - q)/(1 - q)$. Show that the coefficient of variation $C_v(X)$ of this "shifted" geometric is equal to 1.

9.17. Consider the following partial ordering between random variables X_1 and X_2. The random variable X_1 is said to be smaller than the random variable X_2 *in the com-*

pletion rate sense if at time t the completion rate of X_1, say $\lambda_1(t)$, is larger than or equal to the completion rate of X_2, say $\lambda_2(t)$, for any t.

(a) Show that this ordering is equivalent to the ratio $(1 - F_1(t))/(1 - F_2(t))$ being monotone decreasing in t.

(b) Show that monotone likelihood ratio ordering $\Rightarrow$ completion rate ordering $\Rightarrow$ stochastic ordering.

9.18. Consider m machines in parallel and n jobs with i.i.d. processing times from distribution F with mean 1. Show that

$$\frac{n}{m} \le E(C_{\max}) < n.$$

Are there distributions for which these bounds are attained?

9.19. Consider a permutation flow shop with m machines in series and n jobs. The processing time of job j on machine i is X_{ij}, distributed according to F with mean 1. Show that

$$n + m - 1 \le E(C_{\max}) \le mn.$$

Are there distributions for which these bounds are attained?

9.20. Consider a single machine and n jobs. The processing time of job j is X_j with mean $E(X_j)$ and variance $\text{Var}(X_j)$. Find the schedule that minimizes $E(\sum C_j)$ and the schedule that minimizes $\text{Var}(\sum C_j)$. Prove your results.

9.21. Assume $X_1 \ge_{st} X_2$. Show through a counterexample that

$$Z_1 = 2X_1 + X_2 \ge_{st} 2X_2 + X_1 = Z_2$$

is not necessarily true.

COMMENTS AND REFERENCES

For a general overview of stochastic scheduling problems, see Dempster, Lenstra and Rinnooy Kan (1982), Möhring and Radermacher (1985b), and Righter (1994).

For an easy to read and rather comprehensive treatment of distributions and classes of distributions based on completion (failure) rates, see Barlow and Proschan (1975), (Chapter 3).

For a lucid and comprehensive treatment of the several forms of stochastic dominance, see Ross (1983a). A definition of the form of stochastic dominance based on symmetric variability appears in Pinedo (1982). For a scheduling application of monotone likelihood ratio ordering, see Brown and Solomon (1973). For a scheduling application of completion rate ordering (Exercise 9.17), see Pinedo and Ross (1980). For an overview of the different forms of stochastic dominance and their importance with respect to scheduling, see Chang and Yao (1993) and Righter (1994).

For the impact of randomness on fixed schedules, see Pinedo and Weber (1984), Pinedo and Schechner (1985), Pinedo and Wie (1986), Shanthikumar and Yao (1991), and Chang and Yao (1993).

Many classes of scheduling policies have been introduced over the years; see, for example, Glazebrook (1981a, 1981b, 1982), Pinedo (1983), and Möhring, Radermacher and Weiss (1984, 1985).

10

Single Machine Models (Stochastic)

Stochastic models, especially with exponential processing times, often contain more structure than their deterministic counterparts and may lead to results that, at first sight, seem surprising. Models that are NP-hard in a deterministic setting often allow a simple priority policy to be optimal in a stochastic setting.

In this chapter, single machine models with arbitrary processing times in a nonpreemptive setting are discussed first. Then the preeemptive cases are considered, followed by models where the processing times are likelihood ratio ordered. Finally, models with exponentially distributed processing times are analyzed.

10.1 ARBITRARY DISTRIBUTIONS WITHOUT PREEMPTIONS

For a number of stochastic problems, finding the optimal policy is equivalent to solving a deterministic scheduling problem. Usually when such an equivalence relationship exists, the deterministic counterpart can be obtained by replacing all random variables with their means. The optimal schedule for the deterministic problem then minimizes the objective of the stochastic version in expectation.

One such case is when the objective in the deterministic counterpart is linear in $p_{(j)}$ and $w_{(j)}$, where $p_{(j)}$ and $w_{(j)}$ denote the processing time and weight of the job in the j-th position in the sequence.

This observation implies that it is easy to find the optimal permutation schedule for the stochastic counterpart of $1 \ || \ \sum w_j C_j$, when the processing time of job j is X_j, from an arbitrary distribution F_j, and the objective is $E(\sum w_j C_j)$. This problem leads to the stochastic version of the WSPT rule, which sequences the jobs in decreasing order of the ratio $w_j / E(X_j)$ or $\lambda_j w_j$. In what follows, this rule is referred to either as the *Weighted Shortest Expected Processing Time first (WSEPT)* rule or as the "λw" rule.

Theorem 10.1.1. *The WSEPT rule minimizes the expected sum of the weighted completion times in the class of nonpreemptive static list policies as well as in the class of nonpreemptive dynamic policies.*

Proof.   The proof for nonpreemptive static list policies is similar to the proof for the deterministic counterpart of this problem. The proof is based on an adjacent pairwise interchange argument identical to the one used in the proof of Theorem 3.1.1. The only difference is that the p_js in that proof have to be replaced by $E(X_j)$s.

The proof for nonpreemptive dynamic policies needs an additional argument. It is easy to show that it is true for $n = 2$ (again an adjacent pairwise interchange argument). Now consider three jobs. It is clear that the last two jobs have to be sequenced according to the λw rule. These last two jobs will be sequenced in this order independent of what happens during the processing of the first job. There are then three sequences that may occur: each of the three jobs starting first and the remaining two jobs sequenced according to the λw rule. A simple interchange argument between the first job and the second shows that all three jobs have to sequenced according to the λw rule. It can be shown by induction that all n jobs have to be sequenced according to the λw rule in the class of nonpreemptive dynamic policies: Suppose it is true for $n - 1$ jobs. If there are n jobs, it follows from the induction hypothesis that the last $n - 1$ jobs have to be sequenced according to the λw rule. Suppose the first job is not the job with the highest $\lambda_j w_j$. Interchanging this job with the second job in the sequence (i.e., the job with the highest $\lambda_j w_j$) results in a decrease in the expected value of the objective function. This completes the proof of the theorem. $\square$

It can be shown that the nonpreemptive WSEPT rule is also optimal in the class of *preemptive* dynamic policies when all n processing time distributions are ICR.

This follows from the fact that at any time when a preemption is contemplated, the λw ratio of the job currently on the machine is actually higher than when it was put on the machine (the expected remaining processing time of an ICR job decreases as processing goes on). If the ratio of the job was the highest among the remaining jobs when it was put on the machine, it remains the highest while it is being processed.

The same cannot be said about jobs with DCR distributions. The expected remaining processing time *increases* while a job is being processed. So the weight divided by the expected remaining processing time of a job, while it is being processed, *decreases* over time. Preemptions may therefore be advantageous when processing times are DCR.

Example 10.1.2 (Optimal Policy with Random Variables That Are DCR)

Consider n jobs with the processing time X_j distributed as follows. The processing time X_j is 0 with probability p_j and it is distributed according to an exponential with rate λ_j with probability $1 - p_j$. Clearly, this distribution is DCR as it is a mixture of two exponentials with rates ∞ and λ_j. The objective to be minimized is the expected sum of the weighted completion times. The optimal preemptive dynamic policy is clear. All n jobs have to be tried out for a split second at time zero to determine which jobs have zero processing times. If a job does not have zero processing time, it is taken immediately off the machine. After having determined in this way which jobs have nonzero processing times, these remaining jobs are sequenced in decreasing order of $\lambda_j w_j$.

Consider the following generalization of the stochastic counterpart of $1 \mid\mid \sum w_j C_j$ described earlier. The machine is subject to breakdowns. The "up" times (i.e., the times that the machine is in operation) are exponentially distributed with rate v. The "down" times are independent and identically distributed (i.i.d.) according to distribution G with mean $1/\mu$. It can be shown that the λw rule is still optimal even in this case. Actually, it can be shown that this stochastic problem with breakdowns is equivalent to a similar deterministic problem without breakdowns. The processing time of job j in the equivalent deterministic problem is determined as follows. Let X_j denote the original random processing time of job j when there are no breakdowns and let Y_j denote the time job j occupies the machine, including the time that the machine is not in operation. The following relationship can be shown easily (see Exercise 10.11):

$$E(Y_j) = E(X_j)\left(1 + \frac{v}{\mu}\right).$$

This relationship holds because of the exponential uptimes of the machines and the fact that all the breakdowns have the same mean. So even with the breakdown process described, the problem is still equivalent to the deterministic problem $1 \mid\mid \sum w_j C_j$ without breakdowns.

The equivalences between the single machine stochastic models and their deterministic counterparts go even further. Consider the stochastic counterpart of $1 \mid chains \mid \sum w_j C_j$. If in the stochastic counterpart the jobs are subject to precedence constraints that take the form of chains, then Algorithm 3.1.4 can be used for minimizing the expected sum of the weighted completion times (in the definition of the ρ-factor the p_j is again replaced by the $E(X_j)$).

Consider now the stochastic version of $1 \mid\mid \sum w_j(1 - e^{-rC_j})$ with arbitrarily distributed processing times. This problem leads to the stochastic version of the WDSPT rule, which sequences the jobs in decreasing order of the ratio

$$\frac{w_j E(e^{-rX_j})}{1 - E(e^{-rX_j})}.$$

This rule is referred to as the *Weighted Discounted Shortest Expected Processing Time first (WDSEPT)* rule. This rule is, in a sense, a generalization of a number of rules considered before (see Figure 10.1).

Theorem 10.1.3. *The WDSEPT rule minimizes the expected weighted sum of the discounted completion times in the class of nonpreemptive static list policies as well as in the class of nonpreemptive dynamic policies.*

Proof. The optimality of this rule can be shown again through a straightforward pairwise interchange argument as in the proof of Theorem 3.1.6. The $w_j e^{-r(t+p_j)}$ is replaced by the $w_j E(e^{-r(t+X_j)})$. Optimality in the class of nonpreemptive dynamic policies follows from an induction argument similar to the one presented in Theorem 10.1.1. □

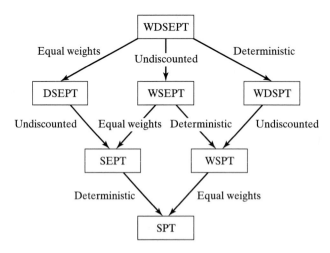

Figure 10.1 Hierarchy of scheduling rules.

Example 10.1.4 (Application of WDSEPT Rule)

Consider two jobs with equal weights—say 1. The processing time distribution of the first job is a continuous time uniform distribution over the interval $[0, 2]$—that is, $f_1(t) = .5$ for $0 \leq t \leq 2$. The processing time distribution of the second job is a discrete time uniform distribution with 0, 1, and 2 as possible outcomes—that is,

$$P(X_2 = 0) = P(X_2 = 1) = P(X_2 = 2) = \tfrac{1}{3}.$$

Clearly, $E(X_1) = E(X_2) = 1$. The discount factor r is $1/10$. To determine the priority indexes of the two jobs, $E(e^{-rX_1})$ and $E(e^{-rX_2})$ have to be computed:

$$E(e^{-rX_1}) = \int_0^\infty e^{-rt} f_1(t)dt = \int_0^2 \frac{1}{2} e^{-0.1t}\, dt = .9063$$

and

$$E(e^{-rX_2}) = \sum_{t=0}^\infty e^{-rt} P(X_2 = t) = \frac{1}{3} + \frac{1}{3}e^{-0.1} + \frac{1}{3}e^{-0.2} = .9078.$$

The priority index of job 1 is therefore 9.678 and the priority index of job 2 is 9.852. This implies that job 2 has to be processed first and job 1 second. If the discount factor would have been zero, any one of the two sequences would have been optimal. Observe that $\text{Var}(X_1) = 1/3$ and $\text{Var}(X_2) = 2/3$. So in this case, it is optimal to process the job with the larger variance first.

In the prior theorem, the optimality of the WDSEPT rule is shown for the class of nonpreemptive static list policies as well as for the class of nonpreemptive dynamic policies. Precedence constraints can be handled in the same way as they are handled for the deterministic counterpart (see Exercise 3.22). The model considered in Theorem 10.1.3, without precedence constraints, is considered again in the next section in an environment that allows preemptions.

The remaining part of this section focuses on due date-related problems. Consider the stochastic counterpart of $1 \parallel L_{\max}$ with processing times that have arbitrary distributions and deterministic due dates. The objective is to minimize the expected maximum lateness.

Theorem 10.1.5.   *The EDD rule minimizes expected maximum lateness for arbitrarily distributed processing times and deterministic due dates in the class of nonpreemptive static list policies, the class of nonpreemptive dynamic policies, and the class of preemptive dynamic policies.*

Proof.   It is clear that the EDD rule minimizes the maximum lateness for any realization of processing times (after conditioning on the processing times, the problem is basically a deterministic problem and Algorithm 3.2.1 applies). If the EDD rule minimizes the maximum lateness for any realization of processing times, it minimizes the maximum lateness of course also in expectation (it actually minimizes the maximum lateness almost surely).   □

Example 10.1.6 (Application of EDD Rule)

Consider two jobs with deterministic due dates. The processing time distributions of the jobs are discrete:

$$P(X_1 = 1) = P(X_1 = 2) = P(X_1 = 4) = \tfrac{1}{3}$$

and

$$P(X_2 = 2) = P(X_2 = 4) = \tfrac{1}{2}.$$

The due date of the first job is $D_1 = 1$ and the due date of the second job is $D_2 = 4$. Now

$$
\begin{aligned}
E(\max(L_1, L_2)) = {} & E\left(\max(L_1, L_2) \mid X_1 = 1, X_2 = 2\right) \times P(X_1 = 1, X_2 = 2) \\
& + E\left(\max(L_1, L_2) \mid X_1 = 1, X_2 = 4\right) \times P(X_1 = 1, X_2 = 4) \\
& + E\left(\max(L_1, L_2) \mid X_1 = 2, X_2 = 2\right) \times P(X_1 = 2, X_2 = 2) \\
& + E\left(\max(L_1, L_2) \mid X_1 = 2, X_2 = 4\right) \times P(X_1 = 2, X_2 = 4) \\
& + E\left(\max(L_1, L_2) \mid X_1 = 4, X_2 = 2\right) \times P(X_1 = 4, X_2 = 2) \\
& + E\left(\max(L_1, L_2) \mid X_1 = 4, X_2 = 4\right) \times P(X_1 = 4, X_2 = 4) \\
= {} & (0 + 1 + 1 + 2 + 3 + 4)\tfrac{1}{6} \\
= {} & \tfrac{11}{6}.
\end{aligned}
$$

It can easily be verified that scheduling job 2 first and job 1 second results in a larger $E(\max(L_1, L_2))$.

Note, however, that for any given sequence, $E(L_{\max}) = E(\max(L_1, L_2))$ does not necessarily have to be equal to $\max(E(L_1), E(L_2))$. Under sequence $1, 2$,

$$E(L_1) = 0 \times \tfrac{1}{3} + 1 \times \tfrac{1}{3} + 3 \times \tfrac{1}{3} = \tfrac{4}{3},$$

whereas

$$E(L_2) = \tfrac{1}{3}\left(\tfrac{1}{2} \times 0 + \tfrac{1}{2} \times 1\right) + \tfrac{1}{3}\left(\tfrac{1}{2} \times 0 + \tfrac{1}{2} \times 2\right) + \tfrac{1}{3}\left(\tfrac{1}{2} \times 2 + \tfrac{1}{2} \times 4\right) = \tfrac{3}{2}.$$

So $\max(E(L_1), E(L_2)) = 3/2$, which is smaller than $E(\max(L_1, L_2))$.

It can be shown that the EDD rule not only minimizes

$$E(L_{\max}) = E(\max(L_1, \dots , L_n)),$$

but also $\max(E(L_1), \dots , E(L_n))$.

It is even possible to develop an algorithm for a stochastic counterpart of the more general $1 \mid prec \mid h_{\max}$ problem considered in Chapter 3. In this problem, the objective is to minimize the maximum of the n expected costs incurred by the n jobs—that is, the objective is to minimize

$$\max \left(E(h_1(C_1)), \dots , E(h_n(C_n)) \right),$$

where $h_j(C_j)$ is the cost incurred by job j being completed at C_j. The cost function h_j is nondecreasing in the completion time C_j. The algorithm is a modified version of Algorithm 3.2.1. The version here is also a backward procedure. Whenever one has to select a schedulable job for processing, it is clear that the distribution of its completion time is the convolution of the processing times of the jobs that have not yet been scheduled. Let f_{J^c} denote the density function of the convolution of the processing times of the set of unscheduled jobs J^c. Job j^* is then selected to be processed last among the set of jobs J^c if

$$\int_0^\infty h_{j^*}(t) f_{J^c}(t)\, dt = \min_{j \in J^c} \int_0^\infty h_j(t) f_{J^c}(t)\, dt.$$

The L.H.S. denotes the expected value of the penalty for job j^* if it is the last job to be scheduled among the jobs in J^c. This rule replaces the first part of Step 2 in Algorithm 3.2.1. The proof of optimality is similar to the proof of optimality in the deterministic case. However, implementation of the algorithm may be significantly more cumbersome because of the evaluation of the integrals.

Example 10.1.7 (Minimizing Maximum Expected Cost)

Consider three jobs with random processing times X_1, X_2, and X_3 from distributions F_1, F_2, and F_3. Particulars of the processing times and cost functions are given in the following table.

jobs	1	2	3
$h_j(C_j)$	$1 + 2C_1$	38	$4C_3$
$E(X_j)$	6	18	12

Note that all three cost functions are linear. This makes the evaluations of all the necessary integrals very easy since the integrals are a linear function of the means of the processing times. If job 1 is the last one to be completed, the expected penalty with regard to job 1 is $1 + 2(6 + 18 + 12) = 73$; the expected penalty with regard to job 2 is 38; and with regard to job 3 is $4(6 + 18 + 12) = 144$. The procedure selects job 2 for the last position. If job 1 would go second, the expected penalty would be $1 + 2(6 + 12) = 37$, and if job 3 would go second, its expected penalty would be $4(6 + 12) = 72$. So job 1 is selected to go second and job 3 goes first. If job 3 goes first, its expected penalty is 48. The maximum of the three expected penalties under sequence 3, 1, 2 is $\max(48, 37, 38) = 48$.

Note that the analysis in Example 10.1.7 is particularly easy since all three cost functions are linear. The only information needed with regard to the processing time of a job is its mean. If any one of the cost functions is nonlinear, the expected penalty of the corresponding job is more difficult to compute; its entire distribution has to be taken into account. The integrals may have to be evaluated via approximation methods.

10.2 ARBITRARY DISTRIBUTIONS WITH PREEMPTIONS: THE GITTINS INDEX

Consider the problem of scheduling n jobs with random processing times $X_1, \ldots, X_n$ from discrete time distributions. The scheduler is allowed to preempt the machine at discrete times $0, 1, 2, \ldots$. If job j is completed at the integer completion time C_j, a reward $w_j \beta^{C_j}$ is received, where β is a discount factor between 0 and 1. The value of β is typically close to 1. It is of interest to determine the policy that maximizes the total expected reward.

Before proceeding with the analysis, it may be useful to relate this problem to another described earlier. It can be argued that this problem is a discrete-time version of the continuous-time problem with the objective

$$E\left(\sum_{j=1}^{n} w_j(1 - e^{-rC_j})\right).$$

The argument goes as follows. Maximizing $\sum w_j \beta^{C_j}$ is equivalent to minimizing $\sum w_j(1 - \beta^{C_j})$. Consider the limiting case where the size of the time unit is decreased and the number of time units is increased accordingly. If the time unit is changed from 1 to $1/k$ with $(k > 1)$, the discount factor β has to be adjusted as well. The appropriate discount factor, which corresponds to the new time unit $1/k$, is then $(\sqrt[k]{\beta})$. If β is relatively close to one, then

$$\sqrt[k]{\beta} \approx 1 - \frac{1 - \beta}{k}.$$

So

$$\beta^{C_j} \approx \left(1 - \frac{1 - \beta}{k}\right)^{kC_j}$$

and

$$\lim_{k\to\infty} \left(1 - \frac{1-\beta}{k}\right)^{kC_j} = e^{-(1-\beta)C_j}.$$

This implies that when β is relatively close to one, it is similar to the $1 - r$ used in earlier models. The current model is then a discrete time stochastic version of the deterministic problem $1 \mid prmp \mid \sum w_j(1 - \exp(-rC_j))$ discussed in Chapter 3. The stochastic version can be used to model the problem described in Example 1.1.3 (the problem described in Example 1.1.3 is actually slightly more general as it assumes that jobs have different release dates).

To characterize the optimal policy for this discrete time scheduling problem with preemptions, it is actually easier to consider a more general reward process. Let $x_j(t)$ denote the state of job j at time t. This state is basically determined by the amount of processing job j has received prior to time t. If the decision is made to process job j during the interval $[t, t + 1]$, a random reward $W_j(x_j(t))$ is received. This random reward is a function of the state $x_j(t)$ of job j. Clearly, this reward process is more general than the one described at the beginning of this section under which a (fixed) reward w_j is received only if job j is *completed* during the interval $[t, t + 1]$. In what follows, the optimal policy is determined for the more general reward process.

The decision to be made at any point in time has two elements. First, a decision has to be made with regard to the job selection. Second, if a particular job is selected, a decision has to be made with regard to the amount of time the job should remain on the machine. The first point in time at which another job is considered for processing is referred to as the *stopping time* and is denoted by τ.

It is shown in what follows that the solution of this problem can be characterized by functions G_j, $j = 1, \ldots, n$, with the property that processing job j^* at time t is optimal if and only if

$$G_{j^*}(x_{j^*}) = \max_{1\le j\le n} G_j(x_j),$$

where x_j is the amount of processing job j already has received by time t—that is, $x_j(t) = x_j$. The function G_j is called the *Gittins index* and is defined as

$$G_j(x_j) = \max_{\pi} \frac{E_\pi\left(\sum_{s=0}^{\tau-1} \beta^{s+1} W_j(x_j(s)) \mid x_j(0) = x_j\right)}{E_\pi\left(\sum_{s=0}^{\tau-1} \beta^{s+1} \mid x_j(0) = x_j\right)},$$

with the stopping time τ being determined by the policy π. However, the value τ is not necessarily the time at which the processing of job j stops. Job j may actually be completed *before* the stopping time.

This is one of the more popular forms in which the Gittins index is presented in the literature. The Gittins index may be described in words as the largest value that can be obtained by dividing the total discounted expected reward over a given time period (determined through the stopping time) by the discounted time.

The next theorem focuses on the optimal policy under the more general reward process described previously.

Theorem 10.2.1. *The policy that maximizes the total discounted expected reward in the class of preemptive dynamic policies prescribes, at each point in time, the processing of the job with the largest Gittins index.*

Proof.   Assume that the scheduler has to pay a fixed charge to the machine if he decides to process job j. Call this charge the *prevailing charge*. Suppose job j is the only one in the system and the scheduler has to decide whether to process it. The scheduler has to decide to process the job for a number of time units, observe the state as it evolves, and then stop processing the moment the prevailing charge does not justify any further processing. If the prevailing charge is very small, it is advantageous to continue processing, whereas if the prevailing charge is too high, any further processing leads to losses.

As a function of the state of job j, say x_j, the so-called fair charge is defined as $\gamma_j(x_j)$, the level of prevailing charge for which the optimal action will neither be profitable nor cause losses. That is, the fair charge is the level of the prevailing charge at which the costs to the scheduler are exactly in balance with the expected outside rewards to be obtained by processing the jobs according to the optimal policy. So

$$\gamma_j(x_j) = \max\left(\gamma : \max_{\pi} E_{\pi}\left(\sum_{s=0}^{\tau-1} \beta^{s+1}\left(W_j(x_j(s)) - \gamma\right) \mid x_j(0) = x_j\right) \geq 0\right),$$

where the policy π determines the stopping time τ. Thus, the fair charge is determined by the optimal action that prescribes processing the job for exactly τ time units or until completion, whichever comes first. Processing the job for less time causes losses, and processing the job for more than τ time units causes losses also. Suppose the prevailing charge is reduced to the fair charge whenever the scheduler would have decided to stop processing the job due to the prevailing charge being too high. Then the scheduler would keep the job on the machine until completion as the process now becomes a fair game. In this case, the sequence of prevailing charges for the job is a nonincreasing function of the time units for which the job has already been processed.

Suppose now that there are n different jobs and at each point in time the scheduler has to decide which one to process during the next time period. Assume that initially the prevailing charge for each job is set equal to its fair charge and the prevailing charges are reduced periodically afterward, as described earlier, every time a stopping time is reached. Thus, the scheduler never pays more than the fair charge and can make sure that his expected total discounted profit is non-negative. However, it is also clear that his total profit cannot be positive because it would have to be positive for at least one job and this cannot happen as the prevailing charges consistently are set equal to the fair charges. The scheduler can only break even if, whenever he selects a job, he processes the job according to the optimal policy. That is, he has to continue processing this job until the optimal stopping time, which determines the level of the fair charge. If he does not act accordingly, he acts suboptimally and incurs an expected discounted loss. So the scheduler acts optimally if, whenever he starts processing a job, he continues to do so as long as the job's fair charge remains greater than its prevailing charge.

The sequence of prevailing charges for each job is a nonincreasing function of the number of time units the job has been processed. By definition, it is a sequence that is independent of the policy adopted. If for each job the sequence of prevailing charges is nonincreasing and if the scheduler adopts the policy of always processing the job with the highest prevailing charge, then he incurs charges in a nonincreasing sequence. This intertwining of sequences of prevailing charges into a single nonincreasing sequence is unique (in terms of charges, not necessarily in terms of jobs). Thus, the policy of processing the job with the largest prevailing charge maximizes the expected total discounted charge paid by the scheduler. This maximum quantity is an upper bound for the expected total discounted reward obtained by the scheduler. This upper bound is achieved by the proposed policy since the policy forces the scheduler to process a job, without interruption, for the time that its fair charge exceeds its prevailing charge (this leads to a fair game in which the scheduler's total expected discounted profit is zero). This completes the proof of the theorem. $\square$

From the expression for the fair charge $\gamma_j(x_j)$, the expression for the Gittins index immediately follows. For the special case in which a fixed reward w_j is received only on completion of job j, the Gittins index becomes

$$G_j(x_j) = \max_{\tau > 0} \frac{\sum_{s=0}^{\tau-1} \beta^{s+1} w_j P(X_j = x_j + 1 + s \mid X_j > x_j)}{\sum_{s=0}^{\tau-1} \beta^{s+1} P(X_j \geq x_j + 1 + s \mid X_j > x_j)}$$

$$= \max_{\tau > 0} \frac{\sum_{s=0}^{\tau-1} \beta^{s+1} w_j P(X_j = x_j + 1 + s)}{\sum_{s=0}^{\tau-1} \beta^{s+1} P(X_j \geq x_j + 1 + s)}.$$

The Gittins index is determined by the maximum of the ratio of the R.H.S. over all possible stopping times. As the expectations of the sums in the numerator and denominator must take into account that the scheduler does not keep the job on the machine for τ time units in case the job is completed early, each element in the sums has to be multiplied with the appropriate probability.

As the computation of Gittins indexes at first sight may seem somewhat involved, an example is in order. The following example considers the reward process where a fixed reward w_j is obtained at the completion of job j.

Example 10.2.2 (Application of the Gittins Index)

Consider three jobs with $w_1 = 60$, $w_2 = 30$, and $w_3 = 40$. Let p_{jk} denote the probability that the processing time of job j takes k time units—that is,

$$p_{jk} = P(X_j = k).$$

The processing times of the three jobs take only values on the integers 1, 2, and 3.

$$p_{11} = \tfrac{1}{6}, \quad p_{12} = \tfrac{1}{2}, \quad p_{13} = \tfrac{1}{3};$$

$$p_{21} = \tfrac{2}{3}, \quad p_{22} = \tfrac{1}{6}, \quad p_{23} = \tfrac{1}{6};$$

$$p_{31} = \tfrac{1}{2}, \quad p_{32} = \tfrac{1}{4}, \quad p_{33} = \tfrac{1}{4}.$$

Assume the discount rate β to be 0.5. If job 1 is put on the machine at time 0, the discounted expected reward at time 1 is $w_1 p_{11}\beta$, which is 5. The discounted expected reward obtained at time 2 is $w_1 p_{12}\beta^2$, which is 7.5. The discounted expected reward obtained at time 3 is $w_1 p_{13}\beta^3$, which is 2.5. The Gittins index for job 1 at time 0 can now be computed easily.

$$G_1(x_1(0)) = G_1(0) = \max\left(\frac{5}{0.5}, \frac{5+7.5}{0.5+0.208}, \frac{5+7.5+2.5}{0.5+0.208+0.042}\right) = 20.$$

Thus, if job 1 is selected as the one to go on the machine at time 0, it will be processed until it is completed. In the same way, the Gittins indexes for jobs 2 and 3 at time zero can be computed:

$$G_2(0) = \max\left(\frac{10}{0.5}, \frac{11.25}{0.5+0.083}, \frac{11.875}{0.5+0.083+0.021}\right) = 20.$$

The computation of the Gittins index of job 2 indicates that job 2, if selected to go on the machine at time 0, may be preempted before being completed; processing is only guaranteed for one time unit.

$$G_3(0) = \max\left(\frac{10}{0.5}, \frac{12.5}{0.5+0.125}, \frac{13.75}{0.5+0.125+0.031}\right) = 20.96.$$

If job 3 would be selected for processing at time 0, it would be processed up to completion.

After comparing the three Gittins indexes for the three jobs at time zero, a decision can be made with regard to the job to be selected for processing. The maximum of the three Gittins indexes is 20.96. So job 3 is put on the machine at time 0 and is kept on the machine until completion. At the completion of job 3, either job 1 or job 2 may be selected. The values of their Gittins indexes are the same. If job 1 is selected for processing, it remains on the machine until it is completed. If job 2 is selected, it is guaranteed processing for only one time unit; if it is not completed after one time unit, it is preempted and job 1 is selected for processing.

What would happen if the processing times have ICR distributions? It can be shown that, in this case, the scheduler never will preempt. The Gittins index of the job being processed increases continuously, whereas the indexes of the jobs waiting for processing remain the same. Consider the limiting case where the length of the time unit goes to 0 as the number of timesteps increases accordingly. The problem becomes a continuous time problem. When the processing times are ICR, the result in Theorem 10.2.1 is equivalent to the result in Theorem 10.1.3. So in one sense Theorem 10.1.3 is more general as it covers the nonpreemptive setting with arbitrary processing time distributions (not just ICR distributions), whereas Theorem 10.2.1 does not give any indication of the form of the optimal policy in a *nonpreemptive* setting when the processing times are *not* ICR. In another sense, Theorem 10.2.1 is more general since Theorem 10.1.3 does not give any indication of the form of the optimal policy in a *preemptive* setting when the processing times are *not* ICR.

The result in Theorem 10.2.1 can be generalized to include jobs arriving according to a Poisson process. In a discrete time framework, this implies that the interarrival times are geometrically distributed with a fixed parameter. The job selected at any point in time is the job with the largest Gittins index among the jobs present. The proof of this result lies beyond the scope of this book.

The result can also be generalized to include breakdowns with up times being i.i.d. geometrically distributed and down times being i.i.d. arbitrarily distributed. For the proof of this result, the reader is also referred to the literature.

10.3 LIKELIHOOD RATIO ORDERED DISTRIBUTIONS

Section 10.1 discussed examples of nonpreemptive stochastic models that are basically equivalent to their deterministic counterparts. In a number of cases, the distributions of the random variables did not matter at all; only their expectations played a role. In this subsection, an example is given of a nonpreemptive stochastic model, which is, to a lesser extent, equivalent to its deterministic counterpart. Its relationship with its deterministic counterpart is not as strong as in the earlier cases since certain conditions on the distribution functions of the processing times are required.

Consider n jobs. The processing time of job j is equal to the random variable X_j with distribution F_j, provided the job is started immediately at time zero. However, over time the machine "deteriorates" (i.e., the later a job begins processing, the greater its processing time). If job j starts with its processing at time t, its processing time is $X_j a(t)$, where $a(t)$ is an increasing concave function. Thus, for any starting time t, the processing time is proportional to the processing time of the job if it had started its processing at time 0. Moreover, concavity of $a(t)$ implies that the deterioration process in the early stages of the process is more severe than in the later stages of the process. The original processing times are assumed to be likelihood ratio ordered in such a way that $X_1 \leq_{lr} \cdots \leq_{lr} X_n$. The objective is to minimize the expected makespan. The following lemma is needed in the subsequent analysis.

Lemma 10.3.1.   *If $g(x_1, x_2)$ is a real valued function satisfying*

$$g(x_1, x_2) \geq g(x_2, x_1)$$

for all $x_1 \leq x_2$, then

$$g(X_1, X_2) \geq_{st} g(X_2, X_1)$$

whenever

$$X_1 \leq_{lr} X_2.$$

Proof.   Let $U = \max(X_1, X_2)$ and $V = \min(X_1, X_2)$. Condition on $U = u$ and $V = v$ with $u \geq v$. The conditional distribution of $g(X_1, X_2)$ is concentrated on the two points $g(u, v)$ and $g(v, u)$. The probability assigned to the smaller value $g(u, v)$ is then

$$P(X_1 = \max(X_1, X_2) \mid U = u, V = v) = \frac{f_1(u) f_2(v)}{f_1(u) f_2(v) + f_1(v) f_2(u)}.$$

In the same way $g(X_2, X_1)$ is also concentrated on two points $g(u, v)$ and $g(v, u)$. The probability assigned to the smaller value $g(u, v)$ in this case is

$$P(X_2 = \max(X_1, X_2) \mid U = u, V = v) = \frac{f_2(u) f_1(v)}{f_2(u) f_1(v) + f_2(v) f_1(u)}.$$

As $u \geq v$ and $X_2 \geq_{lr} X_1$,

$$f_1(u) f_2(v) \leq f_2(u) f_1(v).$$

Therefore, conditional on $U = u$ and $V = v$

$$g(X_2, X_1) \leq_{st} g(X_1, X_2).$$

Unconditioning completes the proof of the Lemma.   □

At first sight, this lemma may seem to provide a very fundamental and useful result. Any pairwise interchange in a deterministic setting can be translated into a pairwise interchange in a stochastic setting with the random variables likelihood ratio ordered. However, the usefulness of this lemma appears to be limited to single machine problems and proportionate flow shops.

The following two lemmas contain some elementary properties of the function $a(t)$.

Lemma 10.3.2.   *If* $0 < x_1 < x_2$, *then for all* $t \geq 0$

$$x_1 a(t) + x_2 a\left(t + x_1 a(t)\right) \geq x_2 a(t) + x_1 a\left(t + x_2 a(t)\right).$$

Proof.   The proof is easy and therefore omitted.   □

From Lemmas 10.3.1 and 10.3.2, it immediately follows that if there are only two jobs, scheduling the job with the larger expected processing time first minimizes the expected makespan.

Lemma 10.3.3.   *The function* $h_{x_1}(t) = t + x_1 a(t)$ *is increasing in* t *for all* $x_1 > 0$.

Proof.   The proof is easy and therefore omitted.   □

Theorem 10.3.4.   *The Longest Expected Processing Time first (LEPT) rule minimizes the expected makespan in the class of nonpreemptive static list policies as well as in the class of nonpreemptive dynamic policies.*

Proof.   Consider first the class of nonpreemptive static list policies. The proof is by induction. It has already been shown to hold for two jobs. Assume the theorem holds for $n - 1$ jobs. Consider any nonpreemptive static list policy and let job k be the job that is scheduled last. From Lemma 10.3.3, it follows that among all schedules that process job k last, the one resulting in the minimum makespan is the one that stochastically minimizes the completion time of the first $n - 1$ jobs. Hence, by the induction hypothesis, of all schedules that schedule job k last, the one with the stochastically smallest makespan is the one that schedules the first $n - 1$ jobs according to LEPT. If k is not the smallest job, then the best schedule is the one that selects the smallest job immediately before this last job k. Let t' denote the time that this smallest job starts its processing and suppose that there are only two jobs remaining to be processed. The problem at this point is a problem with two jobs and an a function that is given by $a_{t'}(t) = a(t' + t)$. Because this function is still concave it follows from the result for two jobs that interchanging these last two jobs reduces the total makespan stochastically. Among all schedules that schedule the smallest job last, the one stated in the theorem, by the induction hypothesis, minimizes the makespan stochastically. This completes the proof for nonpreemptive static list policies.

It remains to be shown that the LEPT rule also stochastically minimizes the makespan in the class of nonpreemptive dynamic policies. Suppose the decision is allowed to depend on what has previously occurred, at most l times during the process (of course, such times occur only when the machine is freed). When $l = 1$, it follows from the optimality proof for static list policies that it is optimal not to deviate from the LEPT schedule. If this remains true when $l - 1$ such opportunities are allowed, it follows from the same argument that it remains true when l such opportunities are allowed (because of the induction hypothesis, such an opportunity would be utilized only once). As the result is true for all l, the proof for nonpreemptive dynamic policies is complete.   □

Example 10.3.5 (Linear Deterioration Function)

Consider two jobs with exponential processing times. The rates are λ_1 and λ_2. The deterioration function $a(t) = 1 + t, t \geq 0$, is linear. If the jobs are scheduled according to sequence 1, 2, then the expected makespan can be computed as follows. If job 1 is completed at time t, the expected time job 2 will occupy the machine is $a(t)/\lambda_2$. The probability job 1 is completed during the interval $[t, t + dt]$ is $\lambda_1 e^{-\lambda_1 t} dt$. So

$$E(C_{max}) = \int_0^\infty \left(t + a(t)\frac{1}{\lambda_2} \right) \lambda_1 e^{-\lambda_1 t} dt = \frac{1}{\lambda_1} + \frac{1}{\lambda_2} + \frac{1}{\lambda_1 \lambda_2}.$$

From this expression, it is clear that the expected makespan in this case does *not* depend on the sequence.

Example 10.3.6 (Increasing Concave Deterioration Function)

Consider two jobs with discrete processing time distributions.

$$P(X_1 = 1) = \tfrac{1}{8}, \quad P(X_1 = 2) = \tfrac{1}{4}, \quad P(X_1 = 3) = \tfrac{5}{8};$$
$$P(X_2 = 1) = \tfrac{1}{4}, \quad P(X_2 = 2) = \tfrac{1}{4}, \quad P(X_1 = 3) = \tfrac{1}{2}.$$

It is clear that $X_1 \geq_{lr} X_2$. $E(X_1) = 2.5$ while $E(X_2) = 2.25$. The deterioration function $a(t) = 1 + t$ for $0 \leq t \leq 2$, and $a(t) = 3$ for $t \geq 2$. Clearly, $a(t)$ is increasing concave. Consider the LEPT sequence (i.e., sequence 1, 2). The expected makespan can be computed by conditioning on the processing time of the first job.

$$E(C_{max}(LEPT)) = \tfrac{1}{8}(1 + 2E(X_2)) + \tfrac{1}{4}(2 + 3E(X_2)) + \tfrac{5}{8}(3 + 3E(X_2)) = \tfrac{287}{32},$$

whereas

$$E(C_{\max}(SEPT)) = \tfrac{1}{4}(1 + 2E(X_1)) + \tfrac{1}{4}(2 + 3E(X_1)) + \tfrac{1}{2}(3 + 3E(X_1)) = \tfrac{292}{32}.$$

Clearly, LEPT is better than SEPT.

Intuitively, it does make sense that if the deterioration function is increasing concave, the longest job should go first. It can also be shown that if the deterioration function is increasing convex, the Shortest Expected Processing Time first (SEPT) rule is optimal; if the deterioration function is linear then any sequence is optimal. However, if the function $a(t)$ is decreasing (i.e., a form of learning takes place that makes it possible to process the jobs faster) then it does not appear that similar results can be obtained.

10.4 EXPONENTIAL DISTRIBUTIONS

In this section, models with exponentially distributed processing times are discussed. Consider the stochastic version of $1 \mid d_j = d \mid \sum w_j U_j$ with job j having an exponentially distributed processing time with rate λ_j and a deterministic due date d. Recall that the deterministic counterpart is equivalent to the *knapsack* problem. The objective is the expected weighted number of tardy jobs.

Theorem 10.4.1. *The WSEPT rule minimizes the expected weighted number of tardy jobs in the classes of nonpreemptive static list policies, nonpreemptive dynamic policies, and preemptive dynamic policies.*

Proof.   First the optimality of the WSEPT rule in the class of nonpreemptive static list policies is shown. Assume the machine is free at some time t and two jobs, with weights w_1 and w_2 and processing times X_1 and X_2, remain to be processed. Consider first the sequence $1, 2$. The probability that both jobs are late is equal to the probability that X_1 is larger than $d - t$, which is equal to $\exp(-\lambda(d - t))$. The penalty for being late is then equal to $w_1 + w_2$. The probability that only the second job is late corresponds to the event where the processing time of the first job is $x_1 < d - t$ and the sum of the processing times $x_1 + x_2 > d - t$. Evaluation of the probability of this event, by conditioning on X_1 (i.e., $X_1 = x$), yields

$$P(X_1 < d - t, X_1 + X_2 > d - t) = \int_0^{d-t} e^{-\lambda_2(d-t-x)} \lambda_1 e^{-\lambda_1 x}\, dx.$$

If $E(\sum wU(1, 2))$ denotes the expected value of the penalty due to jobs 1 and 2, with job 1 processed first, then

$$E\left(\sum wU(1, 2)\right) = (w_1 + w_2)e^{-\lambda_1(d-t)} + w_2 \int_0^{d-t} e^{-\lambda_2(d-t-x)} \lambda_1 e^{-\lambda_1 x}\, dx.$$

The value of the objective function under sequence 2, 1 can be obtained by interchanging the subscripts in the prior expression. Straightforward computation yields

$$E\left(\sum wU(1,2)\right) - E\left(\sum wU(2,1)\right) = (\lambda_2 w_2 - \lambda_1 w_1)\frac{e^{-\lambda_1(d-t)} - e^{-\lambda_2(d-t)}}{\lambda_2 - \lambda_1}.$$

It immediately follows that the difference in the expected values is positive if and only if $\lambda_2 w_2 > \lambda_1 w_1$. Since this result holds for all values of d and t, any permutation schedule that does not sequence the jobs in decreasing order of $\lambda_j w_j$ can be improved by swapping two adjacent jobs, where the first has a lower λw value than the second. This completes the proof of optimality for the class of nonpreemptive static list policies.

Induction can be used to show optimality in the class of nonpreemptive dynamic policies. It is immediate that this is true for two jobs (it follows from the same pairwise interchange argument for optimality in the class of nonpreemptive static list policies). Assume that it is true for $n-1$ jobs. In the case of n jobs, this implies that the scheduler after the completion of the first job will, because of the induction hypothesis, revert to the WSEPT rule among the remaining $n-1$ jobs. It remains to be shown that the scheduler has to select the job with the highest $\lambda_j w_j$ as the first one to be processed. Suppose the decision maker selects a job that does not have the highest $\lambda_j w_j$. Then the job with the highest value of $\lambda_j w_j$ is processed second. Changing the sequence of the first two jobs decreases the expected value of the objective function according to the pairwise interchange argument used for the nonpreemptive static list policies.

To show that WSEPT is optimal in the class of preemptive dynamic policies, suppose a preemption is contemplated at some point in time. The remaining processing time of the job then on the machine is exponentially distributed with the same rate as it had at the start of its processing (because of the memoryless property of the exponential). Since the decision to put this job on the machine did not depend on the value of t at that moment or on the value of d, the same decision remains optimal at the moment a preemption is contemplated. A nonpreemptive policy is therefore optimal in the class of preemptive dynamic policies.   □

This result is in marked contrast to the result in Chapter 3, which states that its deterministic counterpart (i.e., the knapsack problem) is NP-hard.

Consider now the discrete time version of Theorem 10.4.1. That is, the processing time of job j is geometrically distributed with parameter q_j and job j has weight w_j. All jobs have the same due date d. If a job is completed exactly at its due date, it is considered on time. The objective is again $E(\sum w_j U_j)$.

Theorem 10.4.2.   *The WSEPT rule minimizes the expected weighted number of tardy jobs in the classes of nonpreemptive static list policies, nonpreemptive dynamic policies, and preemptive dynamic policies.*

Proof.   Consider two jobs, say jobs 1 and 2, and sequence 1, 2. The probability that both jobs are late is q_1^{d+1} and the penalty is then $w_1 + w_2$. The probability that the first job is on time and the second job is late is

$$\sum_{t=0}^{d} (1 - q_1) q_1^t q_2^{d+1-t}.$$

The penalty is then w_2. So the total penalty under sequence 1, 2 is

$$(w_1 + w_2) q_1^{d+1} + w_2 (1 - q_1) q_2^{d+1} \left(1 - \left(\frac{q_1}{q_2} \right)^{d+1} \right) \bigg/ \left(1 - \frac{q_1}{q_2} \right).$$

The total expected penalty under sequence 2, 1 can be obtained by interchanging the subscripts 1 and 2. Sequence 1, 2 is better than sequence 2, 1 if

$$w_1 q_1^{d+1} q_2 - w_2 q_1^{d+2} - w_2 q_1 q_2^{d+2} + w_2 q_1^{d+2} q_2$$
$$\leq w_1 q_2^{d+2} - w_2 q_1 q_2^{d+1} + w_1 q_1^{d+2} q_2 - w_1 q_1 q_2^{d+2}.$$

After some manipulations, it turns out that sequence 1, 2 is better than 2, 1 if

$$w_1 (1 - q_1)/q_1 \geq w_2 (1 - q_2)/q_2,$$

which is equivalent to

$$\frac{w_1}{E(X_1)} \geq \frac{w_2}{E(X_2)}.$$

That WSEPT is also optimal in the class of nonpreemptive dynamic policies and in the class of preemptive dynamic policies can be shown through the same arguments as the ones used in the proof of Theorem 10.4.1.   □

The WSEPT rule does not necessarily yield an optimal schedule when processing time distributions are not all exponential (or all geometric).

Example 10.4.3 (Optimal Policy with Random Variables That Are ICR)

Consider the case where each one of the processing times is distributed according to an *Erlang(k,λ)* distribution. The rate of an exponential phase of job j is λ_j. This implies $E(X_j) = k/\lambda_j$. The WSEPT rule in general will not yield optimal schedules. Job j having a deterministic processing time p_j is a special case (the number of phases of the *Erlang* for each one of the n jobs approaches ∞ while the mean of each phase approaches zero). It is clear how to construct this way a counterexample for processing time distributions that are ICR.

Example 10.4.4 (Optimal Policy with Random Variables That Are DCR)

Consider the case where each one of the processing times is distributed according to a mixture of exponentials. Assume that the processing time of job j is 0 with probability p_j and exponentially distributed with rate λ_j with probability $1 - p_j$. Clearly,

$$E(X_j) = (1 - p_j)\frac{1}{\lambda_j}.$$

The optimal preemptive policy can be determined easily. Try each job out at time zero for an infinitesimal small period of time. The jobs with zero processing times are then immediately completed. Immediately after time zero, it is known which jobs have nonzero processing times. The remaining processing times of these jobs are then exponentially distributed with probability one. The optimal preemptive policy from that point on is then the nonpreemptive policy described in Theorem 10.4.1.

Theorems 10.4.1 and 10.4.2 can be generalized to include breakdown and repair. Suppose the machine goes through "uptimes," when it is functioning, and "downtimes," when it is being repaired. This breakdown and repair may form an arbitrary stochastic process. Theorem 10.4.1 also holds under these more general conditions since no part of the proof depends on the remaining time till the due date.

Theorem 10.4.1 can also be generalized to include different release dates with arbitrary distributions. Assume a finite number of releases after time 0, say n^*. It is clear from the results presented earlier that at the time of the last release the WSEPT policy is optimal. This may actually imply that the last release causes a preemption (if, at that point in time, the job released is the job with the highest $\lambda_j w_j$ ratio in the system). Consider now the time epoch of the second last release. After this release, a preemptive version of the WSEPT rule is optimal. To see this, disregard for a moment the very last release. All the jobs in the system at the time of the second to last release (*not* including the last release) have to be sequenced according to WSEPT; the last release may in a sense be considered a random "downtime." From the previous results, it follows that all the jobs in the system at the time of the second last release should be scheduled according to preemptive WSEPT independent of the time period during which the last release is processed. Proceeding inductively toward time zero, it can be shown that with arbitrarily distributed releases a preemptive version of WSEPT is optimal in the classes of preemptive static list policies and preemptive dynamic policies.

The WSEPT rule also proves optimal for other objectives. Consider the stochastic counterpart of $1 \mid d_j = d \mid \sum w_j T_j$ with job j again exponentially distributed with rate λ_j. All n jobs are released at time 0. The objective is to minimize the sum of the expected weighted tardinesses.

Theorem 10.4.5. *The WSEPT rule minimizes the expected sum of the weighted tardinesses in the classes of nonpreemptive static list policies, non-preemptive dynamic policies, and preemptive dynamic policies.*

Proof.   The objective $w_j T_j$ can be approximated by the sum of an infinite sequence of $w_j U_j$ unit penalty functions—that is,

$$w_j T_j = \sum_{l=0}^{\infty} w_j U_{jl}.$$

The first unit penalty U_{j0} corresponds to a due date d, the second unit penalty U_{j1} corresponds to a due date $d + \epsilon$, the third corresponds to a due date $d + 2\epsilon$, and so on (see Figure 10.2). From Theorem 10.4.1, it follows that λw rule minimizes each one of these unit penalty functions. If the rule minimizes each one of these unit penalty functions, it also minimizes their sum.   □

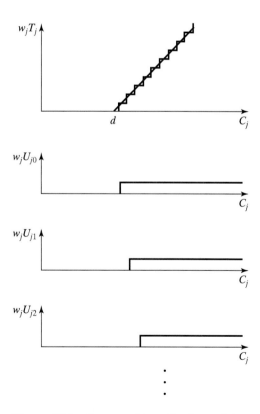

Figure 10.2 Superposition of unit penalties.

This theorem can be generalized along the lines of Theorem 10.4.1 to include arbitrary breakdown and repair processes and arbitrary release processes provided all jobs have due date d (including those released after d).

Actually, a generalization in a slightly different direction is also possible. Consider the stochastic counterpart of the problem $1 \mid\mid \sum w_j h(C_j)$. In this model, the jobs have no specific due dates, but all are subject to the *same* cost function h. The objective is to minimize $E(\sum w_j h(C_j))$. Clearly, $\sum w_j h(C_j)$ is a simple generalization of $\sum w_j T_j$ when all jobs have the same due date d. The function h can again be approximated by a sum of an infinite sequence of unit penalties, the only difference being that the due dates of the unit penalties are not necessarily equidistant as in the proof of Theorem 10.4.5.

Consider now a stochastic counterpart of the problem $1 \mid\mid \sum w_j h_j(C_j)$, with each job having a *different* cost function. Again, all jobs are released at time 0. The objective is to minimize the total expected cost. The following ordering among cost functions is of interest: A cost function h_j is said to be *steeper* than a cost function h_k if

$$\frac{dh_j(t)}{dt} \geq \frac{dh_k(t)}{dt}$$

for every t provided the derivatives exist. This ordering is denoted by $h_j \geq_s h_k$. If the functions are not differentiable for every t, the steepness ordering requires

$$h_j(t + \delta) - h_j(t) \geq h_k(t + \delta) - h_k(t)$$

for every t and δ. Note that a cost function being steeper than another does not necessarily imply that it is higher (see Figure 10.3).

Theorem 10.4.6. *If $\lambda_j w_j \geq \lambda_k w_k \iff h_j \geq_s h_k$, then the WSEPT rule minimizes the total expected cost in the classes of nonpreemptive static list policies, nonpreemptive dynamic policies, and preemptive dynamic policies.*

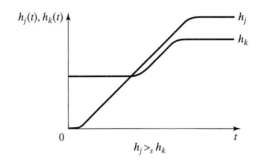

Figure 10.3   One cost function steeper than another.

Proof. The proof follows from the fact that any increasing cost function can be approximated by an appropriate summation of a (possibly infinite) number of unit penalties at different due dates. If two cost functions that may be at different levels go up in the same way over an interval $[t_1, t_2]$, then a series of identical unit penalties go into effect within that interval for both jobs. It follows from Theorem 10.4.1 that the jobs have to be sequenced in decreasing order of λw to minimize the total expected penalties due to these unit penalties. If one cost function is steeper than another in a particular interval, then the steeper cost function has one or more unit penalties going into effect within this interval, which the other cost function has not. To minimize the total expected cost due to these unit penalties, the jobs have to be sequenced again in decreasing order of λw. $\square$

Example 10.4.7 (Application of WSEPT with Due Dates That Are Agreeable)

Consider n jobs with exponentially distributed processing times with rates $\lambda_1, \ldots, \lambda_n$. The jobs have deterministic due dates $d_1 \leq d_2 \leq \cdots \leq d_n$. First, consider $E(\sum w_j T_j)$ as the objective to be minimized. If $\lambda_1 w_1 \geq \lambda_2 w_2 \geq \cdots \geq \lambda_n w_n$, then sequence $1, 2, \ldots, n$ minimizes the objective since $T_1 \geq_s T_2 \geq_s \cdots \geq_s T_n$.

Second, consider $E(\sum w_j U_j)$ with the same due dates $d_1 \leq d_2 \leq \cdots \leq d_n$ as the objective to be minimized. It can be verified easily that the string of inequalities $U_1 \geq_s U_2 \geq_s \cdots \geq_s U_n$ does *not* hold. So sequence $1, 2, \ldots, n$ does not necessarily minimize the objective (see Exercise 10.10).

The result of Theorem 10.4.6 can be easily generalized to include an arbitrary machine breakdown process. It can also be extended, in the case of preemptive static list policies or in the case of preemptive dynamic policies, to include jobs with different release dates as long as the cost functions of the new arrivals satisfy the stated "agreeability" conditions.

The results in this subsection indicate that scheduling problems with exponentially distributed processing times allow for more elegant structural results than their deterministic counterparts. The deterministic counterparts of most of the models discussed in this section are NP-hard. It is intuitively acceptable that a deterministic problem may be NP-hard, whereas its counterpart with exponentially distributed processing times allows for a very simple policy to be optimal. The reason is the following: All data being deterministic (i.e., perfect data) makes it very hard for the scheduler to optimize. To take advantage of all the information available, the scheduler has to spend an inordinately long time doing the optimization. Yet when the processing times are stochastic, the data are fuzzier. The scheduler, with less data at hand, will spend less time performing the optimization. The fuzzier the data, the more likely a simple priority rule minimizes the objective in expectation. Expectation is akin to optimizing for the average case.

EXERCISES (COMPUTATIONAL)

10.1. Consider a single machine and three jobs with i.i.d. processing times with distribution F and mean 1.

   (a) Show that when F is deterministic, $E(\sum C_j) = 6$ under a nonpreemptive schedule and $E(\sum C_j) = 9$ under the processor sharing schedule.

   (b) Show that when F is exponential, $E(\sum C_j) = 6$ under a nonpreemptive schedule and $E(\sum C_j) = 6$ under the processor sharing schedule. (Recall that under a processor sharing schedule all jobs available share the processor equally; i.e., if there are n jobs available, then each job receives $1/n$ of the processing capability of the machine; see Section 5.2).

10.2. Consider the same scenario as in the previous exercise. Assume F is an EME distribution (as defined in Section 9.2) with the parameter p very small.

   (a) Show that $E(\sum C_j) = 6$ under the nonpreemptive schedule.

   (b) Show that $E(\sum C_j) = 3$ under the processor sharing schedule.

10.3. Consider a single machine and three jobs. The distribution of job j, $j = 1, 2, 3$, is discrete uniform over the set $\{10 - j, 10 - j + 1, \ldots, 10 + j\}$. Find the schedule(s) that minimize $E(\sum C_j)$ and compute the value of the objective function under the optimal schedule.

10.4. Consider the same setting as in the previous exercise. Find the schedule that minimizes $E(\sum h_j(C_j))$, where the function $h(C_j)$ is defined as follows.

$$h(C_j) = \begin{cases} 0 & \text{if } C_j \leq 20 \\ C_j - 20 & \text{if } C_j > 20 \end{cases}.$$

Is the *Largest Variance first (LV)* rule or the *Smallest Variance first (SV)* rule optimal?

10.5. Consider the same setting as in the previous exercise. Find the schedule that minimizes $E(\sum h_j(C_j))$, where the function $h(C_j)$ is defined as follows.

$$h(C_j) = \begin{cases} C_j & \text{if } C_j \leq 20 \\ 20 & \text{if } C_j > 20 \end{cases}.$$

Is the *Largest Variance first (LV)* rule or the *Smallest Variance first (SV)* rule optimal?

10.6. Consider two jobs with discrete processing time distributions:

$$P(X_1 = 1) = P(X_1 = 2) = P(X_1 = 4) = \tfrac{1}{3}$$

and

$$P(X_2 = 3) = P(X_2 = 5) = \tfrac{1}{2}.$$

The two jobs have deterministic due dates. The due date of the first job is $D_1 = 2$ and the due date of the second job is $D_2 = 4$. Compute $E(\max(L_1, L_2))$ and $\max(E(L_1), E(L_2))$ under EDD.

10.7. Consider the framework of Section 10.2. There are three jobs, all having a discrete uniform distribution. The processing time of job j is uniformly distributed over the set $\{5 - j, 5 - j + 1, \ldots, 5 + j - 1, 5 + j\}$. The discount factor β is equal to 0.5. The weight of job 1 is 30, the weight of job 2 is 10, and the weight of job 3 is 30. Find the optimal preemptive policy. Determine whether it is necessary to preempt any job at any point in time.

10.8. Redo the instance in Exercise 10.7 with the discount factor $\beta = 1$. Determine all optimal policies. Give an explanation for the results obtained and compare the results with those obtained in Exercise 10.7.

10.9. Consider Example 10.3.5 with the linear deterioration function $a(t) = 1 + t$. Instead of the two jobs with exponentially distributed processing times, consider two jobs with geometrically distributed processing times with parameters q_1 and q_2. Compute the expected makespan under the two sequences.

10.10. Construct a counterexample for the stochastic problem with exponential processing times and deterministic due dates showing that if $\lambda_j w_j \geq \lambda_k w_k \Leftrightarrow d_j \leq d_k$, the λw rule does not necessarily minimize $E(\sum w_j U_j)$.

EXERCISES (THEORETICAL)

10.11. Consider the model in Theorem 10.1.1 with breakdowns. The uptimes are exponentially distributed with rate ν and the downtimes are i.i.d. (arbitrarily distributed) with mean $1/\mu$. Show that the expected time job j spends on the machine is equal to

$$E(Y_j) = \left(1 + \frac{\nu}{\mu}\right) E(X_j),$$

where $E(X_j)$ is the expected processing time of job j. Give an explanation of why this problem is therefore equivalent to the problem without breakdowns (*Hint:* Only the time-axis changes because of the breakdowns).

10.12. Consider the same model as in Exercise 10.11, but assume that the processing time of job j is exponentially distributed with rate λ_j. Assume that the repair time is exponentially distributed with rate μ.

  (a) Show that the number of times the machine breaks down during the processing of job j is geometrically distributed with rate

$$q = \frac{\nu}{\lambda_j + \nu}.$$

  (b) Show that the total amount of time spent on the repair of the machine during the processing of job j is exponentially distributed with rate $\lambda_j \mu / \nu$ provided there is at least one breakdown.

  (c) Show that the total time job j remains on the machine is a mixture of an exponential with rate λ_j and a convolution of two exponentials with rates λ_j and $\lambda_j \mu / \nu$. Find the mixing probabilities.

10.13. Consider the model in Exercise 10.11. Assume that the jobs are subject to precedence constraints that take the form of chains. Show that Algorithm 3.1.4 minimizes the total expected weighted completion time.

10.14. Consider the discrete time stochastic model described in Section 10.2. The continuous time version is a stochastic counterpart of $1 \mid prmp \mid \sum w_j(1 - \exp(-rC_j))$. Show that the Gittins index for this problem is

$$G_j(x_j) = \max_{\tau > 0} \frac{w_j \int_{x_j}^{\tau} f_j(s)e^{-rs} \, ds}{\int_{x_j}^{\tau} (1 - F_j(s))e^{-rs} \, ds}.$$

10.15. Consider the stochastic counterpart of $1 \mid d_j = d \mid \sum w_j U_j$ with the processing time of job j arbitrarily distributed according to F_j. All jobs have a common random due date that is exponentially distributed with rate r. Show that this problem is equivalent to the stochastic counterpart of the problem $1 \mid\mid \sum w_j(1 - \exp(-rC_j))$ (i.e., a problem without a due date but with a discounted cost function) with all jobs having arbitrary distributions. (*Hint:* If, in the counterpart of $1 \mid d_j = d \mid \sum w_j U_j$, job j is completed at time C_j, the probability that it is late is equal to the probability that the random due date occurs before C_j. The probability that this occurs is $1 - e^{-rC_j}$, which is equal to $E(U_j)$).

10.16. Show that if in the model of Section 10.3 the deterioration function is linear—that is, $a(t) = c_1 + c_2 t$, with both c_1 and c_2 constant—the distribution of the makespan is sequence independent.

10.17. Show through a counterexample that LEPT does not necessarily minimize the makespan in the model of Section 10.3 when the distributions are merely ordered in expectation and not in the likelihood ratio sense. Find a counterexample with distributions that are stochastically ordered but not ordered in the likelihood ratio sense.

10.18. Consider the two processing time distributions of the jobs in Example 10.3.6. Assume the deterioration function $a(t) = 1$ for $0 \le t \le 1$ and $a(t) = t$ for $t \ge 1$ (i.e., the deterioration function is increasing convex). Show that SEPT minimizes the makespan.

10.19. Consider the discrete time counterparts of Theorems 10.4.3 and 10.4.4 with geometric processing time distributions. State the results and prove the optimality of the WSEPT rule.

10.20. Generalize the result presented in Theorem 10.4.6 to the case where the machine is subject to an arbitrary breakdown process.

10.21. Generalize Theorem 10.4.6 to include jobs that are released at different points in time.

10.22. Consider the following discrete time stochastic counterpart of the deterministic model $1 \mid d_j = d, prmp \mid \sum w_j U_j$. The n jobs have a common random due date D. When a job is completed before the due date, a discounted reward is obtained. When the due date occurs before its completion, no reward is obtained and it does not pay to continue processing the job. Formulate the optimal policy in the class of preemptive dynamic policies.

10.23. Show that if $w_j = 1$ for all j, and $X_1 \leq_{st} X_2 \leq_{st} \cdots \leq_{st} X_n$, then the WDSEPT rule is equivalent to the SEPT rule for any r, $0 \leq r \leq 1$.

COMMENTS AND REFERENCES

A number of researchers have considered nonpreemptive single machine scheduling problems with arbitrary processing time distributions, see Rothkopf (1966a, 1966b), Crabill and Maxwell (1969), Hodgson (1977), and Forst (1984). For results with regard to the WSEPT rule or equivalently the $c\mu$ rule, see Cox and Smith (1961), Harrison (1975a, 1975b), Buyukkoc, Varaiya, and Walrand (1985), and Nain, Tsoucas, and Walrand (1989). For models that also include stochastic breakdowns, see Glazebrook (1984, 1987), Pinedo and Rammouz (1988), Birge, Frenk, Mittenthal, and Rinnooy Kan (1990), and Frenk (1991).

The Gittins index is due to Gittins and is explained in Gittins (1979). Many researchers have subsequently studied the use of Gittins indexes in single machine stochastic scheduling problems and other applications; see, for example, Whittle (1980, 1981), Glazebrook (1981a, 1981b, 1982), Chen and Katehakis (1986) and Katehakis and Veinott (1987). The proof of optimality of Gittins indexes presented here is due to Weber (1992).

The section on processing time distributions that are likelihood ratio ordered and subject to deterioration is entirely based on the paper by Brown and Solomon (1973). For more results on single machine scheduling subject to deterioration, see Browne and Yechiali (1990).

For an extensive treatment of single machine scheduling with exponential processing time distributions, see Derman, Lieberman, and Ross (1978), Pinedo (1983), and Pinedo and Rammouz (1988). For due date-related objectives with processing time distributions that are not exponential, see Sarin, Steiner, and Erel (1990).

11

Single Machine Models with Release Dates (Stochastic)

In many stochastic environments, the jobs are often not all available at time zero. The jobs come in at different times and randomly. This chapter focuses on single machine stochastic models with the jobs having random processing times and random release dates. The objective is the total expected weighted completion time. Preemptive as well as nonpreemptive models are considered.

An environment with random release dates is somewhat similar to the models considered in queueing theory. In a priority queue, a server (or machine) has to process customers (or jobs) from different classes and each class has it own priority level (or weight).

There are various similarities between stochastic scheduling with random release dates and priority queues. One similarity is that different jobs may have different processing times from different distributions. Another similarity is that different jobs may have different weights. However, there are also various differences. One important difference is that in scheduling, the goal is typically to minimize an objective that involves n jobs, whereas in queueing one usually assumes an infinite stream of customers and focuses on asymptotic results. In scheduling the goal is to find a policy that minimizes the total expected waiting cost of the n jobs or, equivalently, the average expected waiting cost of the n jobs, whereas in queueing, the goal is to quantify the expected waiting time of a typical customer or customer class in steady state and then determine the policy that minimizes the average expected waiting cost per customer or customer class. It pays to draw parallels between stochastic scheduling and priority queues since certain approaches and methodologies are applicable in both areas of research.

The models considered in this chapter are stochastic counterparts of the deterministic models $1 \mid r_j \mid \sum w_j C_j$ and $1 \mid r_j, prmp \mid \sum w_j C_j$. The objective considered is actually not $E(\sum w_j C_j)$, but rather

$$ E\left(\sum_{j=1}^{n} w_j(C_j - R_j)\right). $$

However, the value of the term $E(\sum w_j R_j)$ is, of course, a constant that does not depend on the policy. An equivalent objective is

$$ E\left(\frac{\sum_{j=1}^{n} w_j(C_j - R_j)}{n}\right). $$

If there is an infinite number of customers, then the objective

$$ \lim_{n \to \infty} E\left(\frac{\sum_{j=1}^{n} w_j(C_j - R_j)}{n}\right) $$

is of interest. This last objective is the one typically considered in queueing theory.

The WSEPT rule is optimal in several settings. This chapter focuses on the various conditions under which WSEPT minimizes the objectives under consideration.

11.1 ARBITRARY RELEASES AND ARBITRARY PROCESSING TIMES WITHOUT PREEMPTIONS

The model considered in this section is in one sense more general and in another sense more restricted than the model described in Section 10.1. The generalization lies in the fact that now the jobs have different release dates. The restriction lies in the fact that in Section 10.1 the n jobs have processing times that come from n different distributions, whereas in this section there are only two job classes with two different distributions. The processing times of the two job classes are arbitrarily distributed according to F_1 and F_2 and have means $1/\lambda_1$ and $1/\lambda_2$. The weights of the two job classes are w_1 and w_2, respectively. The release dates of the jobs have an arbitrary joint distribution. Assume that unforced idleness is not allowed; that is, the machine is not allowed to remain idle if there are jobs waiting for processing. Preemptions are not allowed. This model is a stochastic counterpart of $1 \mid r_j \mid \sum w_j C_j$ or, equivalently, $1 \mid r_j \mid \sum w_j (C_j - r_j)$.

Theorem 11.1.1. *Under the optimal nonpreemptive dynamic policy, the decision maker follows the WSEPT rule whenever the machine is freed.*

Proof. The proof is by contradiction and based on a simple adjacent pairwise interchange. Suppose that at a time the machine is freed, jobs from both priority classes are available for processing. Suppose the decision maker starts a job of the lower priority class (even though a job of the higher priority class is available for processing); he schedules a job of the higher priority class immediately after the completion of the job of the lower priority class. Now perform an adjacent pairwise interchange between these two jobs. Note that a pairwise interchange between these two adjacent jobs does not affect the completion times of any one of the jobs processed after this pair of jobs. However, the pairwise interchange does reduce the sum of the weighted expected completion times of the two jobs involved in the interchange. So the original ordering could not have been optimal. It follows that the decision maker must always use the WSEPT rule.   □

The result of Theorem 11.1.1 applies to settings with a finite number of jobs as well as to settings with an infinite arrival stream. The result cannot be generalized to more than two priority classes; with three priority classes, a counterexample can be found.

Example 11.1.2 (Counterexample with Three Priority Classes)

The following counterexample has three jobs and is entirely deterministic.

jobs	1	2	3
p_j	1	4	1
r_j	0	0	1
w_j	1	5	100

At time 0 the job with the highest w_j/p_j ratio is job 2. However, under the optimal schedule, it is job 1 that has to be processed at time 0. After job 1 has been completed at time 1, job 3 starts its processing. Under this schedule, the total weighted completion time is

$$1 + 1 \times 100 + 6 \times 5 = 131.$$

If job 2 would have started its processing at time 0, then the total weighted completion time would be

$$4 \times 5 + 4 \times 100 + 6 \times 1 = 426.$$

The proof of Theorem 11.1.1, for two priority classes, does not go through when unforced idleness is allowed. If unforced idleness is allowed, then it may be optimal to leave the machine idle while a job is waiting in anticipation of an imminent release of a high-priority job.

Example 11.1.3 (Counterexample when Unforced Idleness Is Allowed)

The following counterexample with two jobs is also deterministic.

jobs	1	2
p_j	4	1
r_j	0	1
w_j	1	100

At time 0, there is a job available for processing. However, it is optimal to keep the machine idle until time 1, process job 2 for one time unit and then process job 1. Under this optimal schedule, the total weighted completion time is

$$1 \times 100 + 6 \times 1 = 106.$$

If job 1 would have been put on the machine at time 0, then the total weighted completion time is

$$4 \times 1 + 5 \times 100 = 504.$$

11.2 PRIORITY QUEUES, WORK CONSERVATION, AND POISSON RELEASES

Assume that at the release of job j, at time R_j, the processing time X_j is drawn from distribution F_j. This implies that at any time t the total amount of processing required by the jobs waiting for processing (or, in queueing terminology, the customers waiting in queue) has already been determined. Let $x^r(t)$ denote the remaining processing time of the job that is being processed on the machine at time t. Let $V(t)$ denote the sum of the processing times of the jobs waiting for processing at time t plus $x^r(t)$. In the queueing literature, this $V(t)$ is typically referred to as the amount of work that is present in the system at time t.

At each release date, the $V(t)$ jumps (increases), and the size of the jump is the processing time of the job just released. Between jumps, $V(t)$ decreases continuously with slope -1, as long as the machine is busy processing a job. A realization of $V(t)$ is depicted in Figure 11.1. As long as unforced idleness of the machine is not allowed, the function $V(t)$ does not depend on the priorities of the different job classes nor on the sequence in which the jobs are processed on the machine.

Closed form expressions for the performance measures of interest (e.g., the expected time a typical job spends in the system under a given priority rule) can only be obtained under certain assumptions regarding the release times of the jobs. The release processes considered are those typically used in queueing theory.

Suppose there is a single class of jobs and the jobs have processing times that are i.i.d. and distributed according to F. There is an infinite stream of jobs coming in. The jobs are released according to a Poisson process with rate ν, implying that the probability the number of jobs released by time t, $N(t)$, equals ℓ is

$$P(N(t) = \ell) = \frac{e^{-\nu t}(\nu t)^\ell}{\ell!}.$$

The release times of the jobs are, of course, strongly dependent on one another. The release of any given job occurs a random time after the release of the previous job. Successive interrelease times are independent and exponentially distributed with the same mean.

Poisson release processes have a very useful and important property that, in queueing theory, is often referred to as *Poisson Arrivals See Time Averages (PASTA)*. Consider a single class of jobs that are released according to a Poisson process. The PASTA property implies that an arbitrary job, at its release, encounters an expected number of jobs waiting for processing that is equal to the average number of jobs waiting for processing at any other time that is selected at random. It also implies that the expected number of jobs an arbitrary release finds being

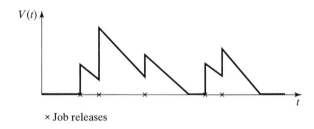

× Job releases

Figure 11.1   Amount of work in system as function of time.

processed on the machine is equal to the average number of jobs being processed at any other time selected at random (note that the number of jobs being processed is a $0 - 1$ random variable; so the expected number being processed is equal to the probability that the machine is processing a job). It also implies that an arbitrary job finds, at its release, an expected amount of work waiting in queue that is equal to the expected amount of work waiting at any other time that is selected at random. The PASTA property is an important characteristic of the Poisson process.

Example 11.2.1 (The PASTA Property)

Consider a Poisson release process with the expected time between consecutive releases being 10 minutes (i.e., $1/\nu = 10$ minutes). Assume the processing times are deterministic and equal to 4 minutes. The expected number of jobs in process (i.e., the time average) is 0.4. The PASTA property implies that this number is equal to the average number of jobs that a new release finds in service. So a Poisson release finds the machine busy with probability 0.4.

Consider now a release process that is not Poisson. A new job is released every 10 minutes (i.e., with deterministic interrelease times). The processing time of each job is again exactly 4 minutes (deterministic). The time average of the number of jobs in the queue is 0 and the time average of the number of jobs in service is 0.4 since 40% of the time a job is being processed and 60% of the time the machine is idle. Since the release process as well as the service times are deterministic, the number of jobs each new release finds in queue is 0. So this average number in queue seen by a new release happens to be equal to the time average (since both averages are zero). However, the average number of jobs that a new release finds in service is also 0, whereas the time average of the number of jobs in service is 0.4; so the average number in service seen by a release is not equal to the time average.

The operating characteristics of a machine subject to Poisson releases can be analyzed within the following framework. Suppose that in each time unit every job in the system (either in queue or in service) has to pay \$1.00 for every unit of processing time it still needs. So if a job needs processing for x time units, then for each unit of time it is waiting in queue it has to pay x dollars. However, while in service, the job's payout rate steadily declines as its remaining processing time decreases.

The average rate at which the system is earning its money is the rate at which the jobs are released, ν, times the expected amount paid by a job. However, the rate at which the system is earning money is also equal to the expected amount of work in the system at a random point in time. Let the random variable V denote the amount of work in the system in steady state. So,

$$E(V) = \lim_{t \to \infty} E(V(t))$$

and

$$E(V) = \nu E(\text{amount paid by a job}).$$

Let W^q and X denote the amount of time the job spends waiting in queue and the amount of time being processed. So the total time between release and completion is

$$W^s = W^q + X.$$

Since the job pays at a constant rate of X per unit time while it waits in queue and at a rate $X - x$ after spending an amount of time x in service,

$$E(\text{amount paid by a job}) = E\left(XW^q + \int_0^X (X - x)\, dx\right).$$

So

$$E(V) = \nu E(XW^q) + \frac{\nu E(X^2)}{2}.$$

In addition, if a job's wait in queue is independent of its processing time (this is the case when the priority rule is *not* a function of the processing time), then

$$E(V) = \nu E(X)E(W^q) + \frac{\nu E(X^2)}{2}.$$

In a queue that operates according to First Come First Served, a job's wait in queue is equal to the work in the system when it is released. So $E(W^q)$ is equal to the average work as seen by a new release. If the release process is Poisson, then (because of PASTA)

$$E(W^q) = E(V).$$

This identity, in conjunction with the identity

$$E(V) = \nu E(X)E(W^q) + \frac{\nu E(X^2)}{2},$$

yields

$$E(W^q) = \frac{\nu E(X^2)}{2(1 - \nu E(X))},$$

where $E(X)$ and $E(X^2)$ are the first two moments of the processing time distributions. This last expression is well known in queueing theory and often referred to as the Pollaczek–Khintchine formula; it is the expected time in an M/G/1 queue.

Poisson releases of jobs also make it possible to derive a closed form expression for the expected length of a busy period $E(B)$ and the expected length of an idle period $E(I)$ of the machine. The utilization rate of the machine can be expressed as

$$\frac{E(B)}{E(I) + E(B)} = \frac{\nu}{\lambda}.$$

So

$$E(B) = \frac{\nu E(I)}{\lambda - \nu}.$$

Because idle times are exponentially distributed, it follows that $E(I) = 1/\nu$ and

$$E(B) = \frac{1}{\lambda - \nu}.$$

The approach and analysis described earlier can also be applied to a setting with a number of different job classes. Let $x_k^r(t)$ denote the remaining processing time of the job that is being processed on the machine at time t, if the job is of class k, $k = 1, \dots, n$. Let $V_k(t)$ denote the sum of the processing times of all jobs of class k, $k = 1, \dots, n$, that are waiting for processing at time t plus $x_k^r(t)$. So

$$V(t) = \sum_{k=1}^{n} V_k(t).$$

If the classes of jobs are ordered according to a given priority list, then an interchange of priorities between two classes that are adjacent on the list, in the preemptive case, does not affect the $V_k(t)$ of any one of the classes with a higher priority nor does it affect the $V_k(t)$ of any one of the classes with a lower priority. Combining adjacent classes of priorities into a single class has no effect on either the $V_k(t)$ of a class with a lower priority or the $V_k(t)$ of a class with a higher priority under any type of policy.

Closed form expressions for performance measures, such as the expected time a typical job of a given class spends in system under a given priority rule, can again only be obtained under certain assumptions regarding the release times of the jobs.

Suppose there are c different classes of jobs, and jobs from class k, $k = 1, \dots, n$, have processing times that are i.i.d. and distributed according to F_k. There are n infinite streams of job releases; jobs from class k are released according to a Poisson process with rate ν_k, implying that the probability of the number of class k jobs released by time t, $N_k(t)$, is ℓ is

$$P(N_k(t) = \ell) = \frac{e^{-\nu_k t}(\nu_k t)^{\ell}}{\ell!}.$$

Release times of jobs from different classes are independent from one another, but release times of jobs from the same class are, of course, strongly dependent on one another. The release of a given job from any given class occurs a random time (exponentially distributed) after the release of the previous job of the same class.

11.3 ARBITRARY RELEASES AND EXPONENTIAL PROCESSING TIMES WITH PREEMPTIONS

Consider a machine that is subject to random releases of n jobs. The processing time of job j is X_j, exponentially distributed with rate λ_j. The weight of job j is w_j and its release date is R_j. The n release dates may have an arbitrary joint distribution; they may either be independent or have some form of dependence. If at the release of a new job the machine is processing a job of lower priority, then the job just released is allowed to preempt the job being processed. The goal is to find the optimal preemptive dynamic policy that minimizes the total expected weighted completion time.

The model considered is a stochastic counterpart of the deterministic model $1 \mid r_j, prmp \mid \sum w_j C_j$ (which is known to be strongly NP-hard). The model is also closely related to the so-called G/M/1 priority queue (i.e., a single server with exponential processing times subject to an arbitrary arrival process that is not necessarily renewal).

Theorem 11.3.1.   *The optimal preemptive dynamic policy is the preemptive WSEPT rule. That is, at any point in time, among the jobs available for processing, the one with the highest $\lambda_j w_j$ must be processed.*

Proof.   Consider a fixed (but arbitrary) time t. At time t, it is known which jobs have already been released. Let q_j denote the probability that, under a certain policy, job j, which was released before time t, has not completed its processing yet. Let $E(V(t))$ denote the total expected time that the machine at time t needs to finish all the jobs that have been released prior to t. Note that, assuming that preemptions generate no additional work and the machine is always kept busy when there are jobs waiting for processing, $E(V(t))$ is independent of the policy used.

Clearly, the amount of work still to be done increases at a release date by the amount of processing the newly released job requires, whereas at other instances the amount of work decreases linearly as long as there is still work to be done.

Because of the memoryless property of the exponential distribution, the expected remaining processing time at time t of the uncompleted job j is $1/\lambda_j$ independent of the amount of processing that job j has received prior to t.

Therefore,

$$E(V(t)) = \sum_{j=1}^{n} \frac{q_j}{\lambda_j}.$$

Consider an arbitrary preemptive static list policy. Reverse the priorities between jobs k and l with adjacent priorities in this static list policy. It is easy to see why this reversal does not affect the waiting times of other jobs. Interchanging the priorities of jobs k and l does not affect the waiting times of higher priority jobs as these jobs have preemptive rights over both k and l; in the same way, jobs not having priority over k and l do not depend on how higher priority jobs are processed.

The expected rate at which the objective function is increasing at time t is

$$w_k q_k + w_l q_l + \sum_{j \neq k,l} w_j q_j.$$

This rate has to be minimized. Because of work conservation,

$$\frac{q_k}{\lambda_k} + \frac{q_l}{\lambda_l} = E(V(t)) - \sum_{j \neq k,l} \frac{q_j}{\lambda_j}.$$

Rewriting this equation yields

$$q_l = \lambda_l \left(E(V(t)) - \sum_{j \neq k,l} \frac{q_j}{\lambda_j} - \frac{q_k}{\lambda_k} \right).$$

The expected rate at which the total cost function increases becomes now, after substitution,

$$q_k \left(w_k - w_l \frac{\lambda_l}{\lambda_k} \right) + \lambda_l w_l \left(E(V(t)) - \sum_{j \neq k,l} \frac{q_j}{\lambda_j} \right) + \sum_{j \neq k,l} w_j q_j.$$

Because reversing the priorities of jobs k and l only affects q_k and q_l, the last two terms of this expected rate do not change. The coefficient of q_k (the first term) is positive only when

$$\lambda_l w_l \leq \lambda_k w_k.$$

So when this inequality holds, the q_k should be kept small. This can only be achieved by giving job k a higher priority.

So any preemptive static list policy that is not in decreasing order of $\lambda_j w_j$ can be improved by a number of adjacent pairwise interchanges. This proves that the preemptive WSEPT policy of the theorem is optimal in the class of preemptive static list policies. The policy is optimal in this class for all possible realizations of release dates.

That this policy is also optimal in the class of preemptive dynamic policies can be shown as follows. Consider first the class of dynamic policies that only allow preemption at the release of a new job. That the policy is optimal in this class of policies can be shown by induction on the number of jobs still to be released. When zero jobs remain to be released, the policy is clearly optimal. Suppose that only one job remains to be released and before the release of this last job (but after the release of the second last job) a job is processed that does not have the highest ratio of $\lambda_j w_j$ among the jobs available for processing. By performing pairwise interchanges among the jobs during this time interval, the expected rate of increase of the objective function can be reduced for all time epochs after this interval. Proceeding inductively, it can be shown that when n jobs remain to be released the policy of the theorem is optimal within this particular class of preemptive dynamic policies.

Consider now the class of dynamic policies with preemptions also allowed at time epochs $\delta, 2\delta, 3\delta, \ldots$ That the policy of the theorem is also optimal in this class of preemptive dynamic policies can be shown by induction as well. Consider the last time interval when an action is taken that does not conform to the stated policy. Then a pairwise switch of jobs during this interval decreases the cost rate for any time epoch after this interval. An inductive argument shows that the policy of the theorem has to be adopted throughout. A continuity argument completes the proof of the theorem. □

The prior theorem also applies to a queueing system subject to an arbitrary input process and exponential service times (i.e., the G/M/1 queue). Closed form expressions for the expected waiting times in a G/M/1 queue can only be obtained for certain specific interrelease time distributions. The following example illustrates how the expected waiting time can be computed when the machine is subject to a stationary input process.

Example 11.3.2 (Arbitrary Releases and Exponential Processing with Preemptions)

Consider a machine with two job classes. Both classes have the same exponential processing time distribution with a mean of 2 minutes (i.e., $\lambda = 0.5$ jobs per minute). One class arrives at regular (fixed) intervals of 8 minutes each. The other class also arrives at regular (fixed) intervals of 8 minutes each. The two classes are synchronized in such a way that every 4 minutes a new job is released. It is easy to determine the expected waiting time of a Class 1 job and a Class 2 job in a D/M/1 queue. The expected waiting times of the two job classes separately and the two job classes together can either be determined analytically or can be found in tables. From the tables available in the literature, it can be established that $E(W_1^q) = 0.04$ minutes, $E(W_2^q) = 0.98$ minutes, and $E(W^q) = 0.51$ minutes.

Consider, for comparison purposes, the following scenario with the same two job classes. The two job classes are now released according to two independent Poisson processes with mean interrelease times $1/\nu = 8$ minutes. It is easy to determine the expected waiting time of a Class 1 job and a Class 2 job in such an M/M/1 queue. Class 1 jobs have preemptive priority over Class 2 jobs. So for Class 1 jobs, Class 2 jobs do not even exist. Analyzing the waiting times of the Class 1 jobs as an M/M/1 queue yields

$$E(W_1^q) = \frac{\nu}{\lambda(\lambda - \nu)} = \frac{0.125}{0.5(0.5 - 0.125)} = 0.6667.$$

The total expected waiting time of all jobs can also be analyzed as an M/M/1 queue. The total expected waiting time is

$$E(W^q) = \frac{0.25}{0.5(0.5 - 0.25)} = 2.$$

From the fact that

$$E(W^q) = 0.5E(W_1^q) + 0.5E(W_2^q),$$

it follows that $E(W_2^q) = 3.33$ min.

Note that the time a Class 2 job spends in queue may not be just a single uninterrupted period. The processing of a Class 2 job may be interrupted repeatedly by arrivals of Class 1 jobs.

Comparing the exponential interrelease times with the deterministic interrelease times shows that the expected waiting time of each job class is smaller when the interrelease times are deterministic.

In the prior example, the two job classes have the same processing time distribution. The optimality of the WSEPT rule also holds when job classes have different processing time distributions. However, in general it is not easy to find closed form expressions for the total expected waiting time of each class when the classes have different distributions. However, it is fairly easy when jobs are released according to Poisson processes.

Example 11.3.3 (Poisson Releases and Exponential Processing with Preemptions)

Consider two job classes. Jobs of Class 1 are released according to a Poisson process with rate $\nu_1 = 1/8$ (i.e., the mean interrelease time is 8 minutes). The distribution of the processing times of Class 1 jobs is exponential with mean and rate 1 (i.e., $\lambda_1 = 1$). The weight of Class 1 jobs is $w_1 = 1$. Class 2 jobs are also released according to a Poisson process with rate $\nu_2 = 1/8$. Their processing times are exponentially distributed with rate $\lambda_2 = 1/3$. Their weight is $w_2 = 3$. Because the $\lambda_j w_j$ values of the two classes are the same, both priority orderings have to be optimal. The fact that the release processes are Poisson makes it possible to obtain closed form expressions for all performance measures.

Consider first the case where the jobs of Class 1 have a higher priority than the jobs of Class 2. To compute the expected waiting time of the jobs of Class 1 is relatively easy. Since these jobs have a preemptive priority over the jobs of Class 2, the jobs of Class 2 do not have any impact on the waiting times of the jobs of Class 1. So these waiting times can be computed by just considering an M/M/1 queue with $v_1 = 1/8$ and $\lambda_1 = 1$. From some elementary results in queueing theory, it follows that

$$E(W_1^q) = \frac{v_1}{\lambda_1(\lambda_1 - v_1)} = \frac{0.125}{1(1 - 0.125)} = \frac{1}{7} = 0.1427$$

and

$$E(W_1^s) = E(W_1^q) + \frac{1}{\lambda_1} = \frac{8}{7} = 1.1427.$$

The expected waiting time of a Class 2 job is harder to compute. From queueing theory, it follows that an arbitrary arrival finds, on its arrival at the queue, an amount of work already waiting for the machine that is equal to the expected waiting time in an M/G/1 queue with Poisson arrivals at rate $v = v_1 + v_2$, and a service time distribution that is a mixture of two exponentials with rates 1 and 1/3 and with mixing probabilities of 0.5 and 0.5. The queue that an arbitrary release of Class 2 finds on arrival consists of jobs of Class 1 as well as Class 2. The arbitrary arrival will have to wait at least for all these jobs to be completed before it can even be considered for processing. This wait will be equal to the expected wait in an M/G/1 queue and is known to be

$$\frac{v E(X^2)}{2(1 - v E(X))} = \frac{0.25 \left(0.5(2/\lambda_1^2) + 0.5(2/\lambda_2^2)\right)}{2 \left(1 - 0.25 \left(0.5(1/\lambda_1) + 0.5(1/\lambda_2)\right)\right)} = 2.5.$$

So the expected total time a Class 2 job remains in system must be at least 2.5 plus its own processing time, which is $1/\lambda_2 = 3$ minutes. However, it is possible that during these 5.5 minutes additional Class 1 jobs are released. These Class 1 jobs have a preemptive priority over the Class 2 job under consideration. So these Class 1 jobs that arrive during these 5.5 minutes have to be taken into account as well. The expected number of Class 1 jobs released during this period is $v_1 \times 5.5 = 11/16$. How much additional wait do these releases cause? It is actually more than their own processing times, because, while these Class 1 jobs are being processed, additional Class 1 jobs may be released. So instead of looking at just 11/16 jobs released, one has to look at the busy periods (consisting of only Class 1 jobs) that each one of these releases generate. The busy period that one such job may cause can be analyzed by computing the busy period in an M/M/1 queue with arrival rate $v_1 = 1/8$ and service rate $\lambda_1 = 1$. From elementary queueing theory, it follows that the expected length of such a busy period generated by a single job is

$$\frac{1}{(\lambda_1 - \nu_1)} = \frac{8}{7}.$$

So the total expected time that an arbitrary Class 2 job spends in the system before it completes its processing is

$$E(W_2^s) = \frac{11}{2} + \frac{11}{16} \times \frac{8}{7} = \frac{11}{14} = \frac{44}{7} = 6.28.$$

Therefore,

$$E(W_2^q) = E(W_2^s) - \frac{1}{\lambda_2} = 3.28.$$

The total objective is

$$\sum_{k=1}^{n} w_k \frac{\nu_k}{\nu} E(W_k^q) = 1 \times \frac{1}{2} \times \frac{1}{7} + 3 \times \frac{1}{2} \times \frac{23}{7} = 5.$$

Consider now the same environment with the priority rule reversed: Class 2 jobs have a preemptive priority over Class 1 jobs. Now the performance measure with regard to Class 2 jobs is easy. They can be viewed as a simple M/M/1 queue with $\nu = 1/8$ and $\lambda = 1/3$:

$$E(W_2^q) = \frac{\nu}{\lambda(\lambda - \nu)} = \frac{0.125}{0.33(0.33 - 0.125)} = \frac{9}{5} = 1.8.$$

The computation of the expected waiting time of a Class 1 job, which has a lower priority, is now harder. The procedure is the same as the one followed for computing the expected waiting time of a Class 2 job when it has the lower priority. Consider again an arbitrary release of a Class 1 job. Upon its release, it finds an amount of work waiting in queue that is identical to what an arbitrary release of Class 1 finds. This amount was computed before and is equal to 2.5. So the expected total time that a Class 1 job has to remain in the system is at least $2.5 + 1/\lambda_1 = 3.5$. However, during this time, there may be additional releases of Class 2 jobs that all have a preemptive priority over the Class 1 job under consideration. The expected number of releases of Class 2 jobs over this period is $3.5 \times \nu_2 = 7/16$. However, each one of these Class 2 jobs generates a busy period of Class 2 jobs, which all have priority over the Class 1 job under consideration. The expected length of a Class 2 job busy period is

$$\frac{1}{(\lambda_2 - \nu_2)} = \frac{24}{5}.$$

So the expected time an arbitrary Class 1 job spends in system before it is completed is

$$E(W_1^s) = \frac{7}{2} + \frac{7}{16} \times \frac{24}{5} = \frac{56}{10} = 5.6.$$

Therefore,

$$E(W_1^q) = E(W_1^s) - \frac{1}{\lambda_1} = 4.6.$$

The total objective is

$$\sum_{k=1}^{n} w_k \frac{\nu_k}{\nu} E(W_k^q) = 1 \times \frac{1}{2} \times \frac{46}{10} + 3 \times \frac{1}{2} \times \frac{9}{5} = 5.$$

As expected, the values of the objective function under the two orderings are the same.

11.4 POISSON RELEASES AND ARBITRARY PROCESSING TIMES WITHOUT PREEMPTIONS

In contrast to the assumptions in the previous section, preemptions are not allowed in this section. The results in this section apply to a stochastic counterpart of the deterministic model $1 \mid r_j \mid \sum w_j C_j$.

The release times of the jobs in the previous sections were assumed to be completely arbitrary. The release time of any one job could be completely independent of the release time of any other job. The release time of job j could be, for example, an arbitrarily distributed random variable R_j, and this time could be measured from time 0. The release processes considered in the previous sections could also allow for any form of stochastic dependency among the release times of the various jobs. The results in these sections hold when there is a finite set of n jobs as well as when there is an infinite arrival stream. Therefore, they are valid in a steady state as well as a transient state. In contrast to the results in the previous sections, the results in this section apply only to steady state; the proof of optimality of the WSEPT rule presented in this section is only valid for the steady state case.

Consider two priority classes with release rates ν_1 and ν_2. The two release processes are independent Poisson. The processing time distributions are G_1 and G_2. Class 1 jobs have nonpreemptive priority over Class 2 jobs. Let $E(W_1^q)$ and $E(W_2^q)$ denote the average wait in queue of a Class 1 job and a Class 2 job, respectively.

Note that the total work in system at any time does not depend on the priority rule as long as the machine is always busy when there are jobs waiting for processing. So the work in the system is the same as it would be if the machine were processing the jobs according to the First Come First Served rule. Under the FCFS rule, the system is equivalent to a single class system subject to a Poisson release process with rate $\nu = \nu_1 + \nu_2$ and a processing time distribution

$$G(x) = \frac{\nu_1}{\nu}G_1(x) + \frac{\nu_2}{\nu}G_2(x),$$

which follows from the fact that a combination of two independent Poisson processes is itself Poisson with a rate that is the sum of the rates of the two underlying processes. The processing time distribution G can be obtained by conditioning on which priority class the job is from. It follows that the average amount of work in the system is

$$
\begin{aligned}
E(V) &= \frac{\nu E(X^2)}{2(1 - \nu E(X))}, \\
&= \frac{\nu\left(\frac{\nu_1}{\nu}E(X_1^2) + \frac{\nu_2}{\nu}E(X_2^2)\right)}{2\left(1 - \nu\left(\frac{\nu_1}{\nu}E(X_1) + \frac{\nu_2}{\nu}E(X_2)\right)\right)} \\
&= \frac{\nu_1 E(X_1^2) + \nu_2 E(X_2^2)}{2(1 - \nu_1 E(X_1) - \nu_2 E(X_2))}.
\end{aligned}
$$

To simplify the notation, let

$$\rho_k = \nu_k E(X_k).$$

So

$$E(V) = \frac{\nu_1 E(X_1^2) + \nu_2 E(X_2^2)}{2(1 - \rho_1 - \rho_2)}.$$

Note that X and W^q (i.e., the processing time and the time waiting for processing) of any given job are not independent of one another. Any information with regard to the realization of the processing time of a job may give an indication of the class to which the job belongs and therefore also an indication of its expected waiting time.

Let V_1 and V_2 denote the amount of work of Classes 1 and 2 in the system.

$$
\begin{aligned}
E(V_i) &= \nu_i E(X_i) E(W_i^q) + \frac{\nu_i E(X_i^2)}{2} \\
&= \rho_i E(W_i^q) + \frac{\nu_i E(X_i^2)}{2}, \qquad i = 1, 2.
\end{aligned}
$$

Define

$$
\begin{aligned}
E(V_i^q) &= \rho_i E(W_i^q), \\
E(V_i^s) &= \nu_i E(X_i^2)/2.
\end{aligned}
$$

So one could interpret V_i^q as the average amount of work of Class i waiting in queue and V_i^s as the average amount of Class i work in service. To compute the average amount of waiting time of a class 1 job, $E(W_1^q)$, consider an arbitrary job release of Class 1. Its wait in queue is the amount of Class 1 work in the system (in queue and in service) on its release plus the amount of Class 2 work in service on its release. Taking expectations and using the fact that Poisson Arrivals See Time Averages yields

$$E(W_1^q) = E(V_1) + E(V_2^s)$$
$$= \rho_1 E(W_1^q) + v_1 E(X_1^2)/2 + v_2 E(X_2^2)/2$$

or

$$E(W_1^q) = \frac{v_1 E(X_1^2) + v_2 E(X_2^2)}{2(1 - \rho_1)}.$$

To obtain $E(W_2^q)$, first note that it follows from $V + V_1 + V_2$ that

$$\frac{v_1 E(X_1^2) + v_2 E(X_2^2)}{2(1 - \rho_1 - \rho_2)} = \rho_1 E(W_1^q) + \rho_2 E(W_2^q) + v_1 E(X_1^2)/2 + v_2 E(X_2^2)/2$$
$$= E(W_1^q) + \rho_2 E(W_2^q).$$

From the last three equations, it follows that

$$\rho_2 E(W_2^q) = \frac{v_1 E(X_1^2) + v_2 E(X_2^2)}{2} \left(\frac{1}{1 - \rho_1 - \rho_2} - \frac{1}{1 - \rho_1} \right)$$

or

$$E(W_2^q) = \frac{v_1 E(X_1^2) + v_2 E(X_2^2)}{2(1 - \rho_1 - \rho_2)(1 - \rho_1)}$$
$$= \frac{v E(X^2)}{2(1 - \rho_1 - \rho_2)(1 - \rho_1)}.$$

If there are n different classes of jobs, a solution can be obtained in a similar way for V_j, $j = 1, \ldots, n$.

First, note that the total amount of work in the system due to jobs of Classes $1, \ldots, k$ is independent of the internal priority rule concerning classes $1, \ldots, k$ and depends only on the fact that each one of them has priority over all the jobs from Classes $k + 1, \ldots, n$. So, $V_1 + \cdots + V_k$ is the same as it would be in case Classes $1, \ldots, k$ are coalesced in a single macro-class I and Classes $k + 1, \ldots, n$ are coalesced in a single macro-class II.

Using this concept of macro-classes in a repetitive fashion, it can be shown that

$$E(W_k^q) = \frac{\nu E(X^2)}{2(1 - \rho_1 - \cdots - \rho_k)(1 - \rho_1 - \cdots - \rho_{k-1})}.$$

Assume now that a Class k job has a weight w_k. This implies that each unit of time a Class k job remains in the system costs w_k dollars. In all the scheduling models considered previously in this book, the objective always has been to minimize the total expected weighted completion time. In the model considered here, there is an infinite arrival stream. The objective cannot be the total expected weighted completion time, but must be something different. The appropriate objective now is to minimize the total expected cost *per unit time*—that is, the average expected amount of money that has to be paid out per unit time by the jobs that are present in the system. It can be shown that this objective is equivalent to the objective of minimizing the average cost per job. More formally, this objective is

$$\sum_{k=1}^{n} w_k \frac{\nu_k}{\nu} E(W_k^q).$$

Minimizing this objective is equivalent to minimizing

$$\sum_{k=1}^{n} w_k \nu_k E(W_k^q).$$

The following theorem specifies the static priority list that minimizes this objective.

Theorem 11.4.1.   *Under the optimal nonpreemptive dynamic policy, the decision maker selects, whenever the machine is freed, from among the waiting jobs one with the highest value of $\lambda_j w_j$. This implies that the jobs are scheduled according to the WSEPT rule.*

Proof.   The proof is again based on an adjacent pairwise interchange argument. Assume $n \geq 3$, Classes 2 and 3 do not conform the $\lambda_j w_j$ rule (i.e., Class 2 has priority over Class 3), and

$$\lambda_2 w_2 < \lambda_3 w_3.$$

Let $\sum(2, 3)$ denote the value of

$$\sum_{k=1}^{n} w_k \nu_k E(W_k^q)$$

under this priority assignment, and let $\sum(3, 2)$ denote the value of the objective after interchanging the priorities of Classes 2 and 3. Now,

$$\sum(2, 3) - \sum(3, 2) = \frac{(2 - 2\rho_1 - \rho_2 - \rho_3)\rho_2\rho_3\nu E(X^2)(\lambda_3 w_3 - \lambda_2 w_2)}{2(1 - \rho_1)(1 - \rho_1 - \rho_2)(1 - \rho_1 - \rho_3)(1 - \rho_1 - \rho_2 - \rho_3)}$$

$$\geq 0.$$

This argument applies to Classes k and $k + 1$ when $k > 1$. To see this, combine all classes with priority over Class k as a single macro-class having priority 1 (with the appropriate arrival rate and service time distribution). Treat Classes k and $k + 1$ as Classes 2 and 3, respectively. A similar argument handles the case for Classes 1 and 2.

Thus, any assignment of priorities that differs from the λw rule can be improved by pairwise interchanges and the WSEPT rule is therefore optimal.   □

The following example illustrates the behavior of the WSEPT rule and compares the nonpreemptive setting with the preemptive setting described in the previous section.

Example 11.4.2 (Poisson Releases and Arbitrary Processing without Preemptions)

Consider a single machine with two job classes. The jobs are released according to a Poisson process and a release is a Class 1 job with probability 0.5 and a Class 2 job with probability 0.5. Class 1 jobs have nonpreemptive priority over Class 2 jobs. Consider first the case where the processing time distributions of both job classes are exponential with a mean of 2 minutes. The interrelease times of Class 1 jobs as well as Class 2 jobs are exponentially distributed with a mean of 8 minutes (i.e., $\nu_1 = \nu_2 = 1/8$ and $\nu = 1/4$). Working out the expressions in this section yields the following results for the expected times that the jobs have to wait before their processing can start:

$$E(W_1^q) = \frac{.125 \times 8 + .125 \times 8}{2(1 - .125 \times 2)} = 1.333$$

and

$$E(W_2^q) = \frac{E(W_1^q)}{1 - \rho_1 - \rho_2} = 2.667.$$

Notice that

$$E(W^q) = 0.5E(W_1^q) + 0.5E(W_2^q) = 2,$$

which is equal to $E(W^q)$ when preemptions are allowed. That the expected waiting times of all the jobs are equal in the preemptive and nonpreemptive cases is to be expected (see Example 11.3.2 and Exercise 11.13).

Consider now the case where both job classes have deterministic processing times with a mean of 2 minutes. Working out the expressions yields the following results for the expected times that the jobs have to wait for processing:

$$E(W_1^q) = \frac{.125 \times 4 + .125 \times 4}{2(1 - .125 \times 2)} = 0.667$$

and

$$E(W_2^q) = \frac{E(W_1^q)}{1 - \rho_1 - \rho_2} = 1.333.$$

Notice that

$$E(W^q) = 0.5E(W_1^q) + 0.5E(W_2^q) = 1,$$

which is only half of the total expected waiting time obtained for the case with exponentially distributed processing times. So a smaller variance in the processing times causes a significant reduction in the expected waiting times.

Clearly, this last example can be generalized to arbitrary processing time distributions.

Example 11.4.3 (Poisson Releases, Exponential Processing, without Preemptions)

Consider again a single machine and two job classes. The jobs are released according to a Poisson process with rate $\nu = 1/4$. Each release is a job of Class 1 with probability 0.5 and a job of Class 2 with probability 0.5. So the releases of jobs of Class 1 are Poisson with rate $\nu_1 = 1/8$ and the releases of jobs of Class 2 are Poisson with rate $\nu_2 = 1/8$. The processing times of jobs of Class 1 are exponentially distributed with mean 1 and the processing times of jobs of Class 2 are exponentially distributed with mean 3. The weight of a Class 1 job is 1 and the weight of a Class 2 job is 3.

If Class 1 jobs have a higher priority than Class 2 jobs, then

$$E(W_1^q) = \frac{.125 \times 2 + .125 \times 18}{2(1 - .125 \times 1)} = 1.429$$

and

$$E(W_2^q) = \frac{E(W_1^q)}{1 - \rho_1 - \rho_2} = 2.857.$$

The value of the objective to be minimized is

$$\sum_{k=1}^{n} w_k \frac{\nu_k}{\nu} E(W_k^q) = 1 \times 0.5 \times 1.429 + 3 \times 0.5 \times 2.857 = 5.$$

Consider now the case where the jobs of Class 2 have a higher priority than the jobs of Class 1:

$$E(W_2^q) = \frac{.125 \times 2 + .125 \times 18}{2(1 - .125 \times 3)} = 2$$

and

$$E(W_1^q) = \frac{E(W_2^q)}{1 - \rho_1 - \rho_2} = 4.$$

The value of the objective to be minimized is

$$\sum_{k=1}^{n} w_k \frac{\nu_k}{\nu} E(W_k^q) = 1 \times 0.5 \times 4 + 3 \times 0.5 \times 2 = 5.$$

Note that the value of the objective to be minimized is the same under the two orderings. This was to be expected since the $\lambda_j w_j$ values of the two job classes are the same. Both priority orderings have to be optimal.

11.5 DISCUSSION

Consider the case with many job classes, Poisson releases, and exponential processing time distributions (i.e., a model that satisfies the conditions of Section 11.3 as well as those of Section 11.4). The results in Section 11.3 imply that the preemptive WSEPT rule minimizes the total expected weighted completion time in the class of preemptive dynamic policies, whereas the results of Section 11.4 imply that the nonpreemptive WSEPT rule minimizes the total expected weighted completion time in the class of nonpreemptive dynamic policies. Clearly, the realizations of the process under the two different rules are different.

To obtain some more insight into the results presented in Section 11.4, consider the following limiting case. Suppose that there are many—say 10,000— different job classes. Each class has an extremely low Poisson release rate. The total job release rate will keep the machine occupied—say 40% of the time. The machine will alternate between busy periods and idle periods; during the busy periods, it may process on the average, say, 10 jobs. These 10 jobs are most likely jobs from 10 different classes. The process during such a busy period may be viewed as a nonpreemptive stochastic scheduling problem (with a random, but finite number of jobs). The results in Section 11.4 imply that the nonpreemptive WSEPT rule minimizes the total expected weighted completion time.

The case not considered in this chapter is a generalization of the case considered in Section 10.2 (i.e., the jobs have arbitrary processing time distributions and different release dates with preemptions allowed). When all the jobs are released at the same time, then the Gittins index policy is optimal. It turns out that when the jobs are released according to a Poisson process, an index policy is again optimal. However, the index is then not as easy to characterize as the Gittins index described in Section 10.2. A very special case for which the optimal preemptive policy can be easily characterized is considered in Exercise 11.7.

Some of the results in this chapter can be extended to machines in parallel. For example, the results in Section 11.4 can be generalized to machines in parallel under the condition that the processing time distributions of all classes are the same.

This chapter mainly focuses on the conditions under which the WSEPT rule is optimal assuming that the jobs are released at different times. It does not appear that similar results can be obtained for other priority rules.

However, it can be shown that under certain conditions the preemptive EDD rule minimizes $L_{\max}$. Assume that the jobs are released at random times and the processing times are also random from arbitrary distributions. So the model is a stochastic counterpart of $1 \mid r_j, prmp \mid L_{\max}$. If the due dates are deterministic, then it can be shown fairly easily that the preemptive EDD rule mininimizes $L_{\max}$ (see Exercise 11.20). If the times between the release date and the due date of each job are exponentially distributed with mean $1/\mu$, then the policy that minimizes $E(L_{\max})$ can be determined also.

EXERCISES (COMPUTATIONAL)

11.1. Consider three jobs. The three jobs have exponential processing times with rates $\lambda_1 = \lambda_2 = 1$ and $\lambda_3 = 2$, and the weights are $w_1 = 1$ and $w_2 = w_3 = 2$. Jobs 1 and 2 are released at time zero and job 3 is released after an exponential time with mean 1. Preemptions are allowed.

  (a) Show that the preemptive WSEPT rule minimizes the total expected weighted completion time.

  (b) Compute the total expected weighted completion time under this rule.

11.2. Consider the same three jobs as in Exercise 11.1. However, preemptions are now not allowed.

  (a) Show that the nonpreemptive WSEPT rule minimizes the total expected weighted completion time.

  (b) Compute the total expected weighted completion time under this rule.

  (c) Compare the outcome of part (b) with the outcome of part (b) in the previous exercise and explain the difference.

11.3. Consider a single machine subject to Poisson releases with rate $\nu = 0.5$. All the jobs are of the same class, and the processing time distribution is a mixture of two exponentials. With probability $1 - p$ the processing time is 0, and with probability p the processing time is exponentially distributed with rate p. So the mean of the mixture is 1. The jobs are served according to the First In First Out (FIFO) rule and preemptions are not allowed.

(a) Apply the Pollaczek–Khintchine formula and find an expression for $E(W_q)$.

(b) What happens with $E(W_q)$ when $p \to 0$?

11.4. Consider again a single machine subject to Poisson releases with rate $\nu = 0.5$. All the jobs are of the same class, and the processing time distribution is a mixture of two exponentials. With probability $1 - p$, the processing time is 0 and with probability p it is exponential with rate p. However, now preemptions are allowed.

(a) Formulate the policy that minimizes the long-term average waiting (or flow) time.

(b) Find an expression for $E(W_q)$ as a function of p under this optimal policy.

(c) Compare the expression under (b) with the expression found for $E(W_q)$ in Exercise 11.3. How does the difference depend on p?

11.5. Consider a single machine that is subject to Poisson releases with rate $\nu = 0.5$. All the jobs are of the same class and the processing time distribution is an Erlang(k, λ) distribution with mean 1 (i.e., $k/\lambda = 1$). Preemptions are not allowed.

(a) Find an expression for $E(W_q)$ as a function of k.

(b) How does $E(W_q)$ depend on k?

11.6. Consider the following setting that is somewhat similar to Example 11.4.3. There are two job classes. The two classes are released according to Poisson processes with rates $\nu_1 = \nu_2 = 0.25$. Preemptions are not allowed. The processing time distribution of each one of the two job classes is a mixture of two exponentials with one of the two exponentials having mean 0. The processing time of a Class 1 job is 0 with probability $1 - p_1$ and exponentially distributed with rate p_1 with probability p_1. The processing time of a Class 2 job is 0 with probability $1 - p_2$ and exponentially distributed with rate p_2 with probability p_2. So the means of the processing times of both job classes are 1. Class 1 jobs have nonpreemptive priority over Class 2 jobs. Compute the expected waiting time of a Class 1 job and of a Class 2 job.

11.7. Consider the same setting as in the previous exercise. However, now Class 1 jobs have preemptive priority over Class 2 jobs. Compute the expected waiting time of a Class 1 job and of a Class 2 job and compare your results with the results obtained in the previous exercise.

11.8. There are two job classes. The classes are released according to Poisson processes with rates $\nu_1 = \nu_2 = 0.25$. The processing time distributions of both job classes are mixtures of two exponentials with one of them having mean 0. The processing time of a Class 1 job is 0 with probability 0.5 and exponentially distributed with mean 2 with probability 0.5. The processing time of a Class 2 job is 0 with probability 0.75 and exponentially distributed with mean 4 with probability 0.25. Assume that the weight of Class 1 jobs is $w_1 = 2$ and the weight of Class 2 jobs is $w_2 = 3$. Preemptions are allowed.

(a) Describe the optimal policy when preemptions are not allowed.

(b) Describe the optimal policy when preemptions are allowed.

11.9. Consider a single machine. Jobs are released in batches of two according to a Poisson process. The arrival rate of the batches is $\nu = 0.25$. Each batch contains one job of Class 1 and one job of Class 2. The processing times of all jobs (from both classes) are exponential with mean 1. Class 1 jobs have preemptive priority over Class 2 jobs. Compute the expected waiting time of Class 1 jobs and the expected waiting time of Class 2 jobs.

11.10. Consider a single machine and two job classes. Preemptions are not allowed. The processing times of Class 1 (2) jobs are exponentially distributed with rate $\lambda_1 = 6$ ($\lambda_2 = 12$). The weight of Class 1 (2) jobs are $w_1 = 4$ ($w_2 = 3$). When the machine is processing a job of any one of the two classes, it is subject to breakdowns. The uptimes of a machine when it is processing a Class 1 (2) job are exponentially distributed with rate $v_1 = 6$ ($v_2 = 5$). The repair times after a breakdown when processing a Class 1 (2) job are arbitrarily distributed with mean $1/\mu_1 = 0.5$ ($1/\mu_2 = 1$). Which class of jobs should have a higher priority? (*Hint:* Look at Exercise 10.11.)

EXERCISES (THEORY)

11.11. Consider three jobs. Two are available at time 0 and the third one is released after an exponential amount of time. The three processing times are all deterministic. Preemptions are not allowed. Is it always optimal to start at time 0 the job with the the highest λ_j/w_j ratio?

11.12. Consider three jobs. Two are available at time 0 and the third one is released after a deterministic (fixed) amount of time. The three processing times are all exponential. Preemptions are not allowed. Is it always optimal to start at time 0 the job with the the highest $\lambda_j w_j$ ratio?

11.13. Compare the numerical results of Examples 11.3.2 and 11.4.2. Explain why the value of the total objective in the preemptive case is equal to the value of the total objective in the nonpreemptive case.

11.14. Consider the setting of Exercise 11.6. Assume $p_1 > p_2$. Which class should have nonpreemptive priority if the objective is to minimize the total expected waiting time over both classes?

11.15. In queueing theory, there is a well-known result known as Little's Law that establishes a relationship between the long-term average of the number of jobs waiting in queue and the expected waiting time of a job in steady state. Little's Law states that

$$E(N^q) = \nu E(W^q).$$

(a) Give an intuitive argument why this relationship holds.
(b) Present a proof for this relationship.

11.16. Consider an arbitrary release process (not necessarily Poisson). Job j has with probability p_j a processing time that is 0 and with probability $1 - p_j$ a processing time that is exponentially distributed with mean $1/\lambda_j$. Job j has a weight w_j. Preemptions are allowed. Describe the optimal preemptive dynamic policy that minimizes the total weighted expected completion time.

11.17. Consider a machine subject to Poisson releases of a single class. The processing time distribution is Erlang$(2, \lambda)$.

    (a) Describe the optimal preemptive policy.

    (b) When do preemptions occur?

11.18. Consider a machine that is subject to breakdowns. Breakdowns can occur only when the machine is processing a job. The uptimes of the machine, in between breakdowns, are i.i.d. exponential. The downtimes are i.i.d. (arbitrarily distributed) with mean $1/\mu$. There are two job classes. Both job classes are released according to Poisson processes. The processing times of the jobs of Class 1 (2) are arbitrarily distributed according to G_1 (G_2). The weight of Class 1 (2) is w_1 (w_2). Show that the nonpreemptive WSEPT rule is optimal in the class of nonpreemptive policies.

11.19. Consider a machine that is subject to breakdowns. The machine can break down only when it is processing a job. The uptimes of the machine, when it is processing jobs, are i.i.d. exponential. The downtimes are i.i.d. exponential as well. There are two job classes that are released according to independent Poisson processes. Class 1 jobs have preemptive priority over Class 2 jobs. The processing times of Class 1 (2) jobs are exponentially distributed with rate λ_1 (λ_2).

    (a) Can Theorem 11.3.1 be generalized to this setting?

    (b) Can Theorem 11.3.1 be generalized to include the case where the machine can break down at any time (i.e., also when it is idle)?

11.20. Consider the following stochastic counterpart of $1 \mid r_j, prmp \mid L_{\max}$. The release dates and due dates are deterministic. The processing time of job j has an arbitrary distribution F_j. Show that EDD minimizes $E(L_{\max})$.

11.21. Consider the following stochastic counterpart of $1 \mid r_j, prmp \mid L_{\max}$. The release dates have an arbitrary distribution and the due date $D_j = R_j + X_j$, where X_j is exponentially distributed with rate 1. The processing time of job j has an arbitrary distribution F_j. Formulate the policy that minimizes $E(L_{\max})$.

COMMENTS AND REFERENCES

Most of the results in this chapter have appeared in the queueing literature. Many books on queueing cover priority queues; see Kleinrock (1976), Ross (1981), Heyman and Sobel (1982), and Wolff (1989). The application of work conservation in queueing theory was first considered by Wolff (1970); his paper contains the results described in Sections 11.1 and 11.2.

    The optimality of WSEPT when processing times are exponentially distributed and release times are arbitrary with preemptions allowed is due to Pinedo (1983). The optimality of WSEPT when the jobs are released according to Poisson processes, the processing times are arbitrarily distributed, and no preemptions allowed is due to Cobham (1954).

12

Parallel Machine Models (Stochastic)

This chapter deals with parallel machine models that are stochastic counterparts of the models discussed in Chapter 5. The body of knowledge in the stochastic case is somewhat less extensive than in the deterministic case.

The results focus mainly on the expected makespan, the total expected completion time, and the expected number of tardy jobs. In what follows, the number of machines is usually limited to two. Some of the proofs can be extended to more than two machines, but such extensions usually require more elaborate notation. Since these extensions would not provide any additional insight, they are not presented here. The proofs for some of the structural properties of the stochastic models tend to be more involved than the proofs for the corresponding properties of their deterministic counterparts.

The first section deals with nonpreemptive models; the results in this section are obtained using interchange arguments. The second section focuses on preemptive models; the results in this section are obtained via dynamic programming approaches. The third section deals with due date-related models.

12.1 THE MAKESPAN WITHOUT PREEMPTIONS

This section considers optimal policies in the classes of nonpreemptive static list policies and nonpreemptive dynamic policies. Since preemptions are not allowed, the main technique for determining optimal policies is based on pairwise interchanges.

First, the exponential distribution is considered in detail as its special properties facilitates the analysis considerably.

Consider *two* machines in parallel and n jobs. The processing time of job j is equal to the random variable X_j, which is exponentially distributed with rate λ_j. The objective is to minimize $E(C_{\max})$. Note that this problem is a stochastic counterpart of $P2 \parallel C_{\max}$, which is known to be NP-hard. However, in Section 10.4, it already became clear that scheduling processes with exponentially distributed processing times often have structural properties that their deterministic counterparts do not have. It turns out that this is also the case with machines in parallel.

A nonpreemptive static list policy is adopted. The jobs are put into a list and at time 0 the two jobs at the top of the list begin their processing on the two machines. When a machine becomes free, the next job on the list is put on the machine. It is not specified in advance on which machine each job will be processed, nor is it known a priori which job will be the last one to be completed.

Let Z_1 denote the time when the second to last job is completed (i.e., the first time a machine becomes free with no jobs on the list to replace it). At this time, the other machine is still processing its last job. Let Z_2 denote the time that the last job is completed on the other machine (i.e., Z_2 equals the makespan $C_{\max}$). Let the difference D be equal to $Z_2 - Z_1$. It is clear that the random variable D depends on the schedule. It is easy to see that minimizing $E(D)$ is equivalent to minimizing $E(C_{\max})$. This follows from

$$Z_1 + Z_2 = 2C_{\max} - D = \sum_{j=1}^{n} X_j,$$

which is a constant independent of the schedule.

In what follows, a slightly more general two machine problem is considered for reasons that become clear later. It is assumed that one of the machines is not available at time 0 and becomes available only after a random time X_0, distributed exponentially with rate λ_0. The random variable X_0 may be thought of as the processing time of an additional job that takes precedence and *must* go first. Let $D(X_0, X_1, X_2, \ldots, X_n)$ denote the random variable D, under the assumption that, at time 0, a job with remaining processing time X_0 is being processed on one machine and a job with processing time X_1 is being started on the other. When one of the two machines is freed, a job with processing time X_2 is started, and so on (see

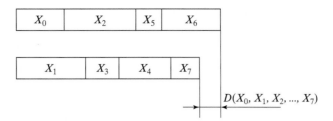

Figure 12.1   The random variable D.

Figure 12.1). The next lemma examines the effect on D of changing a schedule by swapping the two consecutive jobs 1 and 2.

Lemma 12.1.1.   *For any λ_0 and for $\lambda_1 = \min(\lambda_1, \lambda_2, \ldots, \lambda_n)$,*

$$E(D(X_0, X_1, X_2, \ldots, X_n)) \leq E(D(X_0, X_2, X_1, \ldots, X_n)).$$

Proof.   Let q_j (r_j), $j = 0, 1, \ldots, n$, denote the probability that job j is the last job to be completed under schedule $1, 2, \ldots, n$ $(2, 1, \ldots, n)$. The distribution of D may be regarded as a mixture of $n+1$ distributions (all of which are exponential) with either mixing probabilities $q_0, \ldots, q_n$ or mixing probabilities $r_0, \ldots, r_n$. If the last job to be completed is exponential with rate λ_j, then (conditioned on that fact) D is exponentially distributed with rate λ_j. (Recall that if job j is still being processed on a machine while the other machine is idle and no other job remains to be processed, then the remaining processing time of job j is still exponentially distributed with rate λ_j.) So

$$E(D(X_0, X_1, X_2, \ldots, X_n)) = \sum_{j=0}^{n} q_j \frac{1}{\lambda_j}$$

while

$$E(D(X_0, X_2, X_1, \ldots, X_n)) = \sum_{j=0}^{n} r_j \frac{1}{\lambda_j}.$$

To prove the lemma, it suffices to show that

$$q_0 = r_0$$
$$q_1 \leq r_1$$
$$q_j \geq r_j$$

for $j = 2, \ldots , n$. For job 0 to be the last one completed, it has to be larger than the sum of the other n processing times. Clearly, if this is the case, an interchange between jobs 1 and 2 does not affect the probability of job 0 being completed last nor the value of D. To establish the necessary relationships between the mixing probabilities q_j and r_j consider first the case $n = 2$. It can be easily shown that

$$q_0 = \left(\frac{\lambda_1}{\lambda_1 + \lambda_0} \right) \left(\frac{\lambda_2}{\lambda_2 + \lambda_0} \right)$$

$$q_1 = \left(\frac{\lambda_0}{\lambda_0 + \lambda_1} \right) \left(\frac{\lambda_2}{\lambda_1 + \lambda_2} \right)$$

$$q_2 = 1 - q_0 - q_1$$

and

$$r_0 = \left(\frac{\lambda_1}{\lambda_1 + \lambda_0} \right) \left(\frac{\lambda_2}{\lambda_2 + \lambda_0} \right)$$

$$r_1 = 1 - r_0 - r_2$$

$$r_2 = \left(\frac{\lambda_0}{\lambda_0 + \lambda_2} \right) \left(\frac{\lambda_1}{\lambda_1 + \lambda_2} \right).$$

So

$$r_1 - q_1 = \frac{2\lambda_1\lambda_2}{\lambda_2^2 - \lambda_1^2} \left(\frac{\lambda_0}{\lambda_0 + \lambda_1} - \frac{\lambda_0}{\lambda_0 + \lambda_2} \right) \geq 0.$$

This proves the lemma for $n = 2$.

Assume the lemma is true for $n - 1$ jobs and let q'_j and r'_j, $j = 1, \ldots , n - 1$, denote the corresponding probabilities under schedules $0, 1, 2, 3, \ldots , n - 1$ and $0, 2, 1, 3, \ldots , n - 1$. Assume as the induction hypothesis that

$$q'_0 = r'_0$$

$$q'_1 \leq r'_1$$

$$q'_j \geq r'_j,$$

for $j = 2, \ldots , n - 1$. Let q_j and r_j now denote the corresponding probabilities with one additional job. Then

$$q_0 = r_0 = P(X_1 + X_2 + \cdots + X_n < X_0) = \prod_{j=1}^{n} \frac{\lambda_j}{\lambda_j + \lambda_0}$$

and

$$q_j = q'_j \frac{\lambda_n}{\lambda_j + \lambda_n}$$

$$r_j = r'_j \frac{\lambda_n}{\lambda_j + \lambda_n}$$

for $j = 1, \ldots, n - 1$. So, from the induction hypothesis, it follows that

$$q_1 \leq r_1$$

$$q_j \geq r_j,$$

for $j = 2, \ldots, n - 1$. Also, because $\lambda_1 \leq \lambda_j$ for all $j = 2, \ldots, n - 1$, it follows that

$$\frac{\lambda_n}{\lambda_1 + \lambda_n} \geq \frac{\lambda_n}{\lambda_j + \lambda_n}$$

for $j = 2, \ldots, n - 1$. So

$$\sum_{j=1}^{n-1} q'_j \frac{\lambda_n}{\lambda_j + \lambda_n} \leq \sum_{j=1}^{n-1} r'_j \frac{\lambda_n}{\lambda_j + \lambda_n}.$$

Therefore,

$$q_n \geq r_n,$$

which completes the proof of the lemma.   □

This lemma constitutes a crucial element in the proof of the following theorem.

Theorem 12.1.2.   *If there are two machines in parallel and the processing times are exponentially distributed, then the LEPT rule minimizes the expected makespan in the class of nonpreemptive static list policies.*

Proof.   By contradiction. Suppose that a different rule is optimal and that according to this presumably optimal rule, the job with the longest expected processing time is not scheduled for processing either as the first or the second job. (Note that the first and second job are interchangeable as they both start at time 0.) Then an improvement can be obtained by performing a pairwise interchange between this longest job and the job immediately preceding this job in the schedule. As by Lemma 12.1.1, this reduces the expected difference between the completion times of the last two jobs. Through a series of interchanges, it can be shown that the longest job has to be one of the first two jobs in the schedule. In the same way, it can be shown that the second longest job has to be among the first two jobs as well. The third longest job can be moved into the third position to improve the objective, and so on. With each interchange, the expected difference, and thus the expected makespan, are reduced.   □

The approach used in proving the theorem is basically an adjacent pairwise interchange argument. However, this pairwise interchange argument is not identical
to the pairwise interchange arguments used in single machine scheduling. In pairwise interchange arguments applied to single machine problems, *no* restrictions
are made on the relation between the interchanged jobs and those that come after them. In Lemma 12.1.1, jobs not involved in the interchange have to satisfy a
special condition (viz., one of the two jobs being interchanged must have a larger
expected processing time than all jobs following it). Requiring such a condition has
certain implications. When no special conditions are required, an adjacent pairwise
interchange argument actually yields two results: It shows that one schedule minimizes the objective while the *reverse* schedule maximizes that same objective. With
a special condition like the one in Lemma 12.1.1, the argument only works in one
direction. It actually can be shown that the SEPT rule does *not* always maximize
$E(D)$ among nonpreemptive static list policies.

The result presented in Theorem 12.1.2 differs from the results obtained for
its deterministic counterpart considerably. One difference is the following: Minimizing the makespan in a deterministic setting only requires an optimal *partition*
of the n jobs over the two machines. After the allocation has been determined, the
set of jobs allocated to a specific machine may be sequenced in any order. With
exponential processing times, a sequence is determined in which the jobs are to
be *released* in order to minimize the expected makespan. No deviation is allowed
from this sequence and it is not specified at time 0 how the jobs will be partitioned
between the machines. This depends on the evolution of the process.

The following example shows that distributions other than exponential may
result in optimal schedules that differ considerably from LEPT.

Example 12.1.3 (Minimizing Expected Makespan with Arbitrary Processing Times)

Consider two machines and n jobs. The processing time of job j is a mixture of two
exponentials. With probability p_j, it is exponential with rate ∞ (i.e., it has a zero
processing time) and with probability $1 - p_j$, it is exponential with rate λ_j. So

$$E(X_j) = \frac{1 - p_j}{\lambda_j}.$$

Assume $\lambda_1 \leq \cdots \leq \lambda_n$. Since there are no assumptions on $p_1, \ldots, p_n$, sequence
$1, \ldots, n$ may not be LEPT. That the sequence $1, \ldots, n$ still minimizes the expected
makespan can be argued as follows. Note that if a job has zero processing time, it basically does not exist and does not have any effect on the process. Suppose it is known
in advance which jobs have zero processing times and which jobs are exponentially
distributed with processing times larger than zero. From Theorem 12.1.2, it follows
that sequencing the nonzero jobs in increasing order of their rates—that is, according to LEPT—minimizes the expected makespan. This implies that the sequence
$1, \ldots, n$ is optimal for *any* subset of jobs having zero processing times. So, se-

quence $1, \ldots, n$, which is not necessarily LEPT, is always optimal. The $p_1, \ldots, p_n$ can be such that it is actually the SEPT rule that minimizes the expected makespan.

In contrast to the results of Section 10.4, which do not appear to hold for distributions other than the exponential, the LEPT rule does minimize the expected makespan for other distributions as well.

Consider the case where the processing time of job j is distributed according to a mixture of two exponentials (i.e., with probability p_{1j} according to an exponential with rate λ_1 and with probability p_{2j} ($= 1 - p_{1j}$) according to an exponential with rate λ_2). Assume $\lambda_1 < \lambda_2$. So

$$P(X_j > t) = p_{1j}e^{-\lambda_1 t} + p_{2j}e^{-\lambda_2 t}.$$

This distribution can be described as follows: When job j is put on the machine, a (biased) coin is tossed. Dependent on the outcome of the toss, the processing time of job j is either exponential with rate λ_1 or exponential with rate λ_2. After the rate has been determined this way, the distribution of the remaining processing time of job j does not change while the job is being processed. So each processing time is distributed according to one of the two exponentials with rates λ_1 and λ_2.

The subsequent lemma again examines the effect on D of an interchange between two consecutive jobs 1 and 2 on two machines in parallel. Assume again that X_0 denotes the processing time of a job 0 with an exponential distribution with rate λ_0. This rate λ_0 may be different from either λ_1 or λ_2.

Lemma 12.1.4.   *For arbitrary λ_0, if $p_{11} \geq p_{12}$, i.e., $E(X_1) \geq E(X_2)$, then*

$$E(D(X_0, X_1, X_2, \ldots, X_n)) \leq E(D(X_0, X_2, X_1, \ldots, X_n)).$$

Proof.   Note that $D(X_0, X_1, X_2 \ldots, X_n)$ is exponential with one of three rates—namely, λ_0, λ_1, or λ_2. Denote the probabilities of these three events under schedule $0, 1, 2, \ldots, n$ by q_0, q_1, and q_2, where $q_0 + q_1 + q_2 = 1$. Denote by r_0, r_1, and r_2 the probabilities of the same events under schedule $0, 2, 1, 3, \ldots, n$. It is clear that q_0 is equal to r_0 by a similar argument as in Lemma 12.1.1. To prove the lemma, it suffices to show that $q_1 < r_1$.

For $n = 2$, there are four cases to be considered in the computation of q_1. With probability $p_{21}p_{22}$, both jobs 1 and 2 are exponential with rate λ_2. In this case, the probability that $D(X_0, X_1, X_2)$ is exponentially distributed with rate λ_1 is zero. With probability $p_{11}p_{22}$, job 1 is exponentially distributed with rate λ_1, whereas job 2 is exponentially distributed with rate λ_2. For $D(X_0, X_1, X_2)$ to have rate λ_1, job 1 has to outlast job 0 and, after job 2 is started on the machine on which job 0 is completed, job 1 has to outlast job 2 as well. This happens with probability

$$\left(\frac{\lambda_0}{\lambda_0 + \lambda_1}\right)\left(\frac{\lambda_2}{\lambda_2 + \lambda_1}\right).$$

The other two cases can also be computed easily. Summarizing,

$$q_1 = p_{11}p_{12}\left(\frac{\lambda_0}{\lambda_0 + \lambda_1} + \left(\frac{\lambda_1}{\lambda_0 + \lambda_1}\right)\left(\frac{\lambda_0}{\lambda_0 + \lambda_1}\right)\right)$$

$$+ p_{11}p_{22}\left(\frac{\lambda_0}{\lambda_0 + \lambda_1}\right)\left(\frac{\lambda_2}{\lambda_1 + \lambda_2}\right)$$

$$+ p_{21}p_{12}\left(\left(\frac{\lambda_0}{\lambda_0 + \lambda_2}\right)\left(\frac{\lambda_2}{\lambda_1 + \lambda_2}\right) + \left(\frac{\lambda_2}{\lambda_0 + \lambda_2}\right)\left(\frac{\lambda_0}{\lambda_0 + \lambda_1}\right)\right)$$

$$+ p_{21}p_{22}(0).$$

So, the first term on the R.H.S. of this expression corresponds to the event where both jobs 1 and 2 are exponentially distributed with rates λ_1, which happens with probability $p_{11}p_{12}$; the second term corresponds to the event where job 1 is distributed according to an exponential with rate λ_1 and job 2 according to an exponential with rate λ_2, and the third term corresponds to the event where job 1 is distributed according to an exponential with rate λ_2 and job 2 according to an exponential with rate λ_1. The fourth term corresponds to the event where both jobs 1 and 2 are exponentially distributed with rate λ_2. A similar expression can be obtained for r_1. Subtracting yields

$$r_1 - q_1 = (p_1 - p_2)\frac{2\lambda_1\lambda_2}{\lambda_2^2 - \lambda_1^2}\left(\frac{\lambda_0}{\lambda_0 + \lambda_1} - \frac{\lambda_0}{\lambda_0 + \lambda_2}\right) \geq 0,$$

which shows that $r_1 \geq q_1$ for $n = 2$.

Let q_0', q_1', and q_2' and r_0', r_1' and r_2' denote the corresponding probabilities when there are only $n - 1$ jobs under schedule $1, 2, 3, \ldots, n - 1$ and schedule $2, 1, 3, \ldots, n - 1$, respectively.

Now let $n > 2$, and assume inductively that $r_1' \geq q_1'$. Then

$$q_1 = q_1'p_{1n} + q_1'p_{2n}\frac{\lambda_2}{\lambda_1 + \lambda_2} + q_2'p_{1n}\frac{\lambda_2}{\lambda_1 + \lambda_2} + q_0'p_{1n}\frac{\lambda_0}{\lambda_0 + \lambda_1},$$

where the last term on the R.H.S. is independent of the schedule $1, 2, \ldots, n - 1$. A similar expression holds for r_1, and

$$r_1 - q_1 = (r_1' - q_1')\left(p_{1n}\frac{\lambda_1}{\lambda_1 + \lambda_2} + p_{2n}\frac{\lambda_2}{\lambda_1 + \lambda_2}\right) \geq 0.$$

This completes the proof of the lemma.   □

Note that there are no conditions on λ_0; the rate λ_0 may or may not be equal to either λ_1 or λ_2. With this lemma, the following theorem can be shown rather easily.

Theorem 12.1.5.   *The LEPT rule minimizes the expected makespan in the class of nonpreemptive static list policies when there are two machines in parallel and when the processing times are distributed according to a mixture of two exponentials with rates λ_1 and λ_2.*

Proof.   Any permutation schedule can be transformed into the LEPT schedule through a series of adjacent pairwise interchanges between a longer job and a shorter job immediately preceding it. With each interchange, $E(D)$ decreases because of Lemma 12.1.4.   □

Showing that LEPT minimizes the expected makespan can be done in this case without any conditions on the jobs that are not part of the interchange. This is in contrast to Theorem 12.1.2, where the jobs following the jobs in the interchange had to be smaller than the largest job involved in the pairwise interchange. The additional condition, which requires the other expected processing times to be smaller than the expected processing time of the larger of the two jobs in the interchange, is *not* required in this case.

Just as Example 12.1.3 extends the result of Theorem 12.1.2 to a mixture of an exponential and zero, Theorem 12.1.5 can be extended to include mixtures of three exponentials, with rates λ_1, λ_2, and ∞. The next example shows that the LEPT rule, again, does not necessarily minimize the expected makespan.

Example 12.1.6 (Minimizing Expected Makespan with Arbitrary Processing Times)

Let p_{1j} denote the probability job j is exponentially distributed with rate λ_1 and p_{2j} the probability it is distributed with rate λ_2. Assume $\lambda_1 < \lambda_2$. The probability that the processing time of job j is zero is $p_{0j} = 1 - p_{1j} - p_{2j}$. Through similar arguments as those used in Lemma 12.1.4 and Theorem 12.1.5, it can be shown that, to minimize the expected makespan, the jobs in the optimal nonpreemptive static list policy have to be ordered in decreasing order of p_{1j}/p_{2j}. The jobs with the zero processing times again do not play a role in the schedule. As in Example 12.1.3, this implies that the optimal sequence is not necessarily LEPT.

The following example is a continuation of the previous example and is an illustration of the Largest Variance first (LV) rule.

Example 12.1.7 (Application of Largest Variance First [LV] Rule)

Consider the special case of the previous example with

$$\frac{1}{\lambda_1} = 2$$

and

$$\frac{1}{\lambda_2} = 1.$$

Let

$$p_{0j} = a_j$$
$$p_{1j} = a_j$$
$$p_{2j} = 1 - 2a_j.$$

So,

$$E(X_j) = \frac{p_{1j}}{\lambda_1} + \frac{p_{2j}}{\lambda_2} = 1,$$

for all j and

$$\text{Var}(X_j) = 1 + 4a_j.$$

From the previous example, it follows that sequencing the jobs in decreasing order of p_{1j}/p_{2j} minimizes the expected makespan. This rule is equivalent to scheduling the jobs in decreasing order of $a_j/(1 - 2a_j)$. Since $0 \le a_j \le 1/2$, scheduling the jobs in decreasing order of $a_j/(1 - 2a_j)$ is equivalent to scheduling in decreasing order of a_j, which in turn is equivalent to the *Largest Variance first* rule.

The methodology used in proving that LEPT is optimal for the expected makespan on two machines does not easily extend to problems with more than two machines or problems with other processing time distributions. Consider the following generalization of this approach for m machines. Let Z_1 denote the time that the first machine becomes idle with no jobs waiting for processing, Z_2 the time the second machine becomes idle, and so on, and let Z_m denote the time the last machine becomes idle. Clearly, Z_m equals the makespan. Let

$$D_i = Z_{i+1} - Z_i \quad i = 1, \ldots, m - 1.$$

From the fact that the sum of the processing times is

$$\sum_{j=1}^{n} X_j = \sum_{i=1}^{m} Z_i = mC_{\max} - D_1 - 2D_2 - \cdots - (m-1)D_{m-1},$$

which is independent of the schedule, it follows that minimizing the makespan is equivalent to minimizing

$$\sum_{i=1}^{m-1} iD_i.$$

A limited number of processing time distributions can be handled this way. For example, the setting of Theorem 12.1.5 can be extended relatively easily through this approach. However, the scenario of Theorem 12.1.2 cannot be extended that easily.

So far, only the class of nonpreemptive static list policies has been considered in this section. It turns out that most optimal policies in the class of nonpreemptive static list policies are also optimal in the classes of nonpreemptive dynamic policies and preemptive dynamic policies. The proof that a nonpreemptive static list policy is optimal in these other two classes of policies is based on induction arguments very similar to the ones described in the second and third parts of the proof of Theorem 10.4.1. In Section 12.2, an entirely different approach is presented that proves optimality in the class of preemptive dynamic policies. As the optimal preemptive policy is a nonpreemptive static list policy, the policy is also optimal in the classes of nonpreemptive dynamic policies and nonpreemptive static list policies.

Only the expected makespan has been considered in this section. In a nonpreemptive setting, the total expected completion time $E(\sum C_j)$ is more difficult to deal with than the expected makespan. Indeed, an approach similar to the one used to show that LEPT minimizes the expected makespan for exponential processing times has not been found to show that SEPT minimizes the total expected completion time. However, if the processing times are distributed as in Theorem 12.1.5, it can be shown that SEPT minimizes the total expected completion time. If the processing times are distributed as in Example 12.1.7, it can be shown that LV minimizes the total expected completion time.

12.2 THE MAKESPAN AND TOTAL COMPLETION TIME WITH PREEMPTIONS

Pairwise interchange arguments can basically determine the optimal policies in the class of nonpreemptive static list policies. After determining an optimal nonpreemptive static list policy, it can often be argued that this policy is also optimal in the class of nonpreemptive dynamic policies and possibly in the class of preemptive dynamic policies.

In this section, an alternative proof for Theorem 12.1.2 is presented. The approach is entirely different. A dynamic programming type proof is constructed within the class of *preemptive* dynamic policies. After obtaining the result that the nonpreemptive LEPT policy minimizes the expected makespan in the class of preemptive dynamic policies, it is concluded that it is also optimal in the class of nonpreemptive dynamic policies as well as in the class of nonpreemptive static list policies.

This approach can be used for proving that LEPT minimizes the expected makespan for m machines in parallel. It is illustrated for two machines in parallel since the notation is much simpler.

Suppose $\lambda_1 \leq \lambda_2 \leq \cdots \leq \lambda_n$. Let $V(J)$ denote the expected value of the minimum remaining time needed (i.e., under the optimal policy) to finish all jobs given that all the jobs in the set $J = j_1, \ldots, j_l$ already have been completed. If $J = \emptyset$, then $V(J)$ is simply denoted by V. Let $V^*(J)$ denote the same time under the LEPT policy. Similarly, V^* denotes the expected value of the remaining completion time under LEPT when no job has yet been completed.

Theorem 12.2.1. *The nonpreemptive LEPT policy minimizes the expected makespan in the class of preemptive dynamic policies.*

Proof. The proof is by induction on the number of jobs. Suppose that the result is true when there are less than n jobs.

It has to be shown that it is also true when there are n jobs. That is, a policy that at time 0 (when there are n jobs waiting for processing) does not act according to LEPT but at the first job completion (when there are $n-1$ jobs remaining to be processed) switches over to LEPT results in a larger expected makespan than LEPT adopted immediately from time 0 on. In a sense, the structure of this proof is somewhat similar to the proof of Theorem 5.2.7.

Conditioning on the first job completion yields

$$V = \min_{j,k} \left(\frac{1}{\lambda_j + \lambda_k} + \frac{\lambda_j}{\lambda_j + \lambda_k} V^*(\{j\}) + \frac{\lambda_k}{\lambda_j + \lambda_k} V^*(\{k\}) \right).$$

The expected time until the first job completion is the first term on the RHS. The second (third) term is equal to the probability of job j (k) being the first job to be completed, multiplied by the expected remaining time needed to complete the $n-1$ remaining jobs under LEPT. This last equation is equivalent to

$$0 = \min_{j,k} \left(1 + \lambda_j \left(V^*(\{j\}) - V^* \right) + \lambda_k \left(V^*(\{k\}) - V^* \right) + (\lambda_j + \lambda_k) \left(V^* - V \right) \right).$$

Since λ_1 and λ_2 are the two smallest λ_j values and supposedly $V^* \geq V$, the fourth term on the R.H.S. is minimized by $\{j, k\} = \{1, 2\}$. Hence, to show that LEPT is optimal, it suffices to show that $\{j, k\} = \{1, 2\}$ also minimizes the sum of the second and third terms. To simplify the presentation, let

$$A_j = \lambda_j \left(V^*(\{j\}) - V^* \right)$$

and let

$$D_{jk} = A_j - A_k.$$

To show that

$$\lambda_j \left(V^*(\{j\}) - V^*\right) + \lambda_k \left(V^*(\{k\}) - V^*\right) = A_j + A_k$$

is minimized by $\{j, k\} = \{1, 2\}$, it suffices to show that $\lambda_j < \lambda_k$ implies $A_j \leq A_k$ or, equivalently, $D_{jk} \leq 0$. To prove that $D_{jk} \leq 0$ is done in what follows by induction.

Throughout the remaining part of the proof V^*, A_j and D_{jk} are considered functions of the parameters $\lambda_1, \ldots, \lambda_n$. Define $A_j(J)$ and $D_{jk}(J)$ assuming jobs j and k are not in J, in the same way as A_j and D_{jk}—for example,

$$A_j(J) = \lambda_j \left(V^*(J \cup \{j\}) - V^*(J)\right).$$

Before proceeding with the induction, a number of identities have to be established. If j and k are the two smallest jobs not in the set J, then LEPT processes jobs j and k first. Conditioning on the first job completion results in the identity

$$V^*(J) = \frac{1}{\lambda_j + \lambda_k} + \frac{\lambda_j}{\lambda_j + \lambda_k} V^*(J \cup \{j\}) + \frac{\lambda_k}{\lambda_j + \lambda_k} V^*(J \cup \{k\})$$

or

$$(\lambda_j + \lambda_k)V^*(J) = 1 + \lambda_j V^*(J \cup \{j\}) + \lambda_k V^*(J \cup \{k\}).$$

Similarly,

$$
\begin{aligned}
(\lambda_1 + \lambda_2 + \lambda_3)A_1 &= \lambda_1(\lambda_1 + \lambda_2 + \lambda_3)V^*(\{1\}) - \lambda_1(\lambda_1 + \lambda_2 + \lambda_3)V^* \\
&= \lambda_1 \left(1 + \lambda_1 V^*(\{1\}) + \lambda_2 V^*(\{1, 2\}) + \lambda_3 V^*(\{1, 3\})\right) \\
&\quad - \lambda_1 \left(1 + \lambda_1 V^*(\{1\}) + \lambda_2 V^*(\{2\}) + \lambda_3 V^*\right) \\
&= \lambda_1 \left(\lambda_3 V^*(\{1, 3\}) - \lambda_3 V^*(\{1\})\right) \\
&\quad + \lambda_2 \left(\lambda_1 V^*(\{1, 2\}) - \lambda_1 V^*(\{2\})\right) + \lambda_3 A_1
\end{aligned}
$$

or

$$(\lambda_1 + \lambda_2)A_1 = \lambda_1 A_3(\{1\}) + \lambda_2 A_1(\{2\}).$$

The following identities can be established in the same way:

$$(\lambda_1 + \lambda_2)A_2 = \lambda_1 A_2(\{1\}) + \lambda_2 A_3(\{2\})$$

and

$$(\lambda_1 + \lambda_2)A_j = \lambda_1 A_j(\{1\}) + \lambda_2 A_j(\{2\}),$$

for $j = 3, \ldots, n$. Thus, with $D_{12} = A_1 - A_2$, it follows that

$$D_{12} = \frac{\lambda_1}{\lambda_1 + \lambda_2} D_{32}(\{1\}) + \frac{\lambda_2}{\lambda_1 + \lambda_2} D_{13}(\{2\})$$

and

$$D_{2j} = \frac{\lambda_1}{\lambda_1 + \lambda_2} D_{2j}(\{1\}) + \frac{\lambda_2}{\lambda_1 + \lambda_2} D_{3j}(\{2\}),$$

for $j = 3, \dots, n$.

Assume now as induction hypothesis that if $\lambda_j < \lambda_k$ and $\lambda_1 \le \cdots \le \lambda_n$, then

$$D_{jk} \le 0$$

and

$$\frac{dD_{12}}{d\lambda_1} \ge 0.$$

In the remaining part of the proof, these two inequalities are shown by induction on n. When $n = 2$,

$$D_{jk} = \frac{\lambda_j - \lambda_k}{\lambda_j + \lambda_k},$$

and the two inequalities can be established easily.

Assume that the two inequalities of the induction hypothesis hold when there are less than n jobs remaining to be processed. The induction hypothesis now implies that $D_{13}(\{2\})$ as well as $D_{23}(\{1\})$ are nonpositive when there are n jobs remaining to be completed. It also provides

$$\frac{dD_{13}(\{2\})}{d\lambda_1} \ge 0.$$

This last inequality has the following implication: If λ_1 increases, then $D_{13}(\{2\})$ increases. The moment λ_1 reaches the value of λ_2, jobs 1 and 2 become interchangeable. Therefore,

$$D_{13}(\{2\}) \le D_{23}(\{1\}) = -D_{32}(\{1\}) \le 0.$$

From the fact that $\lambda_1 < \lambda_2$, it follows that D_{12} is nonpositive. The induction hypothesis also implies that $D_{2j}(\{1\})$ and $D_{3j}(\{2\})$ are nonpositive. So D_{2j} is nonpositive. This completes the induction argument for the first inequality of the induction hypothesis. The induction argument for the second inequality can be established by differentiating

$$\frac{\lambda_1}{\lambda_1 + \lambda_2} D_{32}(\{1\}) + \frac{\lambda_2}{\lambda_1 + \lambda_2} D_{13}(\{2\})$$

with respect to λ_1 and then using induction to show that every term is positive.   □

This proof shows that LEPT is optimal in the class of preemptive dynamic policies. As the optimal policy is a nonpreemptive static list policy, it also has to be optimal in the class of nonpreemptive static list policies as well as in the class of preemptive dynamic policies. In contrast to the first proof of the same result, this approach also works for an arbitrary number of machines in parallel. The notation, however, becomes significantly more involved.

The interchange approach described in Section 12.1 is not entirely useless. To show that the nonpreemptive LEPT policy is optimal when the processing time distributions are ICR, one has to adopt a pairwise interchange type argument. The reason is obvious. In a preemptive framework, the remaining processing time of an ICR job that has received a certain amount of processing may become less (in expectation) than the expected processing time of a job that is waiting for processing. This then would lead to a preemption. The approach used in Lemma 12.1.1 and Theorem 12.1.2 can be easily applied to a number of different classes of distributions for which the approach used in Theorem 12.2.1 does not appear to yield the optimal nonpreemptive schedule.

Now consider minimizing the total expected completion time of the n jobs. The processing time of job j is exponentially distributed with rate λ_j. In Chapter 5, it was shown that the SPT rule is optimal for the deterministic counterpart of this problem. This gives an indication that in a stochastic setting the *Shortest Expected Processing Time first (SEPT)* rule may minimize the total expected completion time under suitable conditions. Consider again two machines in parallel with n jobs. The processing time of job j is exponentially distributed with rate λ_j. Assume now $\lambda_1 \geq \lambda_2 \geq \cdots \geq \lambda_n$. An approach similar to the one followed in Theorem 12.2.1 for the expected makespan can be followed for the total expected completion time. Let $W(J)$ denote the expected value of the minimum total remaining completion time needed to finish the jobs still in the system under the optimal policy, given that the jobs in set J already have been completed. Again, if $J = \emptyset$, then $W(J)$ is simply denoted by W. Let $W^*(J)$ denote the same quantity under the SEPT policy.

Theorem 12.2.2.   *The nonpreemptive SEPT policy minimizes the total expected completion time in the class of preemptive dynamic policies.*

Proof.   The approach is similar to the one used in Theorem 12.2.1. The proof is again by induction on the number of jobs. Suppose that the result is true when there are less than n jobs. It has to be shown that it is also true when there are n jobs. That is, a policy that at time 0 (when there are n jobs waiting for processing) does not act according to SEPT but at the first job completion (when there are $n - 1$ jobs remaining to be processed) switches over to SEPT results in a larger total expected completion time than SEPT adopted immediately from time 0 on.

Conditioning on the event that occurs at the first job completion yields

$$W = \min_{j,k} \left(\frac{n}{\lambda_j + \lambda_k} + \frac{\lambda_j}{\lambda_j + \lambda_k} W^*(\{j\}) + \frac{\lambda_k}{\lambda_j + \lambda_k} W^*(\{k\}) \right).$$

The increase in the total expected completion time until the first job completion is the first term on the RHS. (Recall that if during a time interval $[t_1, t_2]$ there are k jobs in the system that have not yet been completed, then the total completion time—or, equivalently, total flow time—increases by an amount $k(t_2 - t_1)$.) The second (third) term is equal to the probability of job j (k) being the first job to be completed, multiplied by the total expected remaining completion time under SEPT. This last equation is equivalent to

$$0 = \min_{j,k} \left(n + \lambda_j (W^*(\{j\}) - W^*) + \lambda_k (W^*(\{k\}) - W^*) + (\lambda_j + \lambda_k)(W^* - W) \right).$$

It has to be shown now that if $\lambda_k > \lambda_j$, $\lambda_1 \geq \cdots \geq \lambda_n$, then

$$-1 \leq D_{jk} \leq 0$$

and

$$\frac{dD_{12}}{d\lambda_1} \leq 0.$$

This can be shown, as in Theorem 12.2.1, by induction. If $n = 2$, then $D_{12} = 0$ and the result holds. Doing similar manipulations as in Theorem 12.2.1 yields the following equations:

$$D_{12} = \frac{\lambda_1}{\lambda_1 + \lambda_2} (D_{32}(\{1\}) - 1) + \frac{\lambda_2}{\lambda_1 + \lambda_2} (D_{13}(\{2\}) + 1)$$

and

$$D_{2j} = \frac{\lambda_1}{\lambda_1 + \lambda_2} D_{2j}(\{1\}) + \frac{\lambda_2}{\lambda_1 + \lambda_2} (D_{3j}(\{2\}) - 1),$$

for $j = 3, \ldots, n$. Using these two equations, it is easy to complete the proof.   □

Although the proof of the theorem is presented only for the case of two machines in parallel, the approach does work for an arbitrary number of machines in parallel, but again the notation gets more involved.

More can be said about the total expected completion time on m machines. Consider n processing times $X_1, \ldots, X_n$ from arbitrary distributions $F_1, \ldots, F_n$ such that $X_1 \leq_{st} X_2 \leq_{st} \cdots \leq_{st} X_n$. It has been shown in the literature that SEPT minimizes the total completion time in expectation and even stochastically. Recall that LEPT does *not* minimize the expected makespan when the $X_1, \ldots, X_n$ are arbitrarily distributed and stochastically ordered.

Consider now two machines in parallel, with job processing times exponentially distributed with mean 1, with precedence constraints that have the form of an intree and the expected makespan to be minimized in the class of preemptive dynamic policies (i.e., a stochastic counterpart of $P2 \mid p_j = 1, intree \mid C_{\max}$). For the deterministic version of this problem, the *Critical Path (CP)* rule (sometimes also referred to as the *Highest Level first (HL)* rule) is optimal. The CP rule is in the deterministic case optimal for an arbitrary number of machines in parallel, not just two.

For the stochastic version, the following notation is needed. The root of the intree is level 0. A job is at level k if there is a chain of $k - 1$ jobs between it and the root of the intree. A precedence graph G_1 with n jobs is said to be *flatter* than a precedence graph G_2 with n jobs if the number of jobs at or below any given level k in G_1 is larger than the number of jobs at or below level k in graph G_2. This is denoted by $G_1 \prec_{fl} G_2$. In the following lemma, two scenarios—both with two machines and n jobs but with different intrees—are compared. Let $E(C_{\max}^{(i)}(CP))$ denote the expected makespan under the CP rule when the precedence constraints graph takes the form of intree $G_i, i = 1, 2$.

In the subsequent lemma and theorem, preemptions are allowed. However, it becomes clear afterward that for intree precedence constraints the CP rule does not require any preemptions. Also, recall that whenever a job is completed on one machine, the remaining processing time of the job being processed on the other machine is still exponentially distributed with mean one.

Lemma 12.2.3.   *If $G_1 \prec_{fl} G_2$, then*

$$E(C_{\max}^{(1)}(CP)) \leq E(C_{\max}^{(2)}(CP)).$$

Proof.   The proof is by induction. If $n \leq 2$, it is clear that the lemma holds since graph G_1 must consist of two jobs with no precedence relationship and graph G_2 must consist of a chain of two jobs.

Assume that the lemma holds for graphs of $n - 1$ jobs. Assume that both graphs of n jobs allow two jobs to be started simultaneously at time 0 on the two machines. If this is not the case, then graph G_2 has to be a chain of n jobs, and in this special case the lemma does hold.

For both graphs, the two jobs at the highest levels are selected. Suppose that in both graphs, the job at the very highest level is put on machine 1 and the other job is put on machine 2. Clearly, the job from G_1 on machine 1 is at a lower level than the job from G_2 on machine 1. The job from G_1 on machine 2 can be at either a higher or at a lower level than the job from G_2 on machine 2.

Suppose machine 1 is the first one to complete its job. The remaining time that the job on machine 2 needs to complete its processing is exponential with mean 1.

So at the completion of the job on machine 1, both remaining graphs have $n - 1$ jobs exponentially distributed with mean 1. It is clear that the remaining graphs are ordered in the same way as they were ordered originally (i.e., $G_1 \prec_{fl} G_2$), and the expected remaining processing time is less with G_1 than with G_2 because of the induction hypothesis for $n - 1$ jobs.

Suppose machine 2 is the first one to complete its job. To show that the flatness property of the two remaining graphs is maintained, consider two subcases. In the first subcase, the job from G_2 on machine 2 is at a lower level than the job from G_1 on machine 2. It is clear that when machine 2 is the first one to complete its processing, the remaining part of G_1 is still flatter than the remaining part of G_2. The second subcase is a little bit more involved. Assume the job of G_2 on machine 2 is at level l and the job of G_1 on machine 2 is at level k, $k < l$. The number of jobs below or at level l in G_1 is larger than the number of jobs below or at level l in G_2. Actually, the number of jobs below or at level k in G_1 plus $l - k$ is larger than or equal to the number of jobs below or at level l in G_2. (This statement holds because of the flatness inequality, the structure of the intree, and the fact that there are two machines.) So, at the completion of the job on machine 2, the number of jobs below or at level k in G_1 plus $l - k$ is still less than or equal to the number of jobs below or at level l in G_2 ($k < l$). So the remaining graph of G_1 is still flatter than the remaining graph of G_2 Thus, because of the induction hypothesis, the expected time remaining until the completion of the last job is less under G_1 than under G_2. □

Theorem 12.2.4. *The nonpreemptive CP rule minimizes the expected makespan in the class of nonpreemptive dynamic policies and in the class of preemptive dynamic policies.*

Proof. The theorem is first shown to hold in the class of preemptive dynamic policies. The proof is by contradiction as well as by induction. Suppose the CP rule is optimal with $n - 1$ jobs, but another policy, say policy π, is optimal with n jobs. This policy, at the first job completion, must switch over to the CP rule as the CP rule is optimal with $n - 1$ jobs. In the comparison between π and the CP rule, assume that under both policies the job at the higher level is put on machine 1 and the job at the lower level is put on machine 2. It is easy to check that, if machine 1 is the first one to complete its job, the remaining graph under the CP rule is flatter than the remaining graph under policy π. Because of Lemma 12.2.3, the expected makespan is therefore smaller under the CP rule than under policy π. It is also easy to check that if machine 2 is the first one to complete its job, the remaining graph under the CP rule is flatter than the remaining graph under policy π. Again, the expected makespan is therefore smaller under the CP rule than under π. This proves the optimality of CP in the class of preemptive dynamic policies. It is clear that the CP rule even in the class of preemptive dynamic policies never causes any preemptions. The CP rule is therefore optimal in the class of nonpreemptive dynamic policies as well.

□

As mentioned before, the results presented in Theorems 12.1.2 and 12.2.1, even though they were only proved for $m = 2$, hold for arbitrary m. The CP rule in Theorem 12.2.4 is, however, not necessarily optimal for m larger than two.

Example 12.2.5 (The CP Rule Applied to Three Machines)

Consider three machines and 12 jobs. The jobs are all i.i.d. exponential with mean 1 and subject to the precedence constraints described in Figure 12.2. Scheduling according to the CP rule would put jobs 1, 2, and 3 at time 0 on the three machines. However, straightforward algebra shows that starting with jobs 1, 2, and 4 results in a smaller expected makespan (see Exercise 12.7).

In the deterministic setting discussed in Chapter 5, it was shown that the CP rule is optimal for $Pm \mid p_j = 1, intree \mid C_{max}$ and $Pm \mid p_j = 1, outtree \mid C_{max}$, for any m. One may expect the CP rule to be optimal when all processing times are exponential with mean 1 and precedence constraints take the form of an outtree. However, a counterexample can easily be found for the case of two machines in parallel.

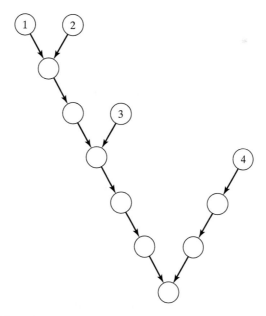

Figure 12.2 The CP rule is not optimal for three machines (Example 12.2.5).

Consider again the problem of two machines in parallel with jobs having i.i.d. exponentially distributed processing times and subject to precedence constraints that take the form of an intree, but now with the total expected completion time as the objective to be minimized.

Theorem 12.2.6. *The nonpreemptive CP rule minimizes the total expected completion time in the class of nonpreemptive dynamic policies and in the class of preemptive dynamic policies.*

Proof.   The proof is similar to the proof of Theorem 12.2.4. A preliminary result similar to Lemma 12.2.3 is again needed.   □

12.3 DUE DATE-RELATED OBJECTIVES

Problems with due dates are significantly harder in a stochastic setting than in a deterministic setting. One of the reasons is that, in a stochastic setting, one cannot work backward from the due dates as can be done in a deterministic setting. The actual realizations of the processing times and the due dates are now not known a priori and working backwards is thus not possible.

However, some results can still be obtained. Consider m parallel machines and n jobs with all processing times deterministic and equal to 1. The weight of job j is equal to w_j and the distribution of the due date of job j is F_j (arbitrary). The problem of determining the schedule that minimizes $E(\sum w_j U_j)$ in the class of nonpreemptive static list policies turns out to be equivalent to a deterministic assignment problem.

Theorem 12.3.1.   *The nonpreemptive static list policy that minimizes $E(\sum w_j U_j)$ can be obtained by solving a deterministic assignment problem with the following cost matrix: If job j is assigned to position i in the permutation schedule, where $km + 1 \leq i \leq (k + 1)m$, then the cost is $w_j F_j(k + 1)$ for $k = 0, 1, 2, \ldots$ The assignment policy that minimizes the expected total cost corresponds to the optimal nonpreemptive static list policy.*

Proof.   The first batch of m jobs in the list complete their processing at time 1, the second batch of m jobs at time 2, and so on. Thus, the probability a job among the first m jobs is overdue is $F_j(1)$, so the expected cost is $w_j F_j(1)$. The expected cost for a job among the second set of m jobs is $w_j F_j(2)$, and so on.   □

Consider now the case where the processing times of the n jobs are i.i.d. exponential with mean 1. Suppose the due date of job j is exponential with rate μ_j, but the due dates are not necessarily independent. Again, the objective is to minimize $E(\sum w_j U_j)$ with m identical machines in parallel.

Theorem 12.3.2. *The nonpreemptive static list policy that minimizes $E(\sum w_j U_j)$ can be obtained by solving a deterministic assignment problem with the following cost matrix: If job j is assigned to position i in the list, where $i = 1, \ldots, m$, then the expected cost is $w_j \mu_j / (1 + \mu_j)$. If job j is assigned to position i, $i = m + 1, \ldots, n$, then the expected cost is*

$$w_j \left(1 - \left(\frac{m}{m + \mu_j} \right)^{i-m} \frac{1}{1 + \mu_j} \right).$$

The assignment that minimizes the expected total cost corresponds to the optimal nonpreemptive static list policy.

Proof. Observe that job j in slot $i = 1, \ldots, m$ starts at time zero. The probability that this job is not completed before its due date is $\mu_j / (1 + \mu_j)$. So its expected cost is $w_j \mu_j / (1 + \mu_j)$. Job j in slot i, $i = m + 1, \ldots, n$, has to wait for $i - m$ job completions before it starts its processing. Given that all machines are busy, the time between successive completions is exponentially distributed with rate m. Thus, the probability that a job in position $i > m$ starts its processing before its due date is $(m/(m+\mu_j))^{i-m}$. Consequently, the probability that it finishes before its due date is

$$\left(\frac{m}{m + \mu_j} \right)^{i-m} \frac{1}{1 + \mu_j}.$$

So the probability that it is not completed by its due date is

$$1 - \left(\frac{m}{m + \mu_j} \right)^{i-m} \frac{1}{1 + \mu_j},$$

and thus has expected cost

$$w_j \left(1 - \left(\frac{m}{m + \mu_j} \right)^{i-m} \frac{1}{1 + \mu_j} \right). \qquad \square$$

EXERCISES (COMPUTATIONAL)

12.1. Consider two machines in parallel and four jobs with exponentially distributed processing times with means $1/5$, $1/4$, $1/3$, and $1/2$.
    (a) Compute the probability of each job being the last one to finish under LEPT.
    (b) Do the same for SEPT.

12.2. Consider the same scenario as in the previous exercise and compute the expected makespan under LEPT and SEPT.

12.3. Consider the same scenario as in Exercise 12.1. Compute the total expected completion time under SEPT and LEPT. (*Hint:* Note that the sum of the expected completion times is equal to the sum of the expected starting times plus the sum of the expected processing times. It suffices to compute the four expected starting times, two of which being zero).

12.4. Consider two machines in parallel and five jobs under the intree precedence constraints depicted in Figure 12.3. All processing times are i.i.d. exponential with mean 1. Compute the expected makespan under the CP rule and under a different schedule.

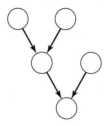

Figure 12.3   Intree (Exercise 12.4).

12.5. Consider two machines and n jobs. The processing times of the n jobs are i.i.d. and distributed according to a mixture of exponentials: $P(X_j = 0) = 1 - p$ and with probability p the processing time is exponentially distributed with rate p (the value p is not necessarily close to 0). Compute the expected makespan under nonpreemptive and preemptive schedules.

12.6. Consider the same scenario as in the previous exercise.
   (a) Compute the total expected completion time for the same set of jobs under nonpreemptive and preemptive schedules.
   (b) Compare the result to the case with all n processing times deterministic and equal to 1.

12.7. Consider Example 12.2.5 with three machines and 12 jobs. Compute the expected makespan under the CP rule and under the rule suggested in Example 12.2.5.

12.8. Find a counterexample showing that the nonpreemptive SEPT rule does not necessarily minimize the total expected completion time in the class of nonpreemptive static list policies when the processing times are merely ordered in expectation and not ordered stochastically.

12.9. Consider the same scenario as in Exercise 12.1. The four jobs are geometrically distributed with means 1/5, 1/4, 1/3, 1/2 (i.e., q is 1/4, 1/3, 1/2, 1).
   (a) Compute the probability of each job being the last one to complete its processing under LEPT (if two jobs finish at the same time, then no job finishes last).
   (b) Compute the expected makespan and compare it with the expected makespan obtained in Exercise 12.2.

12.10. Consider two machines and four jobs. The four jobs are i.i.d. according to the EME distribution with mean 1 (recall that the probability p in the definition of the EME distribution is very close to zero, so $p^2 \ll p$). Compute the expected makespan as a function of p (disregard all terms of p^2 and smaller).

EXERCISES (THEORETICAL)

12.11. Consider m machines in parallel and n jobs. The processing times of the n jobs are i.i.d. exponential with mean 1. Find expressions for $E(C_{\max})$ and $E(\sum C_j)$.

12.12. Consider m machines in parallel and n jobs. The processing times of the n jobs are i.i.d. exponential with mean 1. The jobs are subject to precedence constraints that take the form of chains of different lengths. Find the policy that minimizes the expected makespan and prove that it leads to an optimal schedule.

12.13. Consider n jobs and two machines in parallel. The processing time of job j is with probability p_j exponentially distributed with rate λ. The processing time is with probability q_j distributed according to a convolution of an exponential with rate μ and an exponential with rate λ. The processing time is zero with probability $1 - p_j - q_j$. Show that scheduling the jobs in decreasing order of q_j / p_j minimizes the expected makespan (*Hint:* Every time a machine is freed, the other machine can be in either one of only two states: the remaining processing time of the job on that machine is either exponentially distributed with rate λ or the remaining processing time is distributed according to a convolution of two exponentials with rates μ and λ).

12.14. Consider the processing time distribution in Example 12.1.7. Show that

$$\text{Var}(X_j) = 1 + 4a_j.$$

12.15. Consider m machines in parallel and n jobs. All jobs are available at time $t = 0$. The processing time of job j is drawn from an arbitrary distribution F_j. The objective is to minimize the expected makespan. Show that a rule that prioritizes the jobs at time 0 according to an arbitrary list has as worst case upper bound

$$\frac{E(C_{\max}(LIST))}{E(C_{\max}(LEPT))} \le 2 - \frac{1}{m}.$$

(*Hint:* Use the online scheduling result described in Exercise 5.28.)

12.16. Consider n jobs and two machines in parallel. The processing time of job j is with probability p_j exponentially distributed with rate λ_1 and with probability $1 - p_j$ exponentially distributed with rate λ_2. Job j has a weight w_j. Show that if the WSEPT rule results in the same sequence as the LEPT rule, then the WSEPT rule minimizes the total expected weighted completion time.

12.17. Consider two machines in parallel and five jobs. The processing time of job j is 1 with probability p_j and 2 with probability $1 - p_j$.

    (a) Show that the random variable D (as defined in Section 12.1), can only assume values 0, 1, or 2.

(b) Show that the probability that the random variable D assumes the value 1 is equal to the probability that the sum of the n processing times is odd and therefore independent of the schedule.

(c) Show that minimizing the expected makespan is equivalent to minimizing $P(D = 2)$ and maximizing $P(D = 0)$.

(d) Find the optimal sequence.

12.18. Consider two machines in parallel and n jobs. The processing time of job j is zero with probability p_{0j}, 1 with probability p_{1j}, and 2 with probability $p_{2j} = 1 - p_{1j} - p_{0j}$. Show through a counterexample that the Largest Variance first (LV) rule is not necessarily optimal.

12.19. Consider the two machine setting in Example 12.1.7. Show that the Largest Variance first rule minimizes $E(\sum C_j)$.

12.20. Consider two machines and n jobs. Assume all jobs have the same fixed due date d. Show through counterexamples that there are cases where neither SEPT nor LEPT stochastically maximize the number of jobs completed in time.

12.21. Consider m machines in parallel and n jobs. The processing times of all n jobs are i.i.d. according to a mixture of exponentials. The processing time of job j is zero with probability p and exponential with mean 1 with probability $1 - p$. The due date of job j is exponential with rate μ_j. Show that determining the optimal nonpreemptive static list policy is identical to a deterministic assignment problem. Describe the cost structure of this assignment problem.

12.22. Consider the same setting as in the previous problem. However, the machines are now subject to breakdowns. The uptimes are i.i.d. exponentially distributed with rate ν and the downtimes are also i.i.d. exponential with a different rate. Show that determining the optimal permutation schedule is again equivalent to a deterministic assignment problem. Describe the cost structure of this assignment problem.

12.23. Consider two machines and n jobs. The processing times of the n jobs are i.i.d. exponential with mean 1. To process job j, an amount ρ_j, $0 \leq \rho_j \leq 1$, of a resource is needed. The total amount of that resource available at any point in time is 1. Formulate the policy that minimizes the expected makespan and show that it leads to the optimal schedule.

COMMENTS AND REFERENCES

Many researchers have worked on the scheduling of exponential jobs on identical machines in parallel; see Pinedo and Weiss (1979), Weiss and Pinedo (1980), Bruno, Downey, and Frederickson (1981), Gittins (1981), Van der Heyden (1981), Pinedo (1981a), Weber (1982a, 1982b), Weber, Varaya, and Walrand (1986), Kämpke (1987a, 1987b, 1989), Chang, Nelson, and Pinedo (1992), and Chang, Chao, Pinedo, and Weber (1992). Most of these papers deal with both the makespan and total completion time objectives.

The interchange approach described here, showing that LEPT minimizes the expected makespan when the processing times are exponentially distributed, is based on the paper by Pinedo and Weiss (1979). This paper also discusses hyperexponential distributions that are mixtures of two exponentials. An analysis of the LV rule appears in Pinedo and Weiss (1987).

The dynamic programming type approach showing that LEPT minimizes the expected makespan and SEPT the total expected completion time in a preemptive setting with two machines is based on the paper by Weber (1982b). More general results appear in Weiss and Pinedo (1980), Weber(1982a), Kämpke (1987a, 1987b, 1989), Chang, Chao, Pinedo, and Weber (1992), and Righter (1988, 1992). The most general results with regard to the minimization of the total (unweighted) completion time are probably those obtained by Weber, Varaya, and Walrand (1986). Approximation results appear in Weiss (1990). The fact that the CP rule minimizes the expected makespan in a preemptive as well as a nonpreemptive setting is shown first by Chandy and Reynolds (1975). Pinedo and Weiss (1984) obtain more general results allowing processing times at the different levels of the intree to have different means. Frostig (1988) generalizes this result even further, allowing distributions other than exponential.

There is an extensive literature on the scheduling of nonidentical machines in parallel (i.e., machines with different speeds), which has not been discussed in this chapter. This research focuses on preemptive as well as nonpreemptive models with the objective being either the makespan or total completion time. See, for example, Agrawala, Coffman, Garey, and Tripathi (1984), Coffman, Flatto, Garey, and Weber (1987), Righter (1988), Righter and Xu (1991a, 1991b), Xu, Mirchandani, Kumar, and Weber (1990) and Xu (1991a, 1991b).

The due date-related results in the last section are based on results in a paper by Emmons and Pinedo (1990). For other due date-related results, see Chang, Chao, Pinedo, and Weber (1992) and Xu (1991b).

13

Flow Shops, Job Shops, and Open Shops (Stochastic)

Results for stochastic flow shop, job shop, and open shop models are somewhat limited in comparison with their deterministic counterparts.

For flow shops, nonpreemptive static list policies (i.e., permutation schedules) are considered first. The optimal permutation schedules often remain optimal in the class of nonpreemptive dynamic policies as well as in the class of preemptive dynamic policies. For open shops and job shops, only the classes of nonpreemptive dynamic policies and preemptive dynamic policies are considered.

The results obtained for stochastic flow shops and job shops are somewhat similar to those obtained for deterministic flow shops and job shops. Stochastic open shops are, however, very different from their deterministic counterparts.

The first section discusses stochastic flow shops with unlimited intermediate storage and jobs not subject to blocking. The second section deals with stochastic flow shops with zero intermediate storage; the jobs are subject to blocking.

The third section focuses on stochastic job shops, and the last section goes over stochastic open shops.

13.1 STOCHASTIC FLOW SHOPS WITH UNLIMITED INTERMEDIATE STORAGE

Consider two machines in series with unlimited storage between the machines and no blocking. There are n jobs. The processing time of job j on machine 1 is X_{1j}, exponentially distributed with rate λ_j. The processing time of job j on machine 2 is X_{2j}, exponentially distributed with rate μ_j. The objective is to find the non-preemptive static list policy or permutation schedule that minimizes the expected makespan $E(C_{\max})$.

Note that this problem is a stochastic counterpart of the deterministic problem $F2 \parallel C_{\max}$. The deterministic two machine problem has a very simple solution. It turns out that the stochastic version with exponential processing times has a very elegant solution as well.

Theorem 13.1.1. *Sequencing the jobs in decreasing order of $\lambda_j - \mu_j$ minimizes the expected makespan in the class of nonpreemptive static list policies, the class of nonpreemptive dynamic policies, and the class of preemptive dynamic policies.*

Proof. The proof of optimality in the class of nonpreemptive static list policies is in a sense similar to the proof of optimality in the deterministic case. It is by contradiction. Suppose another sequence is optimal. Under this sequence, there must be two adjacent jobs, say job j followed by job k, such that $\lambda_j - \mu_j < \lambda_k - \mu_k$. It suffices to show that a pairwise interchange of these two jobs reduces the expected makespan. Assume job l precedes job j and let C_{1l} (C_{2l}) denote the (random) completion time of job l on machine 1 (2). Let $D_l = C_{2l} - C_{1l}$.

Perform an adjacent pairwise interchange on jobs j and k. Let C_{1k} and C_{2k} denote the completion times of job k on the two machines under the original, supposedly optimal, sequence, and let C'_{1j} and C'_{2j} denote the completion times of job j under the schedule obtained after the pairwise interchange. Let m denote the job following job k. Clearly, the pairwise interchange does not affect the starting time of job m on machine 1 as this starting time is equal to $C_{1k} = C'_{1j} = C_{1l} + X_{1j} + X_{1k}$. Consider the random variables

$$D_k = C_{2k} - C_{1k}$$

and

$$D'_j = C'_{2j} - C'_{1j}.$$

Clearly, $C_{1k} + D_k$ is the time at which machine 2 becomes available for job m under the original schedule, wheras $C_{1k} + D'_j$ is the corresponding time after the pairwise interchange. First it is shown that the random variable D'_j is stochastically smaller than the random variable D_k. If $D_l \geq X_{1j} + X_{1k}$, then clearly $D_k = D'_j$. The case $D_l \leq X_{1j} + X_{1k}$ is slightly more complicated. Now

$$P(D_k > t \mid D_l \leq X_{1j} + X_{1k}) = \frac{\mu_j}{\lambda_k + \mu_j} e^{-\mu_k t}$$

$$+ \frac{\lambda_k}{\lambda_k + \mu_j} \left(\frac{\mu_k}{\mu_k - \mu_j} e^{-\mu_j t} - \frac{\mu_j}{\mu_k - \mu_j} e^{-\mu_k t} \right).$$

This expression can be explained as follows. Since $D_l \leq X_{1j} + X_{1k}$, then, whenever job j starts on machine 2, job k is either being started or still being processed on machine 1. The first term on the R.H.S. corresponds to the event where job j's processing time on machine 2 finishes before job k's processing time on machine 1, which happens with probability $\mu_j / (\mu_j + \lambda_k)$. The second term corresponds to the event where job j finishes on machine 2 after job k finishes on machine 1; in this case, the distribution of D_k is a convolution of an exponential with rate μ_j and an exponential with rate μ_k.

An expression for $P(D'_j > t \mid D_l \leq X_{1j} + X_{1k})$ can be obtained by interchanging the subscripts j with the subscripts k. Now

$$P(D'_j > t \mid D_l \leq X_{1j} + X_{1k}) - P(D_k > t \mid D_l \leq X_{1j} + X_{1k})$$

$$= \frac{\mu_j \mu_k}{(\lambda_j + \mu_k)(\lambda_k + \mu_j)} \frac{e^{-\mu_j t} - e^{-\mu_k t}}{\mu_k - \mu_j} (\lambda_j + \mu_k - \lambda_k - \mu_j) \leq 0.$$

So D'_j is stochastically smaller than D_k. It can easily be shown, through a straightforward sample path analysis (i.e., fixing the processing times of job m and of all the jobs following job m), that if the realization of D'_j is smaller than the realization of D_k, the actual makespan after the interchange is smaller than or equal to the actual makespan of the original sequence before the interchange. Given that D'_j is stochastically smaller than D_k, the expected makespan is reduced by the interchange. This completes the proof of optimality in the class of nonpreemptive static list (i.e., permutation) policies.

That the rule is also optimal in the class of nonpreemptive dynamic policies can be argued as follows. It is clear that the sequence on machine 2 does not matter. This is because the time machine 2 remains busy processing available jobs is simply the sum of their processing times, and the order in which this happens does not affect the makespan. Consider the decisions that have to be made every time machine 1 is freed. The last decision to be made occurs at that point in time when there are only two jobs remaining to be processed on machine 1. From the pairwise interchange argument described earlier, it immediately follows that the job with the highest $\lambda_j - \mu_j$ has to go first. Suppose that there are three jobs remaining to be processed on machine 1. From the previous argument, it follows that the last two of these three have to be processed in decreasing order of $\lambda_j - \mu_j$. If the first one of the three is not the one with the highest $\lambda_j - \mu_j$, a pairwise interchange between the first and the second reduces the expected makespan. So the last three jobs have to be sequenced in decreasing order of $\lambda_j - \mu_j$. Continuing in this manner, it is shown that sequencing the jobs in decreasing order of $\lambda_j - \mu_j$ is optimal in the class of nonpreemptive dynamic policies.

That the nonpreemptive rule is also optimal in the class of preemptive dynamic policies can be shown in the following manner. It is shown before that in the class of nonpreemptive dynamic policies the optimal rule is to order the jobs in decreasing order of $\lambda_j - \mu_j$. Suppose during the processing of a job on machine 1 a preemption is considered. The situation at this time is essentially no different from the situation at the time the job was started (because of the memoryless property of the exponential distribution). So every time a preemption is contemplated, the optimal decision is to keep the current job on the machine. Thus, the permutation policy is also optimal in the class of preemptive dynamic policies.   □

From the statement of the theorem, it appears that the number of optimal schedules in the exponential case is often smaller than the number of optimal schedules in the deterministic case.

Example 13.1.2 (Exponential Versus Deterministic Processing Times)

Consider n jobs with exponentially distributed processing times. One job has zero processing time on machine 1 and a processing time with a very large mean on machine 2. Assume that this mean is larger than the sum of the expected processing times of the remaining $n - 1$ jobs on machine 1. According to Theorem 13.1.1, these remaining $n - 1$ jobs still have to be ordered in decreasing order of $\lambda_j - \mu_j$ for the sequence to have a minimum expected makespan.

If all the processing times were deterministic with processing times equal to the means of the exponential processing times, it would not have mattered in what order the remaining $n - 1$ jobs were sequenced.

Although at first glance Theorem 13.1.1 does not appear to be similar to Theorem 6.1.4, the optimal schedule with exponential processing times is somewhat similar to the optimal schedule with deterministic processing times. If job k follows job j in the optimal sequence with exponential processing times, then

$$\lambda_j - \mu_j \geq \lambda_k - \mu_k$$

or

$$\lambda_j + \mu_k \geq \lambda_k + \mu_j$$

or

$$\frac{1}{\lambda_j + \mu_k} \leq \frac{1}{\lambda_k + \mu_j},$$

which, with exponential processing times, is equivalent to

$$E(\min(X_{1j}, X_{2k})) \leq E(\min(X_{1k}, X_{2j}))$$

(see Exercise 9.13). This adjacency condition is quite similar to the condition for job k to follow job j in a deterministic setting—namely,

$$\min(p_{1j}, p_{2k}) \leq \min(p_{1k}, p_{2j}).$$

There is another similarity between exponential and deterministic settings. Consider the case where the processing times of job j on both machines are i.i.d. exponentially distributed with the same rate λ_j, for each j. According to the theorem, all sequences must have the same expected makespan. This result is similar to the one for the deterministic proportionate flow shop, where all sequences also have the same makespan.

The remaining part of this section focuses on m machine *permutation* flow shops. For these flow shops, only the class of nonpreemptive static list policies is of interest since the order of the jobs, once determined, is not allowed to change.

Consider an m machine permutation flow shop where the processing times of job j on the m machines are i.i.d. according to distribution F_j with mean $1/\lambda_j$. For such a flow shop, it is easy to obtain a lower bound for $E(C_{\max})$.

Lemma 13.1.3. *Under any sequence,*

$$E(C_{\max}) \geq \sum_{j=1}^{n} \frac{1}{\lambda_j} + (m-1) \max\left(\frac{1}{\lambda_1}, \ldots, \frac{1}{\lambda_n}\right).$$

Proof. The expected time it takes the job with the largest expected processing time to traverse the flow shop is at least $m \max(1/\lambda_1, \ldots, 1/\lambda_n)$. The time that this largest job starts on machine 1 is the sum of the processing times on the first machine of those jobs scheduled before the longest job. After the longest job completes its processing on the last machine, this machine remains busy for a time that is at least as large as the sum of the processing times on the last machine of all those jobs scheduled after the longest job. The lemma thus follows.   □

One class of sequences plays an important role in stochastic permutation flow shops. A sequence $j_1, \ldots, j_n$ is called a SEPT-LEPT sequence if there is a job j_k in this sequence such that

$$\frac{1}{\lambda_{j_1}} \leq \frac{1}{\lambda_{j_2}} \leq \cdots \leq \frac{1}{\lambda_{j_k}}$$

and

$$\frac{1}{\lambda_{j_k}} \geq \frac{1}{\lambda_{j_{k+1}}} \geq \cdots \geq \frac{1}{\lambda_{j_n}}.$$

Both the SEPT and LEPT sequences are examples of SEPT-LEPT sequences.

Theorem 13.1.4.   *If $F_1 \leq_{a.s.} F_2 \leq_{a.s.} \cdots \leq_{a.s.} F_n$, then*

(i) *any SEPT-LEPT sequence minimizes the expected makespan in the class of nonpreemptive static list policies and*

$$E(C_{\max}) = \sum_{j=1}^{n} \frac{1}{\lambda_j} + (m-1)\frac{1}{\lambda_n}.$$

(ii) *the SEPT sequence minimizes the expected total completion time in the class of nonpreemptive static list policies and*

$$E\left(\sum_{j=1}^{n} C_j\right) = m \sum_{j=1}^{n} \frac{1}{\lambda_j} + \sum_{j=1}^{n-1} \frac{j}{\lambda_{n-j}}.$$

Proof.

(i) The first part of the theorem follows from the observation that the jobs in the SEPT segment of the sequence *never* have to wait for a machine while they go through the system. This includes the largest job (i.e., the last job of the SEPT segment of the sequence). The jobs in the LEPT segment of the sequence (excluding the first, i.e., the largest job) have to wait for each machine; that is, they complete their processing on one machine before the next one becomes available. The machines then never have to wait for a job. The makespan is therefore equal to the lower bound established in Lemma 13.1.3.

(ii) The second part follows from the fact that the SEPT sequence is one of the sequences that minimize the expected makespan. To minimize the expected completion time of the job in the kth position in the sequence, the k smallest jobs have to go first and these k jobs have to be sequenced according to a sequence that minimizes the expected completion time of this kth job. So these k smallest jobs may be sequenced according to any SEPT-LEPT sequence including SEPT. This is true for any k. It is clear that under SEPT the expected completion time of the job in each one of the n positions in the sequence is minimized. Therefore, SEPT minimizes the total expected completion time. The actual value of the total expected completion time is easy to compute.

$\square$

It is easy to find examples with $F_1 \leq_{a.s.} F_2 \leq_{a.s.} \cdots \leq_{a.s.} F_n$, where sequences that are not SEPT-LEPT are also optimal (recall that when $F_1, F_2, \ldots, F_n$ are deterministic *any* sequence minimizes the makespan; see Theorem 6.1.8). However, in contrast to deterministic proportionate flow shops, when processing times are stochastic and $F_1 \leq_{a.s.} F_2 \leq_{a.s.} \cdots \leq_{a.s.} F_n$, not *all* sequences are always optimal.

Example 13.1.5 (Optimality of SEPT-LEPT and Other Sequences)

> Consider a flow shop with two machines and three jobs. Job 1 has a deterministic processing time of 11 time units. Job 2 has a deterministic processing time of 10 time units. The processing time of job 3 is zero with probability 0.5 and 10 with probability 0.5. It can be verified easily that *only* SEPT-LEPT sequences minimize the expected makespan. If the processing time of job 1 is changed from 11 to 20, then all sequences have the same expected makespan.

The model in Theorem 13.1.4 assumes that the processing times of job j on the m machines are obtained by m *independent* draws from the same distribution F_j. If the m processing times of job j are all equal to the *same* random variable drawn from distribution F_j, then the model resembles the deterministic proportionate flow shop even more closely. So in this case

$$X_{1j} = X_{2j} = \cdots = X_{mj} = X_j.$$

It is easy to see that in this case, just like in the case of a deterministic proportionate flow shop, the makespan is independent of the job sequence (no particular form of stochastic dominance has to be imposed on the n distributions for this to be true).

Consider the case where the processing times of job j on each one of the m machines are equal to the *same* random variable X_j from distribution F_j and assume

$$F_1 \leq_{cx} F_2 \leq_{cx} \cdots \leq_{cx} F_n.$$

The expectations of the n processing times are therefore all the same, but the variances may be different. The objective to be minimized is the total expected completion time $E(\sum_{j=1}^{n} C_j)$. This problem leads to the application of the so-called *Smallest Variance first (SV)* rule, which selects, whenever machine 1 is freed, among the remaining jobs the one with the smallest variance.

    Theorem 13.1.6. *The SV rule minimizes the total expected completion time in the class of nonpreemptive static list policies.*

    Proof. The proof is by contradiction. Suppose another sequence is optimal. In this supposedly optimal sequence, there has to be a job j followed by a job k such that $X_j \geq_{cx} X_k$. Assume job h precedes job j and job l follows job k. Let

C_{ih} denote the completion time of job h on machine i. Let A denote the sum of the processing times of all the jobs preceding and including job h in the sequence, and let B denote the maximum processing time among the jobs preceding and including job h. Then

$$C_{1h} = A$$

and

$$C_{ih} = A + (i - 1)B.$$

Performing an interchange between jobs j and k does not affect the profile job l encounters on entering the system. So an interchange does not affect the completion times of job l or the jobs following job l in the sequence. The pairwise interchange only affects the sum of the expected completion times of jobs j and k. The completion time of job k before the interchange is equal to the completion time of job j after the interchange. To analyze the effect of the pairwise interchange, it suffices to compare the completion time of job j before the interchange with the completion time of job k after the interchange. Let C_{ij} denote the completion time of job j before the interchange and C'_{ik} the completion time of job k after the interchange. Clearly,

$$C_{mj} = A + X_j + (m - 1) \max(B, X_j)$$

and

$$C'_{mk} = A + X_k + (m - 1) \max(B, X_k).$$

The expected completion time of job k after the pairwise interchange is smaller than the expected completion time of job j before the interchange if

$$\int_0^\infty (A + t + (m - 1) \max(B, t)) \, dF_k(t)$$

$$\leq \int_0^\infty (A + t + (m - 1) \max(B, t)) \, dF_j(t)$$

The integrand is a function that is increasing convex in t. So the inequality indeed holds if $F_k \leq_{cx} F_j$. This implies that the original sequence cannot be optimal and the proof is complete.   □

This result is in contrast to the result stated in Exercise 12.19, where in the case of machines in parallel the Largest Variance first rule minimizes the total expected completion time.

13.2 STOCHASTIC FLOW SHOPS WITH BLOCKING

Consider the stochastic counterpart of $F2 \mid block \mid C_{\max}$ with the processing time of job j on machine 1 (2) being a random variable X_{1j} (X_{2j}) from distribution F_{1j} (F_{2j}). There is zero intermediate storage between the two machines. The objective is to minimize the expected makespan in the class of nonpreemptive static list policies.

When a job starts its processing on machine 1, the preceding job in the sequence starts its processing on machine 2. If job j follows job k in the sequence, then the expected time that job j remains on machine 1, either being processed or blocked, is $E(\max(X_{1j}, X_{2k}))$. If job j is the first job in the sequence, then job j spends only an expected amount of time $E(X_{1j})$ on machine 1 while machine 2 remains idle. If job j is the last job in the sequence, then it spends an expected amount of time $E(X_{2j})$ on machine 2 while machine 1 remains idle. In the same way that the deterministic $F2 \mid block \mid C_{\max}$ problem is equivalent to a deterministic Traveling Salesman Problem, this stochastic model is equivalent to a *deterministic* Traveling Salesman Problem. However, the efficient algorithm described in Section 4.5, which is applicable to the deterministic $F2 \parallel C_{\max}$ problem, is not applicable to the stochastic version of the model. The distance matrix of the Traveling Salesman Problem is determined as follows:

$$d_{0k} = E(X_{1k})$$

$$d_{j0} = E(X_{2j})$$

$$d_{jk} = E(\max(X_{2j}, X_{1k}))$$

$$= E(X_{2j}) + E(X_{1k}) - E(\min(X_{2j}, X_{1k}))$$

$$= \int_0^\infty \bar{F}_{2j}(t)\, dt + \int_0^\infty \bar{F}_{1k}(t)\, dt - \int_0^\infty \bar{F}_{2j}(t)\bar{F}_{1k}(t)\, dt.$$

It is clear that a value for d_{jk} can be computed, but that this value is now not a simple function of two parameters like in Sections 4.5 and 6.2. However, the Traveling Salesman Problem described can be simplified somewhat. The problem is equivalent to a Traveling Salesman Problem with a simpler distance matrix in which the total distance has to be *maximized*. The distance matrix is modified by subtracting the expected sum of the $2n$ processing times from the distances and multiplying the remaining parts by -1:

$$d_{0k} = 0$$

$$d_{j0} = 0$$

$$d_{jk} = E(\min(X_{2j}, X_{1k})) = \int_0^\infty \bar{F}_{2j}(t)\bar{F}_{1k}(t)\, dt.$$

Example 13.2.1 (Flow Shop with Blocking and Exponential Processing Times)

Consider the case where F_{1j} is exponentially distributed with rate λ_j and F_{2j} is exponentially distributed with rate μ_j. The distance

$$d_{jk} = E(\min(X_{2j}, X_{1k})) = \int_0^\infty \bar{F}_{2j}(t)\bar{F}_{1k}(t)\, dt = \frac{1}{\lambda_k + \mu_j}.$$

Although this deterministic Traveling Salesman Problem (in which the total distance must be maximized) has a fairly nice structure, it is not clear whether it can be solved in polynomial time.

It is of interest to study special cases of the problem, with additional structure, to obtain more insight. Consider the case where $F_{1j} = F_{2j} = F_j$. The random variables X_{1j} and X_{2j} are independent draws from distribution F_j. This model is somewhat similar to the deterministic proportionate flow shop model since the distributions of the processing times of any given job on the two machines are identical. However, the actual realizations of the two processing times are not necessarily identical.

Theorem 13.2.2. *If* $F_1 \leq_{st} F_2 \leq_{st} \cdots \leq_{st} F_n$ *then the sequences* $1, 3, 5,$ $\ldots, n, \ldots, 6, 4, 2$ *and* $2, 4, 6, \ldots, n, \ldots, 5, 3, 1$ *minimize the expected makespan in the class of nonpreemptive static list policies.*

Proof. Consider the sequence

$$j_1, \ldots, j_{k-1}, j_k, j_{k+1}, \ldots, j_{l-1}, j_l, j_{l+1}, \ldots, j_n.$$

Partition the sequence into three subsequences, the first being $j_1, \ldots, j_{k-1}$, the second $j_k, \ldots, j_l$, and the third $j_{l+1}, \ldots, j_n$. Construct a new sequence by reversing the second subsequence. The new sequence is

$$j_1, \ldots, j_{k-1}, j_l, j_{l-1}, \ldots, j_{k+1}, j_k, j_{l+1}, \ldots, j_n.$$

If $E(C_{max})$ denotes the expected makespan of the original sequence and $E(C'_{max})$ the expected makespan of the new sequence, then

$$E(C_{max}) - E(C'_{max}) = E(\max(X_{2,j_{k-1}}, X_{1,j_k})) + E(\max(X_{2,j_l}, X_{1,j_{l+1}}))$$
$$- E(\max(X_{2,j_{k-1}}, X_{1,j_l})) - E(\max(X_{2,j_k}, X_{1,j_{l+1}}))$$
$$= -E(\min(X_{2,j_{k-1}}, X_{1,j_k})) - E(\min(X_{2,j_l}, X_{1,j_{l+1}}))$$
$$+ E(\min(X_{2,j_{k-1}}, X_{1,j_l})) + E(\min(X_{2,j_k}, X_{1,j_{l+1}})).$$

So the expected makespan of the second sequence is less than the expected makespan of the first sequence if

$$\int_0^{\infty} \left(\bar{F}_{j_{k-1}}(t) \bar{F}_{j_k}(t) + \bar{F}_{j_l}(t) \bar{F}_{j_{l+1}}(t) \right) dt$$

$$< \int_0^{\infty} \left(\bar{F}_{j_{k-1}}(t) \bar{F}_{j_l}(t) + \bar{F}_{j_k}(t) \bar{F}_{j_{l+1}}(t) \right) dt.$$

Note that the makespan under an arbitrary sequence does not change if two jobs are added, both with zero processing times on the two machines, one scheduled first and the other one last. The processing time distributions of these two jobs are stochastically less than the distribution of any one of the other n jobs. So in the proof, an assumption may be made that there are two additional jobs with zero processing times and that one of these jobs goes first and the other one goes last. In what follows, these two jobs are referred to as jobs 0 and $0'$.

Consider four processing time distributions $F_j \geq_{st} F_k \geq_{st} F_p \geq_{st} F_q$. It can be easily verified that

$$\int_0^{\infty} \left(\bar{F}_j(t) \bar{F}_k(t) + \bar{F}_p(t) \bar{F}_q(t) \right) dt \geq \int_0^{\infty} \left(\bar{F}_j(t) \bar{F}_p(t) + \bar{F}_k(t) \bar{F}_q(t) \right) dt$$

$$\geq \int_0^{\infty} \left(\bar{F}_j(t) \bar{F}_q(t) + \bar{F}_k(t) \bar{F}_p(t) \right) dt.$$

The remaining part of the proof is based on a contradiction argument. An arbitrary sequence not according to the theorem can be improved through a series of successive subsequence reversals until a sequence of the theorem is obtained. Consider a sequence

$$0, j_1, \ldots, j_k, 1, j_{k+1}, \ldots, j_l, 2, j_{l+1}, \ldots, j_{n-2}, 0',$$

where $j_1, \ldots, j_{n-2}$ is a permutation of $3, 4, \ldots, n$. From the previous inequalities, it follows that a subsequence reversal results in the sequence

$$0, 1, j_k, \ldots, j_1, j_{k+1}, \ldots, j_l, 2, j_{l+1}, \ldots, j_{n-2}, 0'$$

with a smaller expected makespan. The makespan can be reduced even further through a second subsequence reversal that results in

$$0, 1, j_k, \ldots, j_1, j_{k+1}, \ldots, j_l, j_{n-2}, \ldots, j_{l+1}, 2, 0'.$$

Proceeding in this manner, it can be easily shown that any sequence can be improved through a series of subsequence reversals until one of the sequences stated in the theorem is obtained. □

Clearly, $1, 3, 5, \ldots, n, \ldots, 6, 4, 2$ is a SEPT-LEPT sequence. That such a sequence is optimal should have been expected. Short jobs should be scheduled in the beginning of the sequence just to make sure that machine 2 does not remain idle for too long, whereas short jobs should also be scheduled toward the end of the sequence to avoid that machine 2 remains busy for a very long time after machine 1 has finished with all its processing. Note that optimal sequences are slightly different when n is even or odd. If n is even, the optimal sequence is $1, 3, 5, \ldots, n-1, n, n-2, \ldots, 6, 4, 2$; if n is odd, the optimal sequence is $1, 3, 5, \ldots, n-2, n, n-1, \ldots, 6, 4, 2$.

Theorem 13.2.2 thus gives an indication of the impact of the means of the processing times on the optimal sequence. Generalizing the result of Theorem 13.2.2 to more than two machines is impossible. Counterexamples have been found.

Consider now the same model with $F_{1j} = F_{2j} = F_j$, $j = 1, \ldots, n$, but with the means of the distributions $F_1, F_2, \ldots, F_n$ being identical and equal to 1. However, the variances of the distributions are now different. Assume that the distributions have symmetric probability density functions and assume that $F_1 \geq_{sv} F_2 \geq_{sv} \cdots \geq_{sv} F_n$. This implies that all random variables lie between 0 and 2. For the two machine model with no intermediate buffers, the following theorem holds.

Theorem 13.2.3.   *If $F_1 \geq_{sv} F_2 \geq_{sv} \cdots \geq_{sv} F_n$, then the sequences $1, 3, 5, \ldots, n, \ldots, 6, 4, 2$ and $2, 4, 6, \ldots, n, \ldots, 5, 3, 1$ minimize the expected makespan in the class of nonpreemptive static list policies.*

Proof.   First it is shown that any sequence can be transformed into a better sequence (with a smaller expected makespan) of the form $1, j_1, \ldots, j_{n-2}, 2$, where $j_1, \ldots, j_{n-2}$ is a permutation of jobs $3, \ldots, n$. Compare sequence

$$j_1, \ldots, j_k, 1, j_{k+1}, \ldots, j_l, 2, j_{l+1}, \ldots, j_{n-2}$$

with sequence

$$1, j_k, \ldots, j_1, j_{k+1}, \ldots, j_l, 2, j_{l+1}, \ldots, j_{n-2}.$$

Subtracting the makespan of the second sequence from that of the first yields:

$$E(\max(X_1, X_{j_{k+1}})) - E(\max(X_{j_1}, X_{j_{k+1}}))$$
$$= E(\min(X_{j_1}, X_{j_{k+1}})) - E(\min(X_1, X_{j_{k+1}}))$$
$$= \int_0^2 \left(\bar{F}_{j_1}(t) \bar{F}_{j_{k+1}}(t) - \bar{F}_1(t) \bar{F}_{j_{k+1}}(t) \right) dt$$
$$= \int_0^2 \left(\bar{F}_{j_{k+1}}(t) \left(\bar{F}_{j_1}(t) - \bar{F}_1(t) \right) \right) dt$$
$$\geq 0.$$

Therefore, it is better to schedule job 1 first. A similar argument shows that job 2 has to be scheduled last.

The next step is to show that any sequence can be transformed into a sequence of the form $1, 3, j_1, \ldots, j_{n-3}, 2$ with a smaller expected makespan. Compare sequence

$$1, j_1, \ldots, j_k, 3, j_{k+1}, \ldots, j_{n-3}, 2$$

with sequence

$$1, 3, j_k, \ldots, j_1, j_{k+1}, \ldots, j_{n-3}, 2.$$

The expected makespan of the first sequence minus the expected makespan of the second sequence is

$$E(\max(X_1, X_{j_1})) + E(\max(X_3, X_{j_{k+1}}))$$

$$- E(\max(X_1, X_3)) - E(\max(X_{j_1}, X_{j_{k+1}}))$$

$$= E(\min(X_1, X_3)) + E(\min(X_{j_1}, X_{j_{k+1}}))$$

$$- E(\min(X_1, X_{j_1})) - E(\min(X_3, X_{j_{k+1}}))$$

$$= \int_0^2 \left(\bar{F}_1(t)\bar{F}_3(t) + \bar{F}_{j_1}(t)\bar{F}_{j_{k+1}}(t) \right) dt$$

$$- \int_0^2 \left(\bar{F}_1(t)\bar{F}_{j_1}(t) + \bar{F}_3(t)\bar{F}_{j_{k+1}}(t) \right) dt$$

$$= \int_0^2 \left(\bar{F}_{j_{k+1}}(t) - \bar{F}_1(t) \right) \left(\bar{F}_{j_1}(t) - \bar{F}_3(t) \right) dt$$

$$\geq 0.$$

So the optimal sequence has to be of the form $1, 3, j_1, \ldots, j_{n-3}, 2$. Proceeding in this manner, the optimality of the two sequences stated in the theorem can be easily verified.   □

This result basically states that the optimal sequence puts jobs with larger variances more toward the beginning and end of the sequence, and jobs with smaller variances more toward the middle of the sequence. Such a sequence could be referred to as an *LV-SV* sequence.

It is not clear whether similar results hold when there are more than two machines in series. Results for problems with more machines are extremely hard to come by since the complexity of the problem increases considerably when going from two to three machines.

Nevertheless, some properties can be shown for m machines in series with blocking (i.e., $Fm \mid block \mid C_{\max}$). Assume that $F_{1j} = F_{2j} = \cdots = F_{mj} = F_j$ with mean $1/\lambda_j$ and that $X_{1j}, \ldots, X_{mj}$ are independent.

Theorem 13.2.4.   *If $F_1 \leq_{a.s.} F_2 \leq_{a.s.} \cdots \leq_{a.s.} F_n$, then a sequence minimizes the expected makespan if and only if it is a SEPT-LEPT sequence.*

Proof.   As the proof of this theorem is straightforward, only a short outline is given. The proof is similar to the proof of Theorem 6.2.4 and consists of two parts. In the first part, it is shown that every SEPT-LEPT sequence attains the lower bound of Lemma 13.1.3. In the second part, it is shown that any sequence that is not SEPT-LEPT leads to a makespan that is strictly larger than the lower bound.

□

13.3 STOCHASTIC JOB SHOPS

Consider now the two machine job shop with job j having a processing time on machine 1 that is exponentially distributed with rate λ_j and a processing time on machine 2 that is exponentially distributed with rate μ_j. Some of the jobs have to be processed first on machine 1 and then on machine 2 while the remaining jobs have to be processed first on machine 2 and then on machine 1. Let $J_{1,2}$ denote the first set of jobs and $J_{2,1}$ the second set of jobs. Minimizing the expected makespan turns out to be an easy extension of the two machine flow shop model with exponential processing times.

Theorem 13.3.1.   *The following nonpreemptive policy minimizes the expected makespan in the classes of nonpreemptive dynamic policies and preemptive dynamic policies: When machine 1 is freed, the decision maker selects from $J_{1,2}$ the job with the highest $\lambda_j - \mu_j$; if all jobs from $J_{1,2}$ have completed their processing on machine 1, the decision maker may take any job from $J_{2,1}$. When machine 2 is freed, the decision maker selects from $J_{2,1}$ the job with the highest $\mu_j - \lambda_j$; if all jobs from $J_{2,1}$ have completed their processing on machine 2, the decision maker may take any job from $J_{1,2}$.*

Proof.   The proof consists of two parts. First, it is shown that jobs from $J_{2,1}$ have a lower priority on machine 1 than jobs from $J_{1,2}$ and jobs from $J_{1,2}$ have a lower priority on machine 2 than jobs from $J_{2,1}$. After that, it is shown that jobs from $J_{1,2}$ are ordered on machine 1 in decreasing order of $\lambda_j - \mu_j$ and jobs from $J_{2,1}$ on machine 2 in decreasing order of $\mu_j - \lambda_j$.

To show the first part, condition on a realization of all $2n$ processing times. The argument is by contradiction. Suppose a schedule is optimal and puts at a certain point in time a job from $J_{2,1}$ on machine 1 *before* a job from $J_{1,2}$. Consider the last job from $J_{2,1}$ processed on machine 1 before the job(s) from $J_{1,2}$. Perform the following change in the schedule: Take this job from $J_{2,1}$ and postpone its processing until the last job from $J_{1,2}$ has been completed. After this change, all jobs from $J_{1,2}$ are completed earlier on machine 1 and are available earlier at machine 2. This implies that machine 1 will finish with all its processing at the same time as before the interchange. However, machine 2 may finish with all its processing earlier than before the interchange because now the jobs from $J_{1,2}$ are available earlier at machine 2. This completes the first part of the proof of the theorem.

To prove the second part, proceed as follows. First, consider $J_{1,2}$. To show that the jobs from $J_{1,2}$ should be scheduled in decreasing order of $\lambda_j - \mu_j$, condition first on the processing times of all the jobs in $J_{2,1}$ on both machines. The jobs from $J_{2,1}$ have a higher priority on machine 2 and a lower priority on machine 1. Assume that two adjacent jobs from $J_{1,2}$ are not scheduled in decreasing order of $\lambda_j - \mu_j$. Performing a pairwise interchange in the same way as in Theorem 13.1.1 results in a smaller expected makespan. This shows that the jobs from $J_{1,2}$ have to be scheduled on machine 1 in decreasing order of $\lambda_j - \mu_j$. A similar argument shows that the jobs from $J_{2,1}$ have to be scheduled on machine 2 in decreasing order of $\mu_j - \lambda_j$. $\square$

The result described in Theorem 13.3.1 is similar to the result described in Section 7.1 with regard to $J2 \parallel C_{\max}$. In deterministic scheduling, the research on the more general $Jm \parallel C_{\max}$ has focused on heuristics and enumerative procedures. Stochastic job shops with more than two machines have not received that much research attention.

13.4 STOCHASTIC OPEN SHOPS

Consider a two machine open shop where the processing time of job j on machine 1 is the random variable X_{1j}, distributed according to F_{1j}, and on machine 2 the random variable X_{2j}, distributed according to F_{2j}. The objective is to minimize the expected makespan. As before, the exponential distribution is considered first. In this case, however, it is not known what the optimal policy is when F_{1j} is exponential with rate λ_j and F_{2j} exponential with rate μ_j. It appears that the optimal policy may not have a simple structure and may even depend on the values of the λs and μs. The special case where $\lambda_j = \mu_j$ can be analyzed. In contrast to the results obtained for the stochastic flow shops, the optimal policy now cannot be regarded as a permutation sequence, but rather as a policy that prescribes a given action that depends on the state of the system.

Theorem 13.4.1.   *The following policy minimizes the expected makespan in the class of preemptive dynamic policies as well as in the class of nonpreemptive dynamic policies: Whenever a machine is freed, the scheduler selects from the jobs that have not yet undergone processing on either one of the two machines, the job with the largest expected processing time. If there are no such jobs remaining, the decision maker may take any job that only needs processing on the machine just freed. Preemptions do not occur.*

Proof.   Just as in the deterministic case, the two machines are continuously busy with the possible exception of at most a single idle period on at most one machine. The idle period can be either an idle period of *Type I* or an idle period of *Type II* (see Figure 8.1). In the case of no idle period at all or an idle period of Type II, the makespan is equal to the maximum of the workloads on the two machines—that is,

$$C_{\max} = \max\left(\sum_{j=1}^{n} X_{1j}, \sum_{j=1}^{n} X_{2j}\right).$$

In the case of an idle period of Type I, the makespan is strictly larger than the R.H.S. of the prior expression. Actually, in this case, the makespan is

$$C_{\max} = \max\left(\sum_{j=1}^{n} X_{1j}, \sum_{j=1}^{n} X_{2j}\right) + \min(I_1, I_2),$$

where I_1 is the length of the idle period and I_2 is the makespan minus the workload on the machine that did not experience an idle period. It is clear that the first term of the R.H.S. of the expression above does not depend on the policy used. To prove the theorem, it suffices to show that the described policy minimizes the expected value of the second term on the R.H.S.—that is, $E(\min(I_1, I_2))$. This term clearly does depend on the policy used.

To obtain some more insight in this second term, consider the following: Suppose job j is the job causing the idle period—that is, job j is the last job to be completed. Given that job j causes an idle period of Type I, it follows from Exercise 9.13 that

$$E(\min(I_1, I_2)) = \frac{1}{2\lambda_j}.$$

If q_j' denotes the probability of job j causing an idle period of Type II under policy π', then

$$E(C_{\max}(\pi')) = E\left(\max\left(\sum_{j=1}^{n} X_{1j}, \sum_{j=1}^{n} X_{2j}\right)\right) + E(H'),$$

where

$$E(H') = \sum_{j=1}^{n} q'_j \frac{1}{2\lambda_j}.$$

From the theory of dynamic programming (see Appendix B), it follows that, to prove optimality of the policy stated in the theorem, say policy π^*, it suffices to show that using π^* from any time t onward results in a smaller expected makespan than acting differently at time t and using π^* from the *next* decision moment onward. Two types of actions at time t violate π^*. First, it is possible to start a job that is not the largest job among jobs not processed on either machine. Second, it is possible to start a job that already has been processed on the other machine while there are still jobs around that have not yet received processing on either machine.

In the remaining part of the proof, the following notation is used: Set J_1 represents the set of jobs that, at time t, have not yet completed their first processing, whereas set J_2 represents the set of jobs that at time t have not yet started with their second processing. Clearly, set J_2 includes set J_1, $J_1 \subset J_2$.

Case 1.   Let π' denote the policy that, at time t, puts job k on the machine freed, with $k \in J_1$ and k not being the largest in J_1, and that reverts back to π^* from the next decision moment onward. Let job 0 be the job that is being processed, at time t, on the busy machine. Let r'_j (r^*_j) denote the probability that job j is the last job to complete its *first* processing under policy π' (π^*) and therefore be a candidate to cause an idle period. Suppose this job j is processed on machine 1. For job j to cause an idle period of Type I, it has to outlast all those jobs that still have to receive their second processing on machine 2. After job j completes its processing on machine 1 and starts its processing on machine 2, it has to outlast all those jobs that have yet to receive their second processing on machine 1. So

$$q^*_j = r^*_j \prod_{l \in \{J_2 - j\}} \frac{\lambda_l}{\lambda_l + \lambda_j}$$

and

$$q'_j = r'_j \prod_{l \in \{J_2 - j\}} \frac{\lambda_l}{\lambda_l + \lambda_j}.$$

Also

$$q'_0 = q^*_0.$$

Note that the expressions for q^*_j and q'_j indicate that q^*_j and q'_j do not depend on the machine on which job l, $l \in \{J_2 - j\}$, receives its second processing: Processing job l the second time on the same machine where it was processed the first time results in values for q_j^* and q_j' that are the same as when job l is processed the second time on the machine on which it was not processed the first time. To show that $E(H^*) \leq E(H')$, it has to be shown that

$$\sum_{j \in \{J_1 - 0\}} \left(r_j^* \left(\prod_{l \in \{J_2 - j\}} \frac{\lambda_l}{\lambda_l + \lambda_j} \right) \frac{1}{2\lambda_j} \right) \leq \sum_{j \in \{J_1 - 0\}} \left(r_j' \left(\prod_{l \in \{J_2 - j\}} \frac{\lambda_l}{\lambda_l + \lambda_j} \right) \frac{1}{2\lambda_j} \right).$$

Note that if $\lambda_a \leq \lambda_b$, then

$$\left(\prod_{l \in \{J_2 - a\}} \frac{\lambda_l}{\lambda_l + \lambda_a} \right) \frac{1}{2\lambda_a} \geq \left(\prod_{l \in \{J_2 - b\}} \frac{\lambda_l}{\lambda_l + \lambda_b} \right) \frac{1}{2\lambda_b}.$$

Suppose the sequence in which the jobs in J_1 start with their first processing under π' is $0, k, 1, 2, \ldots, k-1, k+1, \ldots$ where

$$\lambda_1 \leq \lambda_2 \leq \cdots \leq \lambda_{k-1} \leq \lambda_k \leq \lambda_{k+1} \leq \cdots$$

Performing a pairwise swap in this sequence results in $0, 1, k, 2, \ldots, k-1, k+1, \ldots$ Let this new sequence correspond to policy π''. Now Lemma 12.1.1 can be used whereby π' (π'') corresponds to sequence $X_0, X_2, X_1, \ldots, X_n$ ($X_0, X_1, X_2, \ldots, X_n$) and r_j' (r_j'') corresponds to the r_j (q_j) in Lemma 12.1.1. Using Lemma 12.1.1 and the previous inequalities, it is established that $E(H'') \leq E(H')$. Proceeding in this manner, whereby at each step a pairwise interchange is performed between job k and the job immediately following it, the sequence $1, 2, \ldots, k-1, k, k+1, \ldots$ is obtained. At each step, it is shown that the expected makespan decreases.

Case 2.   Let π' in this case denote the policy that instructs the scheduler at time t to start job l with rate λ_l, $l \in \{J_2 - J_1\}$ and to adopt policy π^* from the next decision moment onward. That is, job l starts at time t with its second processing while there are still jobs in J_1 that have not completed their first processing yet. Let r_j', $j \in J_1$, in this case denote the probability that job j under π' completes its first processing after all jobs in J_1 have completed their first processing and after job l has completed its second processing. Let r_l' denote the probability that job l under π' completes its second processing after all jobs in J_1 have completed their first processing. Assume that when using π^* from t onward the scheduler may, after having started all jobs in J_1, choose job l as the first job to undergo its second processing and may do this on the machine that becomes available first (under Case 1, it became clear that the probability of job j, $j \neq l$, causing a Type I idle period, does not depend on the machine on which job l is processed the second time). Let r_j^* now denote the probability that job j completes its first processing after jobs $J_1 - j$ complete their first processing and after job l completes its second processing. Let r_l^* denote the probability that job l completes its second processing after all jobs in J_1 have completed their first processing. So

$$q_j^* = r_j^* \prod_{i \in \{J_2 - j - l\}} \frac{\lambda_i}{\lambda_i + \lambda_j}$$

for all j in J_1 and

$$q_j' = r_j' \prod_{i \in \{J_2 - j - l\}} \frac{\lambda_i}{\lambda_i + \lambda_j}$$

for all j in J_1. Again

$$q_0' = q_0^*.$$

To show that $E(H^*) \leq E(H')$, it suffices to show that

$$\sum_{j \in \{J_1 - 0\}} \left(q_j^* \left(\prod_{i \in \{J_2 - j - l\}} \frac{\lambda_i}{\lambda_i + \lambda_j} \right) \frac{1}{2\lambda_j} \right) \leq \sum_{j \in \{J_1 - 0\}} \left(q_j' \left(\prod_{i \in \{J_2 - j - l\}} \frac{\lambda_i}{\lambda_i + \lambda_j} \right) \frac{1}{2\lambda_j} \right)$$

From Lemma 12.1.1, it follows that $r_l^* \geq r_l'$ and $r_i^* \leq r_i'$, $i \in J_1$. It then follows that $E(H^*) \leq E(H')$. This completes the proof of the theorem.   □

It appears to be very hard to generalize this result and include a larger class of distributions.

Example 13.4.2 (Processing Times That Are Mixtures of Exponentials)

> Let the processing time of job j on machine i, $i = 1, 2$, be a mixture of an exponential with rate λ_j and zero with arbitrary mixing probabilities. The optimal policy is to process at time 0 all jobs for a very short period on both machines just to check whether their processing times on the two machines are zero or positive. After the nature of all the processing times have been determined, the problem is reduced to the scenario covered by Theorem 13.4.1.

Theorem 13.4.1 states that jobs that still have to undergo processing on both machines have priority over jobs that only need processing on one machine. In a sense, the policy described in Theorem 13.4.1 is similar to the LAPT rule introduced in Section 8.1 for the deterministic $O2 \parallel C_{\max}$ problem.

From Theorem 13.4.1, it follows that the problem is also tractable if the processing time of job j on machine 1 as well as machine 2 is exponentially distributed with rate 1. The policy that minimizes the expected makespan always gives priority to jobs that have not yet undergone processing on either machine. This particular rule does not require any preemptions. In the literature, this rule has been referred to in this scenario as the *Longest Expected Remaining Processing Time first (LERPT)* rule.

Actually, if in the two-machine case all processing times are exponential with mean 1 and if preemptions are allowed, then the total expected completion time can also be analyzed. This model is an exponential counterpart of $O2 \mid p_{ij} = 1, prmp \mid \sum C_j$. The total expected completion time clearly requires a different policy. One particular policy in the class of preemptive dynamic policies is appealing: Consider the policy that prescribes the scheduler to process, whenever possible, on each one of the machines a job that already has been processed on the other machine. This policy may require the scheduler at times to interrupt the processing of a job and start with the processing of a job that has just completed its operation on the other machine. In what follows, this policy is referred to as the *Shortest Expected Remaining Processing Time first (SERPT)* policy.

Theorem 13.4.3.   *In a two machine open shop, the preemptive SERPT policy minimizes the total expected completion time in the class of preemptive dynamic policies.*

Proof.   Let A_{ij}, $i = 1, 2$, $j = 1, \ldots, n$, denote the time that j jobs have completed their processing requirements on machine i. An idle period on machine 2 occurs if and only if

$$A_{1,n-1} \leq A_{2,n-1} \leq A_{1,n},$$

and an idle period on machine 1 occurs if and only if

$$A_{2,n-1} \leq A_{1,n-1} \leq A_{2,n}.$$

Let $j_1, j_2, \ldots, j_n$ denote the sequence in which the jobs leave the system (i.e., job j_1 is the first one to complete both operations, job j_2 the second, etc.). Under the SERPT policy,

$$C_{j_k} = \max(A_{1,k}, A_{2,k}) = \max \left(\sum_{l=1}^{k} X_{1l}, \sum_{l=1}^{k} X_{2l} \right), \qquad k = 1, \ldots, n-1.$$

This implies that the time epoch of the kth job completion, $k = 1, \ldots, n - 1$, is a random variable that is the maximum of two independent random variables, both with *Erlang(k,λ)* distributions. The distribution of the last job completion, the makespan, is different. It is clear that under the preemptive SERPT policy the sum of the expected completion times of the first $n - 1$ jobs that leave the system are minimized. It is not immediately obvious that SERPT minimizes the sum of all n

completion times. Let

$$B = \max(A_{1,n-1}, A_{2,n-1}).$$

The random variable B is independent of the policy. At time B, each machine has at most one more job to complete. A distinction can now be made between two cases.

First, consider the case where, at B, a job remains to be completed on only one of the two machines. In this case, neither the probability of this event occurring nor the waiting cost incurred by the last job that leaves the system (at $\max(A_{1,n}, A_{2,n})$) depends on the policy. Since SERPT minimizes the expected sum of completion times of the first $n - 1$ jobs to leave the system, it follows that SERPT minimizes the expected sum of the completion times of all n jobs.

Second, consider the case where, at time B, a job remains to be processed on both machines. Either (i) there is one job left that needs processing on both machines, or (ii) there are two jobs left, each needing processing on one machine (a different machine for each). Under (i), the expected sum of the completion times of the last two jobs that leave the system is $E(B) + E(B + 2)$, whereas under (ii), it is $E(B) + 1 + E(B) + 1$. In both subcases, the expected sum of the completion times of the last two jobs is the same. As SERPT minimizes the expected sum of the completion times of the first $n - 2$ jobs that leave the system, it follows that SERPT minimizes the expected sum of the completion times of all n jobs.   □

Unfortunately, no results have been reported in the literature for stochastic open shops with more than two machines.

EXERCISES (COMPUTATIONAL)

13.1. Consider a two machine flow shop with unlimited intermediate storage and three jobs. Each job has an exponentially distributed processing time with mean 1 on both machines (the job sequence is therefore immaterial). The six processing times are i.i.d. Compute the expected makespan and the total expected completion time.

13.2. Consider an m machine permutation flow shop without blocking. The processing times of job j on the m machines are identical and equal to the random variable X_j from an exponential distribution with mean 1. The random variables $X_1, \ldots, X_n$ are i.i.d. Determine the expected makespan and the total expected completion time as a function of m and n.

13.3. Compare the expected makespan obtained in Exercise 13.1 with the expected makespan obtained in Exercise 13.2 for $m = 2$ and $n = 3$. Determine which one is larger and give an explanation.

13.4. Consider a two machine flow shop with zero intermediate storage and blocking and n jobs. The processing time of job j on machine i is X_{ij} exponentially distributed with mean 1. The $2n$ processing times are i.i.d. Compute the expected makespan and the total expected completion time as a function of n.

13.5. Consider a two machine flow shop with zero intermediate storage and blocking and n jobs. The processing time of job j on each one of the two machines is equal to the random variable X_j from an exponential distribution with mean 1. The variables $X_1, \ldots, X_n$ are i.i.d. Compute again the expected makespan and the total expected completion time. Compare the results with the results obtained in Exercise 13.4.

13.6. Consider the two machine open shop and n jobs. The processing times X_{ij} are all i.i.d. exponential with mean 1. Assume that the policy in Theorem 13.3.1 is followed (i.e., jobs that still need processing on both machines have priority over jobs that only need processing on one of the machines).

(a) Show that

$$E\left(\max\left(\sum_{j=1}^{n} X_{1j}, \sum_{j=1}^{n} X_{2j}\right)\right) = 2n - \sum_{k=n}^{2n-1} k\binom{k-1}{n-1}\left(\frac{1}{2}\right)^{k}.$$

(b) Show that the probability of the jth job that starts its first processing to cause an idle period of Type I is

$$\left(\frac{1}{2}\right)^{n-1-(j-2-i)}\left(\frac{1}{2}\right)^{n-1-i} = \left(\frac{1}{2}\right)^{2n-j}.$$

(c) Show that the expected makespan is equal to

$$E(C_{\max}) = 2n - \sum_{k=n}^{2n-1} k\binom{k-1}{n-1}\left(\frac{1}{2}\right)^{k} + \left(\frac{1}{2}\right)^{n}.$$

13.7. Consider the same scenario as in Exercise 13.1 with two machines and three jobs. However, now all processing times are i.i.d. according to the EME distribution with mean 1. Compute the expected makespan and the total expected completion time and compare the outcome with the results of Exercise 13.1.

13.8. Consider the same scenario as in Exercise 13.2 with m machines and n jobs. However, now $X_1, \ldots, X_n$ are i.i.d. according to the EME distribution with mean 1. Compute the expected makespan and the total expected completion time as a function of m and n. Compare the results obtained with the results from Exercises 13.2 and 13.7.

13.9. Consider a two machine job shop and three jobs. Jobs 1 and 2 have to be processed first on machine 1 and then on machine 2. Job 3 has to be processed first on machine 2 and then on machine 1. Compute the expected makespan under the assumption that the optimal policy is being followed.

13.10. Consider the following proportionate two machine open shop. The processing time of job j on the two machines is equal to the *same* random variable X_j that is exponentially distributed with mean 1. Assume that the two machine open shop is being operated under the nonpreemptive dynamic policy that always gives priority to a job that has not yet received processing on the other machine. Compute the expected makespan with two jobs and with three jobs.

EXERCISES (THEORETICAL)

13.11. Consider the stochastic counterpart of $F2 \mid block \mid C_{\max}$ that is equivalent to a deterministic TSP with a distance matrix that has to be *minimized*. Verify whether this distance matrix satisfies the triangle inequality (i.e., $d_{jk} + d_{kl} \geq d_{jl}$ for all j, k, and l).

13.12. Consider an m machine flow shop with zero intermediate storages between machines and n jobs. The processing times of $n - 1$ jobs on each one of the m machines are 1. Job n has processing time X_{in} on machine i, and the random variables $X_{1n}, X_{2n}, \ldots, X_{mn}$ are i.i.d. from distribution F_n with mean 1.

    (a) Show that the sequence that minimizes the expected makespan puts the stochastic job either first or last.

    (b) Show that the sequence that puts the stochastic job last minimizes the total expected completion time.

13.13. Consider the two machine flow shop with zero intermediate storage between the two machines. Of $n - 2$ jobs, the processing times are deterministic 1 on each one of the two machines. Of the two remaining jobs, the four processing times are i.i.d. from an arbitrary distribution F with mean 1. Show that, to minimize the expected makespan, one of the stochastic jobs has to go first and the other one last.

13.14. Consider the following stochastic counterpart of $Fm \mid p_{ij} = p_j \mid C_{\max}$. The processing time of job j on each one of the m machines is X_j from distribution F_j with mean 1.

    (a) Find an upper and a lower bound for the expected makespan when F is ICR.

    (b) Find an upper and a lower bound for the expected makespan when F is DCR.

13.15. Consider the following stochastic counterpart of $F2 \mid block \mid C_{\max}$. The processing time of job j on machine i is X_{ij} from distribution F_{ij} with mean 1. The $2n$ processing times are independent.

    (a) Find an upper and a lower bound for the expected makespan when F is ICR.

    (b) Find an upper and a lower bound for the expected makespan when F is DCR.

13.16. Consider a two machine open shop with n jobs. The processing times of job j on machines 1 and 2 are equal to X_j from distribution F. The random variables $X_1, \ldots, X_n$ are i.i.d. Show that, to minimize the expected makespan, the scheduler, whenever a machine is freed, has to select a job that has not been processed yet on the other machine.

13.17. Consider an m machine permutation flow shop with finite intermediate storages and blocking. The processing times of job j on the m machines are $X_{1j}, X_{2j}, \ldots, X_{mj}$, which are i.i.d. from distribution F_j. Assume that

$$F_1 \leq_{a.s.} F_2 \leq_{a.s.} \cdots \leq_{a.s.} F_n.$$

Show that SEPT minimizes the total expected completion time.

13.18. Consider stochastic counterparts of the following five deterministic problems:

    (i) $F2 \mid block \mid C_{\max}$,

    (ii) $F2 \parallel C_{\max}$,

    (iii) $O2 \parallel C_{\max}$,

    (iv) $J2 \parallel C_{\max}$,

    (v) $P2 \mid chains \mid C_{\max}$

Problems (i), (ii), (iii), and (iv) all have n jobs. Problem (iv) has k jobs that have to be processed first on machine 1 and then on machine 2 and $n - k$ jobs that have to be processed first on machine 2 and then on machine 1. Problem (v) has $2n$ jobs in n chains of two jobs each. All processing times are i.i.d. exponential with mean 1. Compare the five problems with regard to the expected makespan and the total expected completion time under the optimal policy.

13.19. Consider a two machine proportionate flow shop with n jobs. If $X_{1j} = X_{2j} = D_j$ (deterministic), the makespan is sequence independent. If X_{1j} and X_{2j} are i.i.d. exponential with rate λ_j, then the expected makespan is sequence independent as well. Consider now a proportionate flow shop with X_{1j} and X_{2j} i.i.d. *Erlang$(2,\lambda_j)$* with each one of the two phases distributed according to an exponential with rate λ_j, $j = 1, \dots, n$. Show via an example that the expected makespan *does* depend on the sequence.

13.20. Consider two machines in series with a buffer storage of size b in between the two machines. The processing times are proportionate. Show that the makespan is decreasing convex in the buffer size b.

13.21. Consider a stochastic counterpart of $Fm \parallel C_{\max}$ with machines that have different speeds—say $v_1, \dots, v_m$. These speeds are fixed (deterministic). Job j requires an amount of work Y_j, distributed according to F_j, on each one of the m machines. The amount of time it takes machine i to process job j is equal to $X_{ij} = Y_j/v_i$. Show that if $v_1 \geq v_2 \geq \cdots \geq v_n$ ($v_1 \leq v_2 \leq \cdots \leq v_n$) and $F_1 \leq_{lr} F_2 \leq_{lr} \cdots \leq_{lr} F_n$, the SEPT (LEPT) sequence minimizes the expected makespan. In other words, show that if the flow shop is decelerating (accelerating), the SEPT (LEPT) sequence is optimal.

COMMENTS AND REFERENCES

The result stated in Theorem 13.1.1 has a very long history. The first publications with regard to this result are by Talwar (1967), Bagga (1970), and Cunningham and Dutta (1973). An excellent paper on this problem is by Ku and Niu (1986). The proof of Theorem 13.1.1, as presented here, is due to Weiss (1982). The SEPT-LEPT sequences for flow shops were introduced and analyzed by Pinedo (1982), and the SV sequences were studied by Pinedo and Wie (1986). For stochastic flow shops with due dates, see Lee and Lin (1991). For models where the processing times of any given job on the m machines are equal to the *same* random variable, see Pinedo (1985). A study of the impact of the randomness of the processing times on the values of the expected makespan and expected total completion time is contained in Pinedo and Weber (1984). Frostig and Adiri (1985) obtain results for flow shops with three machines. Boxma and Forst (1986) obtain results with regard to the expected weighted number of tardy jobs.

The material in the section on stochastic flow shops with blocking is based primarily on the paper by Pinedo (1982). For more results on stochastic flow shops with blocking, see Foley and Suresh (1984a, 1984b, 1986), Kijima, Makimoto, and Shirakawa (1990), Pinedo and Weber (1984), and Suresh, Foley, and Dickey (1985). The analysis of the two machine job shop is from the paper by Pinedo (1981b).

The optimal policy for minimizing the expected makespan in two machine open shops is due to Pinedo and Ross (1982). The result with regard to the minimization of the total expected completion time in two machine open shops is from the paper by Pinedo (1984).

PART 3
Scheduling in Practice

14

General Purpose Procedures for Scheduling in Practice

This chapter describes a number of general purpose procedures that are useful in dealing with scheduling problems in practice and that can be implemented with relative ease in industrial scheduling systems. All the techniques described are heuristics that do not guarantee an *optimal* solution; instead they aim to find reasonably good solutions in a relatively short time. The heuristics tend to be fairly generic and can be adapted easily to a large variety of scheduling problems.

This chapter does not cover real optimization techniques such as branch and bound or dynamic programming. Applications of such optimization techniques tend to be somewhat problem specific and are therefore discussed in detail in the coverage of certain problems in other chapters and in the appendixes.

The first section gives a classification and overview of some of the more elementary priority or dispatching rules such as those described in previous chapters. The second section discusses a method of combining priority or dispatching rules. These composite rules are combinations of multiple elementary dispatching rules. The third technique is a modification of the branch and bound procedure that aims to eliminate branches in an intelligent way so that not all branches have to be examined. The fourth and fifth sections deal with procedures that are based on local search. These techniques tend to be fairly generic and can be applied to a variety of scheduling problems with only minor customization. The fourth section discusses simulated annealing and tabu-search. The fifth section describes a more general local search procedure—namely, genetic algorithms. The last section discusses how different empirical techniques can be combined with one another in a single framework.

14.1 DISPATCHING RULES

Research in dispatching rules has been active for several decades, and many different rules have been studied in the literature. These rules can be classified in various ways. For example, a distinction can be made between *static* and *dynamic* rules. Static rules are not time dependent. They are just a function of the job and/or machine data (e.g., WSPT). Dynamic rules are time dependent. One example of a dynamic rule is the *Minimum Slack (MS) first*, which orders jobs according to $\max(d_j - p_j - t, 0)$ which is time dependent. This implies that at some point in time job j may have a higher priority than job k and at some later point in time jobs j and k may have the same priority.

A second way to classify rules is according to the information on which they are based. A *local* rule uses only information pertaining to either the queue where the job is waiting or the machine where the job is queued. Most of the rules introduced in the previous chapters can be used as local rules. A *global* rule may use information regarding other machines, such as the processing time of the job on the next machine on its route. An example of a global rule is the LAPT rule for the two machine open shop.

In the preceding chapters, many different rules have come up. Of course, there are many more besides the ones discussed. A simple one, often used in practice, is the *Service in Random Order (SIRO)* rule. Under this rule, no attempt is made to optimize anything. Another rule often used is the *First Come First Served* rule, which is equivalent to the *Earliest Release Date first (ERD)* rule. This rule attempts to equalize the waiting times of the jobs (i.e., to minimize the variance of the waiting times). Some rules are only applicable under given conditions in certain machine environments. For example, consider a bank of parallel machines, each with its own queue. According to the *Shortest Queue (SQ) first* rule, every newly released job is assigned to the machine with the shortest queue. This rule is clearly time dependent and therefore dynamic. Many global dynamic rules have been designed for job shops. According to the *Shortest Queue at the Next Operation (SQNO)* rule, every time a machine is freed, the job that has the shortest queue at the next machine on its route is selected for processing.

In Table 14.1, an overview of some of the better known dispatching rules is given. A number of these rules lead to optimal schedules in certain machine environments and are reasonable heuristics in others. All of these rules have variations that can be applied in more complicated settings.

Dispatching rules are useful when one attempts to find a reasonably good schedule with regard to a single objective such as the makespan, the total completion time or the maximum lateness.

However, in real life, objectives are often more complicated. For example, a realistic objective may be a combination of several basic objectives and also a function of time and of the set of jobs waiting for processing. Sorting the jobs on the basis of one or two parameters may not lead to acceptable schedules. More elaborate dispatching rules that take into account several different parameters can address more complicated objective functions. Some of these more elaborate rules are basically a combination of a number of the elementary dispatching rules listed earlier. These more elaborate rules are referred to as composite dispatching rules and are described in the next section.

TABLE 14.1

	RULE	DATA	ENVIRONMENT	SECTION
1	SIRO	—	—	14.1
2	ERD	r_j	$1 \mid r_j \mid \mathrm{Var}(\sum(C_j - r_j)/n)$	14.1
3	EDD	d_j	$1 \parallel L_{\max}$	3.2
4	MS	d_j	$1 \parallel L_{\max}$	14.1
5	SPT	p_j	$Pm \parallel \sum C_j;\ Fm \mid p_{ij} = p_j \mid \sum C_j$	5.3; 6.1
6	WSPT	w_j, p_j	$Pm \parallel \sum w_j C_j$	3.1; 5.3
7	LPT	p_j	$Pm \parallel C_{\max}$	5.1
8	SPT-LPT	p_j	$Fm \mid block,\ p_{ij} = p_j \mid C_{\max}$	6.2
9	CP	$p_j, prec$	$Pm \mid \prec \mid C_{\max}$	5.1
10	LNS	$p_j, prec$	$Pm \mid \prec \mid C_{\max}$	5.1
11	SST	s_{jk}	$1 \mid s_{jk} \mid C_{\max}$	4.5
12	LFJ	M_j	$Pm \mid M_j \mid C_{\max}$	5.1
13	LAPT	p_{ij}	$O2 \parallel C_{\max}$	8.1
14	SQ	—	$Pm \parallel \sum C_j$	14.1
15	SQNO	—	$Jm \parallel \gamma$	14.1

14.2 COMPOSITE DISPATCHING RULES

To explain the structure and construction of these composite dispatching rules, a general framework has to be introduced. Subsequently, two of the more widely used composite rules are described.

A composite dispatching rule is a ranking expression that combines a number of elementary dispatching rules. An elementary rule is a function of attributes of the jobs and/or machines. An attribute may be any property associated with either a job or machine which may be either constant or time dependent. Examples of job attributes are weight, processing time, and due date; examples of machine attributes are speed, the number of jobs waiting for processing, and the total amount of processing that is waiting in queue. The extent to which a given attribute affects the overall priority of a job is determined by the elementary rule that uses it as well as a scaling parameter. Each elementary rule within a composite dispatching rule has its own scaling parameter, which is chosen to properly scale the contribution of the elementary rule to the total ranking expression. The scaling parameters are either fixed by the designer of the rule or variable and a function of time or of the particular job set to be scheduled. If they depend on the particular job set to be scheduled, they require the computation of some job set statistics that characterize the particular scheduling instance at hand as accurately as possible (e.g., whether the due dates in the particular instance are tight). These statistics, which are also called *factors*, usually do not depend on the schedule and can be computed easily from the given job and machine attributes.

The functions that map the statistics into the scaling parameters have to be determined by the designer of the rule. Experience may offer a reasonable guide, but extensive computer simulation is often required. These functions are usually determined only once before the rule is made available for regular use.

Each time the composite dispatching rule is used for generating a schedule, the necessary statistics are computed. Based on the values of these statistics, the values of the scaling parameters are set by the predetermined functions. After the scaling parameters are set, the dispatching rule is applied to the job set.

One example of a composite dispatching rule is a rule that is often used for the $1 \mid\mid \sum w_j T_j$ problem. As stated in Chapter 3, the $1 \mid\mid \sum w_j T_j$ problem is strongly NP-hard. As branch and bound methods are prohibitively time-consuming even with only 30 jobs, it is important to have a heuristic that provides a reasonably good schedule with reasonable computational effort. Some heuristics come immediately to mind; namely, the WSPT rule (that is optimal when all release dates and due dates are zero) and the EDD rule or the MS rule (which are optimal when all due dates are sufficiently loose and spread out). It is natural to seek a heuristic or priority rule that combines the characteristics of these dispatching rules. The *Apparent Tardiness Cost (ATC)* heuristic is a composite dispatching rule that combines the WSPT rule and the MS rule. (Recall that under the MS rule the slack of job j at time t, $\max(d_j - p_j - t, 0)$, is computed and the job with the minimum slack is scheduled.) Under the ATC rule, jobs are scheduled one at a time; that is, every time the machine becomes free, a ranking index is computed for each remaining job. The job with the highest ranking index is then selected to be processed next. This ranking index is a function of the time t at which the machine became free as well as of the p_j, the w_j, and the d_j of the remaining jobs. The index is defined as

$$I_j(t) = \frac{w_j}{p_j} \exp\left(-\frac{\max(d_j - p_j - t, 0)}{K \bar{p}}\right),$$

where K is the scaling parameter that can be determined empirically, and $\bar{p}$ is the average of the processing times of the remaining jobs. If K is very large, the ATC rule reduces to the WSPT rule. If K is very small, the rule reduces to the MS rule when there are no overdue jobs and to the WSPT rule applied to the overdue jobs otherwise.

To obtain good schedules, the value of K (sometimes referred to as the look-ahead parameter) must be appropriate for the particular instance of the problem. This can be done by first performing a statistical analysis of the particular scheduling instance under consideration. There are several statistics that can be used to help characterize scheduling instances. The *due date tightness* factor τ is defined as

$$\tau = 1 - \frac{\sum d_j}{n C_{\max}},$$

where $\sum d_j / n$ is the average of the due dates. Values of τ close to 1 indicate that the due dates are tight and values close to 0 indicate that the due dates are loose. The due date range factor R is defined as

$$R = \frac{d_{\max} - d_{min}}{C_{\max}}.$$

A high value of R indicates a wide range of due dates, whereas a low value indicates a narrow range of due dates. A significant amount of research has been done to establish the relationships between the scaling parameter K and the factors τ and R.

Thus, when one wishes to minimize $\sum w_j T_j$ in a single machine or in a more complicated machine environment (machines in parallel, flexible flow shops), one first characterizes the particular problem instance through the two statistics. Then one determines the value of the look-ahead parameter K as a function of these characterizing factors as well as the particular machine environment. After fixing K, one applies the rule. This rule could be used, for example, in the paper bag factory described in Example 1.1.1.

Several generalizations of the ATC rule have been developed to take release dates and sequence-dependent setup times into account. Such a generalization, the *Apparent Tardiness Cost with Setups (ATCS)* rule, has been designed for the $1 \mid s_{jk} \mid \sum w_j T_j$ problem. The objective is once again to minimize the sum of the weighted tardinesses, but now the jobs are subject to sequence dependent setup times. This implies that the priority of any job j depends on the job just completed on the machine just freed. The ATCS rule combines the WSPT rule, the MS rule, and the SST rule in a single ranking index. The rule calculates the index of job j at time t when job l has completed its processing on the machine as

$$I_j(t, l) = \frac{w_j}{p_j} \exp\left(-\frac{\max(d_j - p_j - t, 0)}{K_1 \bar{p}}\right) \exp\left(-\frac{s_{lj}}{K_2 \bar{s}}\right),$$

where $\bar{s}$ is the average of the setup times of the jobs remaining to be scheduled, K_1 the due date-related scaling parameter, and K_2 the setup time-related scaling parameter. Note that the scaling parameters are dimensionless quantities to make them independent of the units used to express various quantities.

The two scaling parameters, K_1 and K_2, can be regarded as functions of three factors:

(i) the due date tightness factor τ,
(ii) the due date range factor R,
(iii) the setup time severity factor $\eta = \bar{s}/\bar{p}$.

These statistics are not as easy to determine as in the previous case. Even with a single machine, the makespan is now schedule dependent because of the setup times. Before computing the τ and R factors, the makespan has to be estimated. A simple estimate for the makespan on a single machine can be

$$\hat{C}_{max} = \sum_{j=1}^{n} p_j + n\bar{s}.$$

This estimate most likely will overestimate the makespan as the final schedule will take advantage of setup times that are lower than average. The definitions of τ and R have to be modified by replacing the makespan with its estimate.

An experimental study of the ATCS rule, although inconclusive, has suggested some guidelines for the selection of the two parameters K_1 and K_2. The following rule can be used for selecting a proper value of K_1:

$$K_1 = 4.5 + R, \qquad R \leq 0.5$$
$$K_1 = 6 - 2R, \qquad R \geq 0.5.$$

The following rule can be used for selecting a proper value of K_2:

$$K_2 = \tau/(2\sqrt{\eta}).$$

Example 14.2.1 (Application of the ATCS Rule)

Consider an instance of $1 \mid s_{jk} \mid \sum w_j T_j$ with the following four jobs.

jobs	1	2	3	4
p_j	13	9	13	10
d_j	12	37	21	22
w_j	2	4	2	5

The setup times s_{0j} of the first job in the sequence are presented in the following table.

jobs	1	2	3	4
s_{0j}	1	1	3	4

The sequence-dependent setup times of the jobs following the first job are the following:

jobs	1	2	3	4
s_{1j}	–	4	1	3
s_{2j}	0	–	1	0
s_{3j}	1	2	–	3
s_{4j}	4	3	1	–

To use the ATCS rule, the average processing time $\bar{p}$ and the average setup time $\bar{s}$ have to be determined. The average processing time is approximately 11 while the average setup time is approximately 2. An estimate for the makespan is

$$\hat{C}_{\max} = \sum_{j=1}^{n} p_j + n\bar{s} = 45 + 4 \times 2 = 53.$$

The due date range factor $R = 25/53 \approx 0.47$, the due date tightness coefficient $\tau = 1 - 23/53 \approx 0.57$, and the setup time severity coefficient is $\eta = 2/11 \approx 0.18$. Using the given formulae, the parameter K_1 is chosen to be 5 and the parameter K_2 is chosen to be 0.7. To determine which job goes first, $I_j(0, 0)$ has to be computed for $j = 1, \dots, 4$:

$$I_1(0, 0) = \frac{2}{13} \exp\left(-\frac{(12 - 13)^+}{55}\right) \exp\left(-\frac{1}{1.4}\right) \approx 0.15 \times 1 \times 0.51 = 0.075$$

$$I_2(0,0) = \frac{4}{9} \exp\left(-\frac{(37-9)^+}{55}\right) \exp\left(-\frac{1}{1.4}\right) \approx 0.44 \times 0.6 \times 0.47 = 0.131$$

$$I_3(0,0) = \frac{2}{13} \exp\left(-\frac{(21-13)^+}{55}\right) \exp\left(-\frac{3}{1.4}\right) \approx 0.15 \times 0.86 \times 0.103 = 0.016$$

$$I_4(0,0) = \frac{5}{10} \exp\left(-\frac{(22-10)^+}{55}\right) \exp\left(-\frac{4}{1.4}\right) \approx 0.50 \times 0.80 \times 0.05 = 0.020.$$

Job 2 has the highest priority (even though its due date is the latest). As its setup time is 1, its completion time is 10. So at the second iteration, $I_1(10, 2)$, $I_3(10, 2)$, and $I_4(10, 2)$ have to be computed. To simplify the computations, the values of $K_1 \bar{p}$ and $K_2 \bar{s}$ can be kept the same. Continuing the application of the ATCS rule results in the sequence 2, 4, 3, 1 with the sum of the weighted tardinesses equal to 98. Complete enumeration shows that this sequence is optimal. Note that this sequence always selects, whenever the machine is freed, one of the jobs with the smallest setup time.

The ATCS rule can be easily applied to $Pm \mid s_{jk} \mid \sum w_j T_j$ as well. Of course, the look-ahead parameters K_1 and K_2 have to be determined as a function of τ, R, η and m. The ATCS rule has been used in a scheduling system for a paper bag factory such as the one described in Example 1.1.1 (see also Section 16.4).

14.3 FILTERED BEAM SEARCH

Filtered beam search is based on the ideas of branch and bound. Enumerative branch and bound methods are currently the most widely used methods for obtaining optimal solutions to NP-hard scheduling problems. The main disadvantage of branch and bound is that it usually is extremely time-consuming as the number of nodes that have to be considered is very large.

Consider, for example, a single machine problem with n jobs. Assume that for each node at level k jobs have been selected for the first k positions. There is a single node at level 0, with n branches emanating to n nodes at level 1. Each node at level 1 branches out into $n - 1$ nodes at level 2, resulting in a total of $n(n - 1)$ nodes at level 2. At level k there are $n!/(n - k)!$ nodes. At the bottom level, level n, there are $n!$ nodes.

Branch and bound attempts to eliminate a node by determining a lower bound on the objective for all partial schedules that sprout out of that node. If the lower bound is higher than the value of the objective of a known schedule, then the node may be eliminated and its offspring disregarded. If one could obtain a reasonably good schedule through some clever heuristic before going through the branch and bound procedure, then it might be possible to eliminate many nodes. Other elimination criteria (see Chapter 3) may also reduce the number of nodes to be investigated. However, even after these eliminations, there are usually still too many nodes to be evaluated. For example, it may require several weeks on a RISC workstation to find an optimal schedule for an instance of the $1 \mid\mid \sum w_j T_j$ problem with 30 jobs. The main advantage of branch and bound is that, after evaluating all nodes, the final solution is known with certainty to be optimal.

Filtered beam search is an adaptation of branch and bound in which not all nodes at any given level are evaluated. Only the most promising nodes at level k are selected as nodes to branch from. The remaining nodes at that level are discarded *permanently*. The number of nodes retained is called the *beam width* of the search. The evaluation process that determines which nodes are the promising ones is a crucial element of this method. Evaluating each node carefully to obtain an estimate for the potential of its offspring is time-consuming. There is a trade-off here: A crude prediction is quick, but may lead to discarding good solutions, whereas a more thorough evaluation may be prohibitively time-consuming. Here is where the filter comes in. For all the nodes generated at level k, a crude prediction is done. Based on the outcome of this crude prediction, a number of nodes are selected for a thorough evaluation, while the remaining nodes are discarded permanently. The number of nodes selected for a thorough evaluation is referred to as the *filter width*. Based on the outcome of the careful evaluation of all nodes that have passed the filter, a subset of these nodes (the number being equal to the beam width that therefore cannot be greater than the filter width) is selected from where further branches are generated.

A simple example of a crude prediction is the following. The contribution of the partial schedule to the objective and the due date tightness or some other statistic of the jobs remaining to be scheduled are computed. Based on these values, the nodes at a given level are compared to one another and an overall assesment is made.

Every time a node has to undergo a thorough evaluation, all the jobs not yet scheduled are scheduled according to a composite dispatching rule. Such a schedule can still be generated reasonably fast as it only requires sorting. The result of such a schedule is an indication of the promise of the node. If a large number of jobs is involved, nodes may be filtered out by examining the partial schedule obtained by scheduling only a limited number of the remaining jobs with a dispatching rule. This extended partial schedule may be evaluated and based on its value a node may be discarded or retained. If a node is retained, it may be analyzed more thoroughly by having all its remaining jobs scheduled with the composite dispatching rule. The value of this schedule's objective then represents an upper bound on the best schedule among the offspring of that node. The following example illustrates a simplified version of beam search.

Example 14.3.1 (Application of Beam Search)

Consider the following instance of $1 \mid\mid \sum w_j T_j$.

jobs	1	2	3	4
p_j	10	10	13	4
d_j	4	2	1	12
w_j	14	12	1	12

As the number of jobs is rather small, only one type of prediction is made for the nodes at any particular level. No filtering mechanism is used. The beam width is chosen to be 2, which implies that at each level only two nodes are retained. The prediction at a node is made by scheduling the unscheduled jobs according to the ATC rule. With the due date range factor R being $11/37$ and the due date tightness factor τ being $32/37$, the look-ahead parameter is chosen to be 5.

A branch and bound tree is constructed assuming the sequence is developed starting out from $t = 0$. So at the jth level of the tree, jobs are put into the jth position. At level 1 of the tree, there are four nodes: $(1, *, *, *)$, $(2, *, *, *)$, $(3, *, *, *)$, and $(4, *, *, *)$ (see Figure 14.1). Applying the ATC rule to the three remaining jobs at each one of the four nodes results in the four sequences: $(1, 4, 2, 3)$, $(2, 4, 1, 3)$, $(3, 4, 1, 2)$, and $(4, 1, 2, 3)$ with objective values $408, 436, 771$, and 440. As the beam width is 2, only the first two nodes are retained.

Each of these two nodes leads to three nodes at level 2. Node $(1, *, *, *)$ leads to nodes $(1, 2, *, *)$, $(1, 3, *, *)$, and $(1, 4, *, *)$ and node $(2, *, *, *)$ leads to nodes

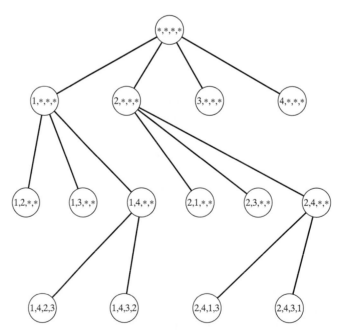

Figure 14.1   Beam search applied to $1 \mid\mid \sum \omega_j T_j$.

$(2, 1, *, *)$, $(2, 3, *, *)$, and $(2, 4, *, *)$. Applying the ATC rule to the remaining two jobs in each one of the six nodes at level 2 results in nodes $(1, 4, *, *)$ and $(2, 4, *, *)$ being retained and the remaining four being discarded.

The two nodes at level 2 lead to four nodes at level 3 (the last level)—namely, nodes $(1, 4, 2, 3)$, $(1, 4, 3, 2)$, $(2, 4, 1, 3)$, and $(2, 4, 3, 1)$. Of these four sequences, sequence $(1, 4, 2, 3)$ is the best with a total weighted tardiness of 408. It can be verified through complete enumeration that this sequence is optimal.

14.4 LOCAL SEARCH: SIMULATED ANNEALING AND TABU-SEARCH

So far, all the algorithms and heuristics described in this chapter have been of the constructive type. They start without a schedule and gradually construct a schedule by adding one job at a time.

In this section and the next, algorithms of the improvement type are considered. Algorithms of the improvement type are conceptually completely different from algorithms of the constructive type. They start out with a complete schedule, which may be selected arbitrarily, and then try to obtain a better schedule by manipulating the current schedule. An important class of improvement type algorithms are the local search procedures. A local search procedure does not guarantee an optimal solution. It usually attempts to find a better schedule than the current one in the *neighborhood* of the current one. Two schedules are *neighbors* if one can be obtained through a well-defined modification of the other. At each iteration, a local search procedure performs a search within the neighborhood and evaluates the various neighboring solutions. The procedure either accepts or rejects a candidate solution as the next schedule to move to based on a given acceptance–rejection criterion.

One can compare the various local search procedures with respect to the following four design criteria:

  (i) The schedule representation needed for the procedure.
 (ii) The neighborhood design.
(iii) The search process within the neighborhood.
 (iv) The acceptance–rejection criterion.

The representation of a schedule may be at times nontrivial. A nonpreemptive single machine schedule can be specified by a simple permutation of the n jobs. A nonpreemptive job shop schedule can be specified by m consecutive strings, each one representing a permutation of n operations on a specific machine. Based on this information, the starting and completion times of all operations can be computed. However, when preemptions are allowed, the format of the schedule representation becomes significantly more complicated.

The design of the neighborhood is a very important aspect of a local search procedure. For a single machine, a neighborhood of a particular schedule may be simply defined as all schedules that can be obtained by doing a single adjacent pairwise interchange. This implies that there are $n - 1$ schedules in the neighborhood of the original schedule. A larger neighborhood for a single machine schedule may be defined by taking an arbitrary job in the schedule and inserting it in another position in the schedule. Clearly, each job can be inserted in $n - 1$ other positions. The entire neighborhood contains less than $n(n - 1)$ neighbors as some of these neighbors are identical. The neighborhood of a schedule in a more complicated machine environment is usually more complex.

An interesting example is a neighborhood designed for the job shop problem with the makespan as objective. To describe this neighborhood, the concept of a critical path has to be used. A critical path in a job shop schedule consists of a set of operations of which the first one starts out at time $t = 0$ and the last one finishes at time $t = C_{\max}$. The completion time of each operation on a critical path is equal to the starting time of the next operation on that path; two successive operations either belong to the same job or are processed on the same machine (see Chapter 7). A schedule may have multiple critical paths that may overlap. Finding the critical path(s) in a given schedule for a job shop problem with the makespan as objective is relatively straightforward. It is clear that, to reduce the makespan, changes have to be made in the sequence(s) of the operations on the critical path(s). A simple neighborhood of an existing schedule can be designed as follows: The set of schedules whose corresponding sequences of operations on the machines can be obtained by interchanging a pair of adjacent operations on the critical path of the current schedule. Note that, to interchange a pair of operations on the critical path, the operations must be on the same machine and belong to different jobs. If there is a single critical path, then the number of neighbors within the neighborhood is at most the number of operations on the critical path minus 1.

Experiments have shown that this type of neighborhood for the job shop problem is too simple to be effective. The number of neighboring schedules that are better than the existing schedule tends to be very limited. More sophisticated neighborhoods have been designed that perform better. One of these is referred to as the *One Step Look-Back Adjacent Interchange*.

Example 14.4.1 (Neighborhood of a Job Shop Schedule)

A neighbor of a current schedule is obtained by first performing an adjacent pairwise interchange between two operations (i, j) and (i, k) on the critical path. After the interchange, operation (i, k) is processed before operation (i, j) on machine i. Consider job k to which operation (i, k) belongs and refer to the operation of job k immediately preceding operation (i, k) as operation (h, k) (it is processed on machine h). On machine h, interchange operation (h, k) and the operation preceding (h, k) on machine h, say operation (h, l) (see Figure 14.2). From the figure it is clear

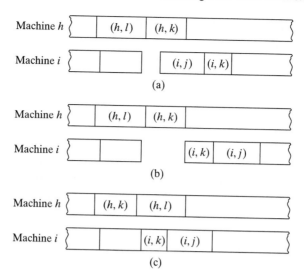

Figure 14.2 One-step look-back interchange for $Jm \parallel C_{\max}$: (a) current schedule, (b) schedule after interchange of (i, j) and (i, k), (c) schedule after interchange of (h, l) and (h, k).

that, even if the first interchange (between (i, j) and (i, k)) does not result in an improvement, the second interchange between (h, k) and (h, l) may lead to an overall improvement.

Actually, this design can be made more elaborate by backtracking more than one step. These types of interchanges are referred to as *Multi-Step Look-Back Interchanges*.

In exactly the same way, one can construct the *One Step Look-Ahead Interchange* and the *Multi-Step Look-Ahead Interchanges*.

The search process within a neighborhood can be done in a number of ways. A simple way is to select schedules in the neighborhood at random, evaluate these schedules, and decide which one to accept. However, it may pay to do a more organized search and select first schedules that appear promising. One may want to consider swapping those jobs that affect the objective the most. For example, when the total weighted tardiness has to be minimized, one may want to move jobs that are very tardy more toward the beginning of the schedule.

The acceptance–rejection criterion is usually the design aspect that distinguishes a local search procedure the most. The difference between the two procedures discussed in the remaining part of this section, simulated annealing and tabu-search, lies mainly in their acceptance–rejection criteria. In simulated annealing, the acceptance-rejection criterion is based on a probabilistic process, whereas in tabu-search it is based on a deterministic process.

Simulated annealing is a search process that has its origin in the fields of material science and physics. It was first developed as a simulation model for describing the physical annealing process of condensed matter.

The simulated annealing procedure goes through a number of iterations. At iteration k of the procedure, there is a current schedule S_k as well as a best schedule found so far, S_0. For a single machine problem, these schedules are sequences (permutations) of the jobs. Let $G(S_k)$ and $G(S_0)$ denote the corresponding values of the objective function. Note that $G(S_k) \geq G(S_0)$. The value of the best schedule obtained so far, $G(S_0)$, is often referred to as the aspiration criterion. The algorithm, in its search for an optimal schedule, moves from one schedule to another. At iteration k, a search for a new schedule is conducted within the neighborhood of S_k. First, a so-called *candidate* schedule, say S_c, is selected from the neighborhood. This selection of a candidate schedule can be done at random or in an organized, possibly sequential, way. If $G(S_c) < G(S_k)$, a move is made, setting $S_{k+1} = S_c$. If $G(S_c) < G(S_0)$, then S_0 is set equal to S_c. However, if If $G(S_c) \geq G(S_k)$, a move is made to S_c with probability

$$P(S_k, S_c) = \exp\left(\frac{G(S_k) - G(S_c)}{\beta_k}\right);$$

with probability $1 - P(S_k, S_c)$ schedule S_c is rejected in favor of the current schedule, setting $S_{k+1} = S_k$. Schedule S_0 does not change when it is better than schedule S_c. The $\beta_1 \geq \beta_2 \geq \beta_3 \geq \cdots > 0$ are control parameters referred to as cooling parameters or temperatures (in analogy with the annealing process mentioned earlier). Often β_k is chosen to be a^k for some a between 0 and 1.

From the prior description of the simulated annealing procedure, it is clear that moves to worse solutions are allowed. The reason for allowing these moves is to give the procedure the opportunity to move away from a local minimum and find a better solution later. Since β_k decreases with k, the acceptance probability for a nonimproving move is lower in later iterations of the search process. The definition of the acceptance probability also ensures that if a neighbor is significantly worse, its acceptance probability is very low and the move is unlikely to be made.

Several stopping criteria are used for this procedure. One way is to let the procedure run for a prespecified number of iterations. Another is to let the procedure run until no improvement has been achieved during a predetermined number of iterations.

The method can be summarized as follows.

Algorithm 14.4.2. (Simulated Annealing)

Step 1. *Set $k = 1$ and select β_1.*

Select an initial sequence S_1 using some heuristic.
Set $S_0 = S_1$.

Step 2. *Select a candidate schedule S_c from the neighborhood of S_k.*
If $G(S_0) < G(S_c) < G(S_k)$, set $S_{k+1} = S_c$ and go to Step 3.
If $G(S_c) < G(S_0)$, set $S_0 = S_{k+1} = S_c$ and go to Step 3.
If $G(S_c) > G(S_k)$, generate a random number U_k from a Uniform(0,1) distribution;
If $U_k \leq P(S_k, S_c)$, set $S_{k+1} = S_c$; otherwise set $S_{k+1} = S_k$ and go to Step 3.

Step 3. *Select $\beta_{k+1} \leq \beta_k$.*
Increment k by 1.
If $k = N$ then STOP, otherwise go to Step 2.

The effectiveness of simulated annealing depends on the design of the neighborhood as well as on how the search is conducted within this neighborhood. If the neighborhood is designed in a way that facilitates moves to better solutions and moves out of local minima, then the procedure will perform well. The search within a neighborhood can be done randomly or in a more organized way. For example, the contribution of each job to the objective function can be computed and the job with the highest impact on the objective can be selected as a candidate for an interchange.

Over the last two decades, simulated annealing has been applied to many scheduling problems in academia as well as in industry with considerable success.

The remaining part of this section focuses on the tabu-search procedure. Tabu-search is in many ways similar to simulated annealing in that it also moves from one schedule to another with the next schedule being possibly worse than the one before. For each schedule, a neighborhood is defined as in simulated annealing. The search in the neighborhood for a potential candidate to move to is again a design issue. As in simulated annealing, this can be done randomly or in an organized way. The basic difference between tabu-search and simulated annealing lies in the mechanism that is used for approving a candidate schedule. In tabu-search, the mechanism is not probabilistic but rather of a deterministic nature. At any stage of the process, a tabu-list of mutations, which the procedure is *not* allowed to make, is kept. A mutation on the tabu-list can be, for example, a pair of jobs that may not be interchanged. The tabu-list has a fixed number of entries (usually between 5 and 9) that depends on the application. Every time a move is made by a mutation in the current schedule, the *reverse* mutation is entered at the top of the tabu-list; all other entries are pushed down one position and the bottom entry is deleted. The reverse mutation is put on the tabu-list to avoid returning to a local minimum that has been visited before. Actually, at times a reverse mutation that is tabu could actually have led to a new schedule, not visited before, that is better than any one generated so far. This may happen when the reverse mutation is close to the bottom of the tabu-list and a number of moves have already been made since that reverse mutation was entered in the list. Thus, if the number of entries in the tabu-list is too small, cycling may occur; if it is too large, the search may be unduly constrained. The method can be summarized as follows.

Algorithm 14.4.3. (Tabu-Search)

Step 1. *Set $k = 1$.*
Select an initial sequence S_1 using some heuristic.
Set $S_0 = S_1$

Step 2. *Select a candidate schedule S_c from the neighborhood of S_k.*
If the move $S_k \rightarrow S_c$ is prohibited by a mutation on the tabu-list, set $S_{k+1} = S_k$ and go to Step 3.
If the move $S_k \rightarrow S_c$ is not prohibited by any mutation on the tabu-list, set $S_{k+1} = S_c$;
enter reverse mutation at the top of the tabu-list;
push all other entries in the tabu-list one position down;
delete the entry at the bottom of the tabu-list.
If $G(S_c) < G(S_0)$, set $S_0 = S_c$;
Go to Step 3.

Step 3. *Increment k by 1.*
If $k = N$ then STOP,
otherwise go to Step 2.

The following example illustrates the method.

Example 14.4.4 (Application of Tabu-Search)

Consider the instance of the single machine total weighted tardiness problem in Example 14.3.1.

jobs	1	2	3	4
p_j	10	10	13	4
d_j	4	2	1	12
w_j	14	12	1	12

The neighborhood of a schedule is defined as all schedules that can be obtained through adjacent pairwise interchanges. The tabu-list is a list of pairs of jobs (j, k) that were swapped within the last two moves and cannot be swapped again. Initially, the tabu-list is empty.

As a first schedule, the sequence $S_1 = 2, 1, 4, 3$ is chosen, with objective function value

$$\sum w_j T_j(2, 1, 4, 3) = 500.$$

The aspiration criterion is therefore 500. There are three schedules in the neighborhood of S_1—namely, 1, 2, 4, 3; 2, 4, 1, 3; and 2, 1, 3, 4. The respective values of the objective function are 480, 436, and 652. Selection of the best non-tabu sequence results in $S_2 = 2, 4, 1, 3$. The aspiration criterion is changed to 436. The tabu-list is updated and now contains the pair $(1, 4)$. The values of the objective functions of the neighbors of S_2 are:

sequence	4, 2, 1, 3	2, 1, 4, 3	2, 4, 3, 1
$\sum w_j T_j$	460	500	608

Note that the second move is tabu. However, the first move is anyhow better than the second. The first move results in a schedule that is worse than the best one so far. The best one is the current one, which therefore is a local minimum. Nevertheless, $S_3 = 4, 2, 1, 3$ and the tabu-list is updated and now contains $\{(2, 4), (1, 4)\}$. Neighbors of S_3 with the corresponding values of the objective functions are:

sequence	2, 4, 1, 3	4, 1, 2, 3	4, 2, 3, 1
$\sum w_j T_j$	436	440	632

Now, although the best move is to 2, 4, 1, 3 (S_2), this move is tabu. Therefore, S_4 is chosen to be 4, 1, 2, 3. Updating the tabu-list results in $\{(1, 2), (2, 4)\}$ and the pair (1,4) drops from the tabu-list as the length of the list is kept to 2. Neighbors of S_4 and their corresponding objective function values are:

sequence	1, 4, 2, 3	4, 2, 1, 3	4, 1, 3, 2
$\sum w_j T_j$	408	460	586

The schedule 4, 2, 1, 3 is tabu, but the best move is to the schedule 1, 4, 2, 3. So $S_5 = 1, 4, 2, 3$. The corresponding value of the objective is better than the aspiration criterion. So the aspiration criterion becomes 408. The tabu-list is updated by adding $(1, 4)$ and dropping $(2, 4)$. Actually, S_5 is a global minimum, but tabu-search, being unaware of this fact, continues.

The information carried along in tabu-search consists of the tabu-list as well as the best solution obtained so far during the search process. Recently, more powerful versions of tabu-search have been proposed; these versions retain more information. One version uses a so-called tabu-tree. In this tree, each node represents a solution or schedule. While the search process goes from one solution to another (with each solution having a tabu-list), the process generates additional nodes. Certain solutions that appear promising may not be used as a take-off point immediately but are retained for future use. If at a certain point during the search process the current solution does not appear promising as a take-off point, the search process can return within the tabu-tree to another node (solution) that had been retained before and take off again, but now in a different direction.

14.5 LOCAL SEARCH: GENETIC ALGORITHMS

Genetic algorithms are more general and abstract than simulated annealing and tabu-search. Simulated annealing and tabu-search may, in a certain way, be viewed as special cases of genetic algorithms.

Genetic algorithms, when applied to scheduling, view sequences or schedules as *individuals* or members of a *population*. Each individual is characterized by its *fitness*. The fitness of an individual is measured by the associated value of the objective function. The procedure works iteratively, and each iteration is referred to as a *generation*. The population of one generation consists of individuals surviving from the previous generation plus the new schedules or *children* from the previous generation. The population size usually remains constant from one generation to the next. The children are generated through reproduction and mutation of individuals who were part of the previous generation (the *parents*). Individuals are sometimes also referred to as *chromosomes*. In a multi-machine environment, a chromosome may consist of subchromosomes, each one containing the information regarding the job sequence on a machine. A mutation in a parent chromosome may be equivalent to an adjacent pairwise interchange in the corresponding sequence. In each generation, the most fit individuals reproduce while the least fit die. The birth, death, and reproduction processes that determine the composition of the next generation can be complex, and usually depend on the fitness levels of the individuals of the current generation.

A genetic algorithm, as a search process, differs in one important aspect from simulated annealing and tabu-search. At each iterative step, a number of different schedules are generated and carried over to the next step. In simulated annealing and tabu-search, only a single schedule is carried over from one iteration to the next. Hence, simulated annealing and tabu-search may be regarded as special cases of genetic algorithms with a population size equal to 1. This diversification scheme is an important characteristic of genetic algorithms. Therefore, in genetic algorithms, the neighborhood concept is not based on a single schedule, but rather on a set of schedules. The design of the neighborhood of the current population of schedules is based on more general techniques than those used in simulated annealing and tabu-search. A new schedule can be constructed by combining parts of different schedules within the population. For example, in a job shop scheduling problem, a new schedule can be generated by combining the sequence of operations on one machine in one parent schedule with a sequence of operations on another machine in another parent schedule. This is often referred to as the *crossover* effect.

A very simplified version of a genetic algorithm is described next.

Algorithm 14.5.1. (Genetic Algorithm)

Step 1. *Set $k = 1$.*
Select ℓ initial sequences $\mathcal{S}_{1,1}, \ldots, \mathcal{S}_{1,\ell}$ using some heuristic.

Step 2. *Select the two best schedules among $\mathcal{S}_{k,1}, \ldots, \mathcal{S}_{k,\ell}$ and call these $\mathcal{S}_k^+$ and $\mathcal{S}_k^{++}$*
Select the two worst schedules among $\mathcal{S}_{k,1}, \ldots, \mathcal{S}_{k,\ell}$ and call these $\mathcal{S}_k^-$ and $\mathcal{S}_k^{--}$
Select two neighbors $\mathcal{S}^*$ and $\mathcal{S}^{**}$ from the neighborhood of $\mathcal{S}_k^+$ and $\mathcal{S}_k^{++}$.
Replace $\mathcal{S}_k^-$ and $\mathcal{S}_k^{--}$ with $\mathcal{S}^*$ and $\mathcal{S}^{**}$;
Keep all other schedules the same and go to Step 3.

Step 3. *Increment k by 1.*
If $k = N$ then STOP,
otherwise go to Step 2.

The use of genetic algorithms has its advantages and disadvantages. One advantage is that they can be applied to a problem without having to know much about the structural properties of the problem. They can be very easily coded and often give fairly good solutions. However, the amount of computation time needed to obtain such a solution can be relatively long in comparison with the more rigorous problem-specific approaches.

14.6 DISCUSSION

For many scheduling problems, one may design procedures that combine elements of several of the techniques presented in this chapter.

For example, the following three-phase approach has proved fairly useful for solving scheduling problems in practice. It combines composite dispatching rules with simulated annealing or tabu-search.

Phase 1. Values of a number of statistics are computed such as the due date tightness, the setup time severity, and so on.

Phase 2. Based on the outcome of Phase 1, a number of scaling parameters for a composite dispatching rule are determined and the composite dispatching rule is applied on the scheduling instance.

Phase 3. The schedule developed in Phase 2 is used as an initial solution for a tabu-search or simulated annealing procedure that tries to generate a better schedule.

This three-phase framework would only be useful if the routine would be used frequently (a new instance of the same problem has to be solved every day). The reason is that the empirical procedure that determines the functions that map values of the job statistics into appropriate values for scaling parameters constitutes a major investment of time. Such an investment pays off only when the routine is used frequently.

EXERCISES (COMPUTATIONAL)

14.1. Consider the instance in Example 14.2.1.

(a) How many different schedules are there?

(b) Compute the value of the objective in case, whenever the machine is freed, the job with the highest w_j/p_j ratio is selected to go next.

(c) Compute the value of the objective in case, whenever the machine is freed, the job with the minimum slack is selected to go next.

(d) Explain why in this instance under the optimal schedule the job with the latest due date has to go first.

14.2. Consider the instance in Example 14.2.1 and determine the schedule according to the ATCS rule with

(a) $K_1 = 5$ and $K_2 = \infty$;

(b) $K_1 = 5$ and $K_2 = 0.0001$;

(c) $K_1 = \infty$ and $K_2 = 0.7$.

(d) $K_1 = 0.0001$ and $K_2 = 0.7$.

(e) $K_1 = \infty$ and $K_2 = \infty$.

14.3. Consider the instance of $P2 \mid\mid \sum w_j T_j$ with the following five jobs.

jobs	1	2	3	4	5
p_j	13	9	13	10	8
d_j	6	18	10	11	13
w_j	2	4	2	5	4

(a) Apply the ATC heuristic on this instance with the look-ahead parameter $K = 1$.

(b) Apply the ATC heuristic on this instance with the look-ahead parameter $K = 5$.

14.4. Consider the instance in Example 14.4.4. Apply the tabu-search technique once more, starting out with the same initial sequence, under the following conditions.

(a) Make the length of the tabu-list 1 (i.e., only the pair of jobs that was swapped during the last move cannot be swapped again). Apply the technique for four iterations and determine whether the optimal sequence is reached.

(b) Make the length of the tabu-list 3 (i.e., the pairs of jobs that were swapped during the last three moves cannot be swapped again). Apply the technique for four iterations and determine whether the optimal sequence is reached.

14.5. Apply the ATC dispatching rule to the following instance of $F3 \mid prmu, p_{ij} = p_j \mid \sum w_j T_j$.

jobs	1	2	3	4
p_j	9	9	12	3
d_j	10	8	5	28
w_j	14	12	1	12

What is the best value for the scaling parameter?

14.6. Apply tabu-search to the instance of $F3 \mid prmu, p_{ij} = p_j \mid \sum w_j T_j$ in Exercise 14.5. Choose as the neighborhood again all schedules that can be obtained through adjacent pairwise interchanges. Start out with sequence 3, 1, 4, 2 and apply the technique for four iterations. Keep the length of the tabu-list equal to 2. Determine whether the optimal sequence is reached.

14.7. Consider the same instance as in Exercise 14.5. Now apply simulated annealing to this instance. Adopt the same neighborhood structure and select neighbors within the neighborhood at random. Choose $\beta_k = (0.9)^k$. Start with 3, 1, 4, 2 as the initial sequence. Terminate the procedure after two iterations and compare the result with the result obtained in the previous exercise. Use the following numbers as uniform random numbers:

$$U_1 = 0.91, \quad U_2 = 0.27, \quad U_3 = 0.83, \quad U_4 = 0.17.$$

14.8. Consider the same instance as in Exercise 14.5. Now apply the Genetic Algorithm 14.5.1 to the instance.

(a) Start with a population of the three sequences 3, 4, 1, 2; 4, 3, 1, 2; and 3, 2, 1, 4 and perform three iterations.

(b) Replace one of the sequences in the initial population under (a) with the sequence obtained in Exercise 14.5 and perform three iterations.

14.9. Consider the instance in Exercise 14.5 of $F3 \mid prmu, p_{ij} = p_j \mid \sum w_j T_j$. Apply the beam search procedure on this instance without a filter. Let a node in the tree represent a partial schedule that consists of a subset of jobs scheduled starting from $t = 0$. Choose the beam width equal to 2. When analyzing a node, schedule the remaining jobs according to the composite dispatching rule described in Section 14.1. Determine whether the resulting schedule is optimal.

14.10. Consider the following instance of $1 \mid s_{jk} \mid C_{\max}$ with six jobs. The sequence-dependent setup times are specified in the following table.

k	0	1	2	3	4	5	6
s_{0k}	—	1	$1+\epsilon$	D	$1+\epsilon$	$1+\epsilon$	D
s_{1k}	D	—	1	$1+\epsilon$	D	$1+\epsilon$	$1+\epsilon$
s_{2k}	$1+\epsilon$	D	—	1	$1+\epsilon$	D	$1+\epsilon$
s_{3k}	$1+\epsilon$	$1+\epsilon$	D	—	1	$1+\epsilon$	D
s_{4k}	D	$1+\epsilon$	$1+\epsilon$	D	—	1	$1+\epsilon$
s_{5k}	$1+\epsilon$	D	$1+\epsilon$	$1+\epsilon$	D	—	1
s_{6k}	1	$1+\epsilon$	D	$1+\epsilon$	$1+\epsilon$	D	—

Assume D to be very large. Define as the neighborhood of a schedule all schedules that can be obtained through an adjacent pairwise interchange.
(a) Find the optimal sequence.
(b) Determine the makespans of all schedules that are neighbors of the optimal schedule.
(c) Find a schedule, with a makespan less than D of which all neighbors have the same makespan. (The optimal sequence may be described as a "brittle" sequence, whereas the last sequence may be described as a more "robust" sequence.)

EXERCISES (THEORY)

14.11. What does the ATCS rule reduce to
(a) if both K_1 and K_2 go to ∞?
(b) if K_1 is very close to zero and $K_2 = 1$?
(c) if K_2 is very close to zero and $K_1 = 1$?

14.12. Consider $Pm \mid\mid \sum w_j C_j$ and the dispatching rule that releases jobs in decreasing order of $w_j/(p_j^k)$. Give an argument for setting the parameter k between 0.75 and 1. Describe the relationship between an appropriate value of k and the number of machines m.

14.13. Consider $Fm \mid p_{ij} = p_j \mid \sum w_j C_j$ and the dispatching rule that releases jobs in decreasing order of $w_j/(p_j^k)$. Give an argument for setting the k larger than 1. Describe the relationship between an appropriate value of k and the number of machines m.

14.14. Consider the following basic mutations that can be applied to a sequence:
(i) An insertion (a job is selected and put elsewhere in the sequence).
(ii) A pairwise interchange of two adjacent jobs.
(iii) A pairwise interchange of two nonadjacent jobs.
(iv) A sequence interchange of two adjacent subsequences of jobs.

(v) A sequence interchange of two nonadjacent subsequences of jobs.

(vi) A reversal of a subsequence of jobs.

Some of these mutations are special cases of others and some mutations can be achieved through repeated applications of others. Taking this into account, explain how these six types of mutations are related to one another.

14.15. Show that if the optimality of a rule can be shown through an adjacent pairwise interchange argument applied to an arbitrary sequence, then

(a) the sequence that minimizes the objective is *monotone* in a function of the parameters of the jobs, and

(b) the reverse sequence *maximizes* that same objective.

14.16. Determine the number of neighbors of a permutation schedule if the neighborhood consists of all schedules that can be reached through

(a) any adjacent pairwise interchange;

(b) any pairwise interchange.

14.17. Consider the $1 \parallel \sum w'_j E_j + \sum w''_j T_j$ problem. Design a composite dispatching rule for the minimization of the sum of the weighted earliness and tardiness penalties. (Consider first the case where all due dates are equal to $C_{\max}$.)

14.18. Describe a neighborhood and a neighborhood search technique for a local search procedure that is applicable to a permutation flow shop scheduling problem with the makespan as objective.

14.19. Describe a neighborhood and a neighborhood search procedure for the problem $1 \mid r_j, prmp \mid \sum w_j C_j$.

14.20. Design a multiphase procedure for $Fm \mid block \mid \sum w_j T_j$ with zero intermediate storage and blocking. Give proper statistics to characterize instances. Present a composite dispatching rule and design an appropriate neighborhood for a local search procedure. (*Hint:* The goodness of fit of an additional job to be included in a partial sequence may be considered to be similar to a sequence-dependent setup time; the structure of a composite dispatching rule may in some respects look like the ATCS rule).

14.21. Design a scheduling procedure for the problem $Pm \mid r_j, M_j \mid \sum w_j T_j$. Let the procedure consist of three basic steps:

(i) a statistics evaluation step,

(ii) a composite dispatching rule step, and

(iii) a simulated annealing step.

Consider the WSPT, LFJ, LFM, EDD and MS rule as rules for possible inclusion in a composite dispatching rule. What factors should be defined for characterization of scheduling instances? What kind of input can the scheduler provide to the three modules?

COMMENTS AND REFERENCES

An excellent treatise of general purpose procedures and heuristics for scheduling problems is the text by Morton and Pentico (1994); they cover most of the techniques described in this chapter.

One of the first studies on dispatching rules is due to Conway (1965a, 1965b). A fairly complete list of the most common dispatching rules is given by Panwalkar and Iskander (1977), and a detailed description of composite dispatching rules is given by Bhaskaran and Pinedo (1992). Special examples of composite dispatching rules are the COVERT rule developed by Carroll (1965) and the ATC rule developed by Vepsalainen and Morton (1987). The ATCS rule is due to Lee, Bhaskaran and Pinedo (1997). Ow and Morton (1989) describe a rule for scheduling problems with earliness and tardiness penalties. Filtered beam search applied to scheduling is due to Ow and Morton (1988).

A great deal of research has been done on simulated annealing, tabu-search, and genetic algorithms applied to scheduling. For a general overview of search spaces for scheduling problems, see Storer, Wu, and Vaccari (1992). For studies on simulated annealing, see Kirkpatrick, Gelatt, and Vecchi (1983), Van Laarhoven, Aarts, and Lenstra (1992), and Matsuo, Suh, and Sullivan (1988). For tabu-search, see Glover (1990), Dell'Amico and Trubian (1991), and Nowicki and Smutnicki (1996). For genetic algorithms applied to scheduling, see Lawton (1992), Della Croce, Tadei, and Volta (1992), and Bean (1994).

The three-step procedure described in the discussion section is due to Lee, Bhaskaran, and Pinedo (1997).

15

More Advanced General Purpose Procedures

The previous chapter covered the most established and widely used generic procedures. This chapter focuses on techniques that are of more recent vintage and not as widely used yet.

The first section focuses on decomposition procedures. Chapter 7 already covered a very well-known decomposition technique—namely, the shifting bottleneck heuristic. The shifting bottleneck technique is a decomposition procedure of the so-called machine-based decomposition type. The first section of this chapter describes several other types of decomposition techniques. The second section discusses constraint guided heuristic search procedures that have been developed in the artificial intelligence community. Constraint guided heuristic search is often also referred to as constraint-based programming. A constraint guided heuristic search procedure attempts to find a feasible schedule given all the constraints in the scheduling environment. The third section discusses a class of techniques that also originated in the artificial intelligence community. These techniques assume that the scheduling process is based on a market mechanism in which each job has to make bids and pay a price for machine time. The fourth section focuses on procedures for scheduling problems with multiple objectives. In practice, most scheduling systems have to deal with multiple objectives and do some form of parametric or sensitivity analysis. The discussion section explains the role of general purpose procedures in the design and development of engines for scheduling systems.

15.1 DECOMPOSITION METHODS AND ROLLING HORIZON PROCEDURES

There are several classes of decomposition methods. The best known class of decomposition methods is usually referred to as machine-based decomposition. A prime example of machine-based decomposition is the shifting bottleneck technique described in Chapter 7. A second class of decomposition methods is referred to as job-based decomposition. A job-based decomposition method is useful when there are constraints with regard to the timing of the various operations of any given job (e.g., when there are minimum and maximum time delays between consecutive operations of a job). A third class consists of the time-based decomposition methods, which are also known as rolling horizon procedures. According to these methods, a schedule is first determined for all machines up to a given point in time, ignoring everything that could happen afterward. After a schedule has been determined up to that given point, a schedule is generated for the next time period, and so on. A fourth class of decomposition methods consists of the hybrid methods. Hybrid methods may combine either machine- or job-based decomposition with time-based decomposition.

Machine-based decomposition is often used in (flexible) flow shop, (flexible) job shop, and (flexible) open shop environments. The decomposition procedure solves the problem by scheduling the machines one at a time in decreasing order of their criticality indexes. That is, the procedure first schedules the machine that is the most difficult to schedule; after having scheduled that machine, it proceeds with scheduling the second most difficult one, and so on. To determine the degree of difficulty in scheduling a machine (i.e., the criticality of a machine), a subproblem has to be defined and solved for each machine. The objective of the overall problem determines the type of objective in the single machine subproblem. However, the objective of the single machine (or parallel machine) subproblem is typically a more difficult objective than the objective of the main problem. The single machine subproblem is usually NP-hard. In Chapter 7, it is shown how the C_{max} objective in the main problem leads to the L_{max} objective in the subproblem and how the

$\sum w_j T_j$ objective in the main problem leads to the $\sum \sum h_{ij}(C_{ij})$ objective in the subproblem, where h_{ij} is piecewise linear convex. It is often not clear how much of an investment in computing time should be made to determine a solution for the single machine subproblem. It is also not clear how the quality of the solutions of the subproblems affects the quality of the overall solution.

An important aspect of machine-based decomposition is its so-called control structure. The control structure is the framework that determines which subproblem has to be solved when. A control structure may typically lead to a significant amount of reoptimization: After a schedule for an additional machine has been generated, the procedure reoptimizes all the machines that had been scheduled earlier, taking into account the sequence on the machine just scheduled. The sequence of the operations on the machine just scheduled may lead to adjustments in the sequences of operations on the machines scheduled earlier. This reoptimization feature has proved to be crucial with regard to the effectiveness of machine-based decomposition procedures.

In a job-based decomposition method, a subproblem consists of all the operations associated with a particular job. The jobs are prioritized and inserted in the schedule one at a time. So the solution of a given subproblem involves the insertion of the operations of a given job into a partial schedule in such a way that the new schedule is feasible and the increase in the overall objective caused by the insertion of the new job is minimal. If an insertion of the new job is not feasible, then the jobs inserted before have to be rescheduled.

Time-based decomposition can be applied in any machine environment. There are various forms of time-based decomposition methods. First, a problem can be decomposed by introducing time intervals of a fixed length and considering in each iteration only jobs that are released during a particular interval. One can also partition the problem by considering each time a fixed number of jobs that are released consecutively.

However, there are also time-based decomposition methods that are based on more "natural" partitioning schemes. A natural partitioning point can be defined as a time t with the property that the schedule of the jobs completed before t does not have an effect on (i.e., is independent of) the schedule of the jobs completed after t. For example, consider the $1 \mid r_j \mid \sum w_j C_j$ problem with a very large number of jobs. This problem is well known to be NP-hard in the strong sense. Assume, for the time being, that the machine is always processing a job when there are jobs available for processing. Suppose that, at a certain point in time, the machine is idle and no jobs are waiting for processing; the machine remains idle until the next job is released. It is clear that such a point in time is a natural partitioning point. More formally, let $V(t)$ denote the total amount of processing that remains to be done on the jobs released prior to t. If $V(t)$ is zero or close to zero for some t, then such a t would be an appropriate partitioning point.

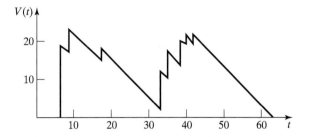

Figure 15.1   Total amount of processing remaining to be done in system (Example 15.1.1).

Example 15.1.1 (Application of Time-Based Decomposition)

Consider the following instance of $1 \mid r_j \mid \sum C_j$ with nine jobs.

jobs	1	2	3	4	5	6	7	8	9
r_j	7	7	9	17	33	35	39	40	42
p_j	9	10	6	3	10	8	6	2	2

The graph of $V(t)$ is presented in Figure 15.1. At $t = 33$, the $V(t)$ jumps from 2 to 12. So just before time 33, the $V(t)$ reaches a minimum. One possible partition would consider jobs 1, 2, 3, 4 as one problem and jobs 5, 6, 7, 8, 9 as a second problem. Minimizing $\sum C_j$ in the first problem results in schedule 1, 3, 4, 2. The second problem starts at time 35 and the optimal schedule is 6, 8, 9, 7, 5.

If the jobs have different weights (i.e., the objective is $\sum w_j C_j$ rather than $\sum C_j$), then the partitioning scheme is not as straightforward anymore. For example, consider a job with weight 0 (or an extremely low weight). Such a job, most likely, will appear more toward the end of the schedule.

Partitioning schemes for problems with due dates may be designed completely differently. Due dates may appear in clusters. Some intervals may have a large number of due dates while others may have a small number. A partitioning of the job set may be done immediately at the end of a cluster of due dates; the subset of jobs to be considered includes all the jobs that have due dates in that cluster.

Example 15.1.2 (Application of Time-Based Decomposition)

Consider the following instance of $1 \mid\mid \sum T_j$ with nine jobs all released at time 0.

jobs	1	2	3	4	5	6	7	8	9
p_j	7	6	7	6	5	5	8	9	4
d_j	22	24	24	25	33	51	52	52	55

There are two clusters of due dates. One cluster lies in the range [22, 25], and the second cluster lies in the range [51, 55]. There is a single due date, d_5, that lies in the middle. According to a partitioning scheme based on clusters of due dates, it makes sense to consider jobs 1, 2, 3, 4 as one subproblem and jobs 5, 6, 7, 8, 9 as a second subproblem.

Any sequence in the first subproblem that schedules job 4 last minimizes $\sum T_j$. Any sequence in the second subproblem that schedules job 9 last and either job 7 or 8 second to last minimizes $\sum T_j$ in that problem.

Control structures can play an important role in time-based decomposition procedures as well. After a time-based decomposition procedure has generated a schedule, a postprocessing procedure can be done. In such a postprocessing procedure, the last couple of jobs of one job set are combined with the first couple of jobs of the next job set. This new set of jobs is then reoptimized (see Figure 15.2).

Example 15.1.3 (Application of Time-Based Decomposition)

Consider the following instance of the $1 \parallel \sum w_j T_j$. The processing times and due dates are the same as those in Example 15.1.2.

jobs	1	2	3	4	5	6	7	8	9
p_j	7	6	7	6	5	5	8	9	4
w_j	0	1	1	2	1	1	2	1	4
d_j	22	24	24	25	33	51	52	52	55

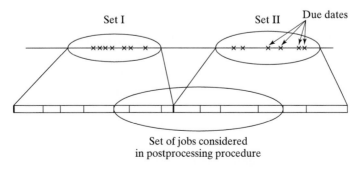

Set of jobs considered
in postprocessing procedure

Figure 15.2   Set of jobs considered in postprocessing procedure.

If the time-based decomposition is done in the same way as in Example 15.1.2, then the first subproblem consists again of jobs 1, 2, 3, 4 and the second subproblem of jobs 5, 6, 7, 8, 9.

In the first subproblem, any schedule that puts job 1 in the fourth position (i.e., the last slot) minimizes the total weighted tardiness. The total weighted tardiness of the first subproblem is then zero. In the second subproblem, any schedule that puts job 8 in the last position and job 5 in the first position minimizes the total weighted tardiness.

What happens now on the borderline between the first and second subproblems? Job 1 is last in the first subproblem and job 5 is first in the second subproblem. It is clear that an interchange between these two jobs results in a schedule with the same objective value. Are these two schedules optimal? Clearly not. If job 1 is processed all the way at the end of the schedule (i.e., completing its processing at $C_1 = 57$), then the total weighted tardiness of the entire schedule is zero. This last schedule is therefore optimal.

Hybrid decomposition methods are more complicated procedures. There are several types of hybrid methods. Some hybrid methods do a machine-based decomposition first and solve the single machine subproblems using a rolling horizon procedure. Other hybrid methods first do a time-based decomposition and then find a schedule for each time interval via a machine decomposition method.

The type of decomposition that has to be done first depends on the characteristics of the instance. For example, if there are significant differences between the criticality indexes of the different machines, it may make sense to do a machine-based decomposition first; the scheduling procedure for the subproblems within the machine-based decomposition procedure may be done via time-based decomposition. However, if there are natural partitioning points that are clear and pronounced, then it may make sense to do a time-based decomposition first and solve the subproblems via a machine-based decomposition procedure.

In practice, decomposition methods are often combined with other methods (e.g., local search procedures; see Section 17.3).

15.2 CONSTRAINT GUIDED HEURISTIC SEARCH

Constraint guided heuristic search is a technique that originated in the artificial intelligence (AI) community. In recent years, it has often been implemented in combination with operations research (OR) techniques to improve its effectiveness.

Constraint guided heuristic search, according to its original design, only tries to find a good solution that is feasible and that satisfies all the given constraints. However, these solutions may not necessarily be optimal. The constraints may include different release dates and due dates of the jobs. It is possible to embed such a technique in a framework designed to minimize any due date-related objective function.

Constraint guided heuristic search applied to $Jm \parallel C_{max}$ works as follows. Suppose that in a job shop a schedule has to be found with the makespan C_{max} less than or equal to a given deadline $\bar{d}$. The constraint satisfaction algorithm has to produce for each machine a sequence of operations such that the overall schedule has a makespan less than or equal to $\bar{d}$.

Before the actual procedure starts, an initialization step has to be done. For each operation, a computation is done to determine its earliest possible starting time and latest possible completion time on the machine in question. After all the time windows have been computed, the time windows of all the operations on each machine are compared with one another. When the time windows of two operations on any given machine do not overlap, a precedence relationship between the two operations can be imposed; in any feasible schedule, the operation with the earlier time window must precede the operation with the later time window. Actually, a precedence relationship may be inferred even when the time windows do overlap. Let S'_{ij} (S''_{ij}) denote the earliest (latest) possible starting time of operation (i, j) and C'_{ij} (C''_{ij}) the earliest (latest) possible completion time of operation (i, j) under the current set of precedence constraints. Note that the earliest possible starting time of operation (i, j) (i.e., S'_{ij}) may be regarded as a local release date of the operation and may be denoted by r_{ij}, whereas the latest possible completion time (i.e., C''_{ij}) may be considered a local due date denoted by d_{ij}. Define the slack between the processing of operations (i, j) and (i, k) on machine i as

$$\sigma_{(i,j) \to (i,k)} = S''_{ik} - C'_{ij}$$

or

$$\sigma_{(i,j) \to (i,k)} = C''_{ik} - S'_{ij} - p_{ij} - p_{ik}$$

or

$$\sigma_{(i,j) \to (i,k)} = d_{ik} - r_{ij} - p_{ij} - p_{ik}.$$

If

$$\sigma_{(i,j) \to (i,k)} < 0,$$

then there does not exist, under the current set of precedence constraints, a feasible schedule in which operation (i, j) precedes operation (i, k) on machine i; so a precedence relationship can be imposed that requires operation (i, k) to appear before operation (i, j). In the initialization step of the procedure, all pairs of time windows are compared with one another and all implied precedence relationships are inserted in the disjunctive graph. Because of these additional precedence constraints, the time windows of each one of the operations can be adjusted (narrowed; i.e., this involves a recomputation of the release date and the due date of each operation).

Constraint satisfaction techniques in general rely on constraint propagation. A constraint satisfaction technique typically attempts, in each step, to insert new precedence constraints (disjunctive arcs) that are implied by the precedence constraints inserted before. With the new precedence constraints in place, the technique recomputes the time windows of all operations. For each pair of operations that have to be processed on the same machine, it has to be verified which one of the following four cases holds.

Case 1. If $\sigma_{(i,j)\to(i,k)} \geq 0$ and $\sigma_{(i,k)\to(i,j)} < 0$,
then the precedence constraint $(i, j) \to (i, k)$ has to be imposed.

Case 2. If $\sigma_{(i,k)\to(i,j)} \geq 0$ and $\sigma_{(i,j)\to(i,k)} < 0$,
then the precedence constraint $(i, k) \to (i, j)$ has to be imposed.

Case 3. If $\sigma_{(i,j)\to(i,k)} < 0$ and $\sigma_{(i,k)\to(i,j)} < 0$,
then no schedule satisfies the precedence constraints already in place.

Case 4. If $\sigma_{(i,j)\to(i,k)} \geq 0$ and $\sigma_{(i,k)\to(i,j)} \geq 0$,
then either ordering between the two operations is still possible.

In one of the steps of the algorithm described in this section, a pair of operations has to be selected that satisfies Case 4 (i.e., either ordering between the operations is still possible). In this step of the algorithm, many pairs of operations may still satisfy Case 4. If there is more than one pair of operations that satisfies Case 4, then a search control heuristic has to be applied. The selection of a pair is based on the sequencing flexibility this pair still provides. The pair with the lowest flexibility is selected. The reasoning behind this approach is straightforward. If a pair with low flexibility is not scheduled early in the process, then it may be that later in the process the pair may not be schedulable at all. So it makes sense to give priority to those pairs with a low flexibility and postpone pairs with a high flexibility. Clearly, the flexibility depends on the amounts of slack under the two orderings. One simple estimate of the sequencing flexibility of a pair of operations, $\phi((i, j)(i, k))$, is the minimum of the two slacks—that is,

$$\phi((i, j)(i, k)) = \min(\sigma_{(i,j)\to(i,k)}, \sigma_{(i,k)\to(i,j)}).$$

However, relying on this minimum may lead to problems. For example, suppose one pair of operations has slack values 3 and 100, whereas another pair has slack values 4 and 4. In this case, there may be only limited possibilities for scheduling the second pair, and postponing a decision with regard to the second pair may well eliminate them. A feasible ordering with regard to the first pair may not really be in jeopardy. Instead of using $\phi((i, j)(i, k))$, the following measure of sequencing flexibility has proved to be more effective:

$$\phi'((i, j)(i, k)) = \sqrt{\min(\sigma_{(i,j)\to(i,k)}, \sigma_{(i,k)\to(i,j)}) \times \max(\sigma_{(i,j)\to(i,k)}, \sigma_{(i,k)\to(i,j)})}.$$

So if the max is large, then the flexibility of a pair of operations increases and the urgency to order the pair goes down. After the pair of operations with the lowest sequencing flexibility $\phi'((i, j)(i, k))$ has been selected, the precedence constraint that retains the most flexibility is imposed—that is, if

$$\sigma_{(i,j)\to(i,k)} \geq \sigma_{(i,k)\to(i,j)},$$

operation (i, j) must precede operation (i, k).

In one of the steps of the algorithm, it also can happen that a pair of operations satisfies Case 3. When this is the case, the partial schedule under construction cannot be completed and the algorithm has to backtrack. Backtracking can take any one of a number of forms. Backtracking may imply that one or more of the ordering decisions made in earlier iterations has to be annulled. It may imply that there does not exist a feasible solution for the problem in the way that it has been presented and that certain constraints have to be relaxed.

The constraint guided heuristic search procedure can be summarized as follows.

Algorithm 15.2.1. (Constraint Guided Heuristic Search)

Step 1. *Compute for each unordered pair of operations*
$\sigma_{(i,j)\to(i,k)}$ *and* $\sigma_{(i,k)\to(i,j)}$.

Step 2. *Check dominance conditions and classify remaining ordering decisions.*
If any ordering decision is either of Case 1 or Case 2 go to Step 3.
If any ordering decision is of Case 3, then backtrack;
otherwise go to Step 4.

Step 3. *Insert new precedence constraint and go to Step 1.*

Step 4. *If no ordering decision is of Case 4, then a solution is found. STOP.*
Otherwise go to Step 5.

Step 5. *Compute $\phi'((i, j)(i, k))$ for each pair of operations not yet ordered.*
Select the pair with the minimum $\phi'((i, j)(i, k))$.
If $\sigma_{(i,j)\to(i,k)} \geq \sigma_{(i,k)\to(i,j)}$, then operation (i, k) must follow (i, j);
otherwise operation (i, j) must follow operation (i, k).
Go to Step 3.

To apply the constraint guided heuristic search procedure to $Jm \parallel C_{\max}$, it has to be embedded in the following framework. First, an upper bound d_u and a lower bound d_l have to be found for the makespan.

Algorithm 15.2.2. (Framework for Constraint Guided Heuristic Search)

Step 1. *Set $d = (d_l + d_u)/2$.*
   Apply Algorithm 15.2.1.
Step 2. *If $C_{\max} < d$, set $d_u = d$.*
   If $C_{\max} > d$, set $d_l = d$.
Step 3. *If $d_u - d_l > 1$ return to Step 1.*
   Otherwise STOP.

The following example illustrates the use of the contraint satisfaction technique.

Example 15.2.3 (Constraint Guided Heuristic Search Applied to a Job Shop)

Consider the instance of the job shop problem described in Example 7.1.1 in Chapter 7.

jobs	machine sequence	processing times		
1	1, 2, 3	$p_{11} = 10$, $\quad p_{21} = 8$, $\quad p_{31} = 4$		
2	2, 1, 4, 3	$p_{22} = 8$, $p_{12} = 3$, $\quad p_{42} = 5$, $\quad p_{32} = 6$		
3	1, 2, 4	$p_{13} = 4$, $\quad p_{23} = 7$, $\quad p_{43} = 3$		

Consider a due date $d = 32$ by when all jobs have to be completed. Consider again the disjunctive graph but disregard all disjunctive arcs (see Figure 15.3). By doing all longest path computations, the local release dates and local due dates for all operations can be established.

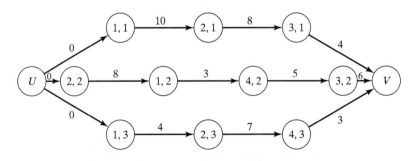

Figure 15.3   Disjunctive graph without disjunctive arcs.

operations	r_{ij}	d_{ij}
(1, 1)	0	20
(2, 1)	10	28
(3, 1)	18	32
(2, 2)	0	18
(1, 2)	8	21
(4, 2)	11	26
(3, 2)	16	32
(1, 3)	0	22
(2, 3)	4	29
(4, 3)	11	32

Given these time windows for all the operations, it has to be verified whether these constraints already imply any additional precedence constraints. Consider, for example, the pair of operations (2, 2) and (2, 3), which both have to go on machine 2. Computing the slack yields

$$\sigma_{(2,3)\to(2,2)} = d_{22} - r_{23} - p_{22} - p_{23}$$
$$= 18 - 4 - 8 - 7$$
$$= -1,$$

which implies that the ordering (2, 3) $\to$ (2, 2) is not feasible. So the disjunctive arc (2, 2) $\to$ (2, 3) has to be inserted. In the same way, it can be shown that the disjunctive arcs (2, 2) $\to$ (2, 1) and (1, 1) $\to$ (1, 2) have to be inserted as well.

Given these additional precedence constraints, the release and due dates of all operations have to be updated. The updated release and due dates are presented in the following table.

operations	r_{ij}	d_{ij}
(1, 1)	0	18
(2, 1)	10	28
(3, 1)	18	32
(2, 2)	0	18
(1, 2)	10	21
(4, 2)	13	26
(3, 2)	18	32
(1, 3)	0	22
(2, 3)	8	29
(4, 3)	15	32

These updated release and due dates do not imply any additional precedence constraints. Going through Step 5 of the algorithm requires the computation of the factor $\phi'((i, j)(i, k))$ for every unordered pair of operations on each machine.

pair	$\phi'((i, j)(i, k))$
$(1, 1)(1, 3)$	$\sqrt{4 \times 8} = 5.65$
$(1, 2)(1, 3)$	$\sqrt{5 \times 14} = 8.36$
$(2, 1)(2, 3)$	$\sqrt{4 \times 5} = 4.47$
$(3, 1)(3, 2)$	$\sqrt{4 \times 4} = 4.00$
$(4, 2)(4, 3)$	$\sqrt{3 \times 11} = 5.74$

The pair with the least flexibility is $(3, 1)(3, 2)$. Since the slacks are such that

$$\sigma_{(3,2) \to (3,1)} = \sigma_{(3,1) \to (3,2)} = 4,$$

either precedence constraint can be inserted. Suppose the precedence constraint $(3, 2) \to (3, 1)$ is inserted. This precedence constraint causes significant changes in the time windows during which the operations have to be processed.

operations	r_{ij}	d_{ij}
$(1, 1)$	0	14
$(2, 1)$	10	28
$(3, 1)$	24	32
$(2, 2)$	0	14
$(1, 2)$	10	17
$(4, 2)$	13	22
$(3, 2)$	18	28
$(1, 3)$	0	22
$(2, 3)$	8	29
$(4, 3)$	15	32

However, this new set of time windows imposes an additional precedence constraint—namely, $(4, 2) \to (4, 3)$. This new precedence constraint causes the following changes in the release and due dates of the operations.

operations	r_{ij}	d_{ij}
$(1, 1)$	0	14
$(2, 1)$	10	28
$(3, 1)$	24	32
$(2, 2)$	0	14
$(1, 2)$	10	17
$(4, 2)$	13	22
$(3, 2)$	18	28
$(1, 3)$	0	22
$(2, 3)$	8	29
$(4, 3)$	18	32

These updated release and due dates do not imply additional precedence constraints. Going through Step 5 of the algorithm requires one to compute for every unordered pair of operations on each machine the factor $\phi'((i, j)(i, k))$.

pair	$\phi'((i, j)(i, k))$
$(1, 1)(1, 3)$	$\sqrt{0 \times 8} = 0.00$
$(1, 2)(1, 3)$	$\sqrt{5 \times 10} = 7.07$
$(2, 1)(2, 3)$	$\sqrt{4 \times 5} = 4.47$

The pair with the least flexibility is $(1, 1)(1, 3)$, and the precedence constraint $(1, 1) \rightarrow (1, 3)$ has to be inserted.

Inserting this last precedence constraint enforces one more constraint—namely, $(2, 1) \rightarrow (2, 3)$. Now only one unordered pair of operations remains—namely, pair $(1, 3)(1, 2)$. These two operations can be ordered in either way without violating any due dates. A feasible ordering is $(1, 3) \rightarrow (1, 2)$. The resulting schedule with a makespan of 32 is depicted in Figure 15.4. This schedule meets the due date originally set but is not optimal.

When the pair $(3, 1)(3, 2)$ had to be ordered, it could have been ordered in either direction because the two slack values were equal. Suppose at that point the opposite ordering was selected—that is, $(3, 1) \rightarrow (3, 2)$. Restarting the process at that point yields the following release and due dates:

operations	r_{ij}	d_{ij}
$(1, 1)$	0	14
$(2, 1)$	10	22
$(3, 1)$	18	26

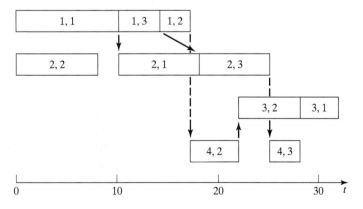

Figure 15.4   Final schedule in Example 15.2.3.

operations	r_{ij}	d_{ij}
(2, 2)	0	18
(1, 2)	10	21
(4, 2)	13	26
(3, 2)	18	32
(1, 3)	0	22
(2, 3)	8	29
(4, 3)	15	32

These release and due dates enforce a precedence constraint on the pair of operations (2, 1)(2, 3) and the constraint is (2, 1) → (2, 3). This additional constraint has the following effect on the release and due dates:

operations	r_{ij}	d_{ij}
(1, 1)	0	14
(2, 1)	10	22
(3, 1)	18	26
(2, 2)	0	18
(1, 2)	10	21
(4, 2)	13	26
(3, 2)	22	32
(1, 3)	0	22
(2, 3)	18	29
(4, 3)	25	32

These new release and due dates have an effect on the pair (4, 2)(4, 3), and the arc (4, 2) → (4, 3) has to be included. This additional arc does not cause any additional changes in the release and due dates. At this point, only two pairs of operations remain unordered—namely, the pair (1, 1)(1, 3) and the pair (1, 2)(1, 3).

pair	$\phi'((i, j)(i, k))$
(1, 1)(1, 3)	$\sqrt{0 \times 8} = 0.00$
(1, 2)(1, 3)	$\sqrt{5 \times 14} = 8.36$

So the pair (1, 1)(1, 3) is more critical and has to be ordered (1, 1) → (1, 3). It turns out that the last pair to be ordered, (1, 2)(1, 3), can be ordered either way.

The resulting schedule turns out to be optimal and has a makespan of 28.

As stated before, constraint satisfaction is not only suitable for makespan minimization. It can also be applied to problems with due date-related objectives and with each job having its own release date.

15.3 MARKET-BASED AND AGENT-BASED PROCEDURES

Consider a flexible job shop with c workcenters. Each workcenter has a number of identical machines in parallel. The jobs have different release dates and the objective is to minimize the total weighted tardiness $\sum w_j T_j$. This problem can be referred to as $FJc \mid r_j \mid \sum w_j T_j$.

Each job has a Job Agent (JA) who receives a budget at the release of the job. The budget of the JA is a function of the weight of the job, the tightness of its due date, and the total amount of processing it requires at the various workcenters. At each point in time, when an operation of a job has been completed at a workcenter, the JA sends out a call for bids from the machines at the next workcenter on the job's route. The information specified in this call for bids includes the preferred time frame in which the JA wants the next operation to be done. The JA determines the desired time frame by taking the job's final due date into account, as well as an estimate of the remaining time it needs to complete all remaining operations.

Each machine has a Machine Agent (MA). The cost structure of the machine consists of a fixed cost per unit time that is incurred at all times (regardless of whether the machine is processing a job). The cost per unit time of each machine is known to all parties. If a JA issues a call for bids, then the machines that can process the particular operation send in their bids. The information in a bid includes the time that the machine intends to process that operation as well as the price. The MA, of course, would like the price to be higher than the fixed cost because then he makes a profit. A machine is allowed to send multiple bids to the same JA for different time periods with different prices. The rules and heuristics a machine uses to generate bids can be complicated.

It may happen that an MA receives at the same point in time several calls for bids from different JAs. The MA may bid on any subset as long as the time periods specified in the different bids do not overlap. The MA's goal is to make as much money as possible on all the processing it does.

The JA receives, as a result of his call for bids, a number of bids and compares them according to a decision rule that takes the timing of the processing as well as the pricing into account. It then decides to whom to give the award. The decision rules and heuristics the JA uses when making awards not only take into account the times and the prices stated in the bids, but also the characteristics of the job (its weight, due date, and processing times of all remaining operations). The goal the JA tries to optimize when making an award is a composite objective that includes the estimated weighted tardiness of the job if the next operation is done in the given time frame as well as the amount it has to pay for that operation and estimates for the amounts it has to pay to other machines for the remaining operations. The JA cannot spend more on awards than what is in his budget. The total budget of a JA is the total cost of processing all the operations on the various machines (including

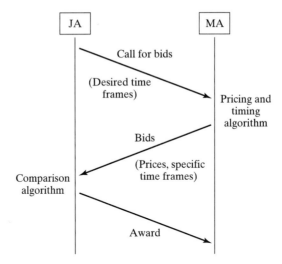

Figure 15.5   Overview of bidding and pricing process.

fixed costs as well as running costs) multiplied by a constant that depends on the weight of the job.

However, it may occur that a JA, after sending out a call, does not receive any bid at all. In such a case, the JA must issue a new call for bids and change the relevant information in the call. For example, the JA may ask for a time frame that is later than the time frame originally requested. This implies that at one point in time, several iterations may take place going back and forth between the JA and the various MAs. An overview of the bidding and pricing process is depicted in Figure 15.5.

Scheduling based on contract negotiations is characterized by the rules used by the various Machine Agents to determine their bids as well as by the rules used by the Job Agents to make the awards. These decision rules depend strongly on budgetary constraints as well as on the level of information available to the JA and MA.

Summarizing, each bidding and pricing mechanism has its own characteristics, including

  (i) the Budgetary Constraints,
 (ii) the Information Infrastructure, and
(iii) the Bidding and Pricing Rules.

The Budgetary Constraints: The budget of the JA of job j, B_j, is a function of the total amount of processing of all the operations of the job, the amount of slack there is with regard to the due date, and the weight of the job—that is,

$$B_j = f \left(w_j, \left(d_j - r_j - \sum_{i=1}^{n} p_{ij} \right), \sum_{i=1}^{n} p_{ij} \right)$$

The B_j may also be regarded as a value assigned to the job by a higher level manager or coordinator and that gives an indication regarding the overall priority of job j.

The Information Infrastructure: A framework with the highest level of information assumes that the JAs as well as the MAs have complete information: They know exactly what has happened up to now anywhere in the entire shop, which jobs are still circulating in the shop, their remaining operations, and their due dates. They also know everything about the job budgets and the amounts already spent. However, nobody has any information about future contracts.

A framework with a very low level of information is a setting where a JA only has information concerning the processing of its own operations in the past (pricing, awards, processing times, etc.), but no information concerning the processing of operations of other jobs. An MA only has information concerning the processing of operations on his own machine in the past and the commitments he has made for the future (including the prices agreed on). An MA does not have any detailed information concerning the processing of operations on other machines. (However, an MA may have some limited information concerning past processing on other machines; he may have lost a bid for a job to another machine in the past and may deduce that the other machine must have processed the operation in question at a lower price than he had quoted in his bid.) In such an environment, the levels of the bids depend on past experiences of both the JA and MA. This implies that in the beginning of the scheduling process the prices may not reflect true market conditions since little information is available. If the number of jobs is small, then this could be a problem. It may then be hard to make good decisions since the total level of information is very low. For problems with a small number of jobs, it makes sense for the JAs and MAs to already have some information at the outset.

An intermediate level of information is a situation in which the JA and MA have some general statistics with regard to the problem (i.e., the average amount of processing), the distribution of the processing times, the distribution of the weights of the jobs, the due date tightness factors, and so on. If the scheduling problem has a very large number of jobs and a learning process takes place, then both the JA and MA will have, after a while, some forms of estimation procedures and statistics.

The Bidding and Pricing Rules: If the time frame requested in a call for bids is such that the MA believes that the machine will end up with idle periods before and after the processing of this operation (because the idle periods may not be large enough to accommodate any other operation), then the MA may price its bid somewhat higher just to receive some compensation for the anticipated idle periods (see Figure 15.6.a). If the MA expects more calls for bids that may result in better

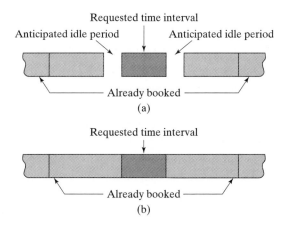

Figure 15.6   Pricing based on goodness of fit.

fitting jobs, he may not bid at all (when the MA has complete information, he may have a pretty good idea of future calls for bids).

If an MA expects many calls for bids in the immediate future, he may price his bids on the higher side because he is not afraid of losing. In the same vein, toward the end of the process, when the number of calls for bids dwindles, prices may come down.

Yet if the requested time period in the call for bids is a time period that happens to be exactly equal to an idle period in the machine's current schedule (see Figure 15.6.b), the MA may price his bid on the lower side because a perfect fit is advantageous for the machine and the MA would like to be sure that no other MA will come up with a lower bid. So the pricing in an MA's bid is determined by a number of factors. The three most important ones are:

(i)   The goodness of fit of the operation in the current machine schedule.
(ii)   The anticipated supply of machine capacity and the level of competition.
(iii)   The anticipated demand for machine capacity.

Expressions of goodness of fit can be developed for most information infrastructure settings. Even in a framework with a very low level of information, a reasonable, accurate estimate for the goodness of fit can be made since the MA should have, even in such an environment, all the necessary information. Yet estimates for the anticipated supply of machine capacity and demand for machine capacity may depend on the information infrastructure. The accuracy of such an assessment is, of course, higher when more information is available.

The selection heuristic a JA uses when receiving multiple bids depends, of course, on the timing and cost. Certain simple dominance rules are evident: A bid at a lower cost at an earlier time dominates a bid at a higher cost at a later time. The JA makes its selection based on remaining slack and on the expectation of future delays at subsequent workcenters. When more exact information is available, the JA may be more willing to postpone the processing.

To determine the optimal decision rules for both the JAs and MAs is not easy. The decision rules of the MAs depend heavily on the level of information available to the JAs and MAs. In general, it tends to be easier to determine a fair price when more information is available.

Example 15.3.1 (Bidding and Pricing in a Flexible Flow Shop)

Consider the last stage in a flexible flow shop. This last stage has two identical machines. All jobs, after having been processed at that stage, leave the system. Suppose the current time t is 400. Machine 1 is idle and does not have any commitments for the future, and machine 2 is busy until time 450 (machine 2 is processing a job that started at time 380 and that has a duration of 70). At time 400, a JA sends out a call for bids. The processing time of the last operation of this job is 10, the due date of the job is 390 (i.e., it is already overdue), and the job's weight is 5. So it is incurring for each additional time unit in the system a penalty of 5. The fixed costs of operating each machine is 4 dollars per unit time (i.e., a machine has to charge at least $4 per unit time that it is processing a job). Which price should machine 1 quote the JA for processing the job starting immediately at time 400 assuming machine 1 has complete information? The only competitor of machine 1 is machine 2, which can only start processing this operation at time 450. Machine 2 has to charge at least $4 \times 10 = 40$ to make a profit. The job would be completed at 460; the tardiness of the job is 70, and the total penalty of this job is $70 \times 5 = 350$. If machine 1 would process the job, then the job would be completed at time 410, its penalty being 20×5 is 100. So processing on machine 1 reduces the penalty by 250. To get the award, the MA of machine 1 has to quote a price that is less than $350 + 40 - 100 = 290$.

What happens now if there is incomplete information? The MA of machine 1 has to charge at least 40 to cover its cost. If machine 1 had received at time 380 a call for bids on the processing of the operation that went at that point to machine 2, then machine 1 knows that machine 2 has to be busy until time 450 and the case reduces to the case considered before. However, if machine 1 did not receive that call for bids at time 380, then it does not have any idea about what is being processed on machine 2. The price that machine 1 will charge may depend on completely different factors— for example, on the prices at which it has been awarded contracts in the past.

Example 15.3.2 (Market-Based Procedure Applied to a Job Shop)

Consider the following instance of $Jm \mid r_j \mid \sum w_j T_j$.

job	w_j	r_j	d_j	machine sequence	processing times		
1	1	5	24	1, 2, 3	$p_{11} = 5,$	$p_{21} = 10,$	$p_{31} = 4$
2	2	0	18	3, 1, 2	$p_{32} = 4,$	$p_{12} = 5,$	$p_{22} = 6$
3	2	0	16	3, 2, 1	$p_{33} = 5,$	$p_{23} = 3,$	$p_{13} = 7$

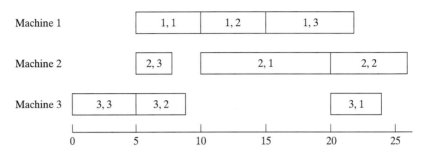

Figure 15.7   Final schedule in Example 15.3.2.

At time 0, both jobs 2 and 3 are interested in machine 3. Both jobs have the same weight and the same total amount of processing. However, job 3 has a tighter due date than job 2. So it is likely that $B_3 > B_2$ and that job 3 would put in a higher bid and win the bidding contest. The next decision moment occurs at time 5. Job 3 would like to go on machine 2, job 1 would like to go on machine 1, and job 2 still wants to go on machine 3. None of the three jobs has any competition. So the JMs and MMs should be able to come to an agreement. The next decision moment occurs at time 10. Both jobs 3 and 2 want to go on machine 1; job 1 wants to go on machine 2. So jobs 3 and 2 again have to compete with one another, now for machine 1. Since the remaining slack of job 2 is $(18 - 10 - 11) = -3$ and the remaining slack of job 3 is $(16 - 10 - 7) = -1$, it appears that job 2 will win the contest. So job 2 goes on machine 1 and job 1 goes on machine 2. At subsequent decision moments, there are no contests between jobs for machines. So the resulting schedule is the one depicted in Figure 15.7. The total weighted tardiness of this schedule is 28.

Note that this schedule is the same as the one generated by the shifting bottleneck procedure described in Chapter 7. However, this schedule is not optimal. The optimal schedule has a total weighted tardiness of 18.

Example 15.3.3 (Bidding Procedure Applied in Practice)

A simple bidding procedure has been implemented in the paint shop of an automobile manufacturing facility and the system has been working quite effectively. There is a dispatcher, which functions as JA, and there are two paint booths. Each paint booth has its own space where a fixed number of trucks can queue up. Each one of the paint booths has an MA. The dispatcher sends out an announcement of a new truck. The dispatcher functions as a JA and his announcement is equivalent to a call for bids. A paint booth has three bidding parameters—namely, the color of the last truck that has gone through, whether or not the queue is empty, and (in case the queue is not empty) whether there is space in the queue. Each MA sends the values of the three parameters to the dispatcher. The dispatcher uses the following decision rules when he makes an award: If the color of the last truck at a paint booth is the same as the color of the truck to be dispatched, then the dispatcher makes the award. If the last colors at both paint booths are different and one of the two queues is empty, the dispatcher makes the award to the booth with the empty queue. If both queues have trucks waiting, then the dispatcher makes the award to a booth that has space in the queue. Even though this bidding process is very simple, it performs well.

A careful study of Examples 15.3.2 and 15.3.3 may lead to a conclusion that the bidding and pricing mechanisms described in this section may behave somewhat myopically. When a JA sends out a call for bids only for the next operation to be processed, then the results of the bidding and pricing rules may be comparable to dispatching rules. Of course, if a bidding and pricing rule is made more elaborate, then the mechanism performs less myopically. It is easy to construct a more sophisticated framework. For example, suppose a JA is allowed to simultaneously issue several calls for bids to various workcenters for a number of operations that have to be processed one after another. If the JAs issue calls for bids for a number of their operations the moment they enter the shop, then the scheduling process of each one of the MAs becomes considerably more complicated. The bidding and pricing rules will be, of course, significantly more involved. In such a framework, it may be more likely that a JA has to go back and forth negotiating with several machines at the same time just to make sure that the time commitments with the various machines do not overlap. In such frameworks, it may also be possible that a job may not be ready at the time of a commitment.

The mechanisms described in this section can be easily generalized to more complicated machine environments. Consider a flexible job shop in which each workcenter does not consist of a bank of *identical* machines in parallel, but rather of a bank of *unrelated* machines in parallel. In this case, if a machine makes a bid for a certain operation, then it may let its pricing depend on how fast it can do the operation relative to the other machines in the same workcenter. For example, if a machine knows it can do an operation much faster than any other machine in the workcenter and it can also do this operation much faster than other operations that will come up for bidding subsequently, then it may take this into account in its pricing.

Bidding and pricing mechanisms, when applied to a completely deterministic problem, cannot perform as well as an optimization technique (e.g., a shifting bottleneck procedure) designed to search for a global optimum. However, distributed scheduling may play an important role in practice in more complicated machine environments with a long horizon that are subject to various forms of randomness (in settings for which it would be hard to develop optimization algorithms). In settings with long time horizons, one more concept starts playing a role—namely, the concept of learning. For both the JM and MM, the historical information starts playing a more important role in their decision making.

15.4 PROCEDURES FOR SCHEDULING PROBLEMS WITH MULTIPLE OBJECTIVES

Chapter 4 presents solutions to a few single machine scheduling problems with two objectives. However, the problems considered in Chapter 4 are relatively simple. In practice, the machine environment is usually more complex, and the scheduler often has to deal with a weighted combination of many objectives. The weights may be time or situation dependent. Typically, a scheduler may not know the exact weights and may want to perform a parametric analysis to get a feeling for the trade-offs. When a scheduler creates a schedule that is better with respect to one objective, he may want to know how other objectives are affected.

When there are multiple objectives, the concept of Pareto-optimality plays a role. A schedule is Pareto-optimal if it is impossible to improve on one of the objectives without making at least one other objective worse (see Chapter 4). Usually, only Pareto-optimal schedules are of interest. (However, in practice, a schedule may be Pareto-optimal only with respect to the set of schedules generated by the heuristic and not with respect to the set of all possible schedules.) When there are multiple objectives, the scheduler may want to view a set of Pareto-optimal schedules before deciding which schedule to select. So a system must retain at all times multiple schedules in memory.

There are major differences between multi-objective problems that allow preemptions and those that do not allow preemptions. Problems that allow preemptions are often mathematically easier than those that do not allow preemptions. In a preemptive environment, there are typically an infinite number of feasible schedules, whereas in a nonpreemptive environment, there are usually only a finite number of feasible schedules. The trade-off curve in a preemptive environment is often a continuous function that is piece-wise linear and convex. The trade-off curve in a nonpreemptive environment consists of a set of points; the envelope of such a set of points is always decreasing but not necessarily convex.

Efficient (polynomial time) algorithms exist only for the simplest multi-objective scheduling problems. Most multi-objective scheduling problems are NP-hard. However, in practice, it is still necessary to find good schedules in a fast and effective manner. Various adaptations of conventional heuristic procedures can be applied to multi-objective problems. The conventional approaches that can be used include:

 (i) procedures based on dispatching rules,
 (ii) branch and bound and filtered beam search, and
(iii) local search techniques.

One approach that is particularly useful for a parametric analysis of a multi-objective problem is

(iv) perturbation analysis.

The remaining part of this section discusses these four approaches in more detail.

Procedures based on dispatching rules. It is often the case that a dispatching rule that is effective with regard to one objective does not perform particularly well with regard to a second objective, whereas a different dispatching rule that is effective with regard to the second objective does not perform particularly well with regard to the first one. There are many ways of combining different dispatching rules within a single framework.

One way is described in Chapter 14 and is exemplified by the composite ATC and ATCS dispatching rules. A composite dispatching rule combines two (or more) dispatching rules within a single framework. Job j has a single index function $I_j(t)$ to which each rule contributes in a certain way. Every time a machine is freed, a job is selected based on the values of the indexes of the waiting jobs. Adjustments of the scaling parameters may result in different Pareto-optimal schedules. This is to be expected since a particular combination of scaling parameters will emphasize a certain subset of the basic rules within the composite rule and each one of the basic rules within the composite rule may favor a different objective. So by adjusting the scaling parameters in a composite dispatching rule, it may be possible to generate schedules for different parts of the trade-off curve.

Another approach for building a procedure based on dispatching rules is by analyzing the partial schedule and the status of the scheduling process. Through such an analysis, it can be determined at any point in time which dispatching rule is the most appropriate one to use. The following example illustrates how two dispatching rules can be combined in a preemptive environment with multiple objectives.

Example 15.4.1 (A Trade-Off Curve in a Preemptive Setting)

Consider the following instance of $P3 \mid prmp \mid \theta_1 C_{\max} + \theta_2 \sum C_j$ with three identical machines in parallel and five jobs. The processing times of the jobs are 5, 6, 8, 12, and 14. Preemptions are allowed and there are two objectives: the makespan and the total completion time. The trade-off curve is piece-wise linear decreasing convex (see Figure 15.8). The minimum total completion time is 56. The minimum makespan when the total completion time is 56 is 17 (see Exercise 5.22). The overall minimum makespan is 15. The minimum total completion time when the makespan is 15 is 58 (see Exercise 5.19). The coordinates (56, 17) and (58, 15) are breakpoints in the trade-off curve. One endpoint of the trade-off curve is achieved with a nonpreemptive schedule while the other endpoint is achieved with a preemptive schedule. All the points on the trade-off curve can be obtained by switching, in an appropriate manner, back and forth between the SPT and LRPT rules. Actually, the algorithm for generating a schedule that corresponds to any given point on the trade-off curve is polynomial (see Exercise 5.19).

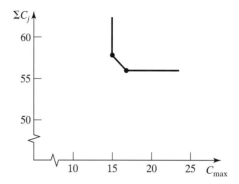

Figure 15.8 Trade-off curve in a preemptive environment (Example 15.4.1).

From Figure 15.8, it is clear that if $\theta_1 = \theta_2 = 0.5$, both breakpoints (and all the points in between) are optimal.

From a mathematical point of view, a scheduler can limit himself to those schedules that correspond to the breakpoints in the trade-off curve. However, it may be the case that other considerations also play a role. For example, the scheduler may prefer a nonpreemptive schedule over a preemptive one.

As stated before, nonpreemptive problems are usually harder than their pre-emptive counterparts. The following example illustrates the application of a procedure based on dispatching rules in a nonpreemptive environment.

Example 15.4.2 (Two Objectives in a Nonpreemptive Setting)

Consider an instance of $Pm \;||\; \theta_1 \sum w_j T_j + \theta_2 C_{max}$. There are m machines in parallel and n jobs. Job j has a processing time p_j, a due date d_j, and a weight w_j. The objectives are the total weighted tardiness $\sum w_j T_j$ and the makespan C_{max}. The composite objective is $\theta_1 \sum w_j T_j + \theta_2 C_{max}$, where θ_1 and θ_2 are the weights of the two objectives. An appropriate rule for the first objective is the Apparent Tardiness Cost first (ATC) rule, which, when a machine is freed at time t, selects the job with the highest index

$$I_j(t) = \frac{w_j}{p_j} \exp \left(-\frac{\max(d_j - p_j - t, 0)}{K \bar{p}} \right),$$

where $\bar{p}$ is the average of the remaining processing times and K is a scaling parameter (see Chapter 14). In contrast, a rule suitable for minimizing the makespan is the Longest Processing Time first (LPT) rule. Clearly, it is better to balance the m machines at the end of the process with shorter jobs than with longer jobs.

Combining the two rules for the joint objective can be done as follows. Since the makespan is determined mainly by the assignment of the jobs to the machines at the end of the process, it makes sense to schedule the jobs early in the schedule according to the ATC rule and switch at some point in time (or after a certain number of jobs) to the LPT rule. The timing of the switch-over depends on several factors, including the relative weights of the two objectives θ_1 and θ_2, the number of machines, and the number of jobs. So a rule can be devised that schedules the jobs in the beginning according to ATC and then switches over to LPT.

Examples 15.4.1 and 15.4.2 are, of course, very stylized illustrations of how different dispatching rules can be combined within a single framework. A procedure based on dispatching rules is usually more complicated than those described above.

Branch and bound and filtered beam search. These methods are applicable to problems with multiple objectives in the same way as they are applicable to problems with a single objective.

The manner in which the branching tree is constructed is similar to the way it is constructed for problems with a single objective. However, the bounding techniques are often different. In filtered beam search, the subroutines at each one of the nodes in the branching tree may use either composite dispatching rules or local search techniques.

Local search techniques. The three most popular local search techniques (i.e., simulated annealing, tabu-search, and genetic algorithms) are discussed in Chapter 14. These techniques are just as suitable for multi-objective problems as they are for single objective problems. However, there are differences. With a single objective, a local search procedure must retain in memory only the very best schedule found so far. With multiple objectives, a procedure must keep in memory all schedules that are Pareto-optimal among the schedules generated so far. The criteria according to which a new schedule is accepted depends on whether the new schedule is Pareto-optimal among the set of retained schedules. If a new schedule is found that is Pareto-optimal with respect to the set of schedules that are currently in memory, then the current schedules have to be checked whether they remain Pareto-optimal after the new schedule has been included. Genetic algorithms are particularly well suited for multi-objective problems since a genetic algorithm is already designed to carry a population of solutions from one iteration to the next. The population could include Pareto-optimal schedules and other schedules. The way the children of a population of schedules are generated may depend on the weights of the various individual objectives within the overall objective. The procedure may focus on those schedules that are the best with respect to the objectives with the greatest weight and use those schedules for cross-overs and other manipulations.

Perturbation analysis. Perturbation analysis is important when a parametric analysis has to be done. The weights of the various objectives may not be fixed and may be allowed to vary. In practice, this may happen when a scheduler simply does not know the exact weights of the objectives and would like to see optimal schedules that correspond to various sets of weights. Suppose that a scheduler has a schedule that is very good with respect to one set of weights but would like to increase the weight of one objective and decrease the weight of another. He may want to consider the contribution that each job makes to the objective that now has been assigned a larger weight and select those jobs that contribute the most to this objective. The positions of these jobs can now be changed. This can be done via a local search heuristic in such a way that their contribution to the objective, which has become more important, decreases while attempting to ensure that their contributions to the objectives, which have become less important, do not increase substantially.

The next example illustrates the effectiveness of combining dispatching rules with local search or perturbation analysis in a nonpreemptive environment.

Example 15.4.3 (Three Objectives in a Nonpreemptive Setting)

Consider the nonpreemptive scheduling problem described in Example 14.2.1. However, consider now three objectives: the total weighted tardiness, the makespan, and the maximum lateness (i.e., $1 \mid s_{jk} \mid \theta_1 \sum w_j T_j + \theta_2 C_{\max} + \theta_3 L_{\max}$).

sequence	rule	$\sum w_j T_j$	$C_{\max}$	$L_{\max}$	Pareto-optimal
1, 2, 3, 4		204	54	32	no
1, 2, 4, 3		139	51	30	no
1, 3, 2, 4	(SST)	162	49	27	yes
1, 3, 4, 2	(MS, EDD)	177	53	19	yes
1, 4, 2, 3		101	53	32	yes
1, 4, 3, 2		129	52	20	yes
2, 1, 3, 4	(SST)	194	50	28	no
2, 1, 4, 3		150	50	29	yes
2, 3, 1, 4		203	51	29	no
2, 3, 4, 1		165	54	42	no
2, 4, 1, 3	(SPT)	110	51	30	yes
2, 4, 3, 1	(SPT, SST, ATCS)	98	48	36	yes
3, 1, 2, 4		215	53	31	no
3, 1, 4, 2		213	55	21	no
3, 2, 1, 4		211	53	31	no
3, 2, 4, 1		159	54	42	no
3, 4, 1, 2		191	59	34	no
3, 4, 2, 1		135	54	42	no
4, 1, 2, 3		140	58	37	no
4, 1, 3, 2		162	56	24	no
4, 2, 1, 3	(WSPT)	118	53	32	no
4, 2, 3, 1	(WSPT)	122	54	42	no
4, 3, 1, 2		146	55	30	no
4, 3, 2, 1		102	52	40	no

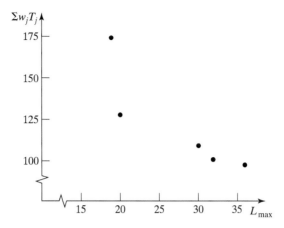

Figure 15.9   Trade-off curve in a nonpreemptive environment (Example 15.4.3)

One can plot the Pareto-optimal points for any two of the three objectives. The trade-offs between the total weighted tardiness and the makespan objective are particularly simple. Sequence 2, 4, 3, 1 minimizes both the makespan and total weighted tardiness; there are no trade-offs. The trade-offs between the total weighted tardiness and the maximum lateness are less simple. If these two objectives are considered, there are five Pareto-optimal schedules—namely,

$$(2, 4, 3, 1), (2, 4, 1, 3), (1, 3, 4, 2), (1, 4, 3, 2), (1, 4, 2, 3)$$

(see Figure 15.9). However, note that although schedule 2, 4, 1, 3 is Pareto-optimal, it can never be optimal when the objective is $\theta_1 L_{max} + \theta_2 \sum w_j T_j$. For any set of weights θ_1 and θ_2, either schedule 1, 4, 3, 2 or schedule 1, 4, 2, 3 dominates schedule 2, 4, 1, 3.

If the objectives C_{max} and L_{max} are considered, then there are four Pareto-optimal schedules—namely,

$$(2, 4, 3, 1), (1, 3, 4, 2), (1, 4, 3, 2), (1, 3, 2, 4).$$

From the two sets of Pareto-optimal schedules, it is clear how useful local search techniques and perturbation analysis are in conjunction with dispatching rules. Each set of Pareto-optimal schedules contains schedules that can be obtained with dispatching rules. All other schedules in these two sets are either one or at most two adjacent pairwise interchanges away from one of the schedules obtained with a dispatching rule.

In practice, the best approach for handling multi-objective problems may be a combination of several of the four approaches described earlier. For example, consider a two-phase procedure where in the first phase several (composite) dispatching rules are used to develop a number of reasonably good initial feasible schedules that are all different. One schedule may have a very low value of one objective while a second may have a very low value of another objective. These schedules are then fed into a genetic algorithm that attempts to generate a population of even better schedules. Any schedule obtained in this manner can also be used in either a branch and bound approach or in a beam search approach, where a good initial feasible schedule reduces the number of nodes to be evaluated.

15.5 DISCUSSION

The heuristic approaches described in this chapter are all different. They tend to work well only on certain problems and then often only on instances with certain characteristics. However, their usefulness often lies in their contribution to the overall performance when put in a framework in which they have to operate with one or more other heuristic approaches. For example, in the shifting bottleneck procedure, many different approaches have been used in the optimization of the single machine subproblem—namely,

  (i) composite dispatching rules,
 (ii) dynamic programming, and
(iii) local search.

There is another class of techniques based on *machine pricing* and *time pricing* principles; this class is somewhat similar to the market- and agent-based procedures. Complicated job shop scheduling problems can be formulated as mathematical programs. These mathematical programs usually have several sets of constraints. For example, one set of constraints may enforce the fact that two jobs cannot be assigned to the same machine at the same point in time. Such a set may be regarded as machine capacity constraints. Another set of constraints may have as its purpose that certain precedence constraints are enforced. Disregarding or relaxing one or more sets of constraints may make the solution of the scheduling problem significantly easier. It is possible to incorporate such a set of constraints in the objective function by multiplying it with a so-called *Lagrangian* multiplier. This *Lagrangian* multiplier is in effect the penalty one pays for violating the constraints. If the solution of the modified mathematical program violates any one of the relaxed constraints, the value of the *Lagrangian* multiplier has to be changed at the next iteration of the scheduling process to increase the penalty of violation and encourage the search for a solution that does not violate the relaxed constraints.

The third section of Chapter 18 focuses on the design of scheduling engines and algorithm libraries. The scheduling engine has to be designed in such a way that the scheduler can easily build a framework by combining a number of different techniques in the way most suitable for the particular problem (or instance) at hand.

EXERCISES (COMPUTATIONAL)

15.1. Consider the following instance of $1 \mid r_j \mid \sum w_j C_j$.

jobs	1	2	3	4	5	6	7	8	9
r_j	7	7	9	17	33	35	39	40	42
w_j	1	5	1	5	5	1	5	1	5
p_j	9	10	6	3	10	8	6	2	2

Design a job-based decomposition procedure for this problem and apply the procedure to the given instance. Compare your result with the result in Example 15.1.1 when all weights are equal.

15.2. Consider $1 \mid r_j, prmp \mid \sum w_j T_j$.

jobs	1	2	3	4	5	6	7	8	9
p_j	7	6	7	6	5	5	8	9	4
w_j	6	7	7	6	5	6	8	9	6
r_j	2	8	11	11	27	29	31	33	45
d_j	22	25	24	24	42	51	53	52	55

Design a job-based decomposition technique when the jobs are subject to release dates and preemptions are allowed. Apply your procedure to the given instance.

15.3. Apply the constraint guided heuristic search technique to the following instance of $Jm \parallel L_{max}$.

jobs	machine sequence	d_j	processing times
1	1, 2, 3	28	$p_{11} = 10$, $p_{21} = 8$, $p_{31} = 4$
2	2, 1, 4, 3	22	$p_{22} = 8$, $p_{12} = 3$, $p_{42} = 5$, $p_{32} = 6$
3	1, 2, 4	21	$p_{13} = 4$, $p_{23} = 7$, $p_{43} = 3$

15.4. Consider the instance of $Jm \mid r_j \mid \sum w_j T_j$ described in Example 15.3.2.

job	w_j	r_j	d_j	machine sequence	processing times
1	1	5	24	1, 2, 3	$p_{11} = 5$, $p_{21} = 10$, $p_{31} = 4$
2	2	0	18	3, 1, 2	$p_{32} = 4$, $p_{12} = 5$, $p_{22} = 6$
3	2	0	16	3, 2, 1	$p_{33} = 5$, $p_{23} = 3$, $p_{13} = 7$

Apply the constraint guided heuristic search technique to this instance. (*Hint:* In contrast to the L_{max} objective, the different jobs now may have completely different tardinesses. The following heuristic may be used. Assume that all the jobs end up with approximately the same amount of tardiness penalty $w_j T_j$. Parametrize on this tardiness penalty: Assume that each job has the same weighted tardiness penalty $w_j T_j = z$. The z together with the weight w_j and the due date d_j translates into a deadline—i.e., $\bar{d}_j = d_j + z/w_j$.)

15.5. Consider a flexible flow shop with a number of stages. The last stage is a bank of two unrelated machines in parallel. At time 400, only one more job has to be processed at the last stage. This job has a due date of 390 and a weight of 5. Both machines are idle at time 400; if machine 1 would process the job its processing time is 10 and if machine 2 would process the job its processing time is 40. The operating cost per unit time of machine 1 is 4 and the operating cost per unit time of machine 2 is 1. Each MA has all information.

What prices should the two machines submit and which machine ends up processing the job?

15.6. Consider the following instance of $1 \ || \ \theta_1 \sum w_j C_j + \theta_2 L_{max}$ with six jobs. The objectives are $\sum w_j C_j$ and L_{max} with weights θ_1 and θ_2, respectively.

jobs	1	2	3	4	5	6
d_j	7	7	9	17	33	35
w_j	9	15	12	3	20	24
p_j	9	10	6	3	10	8

(a) Show that the optimal preemptive schedule is nonpreemptive.
(b) Describe a heuristic that gives a good solution with any given combination of weights θ_1 and θ_2.
(c) Describe a branch and bound approach for this problem.

15.7. Apply the constraint guided heuristic search procedure to the following instance of $Jm \ || \ C_{max}$.

jobs	machine sequence	processing times			
1	1, 2, 3, 4	$p_{11} = 9$,	$p_{21} = 8$,	$p_{31} = 4$,	$p_{41} = 4$
2	1, 2, 4, 3	$p_{12} = 5$,	$p_{22} = 6$,	$p_{42} = 3$,	$p_{32} = 6$
3	3, 1, 2, 4	$p_{33} = 10$,	$p_{13} = 4$,	$p_{23} = 9$,	$p_{43} = 2$

15.8. Consider again the instance of $Jm \ | \ r_j \ | \ \sum w_j T_j$ described in Example 15.3.2 and Exercise 15.4. Apply a bidding and pricing heuristic, in which each JA has to negotiate, each time an operation has been completed, for the operation that comes after the next one (not for the next one as in Example 15.3.2).

Note that while the scheduling process goes on, the JA has, when an operation has been completed, already a contract in place for the next operation; the JA then only has to negotiate for the operation after the next one (however, when a job enters the shop for the first time, the JA has to negotiate for both the first and second operations on the job's route).

Compare your results with the results in Example 15.3.2.

15.9. Consider the instance in Example 15.4.3.
   (a) Consider the objective $\theta_1 L_{\max} + \theta_2 \sum w_j T_j$ (assume $\theta_1 + \theta_2 = 1$). Find the ranges of θ_1 and θ_2 for which each one of the schedules is optimal.
   (b) Consider the objective $\theta_1 L_{\max} + \theta_3 C_{\max}$ (assume $\theta_1 + \theta_3 = 1$). Find the ranges of θ_1 and θ_3 for which each one of the schedules is optimal.

15.10. Consider again the instance in Example 15.4.3. Consider now the objective $\theta_1 L_{\max} + \theta_2 \sum w_j T_j + \theta_3 C_{\max}$ (assume $\theta_1 + \theta_2 + \theta_3 = 1$). Find the ranges of $\theta_1, \theta_2,$ and θ_3 for which each one of the schedules is optimal.

EXERCISES (THEORY)

15.11. Consider Example 15.1.1 and Exercise 15.1. Design a hybrid decomposition scheme for $1 \mid r_j \mid \sum w_j C_j$, that takes the function $V(t)$ into account as well as the different weights of the jobs.

15.12. Consider Example 15.1.3 and Exercise 15.2. Design a hybrid decomposition scheme for $1 \mid r_j, prmp \mid \sum w_j T_j$ that takes into account $V(t)$, due date clusters, and the weights of the jobs.

15.13. Consider the problem $1 \mid r_j \mid \sum T_j$. Design a time-based decomposition method that is based on both the release and due dates of the jobs.

15.14. Consider the market-based procedure described in Section 15.3. In the heuristic that the MA uses, he may wish to have an estimate for the future supply and demand for machine capacity. Describe methods to estimate future supply and demand in the two information infrastructures described in Section 15.3.

15.15. Consider the market-based procedure described in Section 15.3. Describe a goodness of fit measure assuming the MA knows the frequency of the calls for bids that are about to come in as well as the distribution of the corresponding processing times (i.e., the mean and the variance).

15.16. Consider the market-based procedure described in Section 15.3. Suppose a JA receives multiple bids after sending out his call for bids. Design a heuristic for the JA to select a bid.

15.17. Consider a machine environment with two uniform machines in parallel with different speeds. Preemptions are allowed. There are two objectives: the makespan and the total completion time. Prove or disprove that the trade-off curve is decreasing convex.

15.18. Consider the ATCS composite dispatching rule and the following combinations of scaling parameters K_1 and K_2.
   (a) Both K_1 and K_2 very low.
   (b) K_1 very low and K_2 very high.

(c) K_1 very high and K_2 very low.

(d) Both K_1 and K_2 very high.

Which objective function(s) is (are) favored in each one of these four cases?

COMMENTS AND REFERENCES

An excellent treatise on decomposition methods is given by Ovacik and Uzsoy (1997). For some more recent results, see the papers by Chand, Traub, and Uzsoy (1996, 1997), Szwarc (1998), and Elkamel and Mohindra (1999).

Constraint guided heuristic search (constraint-based programming) is a development that originated among computer scientists and AI experts; see Fox and Smith (1984) and Fox (1987). The example of constraint guided heuristic search presented in Section 15.2 is an adaptation of the paper by Cheng and Smith (1997). For more applications of constraint-based programming to scheduling, see Nuijten (1994), Baptiste, Le Pape, and Nuijten (1995), and Nuijten and Aarts (1996).

Market-based and agent-based procedures have been a topic of significant research interest in the scheduling community as well as in other communities. Some of the early papers in the scheduling field are by Shaw (1987, 1988a, 1989), Ow, Smith, and Howie (1988), and Roundy, Maxwell, Herer, Tayur, and Getzler (1991). For examples of auction and bidding protocols, see the paper by Sandholm (1993), Kutanoglu and Wu (1999), and Wellman, Walsh, Wurman, and MacKie-Mason (2001). Sabuncuoglu and Toptal (1999a, 1999b) present a clear overview of the concepts in distributed scheduling and bidding algorithms.

A significant amount of research has been done on multi-objective scheduling. Most of it focuses on single machine scheduling; see Chapter 4 and its references. Not as much research has been done on parallel machine problems with multiple objectives. Eck and Pinedo (1993) consider a nonpreemptive parallel machine scheduling problem with makespan and total completion time as objectives. McCormick and Pinedo (1995) consider a preemptive parallel machine scheduling problem with makespan and total completion time as objectives and obtain a polynomial time algorithm.

The pricing procedure described in the discussion is due to Luh, Hoitomt, Max and Pattipati (1990), Hoitomt, Luh and Pattipati (1993), and Luh and Hoitomt (1993).

16

Modeling and Solving Scheduling Problems in Practice

Parts I and II described a number of stylized and (supposedly) elegant mathematical models in detail. The deterministic models led to a number of simple priority rules as well as to many algorithmic techniques and heuristic procedures. The stochastic models provided some insight into the robustness of the priority rules. The results for the stochastic models led to the conclusion that the more randomness there is in a system, the less advisable it is to employ very sophisticated optimization techniques. Equivalently, the more randomness the system is subject to, the simpler the scheduling rules should be.

It is not clear how all this knowledge can be applied to scheduling problems in the real world. Such problems tend to differ considerably from the stylized models studied by academic researchers. The first section of this chapter focuses on the differences between real-world problems and theoretical models. The subsequent four sections deal with examples of scheduling problems that have appeared in industry and for which algorithmic procedures have been developed. The second section illustrates the use of the Profile Fitting heuristic (described in Section 6.2). The third section discusses an application of the LPT heuristic (described in Section 5.1) within an algorithmic framework for flexible flow shops with bypass. The next section illustrates an application of the ATCS heuristic (described in Section 14.2) within a framework for flexible flow shops without bypass. The fifth section contains an application of constraint guided heuristic search (described in Section 15.2). In the last section, a number of modeling issues are discussed.

The applications described in this chapter illustrate that the rules and techniques introduced and analyzed in Parts I and II can be very useful. However, these rules and techniques usually have to be embedded in a more elaborate framework that deals with all aspects of the problem.

16.1 SCHEDULING PROBLEMS IN PRACTICE

Real-world scheduling problems are usually very different from the mathematical models studied by researchers in academia. It is not easy to list all the differences between these problems and the theoretical models as every real-world scheduling problem has its own peculiarities. Nevertheless, a number of differences are common and therefore worth mentioning.

(i) Theoretical models usually assume that there are n jobs to be scheduled and that after scheduling these n jobs the problem is solved. In the real world, there may be n jobs in the system at some point in time, but new jobs are being added continuously. Scheduling the current n jobs has to be done without a perfect knowledge of the near future. Hence, some provisions have to be made to be prepared for the unexpected. The dynamic nature may require, for example, that slack times are built into the schedule to accommodate unexpected rush jobs or machine breakdowns.

(ii) Theoretical models usually do not emphasize the *resequencing* problem. In practice, the following problem often occurs: There exists a schedule, which was determined earlier based on certain assumptions, and an (unexpected) random event occurs that requires either major or minor modifications in the existing schedule. The rescheduling process, which is sometimes referred to as *reactive* scheduling, may have to satisfy certain constraints. For example, one may wish to keep the changes in the existing schedule at a minimum even if an optimal schedule cannot be achieved this way. This implies that it is advantageous to construct schedules that are in a sense *robust*. That is, resequencing brings about only minor changes in the schedule. The opposite of robust is often referred to as *brittle*.

(iii) Machine environments in the real world are often more complicated than the machine environments considered in previous chapters. Processing restrictions and constraints may also be more involved. They may be either machine dependent, job dependent or time dependent.

(iv) In the mathematical models, the weights (priorities) of the jobs are assumed to be fixed (i.e., they do not change over time). In practice, the weight of a job often fluctuates over time and it may do so in a random fashion. A low-priority job may suddenly become a high-priority job.

 (v) Mathematical models often do not take *preferences* into account. In a model, a job either can or cannot be processed on a given machine. That is, whether the job can be scheduled on a machine is a 0–1 proposition. In reality, it often occurs that a job *can* be scheduled on a given machine, but for some reason there is a preference to process it on another one. Scheduling it on the first machine would only be done in case of an emergency and may involve additional costs.

(vi) Most theoretical models do not take machine availability constraints into account; usually it is assumed that machines are available at all times. In practice, machines are usually *not* continuously available. There are many reasons that machines may not be in operation. Some of these reasons are based on a deterministic process, others on a random process. The shift pattern of the facility may be such that the facility is not in operation throughout. At times preventive maintenance may be scheduled. The machines also may be subject to a random breakdown and repair process.

(vii) Most penalty functions considered in research are piecewise linear (e.g., the tardiness of a job, the unit penalty, etc.). In practice, there usually does exist a committed shipping date or due date. However, the penalty function is usually not piecewise linear. In practice, the penalty function may take, for example, the shape of an "S" (see Figure 16.1). Such a penalty function may be regarded as a function that lies somewhere in between the tardiness functon and the unit penalty function.

(viii) Most theoretical research has focused on models with a single objective. In the real world, there are usually a number of objectives. Not only are there several objectives, but their respective weights may vary over time and may even depend on the particular scheduler in charge. One particular combination of objectives appears to occur very often, especially in the process industry—namely, the minimization of the sum of the weighted tardinesses and the minimization of the sum of the sequence-dependent setup times (especially on bottleneck machines). The minimization of the sum of

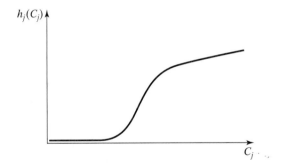

Figure 16.1   A penalty function in practice.

the weighted tardinesses is important since maintaining quality of service is usually an objective that carries weight. The minimization of the sum of the sequence-dependent setup times is important because, to a certain extent, it increases the throughput. When such a combination is the overall objective, the weights given to the two objectives may not be fixed. The weights may depend on the time as well as the current status of the production environment. If the workload is relatively heavy, then minimization of the sequence-dependent setup times is more important; if the workload is relatively light, minimization of the sum of the weighted tardinesses is more important.

(ix) The scheduling process, in practice, is often strongly connected with the assignment of shifts and the scheduling of overtime. Whenever the workload appears to be excessive and due dates appear to be too tight, the decision maker may have the option to schedule overtime or put in extra shifts to meet the committed shipping dates.

(x) The stochastic models studied in the literature usually assume very special processing time distributions. The exponential distribution, for example, is a distribution that has been studied in depth. In reality, processing times are usually not exponentially distributed. Some measurements have shown that processing times may have a density function like the one depicted in Figure 16.2.a. One can think of this density function as the convolution of a deterministic (fixed value) and an Erlang(k, λ). The number of phases of the Erlang(k, λ) tends to be low—say 2 or 3. This density function tends to occur in the case of a manual performance of a given task. That processing times may have such a density function is plausible. One can imagine that there is a certain minimum time that is needed to perform the task to be done. Even the best worker cannot get below this mimimum (which is equal to the fixed value). However, there is a certain amount of variability in the processing times that may depend on the person performing the task. The density func-

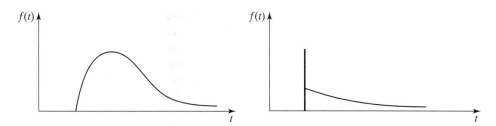

Figure 16.2   Density functions in practice.

tion may have a tail at the right that represents those processing times during which something went wrong. One can easily show that this density function has an increasing completion rate. Another type of density function that does occur in practice is the one depicted in Figure 16.2.b. The processing time is a fixed value with a very high probability, say .98, and with a very low probability, say .02, it is an additional random time that is exponentially distributed with a very large mean. This type of density function often occurs in automated manufacturing or assembly. If a robot performs a task, the processing time is always fixed (deterministic); however, if by accident something goes wrong, the processing time becomes immediately significantly larger.

(xi) Another important aspect of random processing times is correlation. Successive processing times on the same machine tend to be highly positively correlated in practice. In the stochastic models studied in the literature, usually all processing times are assumed to be independent draws from (a) given distribution(s). One of the effects of a positive correlation is an increase in the variance of the performance measures.

(xii) Processing time distributions may be subject to change due to *learning* or *deterioration*. When the distribution corresponds to a manual operation, then the possibility of learning exists. The human performing the operation may be able to reduce the average time he needs for performing the operation. If the distribution corresponds to an operation in which a machine is involved, then the aging of the machine may have as effect that the average processing time increases.

Despite the many differences between the real-world and mathematical models discussed in the previous sections, the general consensus is that the theoretical research done in the past has not been a complete waste of time. It has provided valuable insights into many scheduling problems, and these insights have proved to be useful in the development of algorithmic frameworks for a large number of real-world scheduling systems.

The remaining part of this chapter deals with examples of real-world problems for which scheduling theory has turned out to be useful. In practice, scheduling problems are often tackled with seemingly crude heuristics. The reason for not applying more sophisticated procedures is usually based on the relatively high frequency of random events. Such events often cause schedule changes during the execution of an existing schedule.

16.2 CYCLIC SCHEDULING OF A FLOW LINE

Consider a number of machines in series with a limited buffer between each two successive machines. When a buffer is full, the machine that feeds that buffer cannot release any more jobs; this machine is then blocked. Serial processing is common in assembly lines for physically large items, such as TV sets, copiers, and automobiles, whose size makes it difficult to keep many items waiting in front of a machine. Also, the material handling system does not permit a job to bypass another so that each machine serves the jobs according to the First in First Out (FIFO) discipline. These flow lines with blocking are often used in *Flexible Assembly Systems*.

Since different jobs may correspond to different product types, the processing requirements of one job may be different from those of another. Even if jobs are from the same product family, they may have different option packages. The timing of job releases from a machine may be a function of the queue waiting for processing on that machine as well as the queue at the next machine.

This machine environment with limited buffers is mathematically equivalent to a system of machines in series with *no* buffers between any machines. In such a system with no buffers, a buffer space can be represented by a machine at which all processing times are zero. So any number of machines with limited buffers between machines can be transformed mathematically into a system with a (larger) number of machines in series and *zero* buffers in between the machines. Such a model, with no buffers, is mathematically easier to formulate because there are no differences in the buffer sizes.

Sequences and schedules used in flow lines with blocking are often periodic or cyclic. That is, a number of different jobs are scheduled in a certain way, and this schedule is repeated over and over again. It is not necessarily true that a cyclic schedule has the highest throughput. Often, an acyclic schedule has the highest throughput. However, cyclic schedules have a natural advantage because of their simplicity; they are easy to keep track of and impose a certain discipline. In practice, there is often an underlying basic cyclic schedule from which minor deviations are allowed depending on current orders.

To discuss this class of schedules further, it is useful to define the *Minimum Part Set (MPS)*. Let $\mathcal{N}_k$ denote the number of jobs that correspond to product type k in the overall production target.

$$\bar{\mathcal{N}} = \left(\frac{\mathcal{N}_1}{q}, \ldots, \frac{\mathcal{N}_l}{q} \right)$$

represents the smallest set that has the same proportions of the different product types as the production target. This set is usually referred to as the Minimum Part Set (MPS). Given the vector $\bar{\mathcal{N}}$, the items in an MPS may, without loss of generality, be regarded as n jobs, where

$$n = \frac{1}{q} \sum_{k=1}^{l} \mathcal{N}_k.$$

The processing time of job j, $j = 1, \ldots, n$, on machine i is p_{ij}. Cyclic schedules are specified by the sequence of the n jobs in the MPS. The fact that some jobs within an MPS may correspond to the same product type and have identical processing requirements does not affect the approach described next.

The *Profile Fitting (PF)* heuristic described in Chapter 6 for $Fm \mid block \mid C_{\max}$ is well suited for the machine environment just described and appears to be a good heuristic for minimizing the cycle time in steady state. The cycle time is the time between the first jobs of two consecutive MPSs entering the system. Minimizing the cycle time is basically equivalent to maximizing the throughput. The following example illustrates the cycle time concept.

Example 16.2.1 (MPS Cycle Time)

Consider an assembly line with four machines and no buffers between machines. There are three different product types and they have to be produced in equal amounts—that is,

$$\bar{\mathcal{N}} = (1, 1, 1).$$

The processing times of the three jobs in the MPS are:

jobs	1	2	3
p_{1j}	0	1	0
p_{2j}	0	0	0
p_{3j}	1	0	1
p_{4j}	1	1	0

The second machine (i.e., the second row of processing times) with zero processing times for all three jobs functions as a buffer between the first and third machines. The Gantt charts for this example under the two different sequences are shown in Figure 16.3. Under both sequences, steady state is reached during the second cycle. Under sequence 1, 2, 3, the cycle time is three, whereas under sequence 1, 3, 2, the cycle time is two.

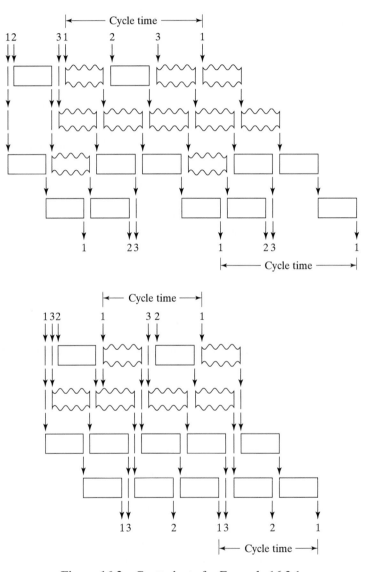

Figure 16.3   Gantt charts for Example 16.2.1.

Heuristics that attempt to minimize the makespan in a flow shop with blocking and a finite number of jobs ($Fm \mid block \mid C_{max}$) also tend to minimize the cycle time of a cyclic schedule for a flow environment with blocking. A variation of the Profile Fitting (PF) heuristic that is suitable for the problem just described works as follows: One job is selected to go first. The selection of the first job in the MPS sequence may be done arbitrarily or according to some scheme. This first job generates a *profile*. For the time being, it is assumed that the job does not encounter any blocking and proceeds smoothly from one machine to the next (in steady state, the first job in an MPS may be blocked by the last job from the previous MPS). The profile is determined by the departure time of this first job, job j_1, from machine i:

$$D_{i,j_1} = \sum_{h=1}^{i} p_{h,j_1}.$$

To determine which is the most appropriate job to go second, every remaining job in the MPS is tried out. For each candidate job, a computation has to be done to determine the times that the machines are idle and the times that the job is blocked at the various machines. The departure times of a candidate job for the second position, say job c, can be computed recursively as follows:

$$D_{1,j_2} = \max(D_{1,j_1} + p_{1c}, D_{2,j_1})$$

$$D_{i,j_2} = \max(D_{i-1,j_2} + p_{ic}, D_{i+1,j_1}), \quad i = 2, \ldots, m-1$$

$$D_{m,j_2} = D_{m-1,j_2} + p_{mc}$$

If candidate c is put into the second position, then the time wasted at machine i, because of either idleness or blocking, is $D_{i,j_2} - D_{i,j_1} - p_{ic}$. The sum of these idle and blocked times over all m machines is computed for candidate c. This procedure is repeated for all remaining jobs in the MPS. The candidate with the smallest total wasted time is then selected for the second position.

After the best fitting job is added to the partial sequence, the new profile (the departure times of this job from all the machines) is computed and the procedure is repeated. From the remaining jobs in the MPS, again the best fitting is selected and so on.

Observe that, in Example 16.2.1 after job 1, the PF heuristic would select job 3 to go next as this would cause only one unit of blocking time (on machine 2) and no idle times. If job 2 were selected to go after job 1, one unit of idle time would be incurred at machine 2 and one unit of blocking on machine 3, resulting in two units of time wasted. So in this example, the heuristic would lead to the optimal sequence.

Experiments have shown that the PF heuristic results in schedules that are close to optimal. However, the heuristic can be refined and made to perform even better. In the description presented earlier the goodness of fit of a particular job was measured by summing all the times wasted on the m machines. Each machine was considered equally important. Suppose one machine is a bottleneck machine, at which more processing has to be done than on any one of the other machines. It is intuitive that lost time on a bottleneck machine is more damaging than lost time on a machine that on the average does not have much processing to do. When measuring the total amount of lost time, it may be appropriate to weigh each inactive time period by a factor proportional to the degree of congestion at the particular machine. The higher the degree of congestion at a machine, the larger the weight. One measure for the degree of congestion of a machine is easy to calculate; simply determine the total amount of processing to be done on all jobs in an MPS at the machine in question. In the numerical example presented before, the third and fourth machines are more heavily used than the first and second machines (the second machine was not used at all and basically functioned as a buffer). Time wasted on the third and fourth machines is therefore less desirable than time wasted on the first and second machines. Experiments have shown that such a weighted version of the PF heuristic works exceptionally well.

Example 16.2.2 (Application of Weighted Profile Fitting Heuristic)

Consider three machines and an MPS of four jobs. There are no buffers between machines 1 and 2 and between machines 2 and 3. The processing times of the four jobs on the three machines are in the following table.

jobs	1	2	3	4
p_{1j}	2	4	2	3
p_{2j}	4	4	0	2
p_{3j}	2	0	2	0

All three machines are actual machines and none is a buffer. The workloads on the three machines are not entirely balanced. The workload on machine 1 is 11, on machine 2 it is 10, and on machine 3 it is 4. So time lost on machine 3 is less detrimental than time lost on the other two machines. If a weighted version of the Profile Fitting heuristic is used, then the weight applied to the time wasted on machine 3 must be smaller than the weights applied to the time wasted on the other two machines. In this example, the weights for machines 1 and 2 are chosen to be 1 while the weight for machine 3 is chosen to be 0.

Assume job 1 is the first job in the MPS. The profile can be easily determined. If job 2 is selected to go second, there will be no idle times or times blocked on machines 1 and 2; however, machine 3 will be idle for 2 time units. If job 3 is selected to go second, machines 1 and 2 are both blocked for two units of time. If job 4 is selected to go second, machine 1 will be blocked for one unit of time. As the weight of machine 1 is 1 and of machine 3 is 0, the weighted PF selects job 2 to go second. It can be easily verified that the weighted PF results in the cycle 1, 2, 4, 3, with a cycle time of 12. If the unweighted version of the PF heuristic were applied and job 1 again was selected to go first, then job 4 would be selected to go second. The unweighted PF heuristic would result in the sequence 1, 4, 3, 2 with a cycle time of 14.

16.3 SCHEDULING OF A FLEXIBLE FLOW LINE WITH LIMITED BUFFERS AND BYPASS

Consider a number of stages in series with a number of machines in parallel at each stage. A job, which in this environment is often equivalent to a batch of relatively small identical items such as Printed Circuit Boards (PCBs), needs to be processed at every stage on only one machine. Usually any of the machines will do, but it may be that not all the machines at a given stage are identical and that a given job has to be processed on a specific machine. A job may not need to be processed at a stage at all; in this case, the material handling system that moves jobs from one stage to the next will usually allow a job to bypass a stage and all the jobs residing at that stage (see Figure 16.4). The buffers at each stage may have limited capacity. If this is the case and the buffer is full, then either the material handling system must come to a standstill or, if there is the option to recirculate, the job bypasses that stage and recirculates. The manufacturing process is repetitive and it is of interest to find a cyclic schedule (similar to the one in the previous section).

The *Flexible Flow Line Loading (FFLL)* algorithm was originally designed at IBM for the machine environment just described. The algorithm was actually conceived for an assembly system used for the insertion of components in PCBs. The two main objectives of the algorithm are

(i) the maximization of throughput, and
(ii) the minimization of WIP.

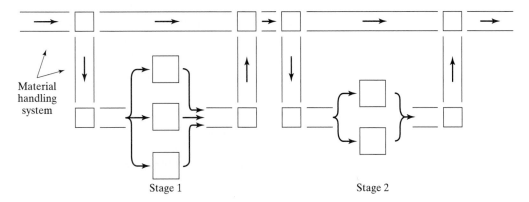

Figure 16.4   Flexible flow line with bypass.

With the goal of maximizing the throughput, an attempt is made to minimize the makespan of a whole day's mix. The FFLL algorithm actually attempts to minimize the cycle time of a Minimum Part Set (MPS). As the amount of buffer space is limited, it is recommended to minimize the WIP to reduce blocking probabilities. The FFLL algorithm consists of three phases:

Phase 1. the machine allocation,

Phase 2. the sequencing, and

Phase 3. the release timing.

The machine allocation phase assigns each job to a particular machine in each bank of machines. Machine allocation is done before sequencing and timing because, to perform the last two phases, the workload assigned to each machine must be known. The lowest conceivable maximum workload at a bank would be obtained if all the machines in that bank were given the same workload. To obtain balanced or nearly balanced workloads over the machines in a bank, the Longest Processing Time first (LPT) heuristic, described in Section 5.1, is used. According to this heuristic, all the jobs are, for the time being, assumed to be available at the same time and are allocated to a bank one at a time on the next available machine in decreasing order of their processing time. After the allocation is determined in this way, the items assigned to a machine may be resequenced, which does not alter the workload balance over the machines in a given bank. The output of this phase is merely the allocation of jobs and not the sequencing of the jobs or the timing of the processing.

The sequencing phase determines the order in which the jobs of the MPS are released into the system, which has a strong effect on the cycle time. The FFLL algorithm uses a *Dynamic Balancing* heuristic for sequencing an MPS. This heuristic is based on the intuitive observation that jobs tend to queue up in the buffer of a machine if a large workload is sent to that machine within a short time interval. This occurs when there is an interval in the loading sequence that contains many jobs with large processing times going to the same machine. Let n be the number of jobs in an MPS and m the number of machines in the entire system. Let p_{ij} denote the processing time of job j on machine i. Note that $p_{ij} = 0$ for all but one machine in a bank. Let

$$W_i = \sum_{j=1}^{n} p_{ij},$$

that is, the workload in an MPS destined for machine i. Let

$$W = \sum_{i=1}^{m} W_i$$

represent the entire workload of an MPS. For a given sequence, let S_j denote the set of jobs loaded into the system up to and including job j. Let

$$\alpha_{ij} = \sum_{k \in S_j} \frac{p_{ik}}{W_i}.$$

The α_{ij} represents the fraction of the total workload of machine i that has entered the system by the time job j is loaded. Clearly, $0 \leq \alpha_{ij} \leq 1$. The Dynamic Balancing heuristic attempts to keep the $\alpha_{1j}, \alpha_{2j}, \dots, \alpha_{mj}$ as close to one another as possible—that is, as close to an ideal target α_j^*, which is defined as follows:

$$\alpha_j^* = \sum_{k \in S_j} \sum_{i=1}^{m} p_{ik} \Bigg/ \sum_{k=1}^{n} \sum_{i=1}^{m} p_{ik}$$

$$= \sum_{k \in S_j} p_k / W,$$

where

$$p_k = \sum_{i=1}^{m} p_{ik},$$

which represents the workload on the entire system due to job k. So α_j^* is the fraction of the total system workload that is released into the system by the time job j is loaded. The cumulative workload on machine i, $\sum_{k \in S_j} p_{ik}$, should be as close as possible to the target $\alpha_j^* W_i$. Now let o_{ij} denote a measure of *overload* at machine i due to job j entering the system

$$o_{ij} = p_{ij} - p_j W_i / W.$$

Clearly, o_{ij} may be negative (in which case there is actually an underload). Now let

$$O_{ij} = \sum_{k \in S_j} o_{ik} = \left(\sum_{k \in S_j} p_{ik} \right) - \alpha_j^* W_i.$$

That is, O_{ij} denotes the cumulative overload (or underload) on machine i due to the jobs in the sequence up to and including job j. To be exactly on target means that machine i is neither overloaded nor underloaded when job j enters the system (i.e., $O_{ij} = 0$). The Dynamic Balancing heuristic now attempts to minimize

$$\sum_{i=1}^{m} \sum_{j=1}^{n} \max(O_{ij}, 0).$$

The procedure is basically a greedy heuristic that selects from among the remaining items in the MPS the one that minimizes the objective at that point in the sequence.

The release timing phase works as follows. From the allocation phase, the MPS workloads at each machine are known. The machine with the greatest MPS workload is the bottleneck since the cycle time of the schedule cannot be smaller than the MPS workload at the bottleneck machine. It is easy to determine a timing mechanism that results in a minimum cycle time schedule. First, let all jobs in the MPS enter the system as rapidly as possible. Consider the machines one at a time. At each machine, the jobs are processed in the order in which they arrive and processing starts as soon as the job is available. The release times are now modified as follows. Assume that the starting and completion times on the bottleneck machine are fixed. First, consider the machines that are positioned before the bottleneck machine (i.e., upstream of the bottleneck machine) and delay the processing of all jobs on each one of these machines as much as possible without altering the job sequences. The delays are thus determined by the starting times on the bottleneck machine. Second, consider the machines that are positioned after the bottleneck machine. Process all jobs on these machines as early as possible, again without altering job sequences. These modifications in release times tend to reduce the number of jobs waiting for processing, thus reducing required buffer space.

This three-phase procedure attempts to find the cyclic schedule with minimum cycle time in steady state. If the system starts out empty at some point in time, it may take a few MPSs to reach steady state. Usually this transient period is very short. The algorithm tends to achieve short cycle times during the transient period as well.

Extensive experiments with the FFLL algorithm indicates that the method is a valuable tool for scheduling flexible flow lines.

Example 16.3.1 (Application of FFLL Algorithm)

Consider a flexible flow shop with three stages. At stages 1 and 3, there are two machines in parallel. At stage 2, there is a single machine. There are five jobs in an MPS. If p'_{kj} denotes the processing time of job j at stage k, $k = 1, 2, 3$, then

jobs	1	2	3	4	5
p'_{1j}	6	3	1	3	5
p'_{2j}	3	2	1	3	2
p'_{3j}	4	5	6	3	4

The first phase of the FFLL algorithm performs an allocation procedure for stages 1 and 3. Applying the LPT heuristic to the five jobs on the two machines in stage 1 results in an allocation of jobs 1 and 4 to one machine and jobs 5, 2, and 3 to the other machine. Both machines have to perform nine time units of processing. Applying the LPT heuristic to the five jobs on the two machines in stage 3 results in an allocation of jobs 3 and 5 to one machine and jobs 2, 1, and 4 to the other machine (the LPT heuristic actually does *not* result in an optimal balance in this case). Note that machines 1 and 2 are at stage 1, machine 3 is at stage 2, and machines 4 and 5 are at stage 3. If p_{ij} denotes the processing time of job j on machine i, then

jobs	1	2	3	4	5
p_{1j}	6	0	0	3	0
p_{2j}	0	3	1	0	5
p_{3j}	3	2	1	3	2
p_{4j}	4	5	0	3	0
p_{5j}	0	0	6	0	4

The workload of machine i due to a single MPS, W_i, can now be easily computed. The workload vector W_i is (9, 9, 11, 12, 10) and the entire workoad W is 51. The workload imposed on the entire system due to job k, p_k, can also be computed easily. The p_k vector is (13, 10, 8, 9, 11). Based on these numbers, all values of the o_{ij} can be computed—for example,

$$o_{11} = 6 - 13 \times 9/51 = +3.71$$

$$o_{21} = 0 - 13 \times 9/51 = -2.29$$

$$o_{31} = 3 - 13 \times 11/51 = +0.20$$

$$o_{41} = 4 - 13 \times 12/51 = +0.94$$

$$o_{51} = 0 - 13 \times 10/51 = -2.55$$

and computing the entire o_{ij} matrix yields

$$
\begin{Vmatrix}
+3.71 & -1.76 & -1.41 & +1.41 & -1.94 \\
-2.29 & +1.23 & -0.41 & -1.59 & +3.06 \\
+0.20 & -0.16 & -0.73 & +1.06 & -0.37 \\
+0.94 & +2.64 & -1.88 & +0.88 & -2.59 \\
-2.55 & -1.96 & +4.43 & -1.76 & +1.84
\end{Vmatrix}.
$$

Of course, the solution also depends on the initial job in the MPS. If the initial job is chosen according to the Dynamic Balancing heuristic, then job 4 is the job that goes first. The O_{i4} vector is then $(+1.41, -1.59, +1.06, +0.88, -1.76)$. There are four jobs that then qualify to go second—namely, jobs 1, 2, 3, and 5. If job j goes second, then the respective O_{ij} vectors are:

$$O_{i1} = (+5.11, -3.88, +1.26, +1.82, -4.31)$$
$$O_{i2} = (-0.35, -0.36, +0.90, +3.52, -3.72)$$
$$O_{i3} = (+0.00, -2.00, +0.33, -1.00, +2.67)$$
$$O_{i5} = (-0.53, +1.47, +0.69, -1.71, +0.08).$$

It is clear that the Dynamic Balancing heuristic then selects job 5 to go second. Proceeding in the same manner, the Dynamic Balancing heuristic selects job 1 to go third and

$$O_{i1} = (+3.18, -0.82, +0.89, -0.77, -2.47).$$

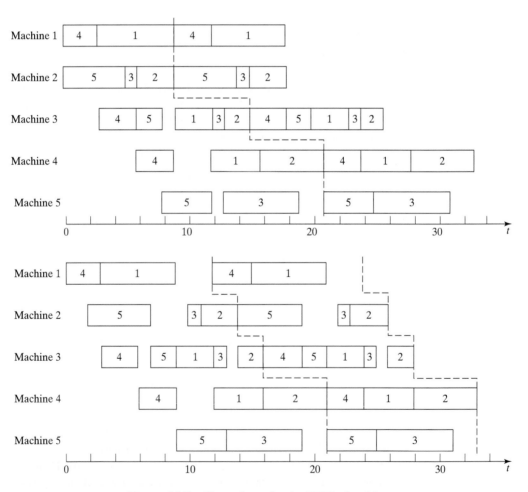

Figure 16.5 Gantt charts for the FFLL algorithm.

In the same manner, it is determined that job 3 goes fourth. Then

$$O_{i3} = (+1.76, -1.23, +0.16, -2.64, +1.96).$$

The final cycle is thus 4, 5, 1, 3, 2.

Applying the release timing phase on this cycle initially results in the schedule depicted in Figure 16.5. The cycle time of 12 is actually determined by machine 4 (the bottleneck machine) and there is therefore no idle time allowed between the processing of jobs on this machine. It is clear that the processing of jobs 3 and 2 on machine 5 can be postponed by three time units.

16.4 SCHEDULING OF A FLEXIBLE FLOW LINE WITH UNLIMITED BUFFERS AND SETUPS

In this section, the environment of Example 1.1.1 in Chapter 1 is considered (the paper bag factory). As in the previous section, there are a number of stages in series with a number of machines in parallel at each stage. The machines at a particular stage may be different for various reasons. For example, the more modern machines can accommodate a greater variety of jobs and can operate at a higher speed than the older machines. One stage (sometimes two) constitutes a bottleneck. The scheduler is usually aware which stage this is. However, unlike the environments considered in Sections 16.2 and 16.3, this is not a repetitive manufacturing process. The set of jobs to be scheduled at one point in time is usually different from the set of jobs to be scheduled at any other point in time.

The jobs that have to be processed are characterized by their processing times, their due dates (committed shipping dates), as well as their physical characteristics. The physical characteristics of a job are usually determined by one or more parameters. In the paper bag factory described in Example 1.1.1, the parameters of a job are the size and shape of the bags as well as the colors that are used in the printing process. When one job is completed on a given machine and another has to start, a setup is required. The duration of the setup depends on both jobs; in particular, on the similarities (or differences) between the physical characteristics of the two jobs.

The machine configuration is a flexible flow shop (see Figure 16.6). The scheduler has to keep three basic objectives in mind. The primary objective is to meet the due dates (committed shipping dates). This objective is more or less equivalent to minimizing the total weighted tardiness $\sum w_j T_j$. Another important objective is the maximization of throughput. This objective is somewhat equivalent to the minimization of the sum of setup times especially with regard to the machines at the bottleneck stage. A third objective is the minimization of the Work-In-Process inventory. The decision-making process is not that easy. For example,

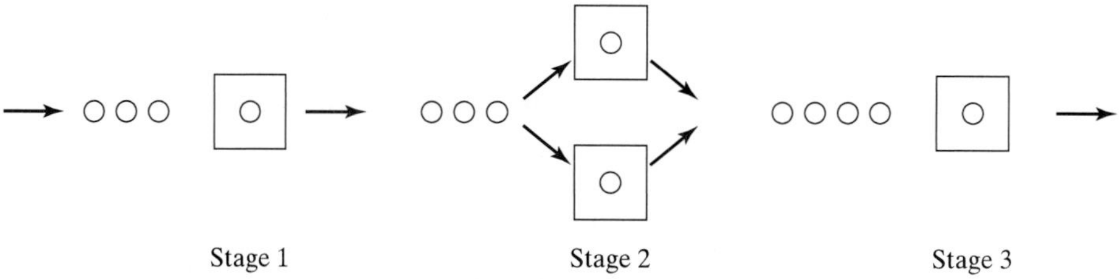

Stage 1   Stage 2   Stage 3

Figure 16.6   Flexible flow line with unlimited buffers.

suppose the decision maker can schedule a job on a particular machine at a particular time in such a way that the setup time is zero. However, the job's committed shipping date is 6 weeks from now. His decision depends on the machine utilization, the available storage space, and other factors. Clearly, the final schedule is the result of compromises among the three objectives.

The setup time between two consecutive jobs on a machine at any one of the stages is a function of the physical characteristics of the two jobs. As said before, the physical characteristics of the jobs are determined by one or more parameters. For simplicity, it is assumed here that only a single parameter is associated with the setup time of job j on machine i, say a_{ij}. If job k follows job j on machine i, then the setup time in between jobs j and k is

$$s_{ijk} = h_i(a_{ij}, a_{ik}).$$

The function h_i may be machine dependent. The algorithmic framework consists of five phases:

Phase 1. Bottleneck Identification,
Phase 2. Computation of Time Windows at Bottleneck Stage,
Phase 3. Computation of Machine Capacities at Bottleneck Stage,
Phase 4. Scheduling of Bottleneck Stage, and
Phase 5. Scheduling of Non-Bottleneck Stages.

The first phase constitutes the bottleneck identification, which works as follows. At least one of the stages has to be the bottleneck. The scheduler usually knows in advance which stage is the bottleneck and schedules this stage first. If two stages are bottlenecks, then the scheduler starts with the bottleneck that is the most downstream. If it is not clear which stage is the bottleneck, then it can be determined from the loading, the number of shifts assigned, and the estimates of the amounts of time that machines are down due to setups. The phenomenon of a moving bottleneck is not taken into account in this phase as the planning horizon is assumed to be short; the bottleneck therefore does not have sufficient time to move.

The second phase computes the time windows for the jobs at the bottleneck stage. For each job, a time window is computed (i.e., a release date and a due date), during which the job should be processed at the bottleneck stage. The due date of the job is computed as follows. The shipping date is the due date at the last stage. We assume that every job, after leaving the bottleneck stage, has a short wait at each one of the subsequent stages. That is, the length of their stay at any one of these stages equals their processing time multiplied by some safety factor. Under this assumption, a (local) due date for a job at the bottleneck stage can be obtained. The release date for a job at the bottleneck stage is computed as follows. For each job, the status is known. The status of job j, σ_j, may be 0 (the raw material for this job has not yet arrived), 1 (the raw material has arrived but no processing has yet taken place), 2 (the job has gone through the first stage but not yet through the second), or 3 (the job has gone through the second but not yet through the third), and so on. If

$$\sigma_j = l, \quad l = 0, 1, 2, \ldots, c - 1,$$

then the release date of job j at the bottleneck stage b is $r_{bj} = f(l)$, where the function $f(l)$ is decreasing in the status l of job j. A high value of σ_j implies that job j already has received some processing and is expected to have an early release at the bottleneck stage b. The function f has to be determined empirically.

The third phase computes the machine capacities at the bottleneck stage. The capacity of each machine over the planning horizon is computed based on its speed, the number of shifts assigned to it, and an estimate of the amount of time spent on setups. If the machines at the bottleneck stage are not identical, then the jobs that go through this stage are partitioned into buckets. For each type of machine, it has to be determined which jobs *must* be processed on that type and which jobs *may* be processed on that type. These statistics are gathered for different time frames (e.g., 1 week ahead, 2 weeks ahead, etc.). Based on these statistics, it can be determined which machine(s) at this stage have the largest loads and are the most critical.

The fourth phase does the scheduling of the jobs at the bottleneck stage. The jobs are scheduled one at a time. Every time a machine is freed, a job is selected to go next. This job selection is based on various factors—namely, the setup time (which depends on the job just completed), the due date, the machine capacity (in case the machines at the bottleneck stage are not identical), and so on. The rule used may be, in its simplest form, equivalent to the ATCS rule described in Section 14.2.

The fifth and last phase does the scheduling of the jobs at the non-bottleneck stages. The sequence in which the jobs go through the bottleneck stage more or less determines the sequence in which the jobs go through the other stages. However, some minor swaps may still be made in the sequences at the other stages. A swap may be able to reduce the setup times on a machine.

The following example illustrates the algorithm.

Example 16.4.1 (A Flexible Flow Shop with Setups)

Consider a flexible flow shop with three stages. Stages 1 and 3 consist of a single machine while stage 2 consists of two machines in parallel. The speed of the machine at stage 1, say machine 1, is 4 (i.e., $v_1 = 4$). The two machines at stage 2, say machines 2 and 3, both have speed 1 (i.e., $v_2 = v_3 = 1$). The speed of the machine at stage 3 is 4 (i.e., $v_4 = 4$). There are six jobs to be processed. Job j is characterized by a processing requirement p_j and the time it spends on machine i is p_j/v_i. All the relevant data regarding job j are presented in the following table.

job	1	2	3	4	5	6
p_j	16	24	20	32	28	22
r_j	1	5	9	7	15	6
d_j	30	35	60	71	27	63
w_j	2	2	1	2	2	1

There are sequence-dependent setup times on each one of the four machines. These sequence-dependent setup times are determined by *machine settings* a_{ij}, which have to be in place for job j when it is processed on machine i. If job k follows job j on machine i, then the setup time

$$s_{ijk} = | a_{ik} - a_{ij} | .$$

(This setup time structure is a special case of the setup time structure discussed in Section 4.5.) The initial setup times on each one of the machines is zero. The machine settings are presented in the following table.

job	1	2	3	4	5	6
a_{1j}	4	2	3	1	4	2
a_{2j}	3	4	1	3	1	3
a_{3j}	3	4	1	3	1	3
a_{4j}	2	2	4	1	2	3

Applying the five-phase procedure to minimize the total weighted tardiness results in the following steps:

The bottleneck identification process is easy. It is clear that the second stage is the bottleneck as its throughput rate is only half the throughput rate of the first and third stages.

The second phase requires a computation of time windows at the bottleneck stage. To compute these time windows, local release times and local due times for stage 2 have to be computed. To compute these times, suppose the total time a job spends at either stage 1 or stage 3 is equal to twice its processing time (i.e., the time waiting for a machine at stage 1 or 3 is at most equal to its processing time). These local release dates and due dates are presented in the following table.

job	1	2	3	4	5	6
r_{2j}	9	17	19	23	29	18
d_{2j}	22	23	50	55	13	52

The third phase of the procedure is not applicable here as both machines are assumed to be identical and all jobs can be processed on either one of the two machines.

The fourth phase requires the scheduling of the jobs at the bottleneck stage by the ATCS rule described in Section 14.2. Applying the ATCS rule results in the following sequence: As jobs 1 and 2 are supposedly the first two jobs to be released at stage 2, they start their processing on machines 2 and 3 at their estimated (local) release times 9 and 17, respectively. The estimated completion time of job 1 on machine 2 is $9 + 16 = 25$ and of job 2 on machine 3 is $17 + 24 = 41$. The ATCS rule has to be applied at time 25. There are three jobs to be considered then—namely, jobs 3, 4 and 6. The ATCS indexes have to be computed for these three jobs. The average processing time of the remaining jobs is approximately 25, and the setup times are between 0 and 3. Assume a K_1 value of 2 and a K_2 value of 0.7. Computing the ATCS indexes results in job 4 being started on machine 2 at time 25. The setup time in between the processing of jobs 1 and 4 on machine 2 is zero. So the estimated completion time of job 4 on machine 2 is $25 + 32 = 57$. The next time a machine is freed occurs at the completion of job 2 on machine 3 and the estimated time is 41. Three jobs have to be considered as candidates at time 41—namely, jobs 3, 5, and 6. Computing the indexes results in the selection of job 5. The setup time in between the processing of jobs 2 and 5 on machine 3 is equal to 3. So the estimated completion time of job 5 on machine 3 is $41 + 3 + 28 = 72$. Machine 2 is the next one to be freed at the estimated time 57. Jobs 3 and 6 remain. The ATCS routine selects job 6, with the larger processing time and the smaller setup time, for processing on

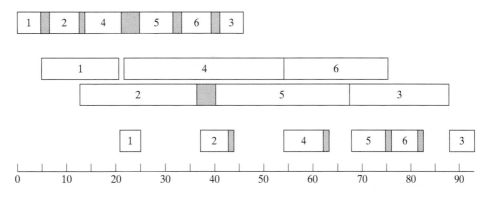

Figure 16.7 Gantt chart for flexible flow shop with six jobs (Example 16.4.1).

machine 2. So job 3 goes on machine 3, starting its processing at 72 and completing it at 92 (see Figure 16.7).

In the fifth and last phase, the schedules on machines 1 and 4 are determined. The sequence in which the jobs start their processing at the first stage is basically identical to the sequence in which they start their processing at the second stage. So the jobs are processed on machine 1 in the sequence 1, 2, 4, 5, 6, 3. After the completion times of the jobs on machine 1 have been determined, the starting and completion times on machines 2 and 3 have to be recomputed. The sequence in which the jobs start their processing at the third stage is basically identical to the sequence in which they complete their processing at the second stage. After computing the completion times at stage 2, the starting and completion times on machine 4 are computed. The completion times on machine 4 and the tardinesses are presented in the following table.

job	1	2	3	4	5	6
C_j	25	43	93	62	75	81.5
$w_j T_j$	0	16	33	0	96	18.5

The total weighted tardiness is 163.5. The final schedule on the four machines is depicted in Figure 16.7.

In more elaborate models, the third and fourth phases of the algorithm can be quite complicated. Suppose that at the bottleneck stage there are a number of nonidentical machines in parallel. The jobs may then be partitioned according to their M_j sets. Jobs that have the same M_j set are put into the same *bucket*. If there are two jobs in the same bucket with exactly the same combination of parameters (which implies a zero setup time), the two jobs may be combined to form a single macrojob (provided the due dates are not too far apart).

Whenever a machine at the bottleneck stage is freed, various machine statistics may be considered and possibly updated to determine which job goes next. First, it has to be determined what the remaining capacity of the machine is over the planning horizon. Second, it has to be determined how much *must* be produced on this machine over the planning horizon. If this number is less than the remaining capacity, other jobs (which can be processed on this machine as well as on others) may be considered as well (i.e., the pool of jobs from which a job can be selected may be enlarged). The job selection process may be determined by rules that are more elaborate than the ATCS rule. The rules may be based on the following criteria:

  (i) the setup time of the job,
 (ii) the due date of the job,
(iii) the flexibility of the job (the number of machines it can go on), and
 (iv) the processing time of the job.

The shorter the setup of a job and the earlier its due date, the higher its priority. The more machines a job can be processed on, the lower its priority. If a machine usually operates at a very high speed (relative to other machines), there is a preference to put larger jobs on this machine. After a job is selected, the statistics with regard to the machine, as well as those with regard to the buckets, have to be updated (i.e., remaining capacity of the machine, contents of the buckets, etc.). In the case of two bottlenecks, the same procedure still applies. The procedure starts with the bottleneck that is most downstream. After scheduling the jobs at this bottleneck, it proceeds with scheduling the jobs at the second bottleneck.

16.5 SCHEDULING OF A BANK OF PARALLEL MACHINES WITH RELEASE AND DUE DATES

This section focuses on the gate assignments at an airport, as described in Example 1.1.2 in Chapter 1.

Consider m machines in parallel and n jobs. Job j has processing time p_j, release date r_j, and due date d_j. Job j may be processed on any machine belonging to set M_j. The goal is to find a feasible assignment of jobs to machines. There may be additional constraints. For example, job j may be processed on machine $i \in M_j$ only if it is started after r_j and is completed before d_j. The entire algorithmic procedure consists of three phases:

> ***Phase 1.*** Constraint guided heuristic search,
>
> ***Phase 2.*** Constraint relaxation, and
>
> ***Phase 3.*** Assignment adjustment.

The constraint guided heuristic search phase uses a generalization of the procedure described in Section 15.2 to search for a feasible schedule or assignment that meets all due dates. More often than not, this phase does not yield a feasible schedule with an assignment for all jobs. Often only a subset of jobs is assigned after the first pass, and with this partial assignment there does not exist a feasible schedule for the remaining jobs. If this is indeed the case, then the procedure has to go through the second phase.

The constraint relaxation phase attempts to find a feasible schedule by relaxing some of the constraints. Constraint relaxation can be done in a number of ways. First, due dates of unscheduled jobs may be postponed and processing times of selected jobs may be shortened. Second, during the first pass, a particular assignment of a job to a machine may have been labeled *impossible*, but in reality this assignment was only *less preferable* because the machine in question was not *perfectly* suited for the particular job. However, such assignment may be made possible in the second pass. Third, one of the additional constraints may be relaxed. Constraint relaxation usually yields a complete schedule, with all jobs assigned to machines. This approach has sometimes been referred to as the *reformulative* approach. The idea is simple: The model has to be reformulated until a feasible schedule is found.

The assignment adjustment phase attempts to improve the schedule via pairwise swaps. Even though the main goal is to obtain a feasible schedule, some objectives may be of importance. Such objectives can include the balancing of the workloads over the various machines and the maximization of the idle times between the processing of consecutive jobs on the various machines (this last objective may be important to be prepared for potential fluctuations in the processing times).

16.6 DISCUSSION

In all four cases described, the scheduling procedures are based on *heuristics* and not on procedures that aim for an *optimal* solution. There are several reasons for this. First, the model is usually only a crude representation of the actual problem; so an optimal solution of the model may not lead to the best solution for the actual problem. Second, almost all scheduling problems in the real world are strongly NP-hard; it would take a very long time to find an optimal solution on a PC or even a workstation. Third, in practice, the scheduling environment is usually subject to a significant amount of randomness; it does not pay therefore to spend an enormous amount of computation time to find a supposedly optimal solution when within a couple of hours, because of some random event, either the machine environment or job set changes.

The solution procedure in each one of the last three sections consisted of a number of phases. There are many advantages in keeping the procedure segmented or modular. The programming effort can be organized more easily and debugging is made easier. Also, if there are changes in the environment and the scheduling procedures have to be changed, a modular design facilitates the reprogramming effort considerably.

EXERCISES (COMPUTATIONAL)

16.1. Consider in the model of Section 16.2 an MPS of six jobs.
   (a) Show that the number of different cyclic schedules is 5! when all jobs are different.
   (b) Compute the number of different cyclic schedules when two jobs are the same (i.e., there are five different job types among the six jobs).

16.2. Consider the model discussed in Section 16.2 with four machines and an MPS of four jobs.

jobs	1	2	3	4
p_{1j}	3	2	3	4
p_{2j}	1	5	2	3
p_{3j}	2	4	0	1
p_{4j}	4	1	3	3

(a) Apply the unweighted PF heuristic to find a cyclic schedule. Choose job 1 as the initial job and compute the cycle time.

(b) Apply again the unweighted PF heuristic. Choose job 2 as the initial job and compute the cycle time.

(c) Find the optimal schedule.

16.3. Consider the same problem as in the previous exercise.

(a) Apply a weighted PF heuristic to find a cyclic schedule. Choose the weights associated with machines $1, 2, 3, 4$ as $2, 2, 1, 2$, respectively. Select job 1 as the initial job.

(b) Apply again a weighted PF heuristic but now with weights $3, 3, 1, 3$. Select again job 1 as the initial job.

(c) Repeat again (a) and (b), but select job 2 as the initial job.

(d) Compare the impact of the weights on the heuristic's performance with the effect of the selection of the first job.

16.4. Consider again the model discussed in Section 16.2. Assume that a system is in steady state if each machine is in steady state. That is, at each machine, the departure of job j in one MPS occurs exactly the cycle time before the departure of job j in the next MPS. Construct an example with three machines and an MPS of a single job that takes more than 100 MPSs to reach steady state assuming the system starts out empty.

16.5. Consider the application of the FFLL algorithm in Example 16.3.1. Instead of letting the Dynamic Balancing heuristic minimize

$$\sum_{i=1}^{m}\sum_{j=1}^{n}\max(O_{ij},0),$$

let it minimize

$$\sum_{i=1}^{m}\sum_{j=1}^{n}|O_{ij}|.$$

Redo Example 16.3.1 and compare the performances of the two Dynamic Balancing heuristics.

16.6. Consider the application of the FFLL algorithm to the instance in Example 16.3.1. Instead of applying LPT in the first phase of the algorithm, find the optimal allocation of jobs to machines (which leads to a perfect balance of machines 4 and 5). Proceed with the sequencing phase and release timing phase based on this new allocation.

16.7. Consider the instance in Example 16.3.1 again.

 (a) Compute in Example 16.3.1 the number of jobs waiting for processing at each stage as a function of time and determine the required buffer size at each stage.

 (b) Consider the application of the FFLL algorithm to the instance in Example 16.3.1 with the machine allocation as prescribed in Exercise 16.6. Compute the number of jobs waiting for processing at each stage as a function of time and determine the required buffer size.

 (Note that with regard to the machines before the bottleneck, the release timing phase in a sense postpones the release of each job as much as possible and tends to reduce the number of jobs waiting for processing at each stage.)

16.8. Consider Example 16.4.1. In the second phase of the procedure, the time windows at the bottleneck stage have to be computed. In the example, job j is estimated to spend twice its processing time at a non-bottleneck stage.

 (a) Repeat the procedure and estimate job j's sojourn time as 1.5 times its processing time.

 (b) Repeat the procedure and estimate job j's sojourn time as 3 times its processing time.

 (c) Compare the results obtained.

16.9. Consider again the instance in Example 16.4.1. Instead of applying the ATCS rule in the fourth phase, select every time a machine is freed a job with the shortest setup time. Compare the sum of the weighted tardinesses under the two rules.

16.10. Consider again the instance in Example 16.4.1. Instead of applying the ATCS rule in the fourth phase, select every time a machine is freed the job with the earliest estimated local due date. Compare the sum of the weighted tardinesses under the two rules.

EXERCISES (THEORY)

16.11. Consider a distribution that is a convolution of a deterministic (i.e., a fixed value) D and an Erlang(k, λ) with parameters k and λ (see Figure 16.2.a). Determine its coefficient of variation as a function of D, k, and λ.

16.12. Consider a random variable X with the following distribution:

$$P(X = D) = p$$

and

$$P(X = D + Y) = 1 - p,$$

where D is a fixed value and the random variable Y is exponentially distributed with rate λ (see Figure 16.2.b). Determine the coefficient of variation as a function of p, D, and λ. Show that this distribution is neither ICR nor DCR.

16.13. Consider a single machine and n jobs. The processing time of job j is a random variable from distribution F. Compare the following two scenarios. In the first scenario, the n processing times are i.i.d. from distribution F and in the second scenario the processing times of the n jobs are all equal to the *same* random variable X from distribution F. Show that the expected total completion time is the same in the two scenarios. Show that the variance of the total completion time is larger in the second scenario.

16.14. Show that the cyclic scheduling problem described in Section 16.2 with two machines and zero intermediate buffer is equivalent to the TSP. Describe the structure of the distance matrix. Determine whether the structure fits the TSP framework considered in Section 4.5.

16.15. Consider the model in Section 16.2. Construct an example where no cyclic schedule of a single MPS maximizes the long-term average throughput rate. That is, to maximize the long-term average throughput rate, one has to find a cyclic schedule of k MPSs, $k \geq 2$.

16.16. The selection of the first job in an MPS in the model of Section 16.2 can be done by choosing the job with the largest total amount of processing. List the advantages and disadvantages of such a selection.

16.17. Consider an MPS with n jobs, which includes jobs j and k where $p_{ij} = p_{ik} = 1$ for all i. Show that there exists an optimal cyclic schedule with jobs j and k adjacent.

16.18. Consider an MPS with n jobs. For any pair of jobs j and k, either $p_{ij} \geq p_{ik}$ for all i or $p_{ij} \leq p_{ik}$ for all i. Describe the structure of the optimal cyclic schedule.

16.19. Consider the FFLL algorithm. Determine whether the largest makespan obtained in Phase 2 is always equal to the cycle time of the cyclic schedule generated by the algorithm.

16.20. Describe an approach for the model in Section 16.4 with two stages being the bottleneck. List the advantages and disadvantages of scheduling the upstream bottleneck first. Do the same with regard to the downstream bottleneck.

16.21. Consider the scheduling problem discussed in Section 16.4. Design alternate ways to compute the time windows (i.e., determining the local release dates and due dates—estimates of the amount of time downstream and the amount of time upstream, etc.). Explain how to take setup times into account.

16.22. Consider the scheduling problem discussed in Section 16.4. If a nonlinear function of the congestion is used for estimating the transit time through one or more stages, should the function be increasing concave or increasing convex? How does the amount of randomness in the system affect the shape of the function?

16.23. Consider the scheduling problem considered in Section 16.5. Design a composite dispatching rule for this problem (*Hint:* Try to integrate the LFJ or LFM rule with the ATC rule). Determine the number of scaling parameters and determine the factors or statistics necessary to characterize scheduling instances.

COMMENTS AND REFERENCES

The differences between real-world scheduling problems and theoretical models have been the subject of a number of papers; see, for example, Mckay, Safayeni, and Buzacott (1988),

Rickel (1988), Oliff (1988), Buxey (1989), Baumgartner and Wah (1991), and Van Dyke Parunak (1991).

The cyclic scheduling of the flow line with limited buffers is analyzed by Pinedo, Wolf, and McCormick (1986) and McCormick, Pinedo, Shenker, and Wolf (1989). Its transient analysis is discussed in McCormick, Pinedo, Shenker, and Wolf (1990). For more results on cyclic scheduling, see Matsuo (1990) and Roundy (1992).

The scheduling problem of the flexible flow line with limited buffers and bypass is studied by Wittrock (1985, 1988).

The flexible flow line with unlimited buffers and setups is considered by Adler, Fraiman, Kobacker, Pinedo, Plotnicoff, and Wu (1993).

The scheduling problem of a bank of parallel machines with release and due dates is considered by Brazile and Swigger (1988, 1991).

Of course, scheduling problems in many other types of industries have been discussed in the literature also. For scheduling problems and solutions in the micro-electronics industry, see, for example, Bitran and Tirupati (1988), Wein (1988), Uzsoy, Lee, and Martin-Vega (1992a, 1992b), Lee, Uzsoy, and Martin-Vega (1992), and Lee, Martin-Vega, Uzsoy, and Hinchman (1993). For scheduling problems and solutions in the automotive industry, see, for example, Burns and Daganzo (1987), Yano and Bolat (1989), Bean, Birge, Mittenthal, and Noon (1991), and Akturk and Gorgulu (1999).

17

Design, Development, and Implementation of Scheduling Systems

Analyzing a scheduling problem and developing a procedure for dealing with the problem on a regular basis is in the real world only part of the story. The procedure has to be embedded in a system that enables the scheduler to actually apply it. The scheduling system has to be incorporated into the information system of the factory or organization in question, which can be a formidable task. This chapter covers a number of system design and system implementation issues.

The first section gives an overview of the general structure of scheduling systems. The second section deals with database and knowledge-base issues. The third section describes schedule generators. The fourth section goes into user interface issues. The fifth section describes the advantages and disadvantages of generic systems and application specific systems. The last section deals with implementation issues.

It is, of course, impossible to cover all the issues concerning the topics just mentioned. Many books have been written on each one of these topics. This chapter focuses only on some of the issues concerning the design, development, and implementation of scheduling systems.

17.1 SYSTEMS ARCHITECTURE

As described in Chapter 1, a scheduling system usually has to interact with several other systems within the organization. It may have to receive information from a higher level production planning system that determines the appropriate actions for the medium and long term with regard to shift schedules, preventive maintenance schedules, and so on. It may receive information from a *Material Requirements Planning (MRP)* system to know the proper release dates of the jobs. The scheduling system may also interact with the shop floor control system to receive up to date information concerning availability of machines, progress of jobs, and so on (see Figures 1.1 and 17.1).

The scheduling system consists of a number of different modules. Those of fundamental importance are:

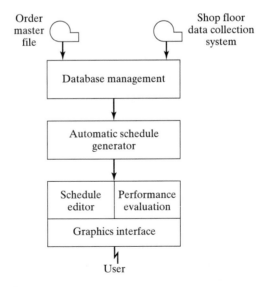

Figure 17.1   Configuration of a scheduling system.

   (i) the database and knowledge-base modules,
  (ii) the schedule generation and regeneration modules, and
 (iii) the user interface modules.

These modules play a crucial role in the functionality of the system. Significant effort is usually required to make a factory's database suitable for input to the system. Making a database accurate, consistent, and complete often involves the design of a series of tests the data must pass before it can be used. A database management module may also have capabilities of manipulating the data, performing various forms of statistical analyses, and enabling the scheduler, through some user interface, to see data in the form of bar charts or pie charts. Some systems have a separate knowledge-base specifically designed for scheduling purposes. A knowledge-base contains, in one format or another, a list of rules to be followed in given situations and possibly a list of objects representing orders and resources. However, not many systems have separate knowledge-bases; in most systems, the knowledge is embedded within the schedule generation module.

A schedule generation module involves the formulation of a suitable model and the formulation of objective functions, constraints and rules, as well as heuristics and algorithms.

User interface modules are important, especially with regard to the implementation process. Without an excellent user interface, there is a good chance that, regardless of its capabilities, the system will be too unwieldy to use. User interfaces often take the form of an electronic Gantt chart with tables and graphs that enable the scheduler to edit the schedule generated by the system and take last minute information into account (see Figure 17.1). When the scheduler edits the schedule, he is usually able to follow the impact of his changes on various performance measures as well as compare different schedules and perform what-if analyses.

17.2 DATABASES AND KNOWLEDGE-BASES

The database management subsystem may be either custom made or a commercial package. A number of commercial database systems available on the market have proved useful for scheduling systems. These systems are usually relational databases incorporating *Stuctured Query Language (SQL)*. Examples of such database management systems are Oracle, Sybase, and Ingres.

Whether a system is custom made or commercial, it needs a number of basic functions. The usual functions include multiple editing, sorting, and searching routines. Before generating a schedule, the scheduler may want to see certain segments of the order masterfile and collect some statistics with regard to the orders and the related jobs. Actually, at times he may not want to feed all jobs into the scheduling routines, but rather a subset. SQL allows fairly complicated job searches to be done with only three key words:

SELECT	fields to be displayed,
FROM	tables containing the fields, and
WHERE	specified conditions hold.

Within the database, a distinction can be made between *static* and *dynamic* data. Static data are all job and machine data that do *not* depend on the schedule. These include job data specified in the customer's order form, such as the ordered product quantity (which is proportional to the processing times of all the operations associated with the job), the committed shipping date (the due date), the time all necessary material is available (the release date), and possibly some processing (precedence) constraints. The priorities (weights) of the jobs are also static data as they are not schedule dependent. Having different weights for different jobs is usually a necessity, but determining their values is not all that easy. In practice, it is usually not necessary to have more than three priority classes; the weights are then, for example, 1, 2 and 4. The three priority classes are sometimes described as *hot*, *very hot*, and *hottest* depending on the level of the manager pushing the job. These weights actually have to be entered manually by the scheduler into the information system database. To determine the priority level, the person who enters the weight may use his own judgment or a formula that takes other data from the information system into account (e.g., total annual sales to the customer or some other measure of customer criticality). The weight of a job may also change from one day to another; a job that was not urgent yesterday may be urgent today. The scheduler may have to go into the file and change the weight of the job before generating a new schedule. The static machine data include machine speeds, scheduled maintenance times, and so on. There may also be static data that are both job and machine dependent (e.g., the setup time between jobs j and k assuming this setup occurs on machine i).

The dynamic data consist of all the data that are schedule dependent: the starting times and completion times of the jobs; the idle times of the machines; the times that a machine is undergoing setups; the sequences in which the jobs are processed on the machines; the number of jobs that are late; the tardinesses of the late jobs; and so on.

The following example illustrates some of these notions.

Example 17.2.1 (Order Master File in a Paper Bag Factory)

Consider the paper bag factory described in Example 1.1.1. The order master file may contain the following data:

ORDER	CUSTOMER	CMT	FC	GS	FBL	QTY	DDT	PRDT
DVN01410	CHEHEBAR CO		16.0	5.0	29.0	55.0	05/25	05/24
DVN01411	CHEHEBAR CO		16.0	4.0	29.0	20.0	05/25	05/25
DVN01412	CHEHEBAR CO		16.0	4.0	29.0	35.0	06/01	
DXY01712	LANSBERG LTD	PR	14.0	3.0	21.0	7.5	05/28	05/23
DXY01713	LANSBERG LTD		14.0	3.0	21.0	45.0	05/28	05/23
DXY01714	LANSBERG LTD		16.0	3.0	21.0	50.0	06/07	
EOR01310	DERMAN INC	HLD	16.0	3.0	23.0	27.5	06/15	

Each order is characterized by an 8-digit alphanumeric order number. A customer may place a number of different orders, each one representing a different type of bag. A bag is characterized by three physical parameters, the so-called face width (FC), the gusset (GS), and the finished bag length (FBL), which correspond to machine settings for a bag of that size. The quantities of bags ordered (QTY) are in multiples of 1,000 (e.g., the first order represents 55,000 bags). The month and day of the committed shipping date are specified in the DDT column. The month and day of the completion date of the order are specified in the PRDT column; the days specified in this column can be either actual completion dates or planned completion dates. The comments (CMT) column is often empty. If a customer calls and puts an order an hold, then HLD is entered in this column and the scheduler knows that this order should not yet be scheduled. If an order has a high priority, then PR is entered in this column. The weights will be a function of these entries (i.e., a job on hold has a low weight, a priority job has a high weight, and the default value corresponds to an average weight).

Setup times may be regarded as either static or dynamic data. This depends on the manner in which they are generated. Setup times may be stored in a table so that whenever a particular setup time needs to be known, the necessary table entry is retrieved. However, this method is not very efficient if the set is very large and if relatively few table look-ups are required. The size of the matrix is n^2 and all entries of the matrix have to be computed beforehand, which may require considerable CPU time as well as memory. An alternative way to compute and retrieve setup times, which is more efficient in terms of storage space used and can be more efficient in terms of computation time, is the following. A number of parameters, say

$$a_{ij}^{(1)}, \ldots, a_{ij}^{(l)},$$

may be associated with job j and machine i. These parameters are static data and may be regarded as given machine settings necessary to process job j on machine i. The setup time between jobs j and k on machine i, s_{ijk}, is a known function of the $2l$ parameters

$$a_{ij}^{(1)}, \ldots, a_{ij}^{(l)}; \quad a_{ik}^{(1)}, \ldots, a_{ik}^{(l)}.$$

The setup time usually is a function of the differences in machine settings for jobs j and k and is determined by production standards.

Example 17.2.2 (Sequence-Dependent Setup Times)

> Assume that to start a job on machine i three machine settings have to be fixed (e.g., the face, the gusset, and the finished bag length in Example 17.2.1). So the total setup time s_{ijk} depends on the time it takes to perform these three tasks and is a function of six parameters—that is,
>
> $$a_{ij}^{(1)}, a_{ij}^{(2)}, a_{ij}^{(3)} \quad a_{ik}^{(1)}, a_{ik}^{(2)}, a_{ik}^{(3)}.$$
>
> If the three tasks are independent of one another and have to be performed sequentially, then the total setup time is
>
> $$s_{ijk} = h_i^{(1)}(a_{ij}^{(1)}, a_{ik}^{(1)}) + h_i^{(2)}(a_{ij}^{(2)}, a_{ik}^{(2)}) + h_i^{(3)}(a_{ij}^{(3)}, a_{ik}^{(3)}).$$
>
> If the three tasks can be done in parallel, the total setup time is
>
> $$s_{ijk} = \max\left(h_i^{(1)}(a_{ij}^{(1)}, a_{ik}^{(1)}), \quad h_i^{(2)}(a_{ij}^{(2)}, a_{ik}^{(2)}), \quad h_i^{(3)}(a_{ij}^{(3)}, a_{ik}^{(3)})\right).$$
>
> Of course, there may be situations where some of the setup tasks can be done in parallel while others have to be done in series.

If the setup times are computed this way, they may be considered dynamic data. The total amount of computation time needed for computing setup times this way depends on the type of algorithm. If a dispatching rule is used to determine a good schedule, this method, based on (static) job parameters, is usually more efficient than the table look-up method mentioned earlier. However, if some kind of local search routine is used, the table look-up method will become more time efficient. The decision on which method to use depends on the relative importance of memory versus CPU time.

The calendar function is often also part of the database system. It contains crucial information with regard to factory holidays, scheduled machine maintenance, number of shifts available, and so on. Calendar data at times are static (e.g., fixed holidays) and at times dynamic (e.g., preventive maintenance shutdowns).

While virtually every scheduling system relies on a database, few scheduling systems have a module that serves specifically as a knowledge-base. However, it appears that in the future knowledge-bases may become more important. In the remainder of this section, a number of knowledge-base issues are discussed.

The design of a knowledge-base, in contrast to the design of a database, has a significant impact on the overall architecture of the system. The most important aspect of a knowledge-base is the knowledge *representation.* One form of knowledge representation is through *rules.* There are several formats for stating rules. One common format is through an *IF–THEN* statement. That is, *IF* a given condition holds, *THEN* a specific action has to be taken.

Example 17.2.3 (IF–THEN Rule for Parallel Machines)

Consider the airport terminal environment described in Example 1.1.2. If a flight is not continuing, it may be acceptable to assign this flight to a remote gate. This implies that arriving passengers may have to walk a certain distance to the terminal, but there are no departing passengers who have to walk that distance. So:

> *IF*   a flight is not continuing,
> *THEN*   the flight may be assigned to a remote gate.

It is easy to code such a rule in most programming languages, such as Fortran, Pascal, or C.

Another format for stating rules is through *predicate logic*, which is based on propositional calculus. An appropriate programming language for dealing with rules in this format is Prolog.

Example 17.2.4 (Logic Rule for Parallel Machines)

Consider the rule in the previous example. A Prolog version of this rule may be:

> GATEOK(G,F):-not(continue(F)), remote-gates(L), member(G,L).

The G refers to a specific gate, the F to a specific flight, and the L to a list of all remote gates. The ":-" may be read as "if"and the "," may be read as "and." A translation of the rule would be: Gate G is suitable for flight F if F is not continuing, if the set L is the set of remote gates, and if gate G is a member of L.

A second form of knowledge representation is through so-called *frames* or *schemata*. A frame or schema provides a structured representation of an object or a class of objects. A frame is a collection of slots and values. Each slot may have a value class as well as a default value. Information can be shared among multiple frames through inheritance. Frames lower in the hierarchy are applicable to more specific operations and resources than the more generic frames at higher levels in the hierarchy.

Example 17.2.5 (Schema for Operation of Job)

A schema that represents a generic operation of a job may be described as follows:

> { { OPERATION
>   { IS = A ACT
>   DURATION: *processing time of this operation*
>   ENABLED-BY: *state that enables this operation*
>   CAUSES: *states caused by this operation*
>   PREVIOUS-OPERATION: *operations that directly precede*
>   NEXT-OPERATION: *operations that follow* } } }.

The first statement specifies that an OPERATION is a type of ACT that is a more generic activity than an operation. A more specific frame in the hierarchy may describe, for example, a printing operation.

Besides activities, constraints and rules can be represented by schemata as well.

Example 17.2.6 (Schema for Due Date Constraint)

A schema that represents a generic due date constraint of a job may be described as follows:

$$
\{\, \{\, \text{DUE-DATE-CONSTRAINT}
$$
$$
\{\, \text{IS} = \text{A CONTINUOUS} - \text{CONSTRAINT}
$$
$$
\text{DOMAIN:} \; \textit{dates}
$$
$$
\text{PIECE} = \text{WISE} = \text{LINEAR} = \text{UTILITY:} \; \textit{cost function} \,\}\, \}\, \}.
$$

The first statement specifies that a DUE-DATE-CONSTRAINT is a continuous constraint. A continuous constraint refers to the form of the domain, whether it is continuous or discrete. The domain slot specifies that the constraint is defined by dates. The schema also specifies a utility function to be used to determine the utility or preference of any particular input value. This utility function is similar to the due date cost functions defined in Chapter 2.

Computer scientists and researchers in artificial intelligence (AI) have been experimenting with other forms of knowledge representation. For example, a third form is *semantic networks*. In academia, some experimentation has been done with this form in conjunction with scheduling system development.

As said before, the design of a knowledge-base in a knowledge-based system has a significant impact on the design of the schedule generation module. This is discussed in more detail in the next section.

17.3 SCHEDULE GENERATION ISSUES

Current schedule generation techniques are an amalgamation of several schools of thoughts that have been converging in recent years. One school of thought that is predominantly followed by industrial engineers and operations researchers may be referred to as the *algorithmic* approach. Another that is often followed by computer scientists and AI experts may be referred to as the *knowledge-based* approach. Recently, these two approaches have started to converge and the differences have become blurred. Some hybrid systems developed in the recent past use a knowledge-base as well as fairly sophisticated heuristics. Certain segments of the procedure are designed according to the algorithmic approach while other segments are designed according to the knowledge-based approach.

Example 17.3.1 (Architecture of Scheduling System for Wafer Fabrication)

For a semiconductor wafer fabrication unit, a hybrid scheduling system has been designed. The system consists of two levels. The higher level operates according to a knowledge-based approach. The lower level is based on an algorithmic approach; it consists of a library of algorithms.

The higher level performs the first phase of the scheduling process. At this level, the current status of the environment is analyzed. This analysis takes into consideration due date tightness, bottlenecks, and so on. The rules embedded in this higher level determine the type of algorithm in the lower level to be used in each situation.

The algorithmic approach usually requires a mathematical formulation of the problem that includes objectives and constraints. The algorithm could be based on any one or a combination of techniques presented in the first part of this book. The quality of the solution is based on the values of the objectives and performance criteria under the given schedule. This form of schedule generation may consist of three phases (see Section 14.6). In the first phase, a certain amount of *preprocessing* is done, where the problem instance is analyzed and a number of statistics are compiled (e.g., the average processing time, the maximum processing time, the due date tightness). The second phase consists of the actual algorithms and heuristics. The structure of the algorithm or heuristic may depend on the statistics compiled in the first phase (e.g., in the way the look-ahead parameter K in the ATC rule may depend on the due date tightness and due date range factors). The third phase may contain a *postprocessor*. The solution that comes out of the second phase is fed into a procedure such as simulated annealing or tabu-search to see if improvements can be obtained. This type of schedule generation is usually coded in a procedural language such as Fortran, Pascal, or C.

The knowledge-based approach is different from the algorithmic approach in various respects. This approach is often more concerned with underlying problem structures which cannot easily be described in an analytical format. To incorporate the scheduler's knowledge into the system, rules or objects are used. This approach is often used when it is only necessary to find a *feasible* solution given the many constraints or rules. However, as some schedules are ranked more preferable than others, heuristics are used at times to obtain a preferred schedule. Through a so-called *inference engine* the approach attempts to find schedules that do not violate prescribed rules and satisfy stated preferences as much as possible. The logic behind the schedule generation process is often a combination of inferencing techniques and search techniques as described in Section 15.2. The inferencing techniques are usually so-called *forward chaining* and *backward chaining* algorithms. A forward chaining algorithm is knowledge driven. It first analyzes the data and the rules and through inferencing techniques attempts to construct a feasible schedule. A backward chaining algorithm is result oriented. It starts out with a promising schedule and attempts to verify whether it is feasible and not in violation of any of the rules. Whenever a satisfactory solution does not appear to exist or when the scheduler judges it too difficult to find, the scheduler may reformulate the problem through a relaxation of the constraints. The relaxation of constraints may either be done automatically (by the system) or by the user. Because of this aspect, the approach has been at times also referred to as the *reformulative* approach.

The programming style used for the development of knowledge-based systems is different from the ones used for systems based on algorithmic approaches. The programming style may depend on the form of the knowledge representation. If the knowledge is represented in the form of *IF-THEN* rules, then the system can be coded using an expert system shell such as *OPS5*. The expert system shell contains an inference engine that is capable of doing forward chaining or backward chaining of the rules to obtain a feasible solution. This approach may have difficulties dealing with conflict resolution and uncertainty. If the knowledge is represented in the form of logic rules (see Example 17.2.4), then an ideal programming language is Prolog. If the knowledge is represented in the form of frames, then a language with object-oriented extensions is required (e.g., *LISP* or *C++*). These languages emphasize user-defined objects, which facilitate a modular programming style.

Algorithmic approaches as well as knowledge-based approaches have their advantages and disadvantages. An algorithmic approach has an edge if

  (i) the problem allows for a crisp and precise mathematical formulation,
 (ii) the number of jobs involved is large,
(iii) the amount of randomness in the environment is minimal,
 (iv) some form of optimization has to be done frequently and in real time, and
  (v) the general rules are consistently being followed without too many exceptions.

A disadvantage of the algorithmic approach is that if the scheduling environment changes, (e.g., certain preferences on assignments of jobs to machines) the reprogramming effort can be substantial.

The knowledge-based approach may have an edge if only feasible schedules are required. Some system developers believe that changes in the scheduling environment or rules can be incorporated more easily in a knowledge-based system than in a system based on the algorithmic approach. Others, however, believe that the effort required to modify any system is mainly a function of how well the code is organized and written; the effort required to modify does not depend that much on the approach used.

A major disadvantage of the knowledge-based approach is that obtaining a reasonable schedule may take substantially more computer time than an algorithmic approach. Certain scheduling systems in practice do have to operate in near-real time (it is very common that schedules must be generated within several minutes).

The amount of available computer time is an important issue with the selection of a schedule generation technique. The time allowed for schedule generation varies from application to application. A large number of applications require real-time performance—that is, the schedule has to be generated in seconds or minutes on the computer at hand. This may be the case if rescheduling is required throughout the day, due to substantial schedule deviations. It would also be true if the scheduler runs iteratively, requiring human interaction between iterations (perhaps for adjustments of workcenter capacities). Some applications do allow for overnight number crunching. For example, the scheduler at the end of the afternoon executes the program and wants an answer by the time he or she arrives at work the next day. A small number of applications allow for virtually unlimited number crunching. When in the airline industry quarterly flight schedules have to be determined, the capital at stake is such that a week of number crunching on a mainframe is fully justified.

As said before, the two approaches have been converging, and most scheduling systems currently developed have elements of both. One language of choice is now *C++* as it incorporates the best of both worlds. It is an easy language for coding algorithmic procedures and it also has object-oriented extensions.

17.4 USER INTERFACES AND INTERACTIVE OPTIMIZATION

The user interfaces are a very important part of the system. These interfaces may determine whether the system is going to be used or not. Most user interfaces, whether the system is based on a workstation or PC, make extensive use of windows and graphics. The user often wants to see two or more different sets of information at the same time. This is not only the case for the static data that are stored in the database, but also for the dynamic data that depend on the schedule.

User interfaces for the database modules often take a fairly conventional form and may be determined by the particular database package used. The database package may also contain its own calendar.

The schedule generation module may provide the user with a number of computational procedures and algorithms. Such a library of procedures within the schedule generation module will require its own user interface, enabling the scheduler to select the appropriate algorithm or even to design an entirely new procedure.

User interfaces that display schedule information take many different forms. The interfaces for schedule *manipulation* determine the basic character of the system as these interfaces are the ones used most extensively by the scheduler. The different forms of schedule manipulation interfaces depend on the level of detail as well as the planning horizon being considered. Four such interfaces are described in detail—namely:

  (i) the Gantt Chart interface,
 (ii) the Dispatch List interface,
(iii) the Capacity Buckets interface, and
 (iv) the Throughput Diagram interface.

The first, and probably most popular, form of schedule manipulation interface is the *Gantt chart* (see Figure 17.2). The Gantt chart is the usual horizontal bar chart with the x-axis representing the time and the y-axis, the various machines. A color and/or pattern code may be used to indicate a characteristic or an attribute of the corresponding job. For example, jobs that under the current schedule are completed after their due date are colored red. The Gantt chart usually has a number of scroll capabilities that allow the user to go back and forth in time or focus on particular

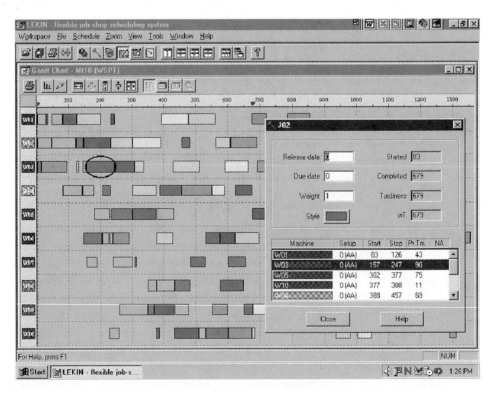

Figure 17.2   Gantt chart of the Leckin system.

machines; it is usually mouse driven. Perhaps the user is not entirely satisfied with the generated schedule and wishes to perform a number of manipulations on his or her own. With the mouse, the user has the ability to "click and drag" a job from one position to another. Providing the interface with a click and drag capability is not a trivial task for the following reason. After changing the position of a particular operation on a machine, other operations on that machine, which belong to other jobs, may have to be pushed either forward or backward in time to maintain feasibility. The fact that other operations have to be processed at different times may also have an effect on other machines. This is often referred to as *cascading* or *propagation* effects. After the scheduler repositions an operation of a job, the system may call a reoptimization procedure embedded in the scheduling routine to control the cascading effects in a proper manner.

Example 17.4.1 (Cascading Effects and Reoptimization)

Consider $F3 \mid\mid \sum w_j T_j$. There is an unlimited storage space between the successive machines and therefore no blocking. Consider a schedule with four jobs as depicted by the Gantt chart in Figure 17.3.a. If the user interchanges jobs 2 and 3 on machine 1 while keeping the order on the two subsequent machines the same, the resulting schedule, because of cascading effects, takes the form depicted in Figure 17.3.b. If the system has reoptimization algorithms at its disposal, the user may decide to reoptimize the operations on machines 2 and 3 while keeping the order on machine 1 the way he constructed it. A reoptimization algorithm then results in the schedule depicted in Figure 17.3.c. To obtain appropriate job sequences for machines 2 and 3, the reoptimization algorithm has to solve an instance of $F2 \mid r_j \mid \sum w_j T_j$.

Gantt charts do have disadvantages especially when there are many jobs and machines. It may be hard to recognize which bar or rectangle corresponds to which job. As the space on the screen (or on the printout) is rather limited, it is hard to attach to each bar the job number associated with it. Gantt chart interfaces usually provide the capability to click the mouse on a given bar and open a window that displays detailed data regarding the corresponding job. Some Gantt charts also have a filter capability, where the user may specify the job(s) that should be exposed on the Gantt chart while disregarding all other jobs. The Gantt chart interface depicted in Figure 17.2 is from the *Lekin* system developed at New York University (see Chapter 19).

The second form of user interface displaying schedule information is the *Dispatch List* interface (see Figure 17.4). Schedulers often want to see a list of the jobs to be processed on each machine in the order in which they are to be processed. With this type of display, the schedulers also want to have editing capabilities. That is, they want to be able to change the sequence in which jobs are processed on a machine or move a job from one machine to another. This sort of interface does not have the disadvantage of the Gantt chart since the jobs are listed with their job

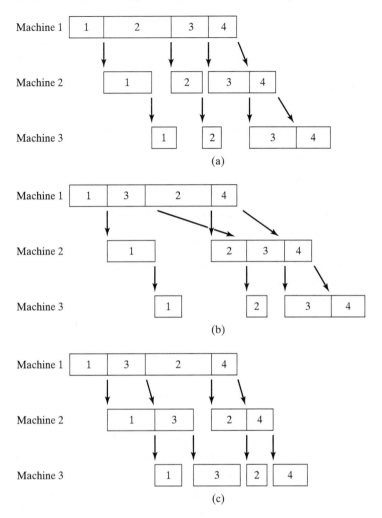

Figure 17.3 Cascading and reoptimization after swap: (a) original schedule, (b) cascading effects after swap of jobs on Machine 1, (c) schedule after reoptimization of Machines 2 and 3.

number and the scheduler knows exactly where each job is in a sequence. If the scheduler would like more attributes (e.g., processing time, due date, completion time under the current schedule, etc.) of the jobs to be listed, then one or more columns can be added next to the job number column, each one with a particular attribute. The disadvantage of the Dispatch List interface is that the scheduler does not have a good overview of the schedule relative to time. The user may not see immediately which jobs are going to be late, which machine is idle most of

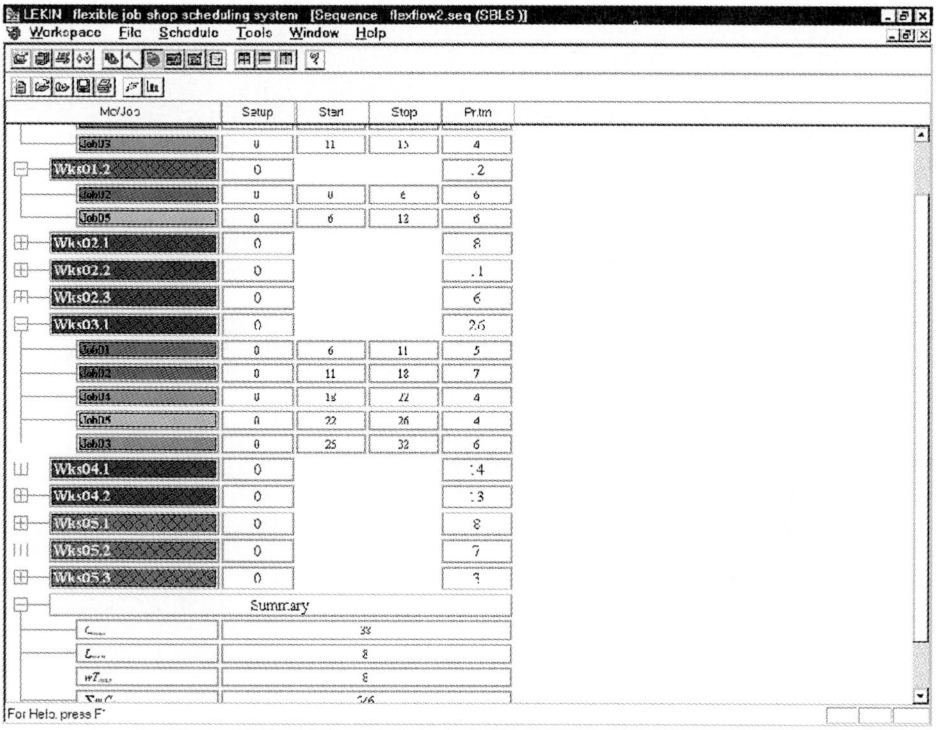

Figure 17.4   The dispatch list interface of the Lekin system.

the time, and so on. The Dispatch List interface in Figure 17.4 is from the Lekin system described in Chapter 19.

The third form of user interface is the *capacity buckets* interface (see Figure 17.5). The time axis is partitioned into a number of time slots or buckets. Buckets may be days, weeks, or months. For each machine, it is known what the processing capacity is for a bucket. The schedule generation, in certain scheduling environments, may be accomplished through the assignment of jobs to machines in given time segments. After such assignments are made, the capacity buckets interface displays the percentages of capacity utilized for each machine during each time segment. If the scheduler sees that a machine is overutilized during a given time period, he or she knows that the jobs in the corresponding bucket have to be rescheduled. The capacity buckets interface contrasts in a sense with the Gantt chart interface. A Gantt chart indicates the number of late jobs as well as their respective tardinesses. The number of late jobs and the total amount of tardinesses gives an indication of the deficiency in capacity. The Gantt chart is thus a good indicator of the available capacity in the short term (days or weeks) when there are a

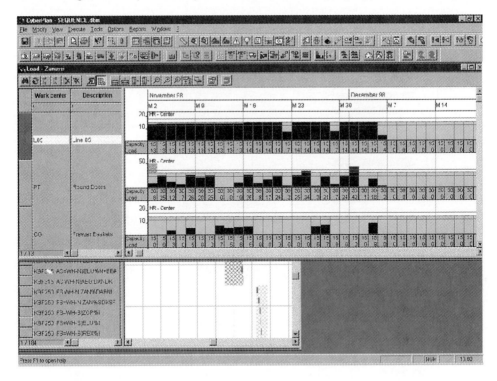

Figure 17.5   The capacity buckets interface of the Cyberplan system.

limited number of jobs (20 or 30). Capacity buckets are useful when the scheduler is performing some form of medium or long-term planning. The bucket size may be either a week or a month and the total period covered 3 or 4 months. Capacity buckets are, of course, a cruder form of information as they do not indicate which jobs are completed on time and which ones are late. The capacity buckets interface depicted in Figure 17.5 is from the Cyberplan System developed by Cybertec.

The fourth form of user interface is the *input–output diagram* or *through-put diagram* interface, which are often of interest when the production is made to stock. These diagrams describe cumulatively over time the total amount of orders received, the total amount produced, and the total amount shipped. The difference at any point in time between the first two curves is the total amount of orders waiting for processing, and the difference between the second and the third curves equals the total amount of finished goods in inventory. This type of interface speci-fies neither the number of late jobs nor their respective tardinesses. It does provide the scheduler with information regarding machine utilization and Work-in-Process (WIP).

Clearly, the different user interfaces for the display of schedule information have to be strongly linked with one another. When the scheduler makes changes in either the Gantt chart interface or the Dispatch List interface, the dynamic data may change considerably due to the cascading effects or reoptimization routines. Changes made in one interface, of course, have to be taken immediately into account in the other interfaces as well.

User interfaces for the display of schedule information have to be linked with other interfaces also (e.g., database management interfaces and schedule generation interfaces). For example, the scheduler may modify an existing schedule in the Gantt chart interface by clicking and dragging; then he may want to freeze certain jobs in their respective positions. After doing this, he may want to reoptimize the remaining jobs that are not frozen using one of the algorithms or procedures available in the schedule generation module. These algorithms are similar to the algorithms described in Part I for situations where machines are not available during given time periods (because of breakdowns or other reasons). Therefore, the schedule manipulation interfaces have to be strongly linked to the interfaces for algorithm selection.

Schedule manipulation interfaces may also have a separate window that displays the values of all the relevant performance measures. If the user has made a change in the schedule, the values before and after the change have to be displayed. Typically, performance measures are displayed in plain alphanumeric format. However, more sophisticated graphical displays may be used as well.

Some schedule manipulation interfaces are sophisticated enough to allow the user to split a job into a number of smaller segments and schedule each of these separately. Splitting an operation is equivalent to (possibly multiple) preemptions.

The more sophisticated schedule manipulation interfaces also allow different operations of the same job to overlap in time. This may occur in many situations in practice. For example, a job may start at a downstream machine of a flow shop before it has completed its processing at an upstream machine. In practice, this occurs when a job constitutes a large batch of identical items. Before the entire batch has been completed at an upstream machine, parts of the batch may already have been transported to the next machine and may have already started their processing there.

17.5 GENERIC SYSTEMS VERSUS APPLICATION-SPECIFIC SYSTEMS

Dozens of software houses have developed systems that they claim can be implemented in many different industrial settings with only minor modifications. It has turned out that the effort involved in customizing such systems can be quite substantial. The code developed for the customization of the system can be more than half the total code of the final version of the system. However, some systems have very sophisticated configurations that allow them to be tailored to different types of industries without much programming effort. These systems are highly modular and have an advantage in the adjustment to specific requirements. A generic system, if it is very modular, can be changed to fit a specific environment by adding specific modules (e.g., special scheduling algorithms). Specialists can develop these algorithms, and the generic scheduling software supplies standard interfaces or hooks that allow integration of the special functions in the package. This concept provides specialists the ability to concentrate on the scheduling problem while the generic software package supplies the functionalities that are less specific (e.g., user interfaces, data management, standard scheduling algorithms for less complex areas, etc.).

Generic systems may be built either on top of a commercial database system, such as Sybase or Oracle, or on top of a proprietary database system developed specifically for the scheduling system. Generic systems use processing data similar to the data presented in the deterministic scheduling framework in Chapter 2. However, the framework in such a scheduling system may be somewhat more elaborate than the framework presented in Chapter 2. For example, the database may allow for an alphanumeric *order* number that refers to the name of a customer. The order number then relates to a *number* of jobs, each one with its own processing time (that often may be referred to as the *quantity* of a batch) and a routing vector that determines the precedence constraints the operations are subject to. The order number has its own due date (committed shipping date), weight (priority factor), and release date (that may be determined by a *Material Requirements Planning* [*MRP*] system connected to the scheduling system). The scheduling system may include procedures that translate the due date of the order into due dates for the different jobs at the different workcenters. Also, the weights of the different jobs belonging to an order may not be exactly equal to the weight of the order. The weights of the different jobs may be a function of the amount of value already added to the product. The weight of the last job pertaining to an order is typically larger than the weight of the first job pertaining to that order.

The way the machine environment is represented in the database is also somewhat more elaborate than the way it is described in Chapter 2. Usually the system allows the specification of workcenters and, within each workcenter, the specification of machines.

Most generic scheduling systems have automatic scheduling routines to generate a first schedule for the user. Of course, such a schedule rarely satisfies the user. This is one of the reasons that scheduling systems often have elaborate user interfaces that allow the scheduler to manually modify an existing schedule. The automated scheduling capabilities generally consist of a number of different dispatching rules, which are basically sorting routines. These rules are usually the same as the priority rules discussed in the previous chapters (SPT, LPT, WSPT, EDD, etc.). Some generic scheduling systems have more elaborate scheduling procedures, such as *forward loading* or *backward loading.* Forward loading implies that the jobs are inserted one at a time starting at the beginning of the schedule—that is, at the current time. Backward loading implies that the schedule is generated starting from the back of the schedule—that is, from the due dates working its way toward the current time (again inserting one job at a time). These insertions, either forward or backward in time, are done according to some priority rule. Some of the more sophisticated automated scheduling procedures first identify the bottleneck workcenter(s) or machine(s); they compute time windows during which jobs have to be processed on these machines and schedule the jobs on these machines through some algorithmic procedure. After the bottlenecks are scheduled, the procedure schedules the remaining machines through either forward or backward loading.

Almost all generic scheduling systems have user interfaces that include Gantt charts and enable the scheduler to manipulate schedules manually. However, these Gantt chart interfaces are not always perfect. For example, most of them do *not* take into account the cascading and propagation effects referred to in the previous section. They may do some automatic rescheduling on the machine or workcenter on which the scheduler has made a change, but they usually do not adapt the schedules on other machines or workcenters to this change. Resulting schedules actually may be infeasible. Some scheduling systems give the scheduler a warning when the resulting schedule, after the scheduler has made modifications, is not feasible for one reason or another.

Besides the Gantt chart interface, most scheduling systems have at least one other type of interface that displays either the actual schedule or important data relating to the schedule. The second interface is typically one of those mentioned in the previous section.

Generic scheduling systems usually have fairly elaborate report generators. These report generators print out the schedule with alphanumeric characters; such printouts can be done fast and on an inexpensive printer. The printout may then resemble what is displayed—for example, in the Dispatch List interface described in the previous section. It is possible to list the jobs in the order in which they will be processed at a particular machine or workcenter. Besides the job number, other relevant data of each job may be printed out as well. Of course, there are also systems that print out entire Gantt charts. But Gantt charts have the disadvantage mentioned before—that it may not be immediately obvious which rectangle or bar corresponds to which job. Usually the bars are too small to display any information in or append any information at.

Generic systems have a number of advantages over application-specific systems. If the scheduling problem is a fairly standard problem and only a minor customization of a generic system suffices, then such an option is usually less expensive than developing an application-specific system from scratch. An additional advantage is that an established company will maintain the system. Yet most software houses that develop scheduling systems do not provide the source code. This makes the user of the system dependent on the software house even for very minor changes.

In many instances, generic systems are simply not suitable and application-specific systems (or modules) have to be developed. There are several good reasons for developing an application-specific system.

One reason is that the scheduling problem is simply so large, because of the number of machines, jobs, and attributes, that a PC-based generic scheduling system cannot handle it. The databases are very large, and the required interface between the shopfloor control system and the scheduling system are of a kind that a generic scheduling system does not have. An example of an environment where this is often the case is semiconductor manufacturing.

A second reason for opting for an application-specific system is that the scheduling environment has so many idiosyncrasies that a generic system simply cannot do the job. The processing environment may have certain restrictions or constraints that are hard to attach to or incorporate in a generic system. For example, certain machines at a workcenter have to start with their processing of different jobs at the same time (for one reason or another) or a group of machines may have to act as a single machine at some times and as separate machines at other times. The order portfolio may also have many idiosyncrasies. That is, there may be a fairly common machine environment used in a fairly standard way (that would fit nicely in a generic system), but with regard to the jobs there are too many exceptions to the rules. Coding in the special situations represents such a large amount of work that it may be advisable to build a system from scratch.

A third reason for developing an application-specific system is that the user may insist on having the source code and being able to maintain the system within his organization.

An important advantage of application-specific systems is that schedule manipulation is usually considerably faster and easier than with generic systems.

17.6 IMPLEMENTATION AND MAINTENANCE ISSUES

During the last two decades, with the advent of the PC in the factory, a large number of scheduling systems have been developed, and many more are under development. This process has made it clear that a large proportion of the theoretical research done during the past decades is of limited use in real-world applications. Fortunately, the system development underway in industry is encouraging theoretical researchers to tackle scheduling problems that are more relevant to the real world. At various academic institutions in Europe, Japan, and North America, research is focused on the development of algorithms as well as systems; significant efforts are being made in the integration of these developments.

In spite of the fact that during the last two decades many companies have made large investments in the development as well as in the implementation of scheduling systems, not that many systems appear to be used on a regular basis. Systems, after being implemented, often remain in use only for a limited time; after a while, they often are ignored altogether.

In those situations where the systems are in use on a more or less permanent basis, the general feeling is that the operations do run smoother. A system in place often does *not* reduce the time the scheduler spends on the scheduling process. However, a system usually does enable the scheduler to produce better schedules. Using an interactive schedule editor, the scheduler is able to compare different schedules and easily monitor the various performance measures. There are other reasons for smoother operations besides simply better schedules. A scheduling system imposes more discipline on the operations. There are compelling reasons now for keeping an accurate database. Schedules are printed out neatly and are visible on monitors. This apparently has an effect on people, encouraging them to follow the schedules.

The system designer should be aware of the reasons that some systems have never been implemented or are never used. In some cases, databases are not sufficiently accurate, and the team implementing the system does not have the patience or time to improve the database (the persons responsible for the database may be different from the people installing the scheduling system). In other cases, the way in which workers' productivity is measured is not in agreement with the performance criteria on which the system is based. User interfaces may not permit the scheduler to resequence quickly enough in the case of unexpected events. Procedures that enable resequencing when the scheduler is absent (e.g., if something unexpected happens during third shift) may not be in place. Finally, systems may not be given sufficient time to settle or stabilize in their environment (this may require many months, if not years).

Even if a system gets implemented and used, the duration during which it remains in use may be limited. Every so often, the machine environment in a factory may change drastically, and the system is not flexible enough to provide suitable schedules for the new environment. Even a change in the scheduler may derail a system.

Summarizing, the following points could be taken into consideration when designing, developing, and implementing a scheduling system:

1. Visualize how the operating environment will evolve over the lifetime of the system before the design process actually starts.

2. Get all the persons affected by the scheduling system involved in the design process. The development process must be a team effort, and all involved have to approve the design specifications.

3. Determine which part of the system can be handled by off-the-shelf software. Using appropriate commercial code speeds up the development process considerably.

4. Keep the design of the code modular. This is necessary not only to facilitate the entire programming effort, but also to facilitate changes in the system after its implementation.

5. Make the objectives of the algorithms embedded in the system consistent with the performance measures by which people who follow the schedules are being judged.

6. Do not take the data integrity of the database for granted. The system has to be able to deal with faulty or missing data and provide the necessary safeguards.

7. Capitalize on potential side benefits of the system (e.g., spin-off reports for distribution to key people). This enlarges the supporters' base of the system.

8. Make provisions to ensure easy rescheduling—not only by the scheduler, but also by others in case the scheduler is absent.

9. Keep in mind that the installment of the system requires patience. It may take months or even years before the system runs smoothly. This period should be a period of continuous improvement.

10. Do not underestimate the necessary maintenance of the system after installation. The effort required to *keep* the system in use on a regular basis is considerable.

It appears that, in the decade to come, even larger efforts will be made in the design, development, and implementation of scheduling systems and that such systems will play a more and more important role in manufacturing as well as services.

EXERCISES

17.1. Consider a job shop with machines in parallel at each workstation (i.e., a flexible job shop). There are hard as well as soft constraints that play a role in the scheduling of the machines. More machines may be installed in the near future. The scheduling process does not have to be done in real time, but can be done overnight. Describe the advantages and disadvantages of an algorithmic approach and of a knowledge-based approach.

17.2. Consider a factory with a single machine with sequence-dependent setup times and hard due dates. It does not appear that changes in the environment are imminent in the near future. Scheduling and rescheduling has to be done in real time.

(a) List the advantages and disadvantages of an algorithmic approach and of a knowledge-based approach.

(b) List the advantages and disadvantages of a commercial system and of an application-specific system.

17.3. Design a schedule generation module that is based on a composite dispatching rule for the $Pm \mid r_j, s_{jk}, M_j \mid \gamma$ environment. There are three objectives—namely, $\sum w_j T_j$, $C_{\max}$, and $L_{\max}$. Each objective has its own weight and the weights are time dependent; every time the scheduler uses the system, he puts in the relative weights of the various objectives. Design the composite dispatching rule and explain how the scaling parameters depend on the relative weights of the objectives.

17.4. Consider the following three measures of machine congestion over a given time period.

(i) the number of late jobs during the period,

(ii) the average number of jobs waiting in queue during the given period, and

(iii) the average time a job has to wait in queue during the period.

How does the selection of congestion measure depend on the objective to be minimized?

17.5. Consider the following scheduling alternatives:

(i) forward loading (starting from the current time),

(ii) backward loading (starting from the due dates), and

(iii) scheduling from the bottleneck stage first.

How does the selection of one of the three alternatives depend on the following factors?

(i) degree of uncertainty in the system,

(ii) workload balance (not one specific stage is a bottleneck), or

(iii) due date tightness.

17.6. Consider the ATC rule. The K factor is usually determined as a function of the due date tightness factor τ and the due date range factor R. However, the process usually requires extensive simulation. Design a learning mechanism that refines the function f that maps τ and R into K during the regular (possibly daily) use of the system's schedule generator. (*Hint:* Consider, for example, a two-dimensional grid of τ and R. For each combination of τ and R, there is a suggested value of K. Whenever the schedule generator is used the suggested value of K is applied. However, during the same run the values $K + \delta$ and $K - \delta$ are also tried. This implies that the system is applied to one instance three times. The values of the objectives under the three different applications are compared with one another; based on the three outcomes, the suggested value of K, for the given combination of τ and R, may be modified.)

17.7. Consider an interactive scheduling system with a user interface for schedule manipulation that allows freezing of jobs. That is, the scheduler can click on a job and freeze the job in a certain position. The other jobs have to be scheduled around the frozen jobs. Freezing can be done with tolerances so that in the optimization process of the remaining jobs the frozen jobs can be moved a little bit. This facilitates the scheduling of the unfrozen jobs. Consider a system that allows freezing of jobs with specified tolerances. Show that freezing in an environment that does not allow preemptions requires tolerances of at least half the maximum processing time in either direction to avoid machine idle times.

17.8. Consider an interactive scheduling system with a user interface that only allows for freezing of jobs with no (zero) tolerances.

(a) Show that in a nonpreemptive environment the machine idle times caused by frozen jobs are always less than the maximum processing time.

(b) Describe how procedures can be designed that minimize in such a scenario machine idle times in conjunction with other objectives, such as the total completion time.

17.9. Consider a user interface of an interactive scheduling system for a bank of parallel machines. Assume that the reoptimization algorithms in the system are designed in such a way that they optimize each machine separately while they keep the current assignment of jobs to machines unchanged. A move of a job (with the mouse) is said to be *reversible* if the move, followed by the reoptimization procedure, followed by the *reverse* move, followed once more by the reoptimization procedure, results in the original schedule.

Suppose a job is moved with the mouse from one machine to another. Show that such a move is reversible if the reoptimization algorithm minimizes the total completion time. Show that the same is true if the reoptimization algorithm minimizes the sum of the weighted tardinesses.

17.10. Consider the same scenario as in the previous exercise. Show that with the type of reoptimization algorithms described in the previous exercise, moves that take jobs from one machine and put them on another are *commutative*. That is, the final schedule does not depend on the sequence in which the moves are done even if all machines are reoptimized after each move.

COMMENTS AND REFERENCES

Many papers and books have been written on the various aspects of production information systems; see, for example, Scheer (1988), Gaylord (1987), and Pimentel (1990).

With regard to the issues concerning the overall development of scheduling systems, a relatively large number of papers have been written, often in proceedings of conferences; for example, Oliff (1988), Karwan and Sweigert (1989), and Interrante (1993). See also Kanet and Adelsberger (1987), Kusiak and Chen (1988), Solberg (1989), Kanet and Sridharan (1990), Adelsberger and Kanet (1991), Fraiman, Pinedo, and Yen (1993), and Pinedo, Samroengraja, and Yen (1994).

For research focusing specifically on knowledge-based scheduling systems, see Smith, Fox, and Ow (1986), Shaw and Whinston (1989), Atabakhsh (1991), Noronha and Sarma (1991), Lefrancois, Jobin, and Montreuil (1991), Smith (1992, 1994), and Kempf (1994).

For work on the design and development of user interfaces for scheduling systems, see Kempf (1989), Woerner and Biefeld (1993), and Wiers (1997).

18

Advanced Concepts in Scheduling System Design

This chapter focuses on a number of issues that have come up in recent years in the design, development, and implementation of scheduling systems. The first section covers issues concerning uncertainty, robustness, and reactive scheduling. In practice, schedules often have to be changed because of random events. To facilitate the rescheduling process, the original schedule has to be *robust*. This first section goes into the generation of robust schedules and the measurement of their robustness. The second section considers machine learning mechanisms. A system cannot consistently generate good schedules that are to the liking of the scheduler. The scheduler often has to tweak the schedule generated by the system to make it usable. A well-designed system can learn over time from the adjustments the scheduler makes in the schedules; the mechanism that allows the system to do so is called a learning mechanism. The third section focuses on the design of scheduling engines. A scheduling engine often contains a library of scheduling routines. Often one procedure may be more appropriate for one type of instance or data set, whereas another procedure may be more appropriate for another type of instance. The scheduler should be able to select, for each instance, which procedure to apply. It may even be the case that a scheduler may want to tackle an instance by combining various procedures. This third section discusses how a scheduling engine should be designed to enable the scheduler to adapt and combine algorithms to achieve maximum effectiveness. The fourth section goes into reconfigurable systems. Experience has shown that the development and implementation of systems is very time-consuming and costly. To reduce the costs, an effort must be made to make the systems as modular as possible. If the modules are well designed and sufficiently flexible, they can be used over and over again without major modifications. The fifth section focuses on design aspects of web-based scheduling systems. This section discusses the effects that networking has on the design of such systems. The last section discusses a number of other issues and presents a view of how scheduling systems may look in the future.

18.1 ROBUSTNESS AND REACTIVE SCHEDULING

In practice, it often happens that, soon after a schedule is generated, an unexpected event occurs that forces the scheduler to change the schedule. Such an event may be a machine breakdown or a rush job that suddenly has to be included. Many schedulers believe that in practice, most of the time, the scheduling process is a *rescheduling* process. Rescheduling is sometimes also referred to as *reactive scheduling*. In a reactive scheduling process, the scheduler tries to accomplish a number of objectives. He tries to accommodate the original objectives; he also tries to make the new schedule look, as much as possible, like the original schedule to minimize confusion.

One way of dealing with the rescheduling process is to put all the operations not yet started back in the hopper and generate a new schedule from scratch, taking into account the disruptions that just occurred. The danger is that the new schedule may be completely different from the original schedule, and too big a difference may cause confusion.

If the disruption is minor (e.g., the arrival of just one unexpected job), then a simple change may suffice. For example, the scheduler may insert the unexpected arrival in the current schedule in such a way that the total additional setup is minimized and no other high-priority job is delayed. A major disruption, like the breakdown of an important machine, often requires substantial changes in the schedule. If a machine goes down for an extended period of time, then the entire workload allocated to that machine over that period has to be transferred to other machines. This may cause extensive delays.

Another way to deal with the rescheduling process is to somehow anticipate the random events. To achieve this, it is necessary for the original schedule to be robust so that when disruptions occur, the necessary changes in the schedule are minimal.

Schedule robustness is a concept that is not easy to measure or even define. Suppose the completion time of a job is delayed by δ (because of a machine break-down or the insertion of a rush job). Let $C'_j(\delta)$ denote the new completion time of job j (i.e., the new time at which job j leaves the system), assuming the sequences of all the operations on all the machines remain the same. Of course, the new completion times of all the jobs are a function of δ. Let Z denote the value of the objective function before the disruption occurred and let $Z'(\delta)$ denote the new value of the objective function. So $Z'(\delta) - Z$ is the difference due to the disruption. One measure of schedule robustness may be

$$\frac{Z'(\delta) - Z}{\delta}.$$

This measure is a function of δ. For small values of δ the ratio may be low, whereas for larger values of δ it may get progressively worse. It is to be expected that this ratio is increasing in δ.

A more accurate measure of robustness can be established when the probabilities of certain events are known in advance. Suppose a perturbation of a random size Δ may occur and the probability that the random variable Δ takes the value δ—that is, $P(\Delta = \delta)$, is known. If Δ can assume only integer values, then

$$\sum_{\delta=0}^{\infty} \left(Z'(\delta) - Z \right) P(\Delta = \delta)$$

is an appropriate measure for the robustness. If the random variable Δ is a continuous random variable with a density function $f(\delta)$, then an appropriate measure is

$$\int_{\delta=0}^{\infty} (Z'(\delta) - Z) f(\delta) d\delta.$$

In practice, it may be difficult to make a probabilistic assessment of random perturbations; one may want to have more practical measures of robustness. For example, one measure could be based on the amount of slack between the completion times of the jobs and their respective due dates. So a possible measure for the robustness of schedule $\mathcal{S}$ is

$$\mathcal{R}(\mathcal{S}) = \frac{\sum_{j=1}^{n} w_j (d_j - C_j)}{\sum w_j d_j}.$$

The larger $\mathcal{R}(\mathcal{S})$, the more robust the schedule. Maximizing this particular measure of robustness is akin to maximizing the total weighted earliness.

When should a scheduler opt for a more robust schedule? This may depend on the probability of a disruption as well as his or her ability to reschedule.

Example 18.1.1 (Measures of Robustness)

Consider a single machine and three jobs. The job data are presented below.

jobs	1	2	3
p_j	10	10	10
d_j	10	22	34
w_j	1	100	100

The schedule that minimizes the total weighted tardiness is schedule 1, 2, 3 with a total weighted tardiness of 0. It is clear that this schedule is not that robust since two jobs with very large weights are scheduled for completion close to their respective due dates. Suppose that immediately after schedule 1, 2, 3 has been fixed a disruption occurs (i.e., at time $0 + \epsilon$), and the machine goes down for $\delta = 10$ time units. The machine can start processing the three jobs at time $t = 10$. If the original job sequence 1, 2, 3 is maintained, then the total weighted tardiness is 1410. The manner in which the total weighted tardiness of sequence 1, 2, 3 depends on the value of δ is depicted in Figure 18.1.

If the original schedule is 2, 3, 1, then the total weighted tardiness, with no disruptions, is 20. However, if a disruption does occur at time $0+\epsilon$, then the impact is considerably less severe than with schedule 1, 2, 3. If $\delta = 10$, then the total weighted tardiness is 30. The way the total weighted tardiness under sequence 2, 3, 1 depends

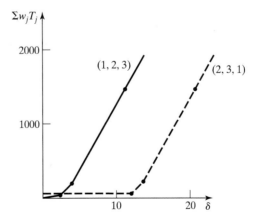

Figure 18.1   Increase in objective value as a function of disruption level (Example 18.1.1).

on δ is also depicted in Figure 18.1. From Figure 18.1, it is clear that schedule 2, 3, 1 (even though originally suboptimal) is more robust than schedule 1, 2, 3.

Under schedule 1, 2, 3, the robustness is

$$\mathcal{R}(1, 2, 3) = \frac{\sum_{j=1}^{n} w_j (d_j - C_j)}{\sum w_j d_j} = \frac{600}{5610} = 0.11,$$

whereas

$$\mathcal{R}(2, 3, 1) = \frac{2580}{5610} = 0.46.$$

So according to this particular measure of robustness, schedule 2, 3, 1 is considerably more robust.

Suppose that with probability 0.01 a rush job with processing time 10 arrives at time $0 + \epsilon$; the scheduler is not allowed, at the completion of this rush job, to change the original job sequence. If at the outset he had selected schedule 1, 2, 3, then the total expected weighted tardiness is

$$0 \times 0.9 + 1410 \times 0.1 = 141.$$

If he had selected schedule 2, 3, 1, then the total expected weighted tardiness is

$$20 \times 0.9 + 30 \times 0.1 = 21.$$

So with a 10% probability of a disruption, it is better to go for the more robust schedule.

Even if a scheduler is allowed to reschedule after a disruption, he still may not select at time 0 a schedule that is optimal with regard to the original data set.

Several other measures of robustness can be defined. For example, assume again that the completion of one job is delayed by δ. However, before computing the effect of the disruption on the objective, each machine sequence is reoptimized separately (i.e., the machine sequences are reoptimized one by one on a machine-by-machine basis). After this reoptimization, the difference in the objective function is computed. The measure of robustness is then the same ratio as the one defined previously. The impact of the disruption is now harder to compute since different values of δ may result in different schedules. This ratio is, of course, less than the ratio without reoptimization. An even more complicated measure of robustness assumes that, after a disruption, a reoptimization is done on a global basis rather than on a local basis (e.g., under this assumption, a disruption may cause an entire job shop to be reoptimized). Other measures of robustness may even allow preemptions in the reoptimization process.

Totally different measures of robustness can be defined based on the capacity utilization of the bottleneck machines (i.e., the percentages of time the machines are utilized) and on the levels of WIP inventory that are kept in front of these machines.

How can one generate robust schedules? Various rules can be followed when creating schedules.

  (i) Insert idle times.
 (ii) Schedule less flexible jobs first.
(iii) Do not postpone the processing of any operation unnecessarily.
 (iv) Keep always a number of jobs waiting in front of highly utilized machines.

The first rule prescribes the insertion of idle periods on given resources at certain points in time. This is equivalent to scheduling the machines below capacity. The durations of the idle periods as well as their timing within the schedule depend on the expected nature of the disruptions. One could argue that the idle periods in the beginning of the schedule may be kept shorter than the idle periods later in the schedule since the probability of an event occurring in the beginning may be smaller than later. In practice, some schedulers follow a rule whereby at any point in time in the current week the machines are scheduled up to 90% of capacity, the next week up to 80%, and the week after that up to 70%. However, one reason for keeping the idle periods in the beginning at the same length may be the following: Although the probability of a disruption is small, its relative impact is more severe than that of a disruption that occurs later on in the process.

The second rule suggests that less flexible jobs should have a higher priority than more flexible jobs. If a disruption occurs, then the more flexible jobs remain to be processed. The flexibility of a job is determined, for example, by the number of machines that can do the processing (e.g., the machine eligibility constraints described in Chapter 2). However, the flexibility of a job may also depend on its setup time structure. Some jobs may require setups that are sequence independent and always short. Other jobs may have sequence-dependent setup times that are highly variable. Their setups are short only when they follow certain other jobs; otherwise their setups are very long. Such jobs are clearly less flexible.

The third rule suggests that the processing of a job should not be postponed unnecessarily. From the point of view of inventory holding costs and earliness penalties, it is desirable to start operations as late as possible. From a robustness point of view, it may be desirable to start operations as early as possible. So there is a trade-off between robustness and earliness penalties.

The fourth rule tries to make sure that a bottleneck machine never starves because of random events occurring upstream. It makes sense to always have a number of jobs waiting for processing at a bottleneck machine. The reason is the following: If no inventory is kept in front of the bottleneck and the machine feeding it suddenly breaks down, then the bottleneck may have to remain idle and may not be able to make up for the lost time later.

Example 18.1.2 (Starvation Avoidance)

Consider a two machine flow shop with 100 identical jobs. Each job has a processing time of 5 time units on machine 1 and 10 time units on machine 2. Machine 2 is therefore the bottleneck. At each job completion on machine 1, the machine may have to undergo a maintenance service for a duration of 45 time units during which it cannot do any processing. The probability that such a service is required is 0.01.

The primary objective is the minimization of the makespan, and the secondary objective is the average amount of time a job remains in the system (i.e., the time in between the start of a job on machine 1 and its completion on machine 2; this secondary objective is basically equivalent to the minimization of the Work-in-Process). However, the weight of the primary objective is 1,000 times the weight of the secondary objective.

Because of the secondary objective, it does not make sense to let machine 1 process the 100 jobs one after another and finish them all by time 500. In an environment in which machine 1 never requires any servicing, the optimal schedule processes the jobs on machine 1 with idle times of five time units in between. In an environment in which machine 1 needs servicing with a given probability, it is necessary to have some jobs wait for machine 2. The optimal schedule is to keep consistently 5 jobs waiting for processing on machine 2. If machine 1 has to be serviced, then machine 2 does not lose any time and the makespan does not go up unnecessarily.

This example illustrates the trade-off between capacity utilization and minimization of Work-in-Process.

Robustness and rescheduling have a strong influence on the design of the user interfaces and the scheduling engine (multi-objective scheduling where one of the performance measures is robustness). Little theoretical research has been done on these issues. This topic may become an important research area in the future.

18.2 MACHINE LEARNING MECHANISMS

In practice, the algorithms embedded in a scheduling system often do not yield schedules that are acceptable to the user. The inadequacy of the algorithms is based on the fact that scheduling problems (which often have multiple objectives) are inherently intractable. It is extremely difficult to develop algorithms that can provide a reasonable and acceptable solution for any instance of a problem in real time.

New research initiatives are focusing on the design and development of learning mechanisms that enable scheduling systems in daily use to improve their schedule generation capabilities. This process requires a substantial amount of experimental work. A number of machine learning methods have been studied with regard to their applicability to scheduling. These methods can be categorized as follows:

- **(i)** rote learning,
- **(ii)** case based reasoning,
- **(iii)** induction methods and neural networks, and
- **(iv)** classifier systems.

These four classes of learning mechanisms are described in what follows in more detail.

Rote learning is a form of brute force memorization. The system saves old solutions that gave good results together with the instances on which they were applied. However, there is no mechanism for generalizing these solutions. This form of learning is only useful when the number of possible scheduling instances is limited (i.e., a small number of jobs of very few different types). It is not very effective in a complex environment, when the likelihood of a reoccurence of an instance is small.

Case-based reasoning attempts to exploit experience gained from past similar problem-solving cases. A scheduling problem requires the identification and generalization of salient features of past schedules, and a mechanism for determining which stored case is the most useful in the current context. Given the large number of interacting constraints inherent in scheduling, existing case indexing schemes are often inadequate for building the case base and subsequent retrieval, and new ways have to be developed. The following example shows how the performance of a composite dispatching rule can be improved through the use of a crude form of case-based reasoning (i.e., the so-called parameter adjustment method).

Example 18.2.1 (Case-Based Reasoning: Parameter Adjustment)

Consider the $1 \mid s_{jk} \mid \sum w_j T_j$ problem (i.e., there is a single machine with n jobs and the total weighted tardiness $\sum w_j T_j$ is the objective to be minimized). Moreover, the jobs are subject to sequence-dependent setup times s_{jk}. This problem has been considered in Example 14.2.1. A fairly effective composite dispatching rule for this scheduling problem is the ATCS rule. When the machine has completed the processing of job l at time t, the ATCS rule calculates the ranking index of job j as

$$I_j(t, l) = \frac{w_j}{p_j} \exp\left(-\frac{\max(d_j - p_j - t, 0)}{K_1 \bar{p}}\right) \exp\left(-\frac{s_{lj}}{K_2 \bar{s}}\right),$$

where $\bar{s}$ is the average setup time of the jobs remaining to be scheduled, K_1 the scaling parameter for the function of the due date of job j, and K_2 the scaling parameter for the setup time of job j. As stated in Chapter 14, the two scaling parameters K_1 and K_2 can be regarded as functions of three factors:

(i) the due date tightness factor τ,

(ii) the due date range factor R, and

(iii) the setup time severity factor $\eta = \bar{s}/\bar{p}$.

However, it is difficult to find appropriate functions that map the three factors into suitable values for the scaling parameters K_1 and K_2.

At this point, a learning mechanism can be of use. Suppose that in the scheduling system there are functions that map combinations of the three factors, τ, R, and η, in two values for K_1 and K_2. These do not have to be algebraic functions; they may be tables of data. When a scheduling instance is considered, the system computes τ, R, and η and looks in the current tables for the appropriate values of K_1 and K_2. (These values may have to be determined by means of an interpolation.) The instance is then solved using the composite dispatching rule with the values K_1 and K_2. The objective value of the generated schedule is also computed. However, in that same step, without any human intervention, the system also solves the same scheduling instance using values $K_1 + \delta$, $K_1 - \delta$, $K_2 + \delta$, $K_2 - \delta$ (various combinations). Since the dispatching rule is very fast, this can be done in real time. The performance measures of the schedules generated with the perturbed scaling parameters are then computed. If any of these schedules turns out to be substantially better than the one generated with the original K_1 and K_2, then there may be a reason for changing the mapping from the characteristic factors onto the scaling parameters. This can be done internally by the system without any input from the scheduler.

The learning mechanism described in the prior example is an online mechanism that operates without supervision. This mechanism is an example of case-based reasoning and can be applied to many multi-objective scheduling problems as well, even when a simple index rule does not exist.

The third class of learning mechanisms are of the induction type. The most common form of an induction type learning mechanism is a neural network. A neural net consists of a number of interconnected neurons or units. The connections among units have weights, which represent the strengths of the connections among the units. A multi-layer feedforward net is composed of input units, hidden units, and output units (see Figure 18.2). An input vector is processed and propagated through the network starting at the input units and proceeding through the hidden units all the way to the output units. The activation level of input unit i is set to the ith component of the input vector. These values are then propagated to the hidden units via the weighted connections. The activation level of each hidden unit is then computed by summing these weighted values and transforming the sum through a function f—that is,

$$a_h = f\left(q_h, \sum_i w_{ih} a_i\right),$$

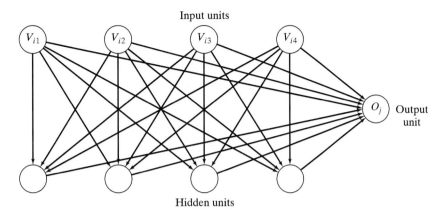

Figure 18.2   A four-layer neural network.

where a_h is the activation level of unit h, q_h is the bias of unit h, and w_{ih} is the connection weight between nodes i and h. These activation levels are propagated to the output units via the weighted connections between the hidden units and the output units and transformed again by means of the function f. The neural net's response to a given input vector is composed of the activation levels of the output units, which is referred to as the *output vector*. The dimension of the output vector does not have to be the same as the dimension of the input vector.

The knowledge of the net is stored in the weights, and there are well-known methods for adjusting the weights on the connections to obtain appropriate responses. For each input vector, there is a most appropriate output vector. A learning algorithm computes the difference between the output vector of the current net and the most appropriate output vector and suggests incremental adjustments to the weights. One such method is the so-called *backpropagation learning algorithm*.

The next example illustrates the application of a neural net to scheduling.

Example 18.2.2 (Neural Net for Parallel Machine Scheduling)

Consider the $Qm \mid r_j, s_{jk} \mid \sum w_j T_j$ problem. There are m nonidentical machines in parallel. Machine i has speed v_i. If job j is processed on machine i, then its processing time is $p_{ij} = p_j/v_i$. The jobs have different release dates and due dates. One of the objectives is to minimize the total weighted tardiness (note that the weights in the objective function are not related to the weights in the neural net). The jobs on a machine are subject to sequence-dependent setup times s_{jk}. There exists a partial schedule of jobs already assigned to each machine and new jobs arrive as time goes by. At each new release, it has to be decided on which machine the job must be processed. The neural net has to provide help with this decision.

Typically, the encoding of the data in the form of input vectors is crucial to the problem-solving process. In the parallel machines application, each input pattern represents a description of the attributes of a sequence on one of the machines. Examples of attributes are:

(i) Increase in the total weighted completion times of all the jobs already scheduled on the machine.

(ii) Number of additional late jobs on the machine.

(iii) Average additional latenesses of jobs already scheduled on the machine.

(iv) Current number of jobs on the machine.

However, the importance of each individual attribute is relative. For example, knowing that the number of jobs on a machine is five does not mean much without knowing the number of jobs on the other machines. Let N_{il} be attribute l with respect to machine i. Two transformations have to be applied to N_{il}.

Translation. The N_{il} value of attribute l under a given sequence on machine i is transformed as follows.

$$N'_{il} = N_{il} - (\min_i N_{il}) \quad i = 1, \ldots, m, \quad l = 1, \ldots, k.$$

In this way, $N'_{i^*l} = 0$ for the best machine, machine i^*, with respect to attribute l and the remaining values N'_{il} correspond to the differences with the best value.

Normalization. The N'_{il} value is transformed by normalizing over the maximum value in the context.

$$N''_{il} = N'_{il} / \max_i N'_{il} \quad i = 1, \ldots, m, \quad l = 1, \ldots, k.$$

These transformations make the comparisons of the input patterns corresponding to the different machines significantly easier. For example, if the value of attribute l is important in the decision-making process, then a machine with the lth attribute close to zero is more likely to be selected.

Attribute values are computed as follows. Each new job is first positioned where its processing time is the shortest (including the setup time immediately preceding and immediately following). After this insertion, the attribute values corresponding to that machine have to be recalculated.

A neural net architecture to deal with this problem can be of the structure described in Figure 18.2. This is a three-layer network with four input nodes (equal to the number of attributes k), four hidden nodes, and one output node. Each input node is connected to all the hidden nodes as well as the output node. The four hidden nodes are connected to the output node as well.

During the training phase of this network, an extensive job release process has to be considered—say 1,000 jobs. Each job release generates m input patterns (m being the number of machines) that have to be fed into the network. During this training process, the *desired* output of the net is set equal to 1 when the machine associated with the given input is selected by the expert for the new release and equal to 0 otherwise. For each input vector, there is a desired output and the learning algorithm has to compute the error or difference between the current neural net output and the desired output to make incremental changes in the connection weights. A well-known learning algorithm for the adjustment of connection weights is the backpropagation learning algorithm; this algorithm requires the choice of a so-called learning rate and a momentum term. At the completion of the training phase, the connection weights are fixed.

   Using the network after the completion of the training phase requires that every time a job is released the m input patterns are fed into the net; the machine associated with the output closest to 1 is then selected for the given job.

In contrast to the learning mechanism described in Example 18.2.1, the mechanism described in Example 18.2.2 requires off-line learning with supervision (i.e., training).

   The fourth class of learning mechanisms consists of the classifier systems. A common form of classifier system can be implemented via a genetic algorithm. However, a chromosome (or string) in such an algorithm now does not represent a schedule, but rather a list of rules (e.g., priority rules) that are to be used in the successive steps (or iterations) within an algorithmic framework designed to generate schedules for the problem at hand. For example, consider a framework for generating job shop schedules that is similar to Algorithm 7.1.3. However, modify this algorithm by replacing Step 3 with a priority rule that selects an operation from set Ω'. A schedule for the job shop can now be generated by doing nm successive iterations of this algorithm, where nm is the total number of operations. Every time a new operation has to be scheduled, a given priority rule is used to select the operation from the current set Ω'. The information that specifies which priority rule should be used in each iteration can be stored in a string of length nm for a genetic algorithm. The fitness of such a string is the value of the objective function that is achieved when all the (local) rules are applied successively in the given framework. This representation of a solution (by specifying rules), however, requires relatively intricate cross-over operators to get feasible off-spring. The genetic algorithm is thus used to search over the space of rules and not over the space of actual schedules. The genetic algorithm actually serves as a meta-strategy that optimally controls the use of priority rules.

18.3 DESIGN OF SCHEDULING ENGINES AND ALGORITHM LIBRARIES

A schedule generator in a scheduling system often contains a library of algorithmic procedures. Such a library may include basic dispatching rules, composite dispatching rules, shifting bottleneck techniques, local search techniques, branch and bound procedures, beam search techniques, mathematical programming routines, and so on. For a specific instance of a problem, one procedure may be more suitable than another. The appropriateness of a procedure may depend on the amount of CPU time available or the length of time the user is willing to wait for a solution.

The user of such a scheduling system may want to have a certain flexibility in the usage of the various types of procedures in the library. The desired flexibility may simply imply an ability to determine which procedure to apply to the given instance of the problem, or it may imply more elaborate ways of manipulating a number of procedures. A scheduling engine may have modules that allow the scheduler to

  (i) analyze the data of an instance and determine algorithmic parameters,

 (ii) set up algorithms in parallel,

(iii) set up algorithms in series,

 (iv) integrate algorithms.

For example, with an algorithm library at his disposal, a user would be able to do the following. The user may have statistical procedures that he can apply to the data set to obtain certain statistics, such as average processing time, range of processing times, due date tightness, setup time severity, and so on. Based on these statistics, the user can then select a procedure and specify appropriate levels for its parameters (e.g., scaling parameters, lengths of tabu-lists, beam widths, number of iterations, etc.).

If a user has a number of computers or processors at his disposal, he may want to apply different procedures simultaneously (i.e., in parallel) since he may not know in advance which one is the most suitable for the instance at hand. The different procedures function then completely independently from one another.

A user may also want to concatenate procedures (i.e., arrange procedures in series). That is, he or she would set the procedures up in such a way that the output of one serves as an input to another (e.g., the outcome of a dispatching rule serves as the initial solution for a local search procedure). The transfer of data from one procedure to the next is usually relatively simple. It may just be a schedule, which, in the case of a single machine, is just a permutation of the jobs. In the case of a parallel machine environment or a job shop environment, it may be a set of sequences, one for each machine.

Example 18.3.1 (Concatenation of Procedures)

Consider a scheduling engine that allows a scheduler to feed the outcome of a composite dispatching rule into a local search procedure. This means that the output of the first stage (i.e., the dispatching rule) is a complete schedule. The schedule is feasible and the starting and completion times of all the operations are determined. The output data of this procedure (and the input data for the next procedure) may contain the following information:

     (i) the sequence of operations on each machine,

     (ii) the start time and completion time of each operation, and

     (iii) the values of specific objective functions.

     The output data do not have to contain all the data listed above. For example, it may include only the sequence of operations on each machine. The second procedure (i.e., the local search procedure) may have a routine that can compute the start and completion times of all the operations given the structure of the problem and the sequences of the operations.

     Another level of flexibility allows the user not only to set procedures up in parallel or in series, but also to integrate procedures in a more complex manner. When different procedures are integrated within one framework, they do not work independently from one another; the effectiveness of one procedure may now depend on the input or feedback received from another procedure. Consider for a scheduling problem a branch and bound procedure or a beam search procedure. At each node in the search tree, one has to obtain either a lower bound or an estimate for the total penalty to be incurred by the remaining jobs. A lower bound can often be obtained as follows. One can assume that the remaining jobs can be scheduled while allowing preemptions. A preemptive version of a problem is often easier to solve than its nonpreemptive counterpart, and the optimal solution of the preemptive problem provides a lower bound for the optimal solution of the nonpreemptive version.

     Another example of an integration of procedures arises in decomposition techniques. A machine-based decomposition procedure is typically a heuristic designed for a complicated scheduling problem with many subproblems. A framework of the procedure for the overall problem can be constructed as in Chapter 7. However, the user may want to specify, knowing the particular problem or instance, which procedure to apply to solve the subproblem.

     If procedures have to be integrated, then one often has to work within a general framework (sometimes also referred to as a control structure) in which one or more specific types of subproblems have to be solved many times. The user may want to have the ability to specify certain parameters within this framework. For example, if the framework is a search tree for a beam search, then the user would like to be able to specify the beam width as well as the filter width. The subproblem that has to be solved at each node of the search tree has to yield (with little computational effort) a good estimate for the contribution of the unscheduled jobs to the overall objective.

     The transfer of data between procedures in an integrated framework may be complicated. It may be that data concerning a subset of jobs or a subset of operations have to be transferred. It may also be that the machines are not available at all times. The positions of the jobs already scheduled on the various machines may be fixed and the procedure that has to schedule the remaining jobs has to know at what times the machines are still available. If there are sequence-dependent setup times, then the procedure also has to know which job was the last one processed on each machine just to be able to compute the sequence-dependent setup time for the next job.

Example 18.3.2 (Integration of Procedures in a Branching Scheme)

Consider a branch and bound approach for $1 \mid s_{jk}, brkdwn \mid \sum w_j T_j$. The jobs are scheduled in a forward manner (i.e., a partial schedule consists of a sequence of jobs that starts at time zero). At each node of the branching tree, a bound has to be obtained for the total weighted tardiness of the jobs still to be scheduled. If a procedure is called to establish a lower bound for all schedules that are descendants of any particular node, then the following input data have to be provided:

   (i) the set of jobs already scheduled and the set of jobs still to be scheduled,

  (ii) the time periods that the machine remains available, and

 (iii) the last job in the current partial schedule (to determine the sequence dependent setup time).

The output data of the procedure may contain a sequence of operations as well as a lower bound. The required output may be just a lower bound; the actual sequence may not be of interest.

If there are no setup times, then a schedule can also be generated in a backward manner (since the value of the makespan is then known in advance).

Example 18.3.3 (Integration of Procedures in a Decomposition Scheme)

Consider a shifting bottleneck framework for a flexible job shop with a number of machines in parallel at each workcenter.

At each iteration, a subset of the workcenters has already been scheduled and an additional workcenter has to be scheduled. The sequences of the operations at the workcenters already scheduled imply that the operations of the workcenter to be scheduled in the subproblem are subject to delayed precedence constraints. When the procedure for the subproblem is called, a certain amount of data has to be transferred. These data may include:

   (i) the release date and due date of each operation,

  (ii) the precedence constraints between the various operations, and

 (iii) the necessary delays that go along with the precedence constraints.

The output data consist of the sequence of the operations as well as their start and completion times. It also contains the values of given performance measures.

It is clear that the type and structure of the information transferred back and forth is now more complicated than in a simple concatenation of procedures.

These forms of integration of procedures have led to the development of so-called *scheduling description languages*. A scheduling description language is a high-level language that enables a scheduler to write the code for a complex integrated algorithm with only a limited number of concise statements or commands. Each statement in a description language involves the application of a relatively powerful procedure. For example, a statement may carry the instruction to apply a tabu-search procedure on a given set of jobs in a given machine environment. The input to such a statement consists of the set of jobs, the machine environment, the processing restrictions and constraints, the length of the tabu-list, an initial schedule, and the total number of iterations. The output consists of the best schedule obtained with the tabu-search procedure. Other statements may be used to set up various different procedures in parallel or concatenate two different procedures.

18.4 RECONFIGURABLE SYSTEMS

The last two decades have witnessed the development of a large number of scheduling systems in industry and academia. Some of these systems are application-specific, whereas others are generic. Application-specific systems typically do somewhat better in the implementation process than customized generic systems. However, application-specific systems are often hard to modify and adapt to changing environments. Generic systems are usually somewhat better designed and more modular. However, any customization of such systems typically requires significant investments.

Considering the experience of the last two decades, it appears useful to provide guidelines that facilitate and standardize the design and development of scheduling systems. Efforts have to be made to provide guidelines as well as tools for the development of scheduling systems. The most recent designs tend to be object oriented.

There are many advantages in following an object-oriented design approach for the development of a scheduling system. First, the design is modular, which makes maintenance and modification of the system relatively easy. Second, large segments of the code are reusable. This implies that two scheduling systems that are substantially different still may share a significant amount of code. Third, the designer thinks in terms of the behavior of objects, not in lower level detail. In other words, the object-oriented design approach can speed up the design process and separate the design process from its implementation.

Object-oriented systems are usually designed around two basic entities—namely, *objects* and *methods*. Objects refer to various types of entities or concepts. The most obvious ones are jobs and machines. However, a schedule is also an object and so are user interface components, such as buttons, menus, and canvases.

There are two basic relationships between object types—namely, the *is–a* relationship and the *has–a* relationship. According to an is–a relationship, an object type is a special case of another object type. According to a has–a relationship, an object type may consist of other object types. Objects usually carry along static information, referred to as attributes, and dynamic information, referred to as the state of the object. An object may have several attributes that are descriptors associated with the object. An object may be in any one of a number of states. For example, a machine may be *busy*, *idle*, or *broken down*. A change in the state of an object is referred to as an *event*.

A method is implemented in a system by means of one or more operators. Operators are used to manipulate the attributes corresponding to objects and may result in changes of object states (i.e., events). However, events may trigger operators as well. The sequence of states of the different objects can be described by a state-transition or event diagram. Such an event diagram may represent the links between operators and events. An operator may be regarded as the way in which a method is implemented in the software. Any given operator may be part of several methods. Some methods may be very basic and can be used for simple manipulations of objects (e.g., a pairwise interchange of two jobs in a schedule). Others may be very sophisticated, such as an intricate heuristic for use on a given set of jobs (objects) in a given machine environment (also objects). The application of a method to an object usually triggers an *event*.

The application of a method to an object may cause information to be transmitted from one object to another. Such a transmission of information is usually referred to as a *message*. Messages represent information (or content) that is transmitted from one object (e.g., a schedule) via a method to another object (e.g., a user interface display). A message may consist of simple attributes or an entire object. Messages are transmitted when events occur (caused by the application of methods to objects). Messages have also been referred to in the literature as memos. The transmission of messages from one object to another can be described by a transition event diagram and requires the specification of protocols.

A scheduling system may be object oriented in its conceptual design and/or in its development. A scheduling system is object oriented in its conceptual design if the design of the system is object oriented throughout. This implies that every concept used and every functionality of the system is either an object or a method of an object (whether it is in the data or knowledge-base, algorithm library, schedule generator, or user interfaces). Even the largest modules within the system are objects, including the algorithm library and user interface modules. A scheduling system is object oriented in its development if only the more detailed design aspects are object oriented and the code is based on a programming language with object-oriented extensions such as C++.

Many scheduling systems developed in the past have object-oriented aspects and tend to be object oriented in their development. A number of these systems also have conceptual design aspects that are object oriented. Some rely on inference engines for the generation of feasible schedules and others are constraint based relying on constraint propagation algorithms and search. These systems usually do not have schedule generators that perform very sophisticated optimization.

Many systems developed in the past are not designed throughout according to an object-oriented philosophy. Some aspects that are typically not object-oriented are:

(i) the design of the schedule generators,

(ii) the design of the user interfaces, and

(iii) the specification of the precedence, routing and layout constraints.

Few existing schedule generators have extensive libraries of algorithms that are easily reconfigurable and would benefit from a modular object-oriented design (an object-oriented design would require a detailed specification of operators and methods). Since most scheduling environments would benefit from highly interactive optimization, schedule generators have to be strongly linked to interfaces that allow schedulers to manipulate schedules manually. Still object-oriented design has not had a major impact on the design of user interfaces for scheduling systems. The precedence constraints, routing constraints, and machine layout constraints are often represented by rules in a knowledge-base, and an inference engine must generate a schedule satisfying the rules. However, these constraints can be modeled conceptually easily using graph and tree objects that then can be used by an object-oriented schedule generator.

18.5 SCHEDULING SYSTEMS ON THE INTERNET

With the ongoing development in information technology, conventional single-user, stand-alone systems have become available on networks and even on the Internet. Basically there are three classes of Internet-based systems:

(i) information access systems,

(ii) information coordination systems, and

(iii) information processing systems.

In information access systems, information can be retrieved and shared through the Internet, EDI, or other electronic systems. The server acts as an information repository and distribution center, such as a homepage on the Internet.

In information coordination systems, information can be generated as well as retrieved by many users (clients). The information flows go in many directions, and the server can synchronize and manage the information, such as in project management and electronic markets.

In information processing systems, the servers can process the information and return the results of this processing to the clients. In this case, the servers function as application programs that are transparent to the users.

Internet scheduling systems are information-processing systems that are similar to the interactive scheduling systems described in previous sections, except that an Internet scheduling system is usually a strongly distributed system. Because of the client–server architecture of the Internet, the three major components of a scheduling system (i.e., its database, scheduling engine, and user interface), have to be adapted. In what follows, we discuss the typical design features of Internet scheduling systems.

The advantages of scheduling system servers on the Internet are the following. First, the input–output interfaces (used for the graphical displays) can be supported by local hosts rather than servers at remote sites. Second, the scheduling server as well as the local clients can handle the data storage and manipulation. This may alleviate the workload at the server sites and give local users the capability and flexibility to manage the database. Third, multiple scheduling servers can jointly solve large-scale and complicated scheduling problems either partially or in their entirety using their distributed computational resources.

While retaining all the functions inherent in an interactive scheduling system, the main components of the scheduling system have to be restructured to comply with the client–server architecture and to achieve the advantages listed before. This restructuring affects the design of the database, scheduling engine, and user interface.

The design of the database has the following characteristics. Both the process manager and schedule manager reside at the servers. However, some data can be kept at the client for display or further processing. Both the Gantt chart and the dispatch lists are representations of the schedule generated by the scheduling engine. The local client can cache the results for fast display and further processing, such as editing. Similarly, both the server and client can process the information. Figure 18.3 exhibits the information flow between the scheduling server and local clients. A client may have a general purpose database management system (such as Sybase or Excel) or an application-specific scheduling database for data storage and manipulation.

The design of the scheduling engine has the following characteristics. A local client can select a suitable algorithm from a library that resides at a server to solve his problem. Typically, there is not a single algorithm that can solve his scheduling problem satisfactorily, and a user may want to construct a composite procedure

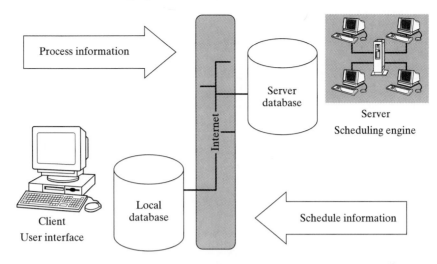

Figure 18.3   Information flow between client and server.

based on several of the algorithms available in that library. The server or client algorithm generator is a workplace for the users to construct new composite procedures. Figure 18.4 shows how a new composite procedure can result in a new algorithm that can then be included in both the server and client libraries. This local workplace can speed up the process of constructing intermediate and final composite methods and extend the server and client libraries at the same time.

The Internet also has an effect on the design of the user interfaces. Using existing Internet support, such as HTML (HyperText Markup Language), Java, Java script, Perl, and CGI (Common Gateway Interface) functions, the graphical user interfaces of scheduling systems can be implemented as library functions at the server sites. Through the use of appropriate browsers (such as Netscape), users can enter data or view schedules with a dynamic hypertext interface. Moreover, a user can also develop interface displays that link server interface functions to other applications. In Figure 18.3 it is shown how display functions can be supported either by remote servers or local clients.

Thus, it is clear that servers can be designed to fully support local clients solving scheduling problems. The local clients can manipulate data and construct new scheduling methods. The scheduling servers function as regular interactive scheduling systems except that they can be used in a multi-user environment on the Internet.

Internet scheduling systems can be used in several ways: unionization and personal customization. Unionization means that an Internet scheduling system can be used in a distributed environment. A distributed scheduling system can exchange information efficiently and collaborate effectively in the solution of hard

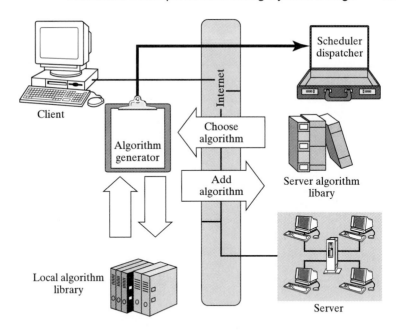

Figure 18.4   Process of constructing new procedures.

scheduling problems. Personal customization implies that a scheduling system can be customized to satisfy an individual user's requirements. Different users may have different requirements since each has his own way of using information, applying scheduling procedures, and solving problems. Personalized scheduling systems can provide shortcuts and improve system performance.

With the development of Internet technology and client–server architectures, new tools can be incorporated in scheduling systems for solving large-scale and complicated problems. It appears that Internet scheduling systems may lead to viable personalized interactive scheduling systems.

18.6 DISCUSSION

Many teams in industry and academia are currently developing scheduling systems. The database (or object base) management systems are usually off the shelf, developed by companies that specialize in these systems (e.g., Oracle). These commercial databases are typically not specifically geared for scheduling applications; they are of a generic nature.

Dozens of software development and consulting companies specialize in scheduling applications. They may even specialize in certain niches (e.g., scheduling applications in the process or microelectronics industries). Each of these companies has its own systems with elaborate user interfaces and its own way of doing interactive optimization.

Research and development in scheduling algorithms and learning mechanisms will most likely only take place in academia or in large industrial research centers. This type of research needs extensive experimentation; software houses often do not have the time for such developments.

In the future, the Internet may allow for the following types of interaction between software companies and universities that develop scheduling systems on the one side and companies that need their scheduling services (the clients) on the other side. A client may use a scheduling system that is made available on the web, enter its data, and run the system. The system gives the client the values of all the performance measures of the schedule it has generated. However, the client cannot yet see the schedule. If the performance measures of the schedule are acceptable to the client, then he may decide to purchase the schedule from the company that owns the system.

EXERCISES

18.1. One way to construct robust schedules is by inserting idle times. Describe all the factors that influence the timing, frequency, and duration of the idle periods.

18.2. Consider all nonpreemptive schedules of n jobs on a single machine. Define a measure for the distance (or the difference) between two schedules.
  (a) Apply the measure when the two schedules consist of the same set of jobs.
  (b) Apply the measure when one set of jobs has one more job than the other set.

18.3. Consider the same set of jobs as in Example 18.1.1. Assume that there is a probability p that the machine needs servicing starting at time 2. The servicing takes 10 time units.
  (a) Assume that neither preemption nor resequencing is allowed (i.e., after the servicing has been completed, the machine has to continue processing the job it was processing before the servicing). Determine the optimal sequence(s) as a function of p.
  (b) Assume preemption is not allowed, but resequencing is allowed. That is, after the first job has been completed, the scheduler may decide not to start the job he originally scheduled to go second. Determine the optimal sequence(s) as a function of p.
  (c) Assume preemption as well as resequencing are allowed. Determine the optimal sequence(s) as a function of p.

18.4. Consider an instance of $P2 \mid p_j = 1, M_j, brkdwn \mid C_{\max}$ with two jobs. At each point in time, each machine has a probability of 0.5 of breaking down for one time unit. Job 1 can only be processed on machine 1, whereas job 2 can be processed on either one of the two machines. Compute the expected makespan under the Least Flexible Job first (LFJ) rule and under the Most Flexible Job first (MFJ) rule.

18.5. Consider the problem $1 \mid s_{jk} \mid \gamma$. Define a measure of job flexibility that is based on the setup time structure.

18.6. Consider a flow shop with limited intermediate storages that is subject to a cyclic schedule as described in Section 16.2. Machine i now has at the completion of each operation a probability p_i that it goes down for an amount of time x_i.

    (a) Define a measure for the congestion level of a machine.

    (b) Suppose that originally there are no buffers between machines. Now k buffer spaces can be inserted between the m machines and the allocation has to be done in such a way that the schedules are as robust as possible. How does the allocation of the buffer spaces depend on the congestion levels at the various machines?

18.7. Explain why rote learning is an extreme form of case-based reasoning.

18.8. Describe the input attributes in a neural net approach for $Rm \mid r_j, M_j \mid \sum w_j T_j$.

18.9. Describe how a branch and bound approach can be implemented for $Pm \mid s_{jk} \mid \sum w_j T_j$. That is, generalize the discussion in Example 18.3.2 to parallel machines.

18.10. Consider Example 18.3.3 and Exercise 18.9. Integrate the ideas presented in an algorithm for the flexible job shop problem.

18.11. Consider a scheduling description language that includes statements that can call different scheduling procedures for the problem $Pm \mid r_j \mid \sum w_j T_j$. Write the specifications for the input and output data for three statements that correspond to three procedures of your choice. Develop also a statement for setting the procedures up in parallel and a statement for setting the procedures up in series. Specify for each one of these last two statements the appropriate input and output data.

18.12. Suppose a scheduling description language is used for the coding of the shifting bottleneck procedure. Describe the type of statements that are required for such a code.

COMMENTS AND REFERENCES

As is evident from Part II of this book, there is an extensive literature on scheduling under uncertainty (i.e., stochastic scheduling). However, the literature on stochastic scheduling, in general, does not address the issue of robustness per se. However, robustness concepts have received some special attention in the literature; see, for example, the work by Leon and Wu (1994), Leon, Wu and Storer (1994), Metha and Uzsoy (1999), Wu, Storer, and Chang (1991), and Wu, Byeon, and Storer (1999). For an overview of research in reactive scheduling, see the excellent survey by Smith (1992). For a more recent analysis of reactive scheduling in job shops, see Sabuncuoglu and Bayiz (2000). For an industrial application of reactive scheduling, see Elkamel and Mohindra (1999).

    Research on learning mechanisms in scheduling systems started in the late 1980s; see, for example, Shaw (1988b), Shaw and Whinston (1989), Yih (1990), Shaw, Park, and Raman (1992), and Park, Raman, and Shaw (1997). The parametric adjustment method for the ATCS rule in Example 18.2.1 is due to Chao and Pinedo (1992). An excellent overview of learning mechanisms in scheduling systems is presented in Aytug, Bhattacharyya, Koehler, and Snowdon (1994). The book by Pesch (1994) focuses on learning in scheduling through genetic algorithms (classifier systems).

A fair amount of development work has been done recently on the design of adaptable scheduling engines. Akkiraju, Keskinocak, Murthy, and Wu (1998a, 1998b) discuss the design of an agent-based approach for a scheduling system developed at IBM. Feldman (1999) describes in detail how algorithms can be linked and integrated, and Webster (2000) presents two frameworks for adaptable scheduling algorithms.

The design, development, and implementation of modular or reconfigurable scheduling systems is often based on objects and methods. For objects and methods, see Booch (1994), Martin (1993), and Yourdon (1994). For modular design with regard to databases and knowledge bases, see, for example, Collinot, LePape, and Pinoteau (1988), Fox and Smith (1984), Smith (1992), and Smith, Muscettola, Matthys, Ow, and Potvin (1990). For an interesting design of a schedule generator, see Sauer (1993). A system design proposed by Smith and Lassila (1994) extends the modular philosophy for scheduling systems farther than any previous system. This is also the case with the approach by Yen (1995) and Pinedo and Yen (1997).

The material concerning Internet scheduling systems is based on the paper by Yen (1997).

19

Examples of System Designs
and Implementations

From the previous chapters, it is evident that there are many different types of scheduling problems. It is not likely that a single system can be designed that is applicable to just about any scheduling problem with only minor customization. This suggests that there is room as well as a need for many different scheduling systems. The variety of available platforms, databases, graphic user interfaces (GUIs), and networking capabilities enlarges the number of possibilities even more.

This chapter describes the architectures and implementations of six scheduling systems. The first system is the Advanced Planner and Optimizer (APO) of SAP from Germany. The APO system is a flexible system that can be easily adapted to many industrial settings. The second system is developed at IBM's T.J. Watson Research Center. This system has been installed at a number of sites, primarily in the paper industry. The third system is the TradeMatrix Production Scheduler system developed by i2 Technologies based in Dallas, Texas. At the moment, this system may actually be the most widely used in industry. The fourth system is an application-specific system developed by Cybertec from Italy for a facility that produces jars of yogurt. The fifth system is the virtual production engine by SynQuest. This system covers various levels of planning and scheduling, including strategic, tactical, and operational. The sixth and last system is an academic system that has been developed at New York University (NYU) for educational purposes. This system has been in use at many universities all over the world for several years.

19.1 THE SAP–APO SYSTEM

The main product of the software company SAP is its ERP system, R/3. In 1998, the company decided to develop its own planning and scheduling systems for supply chain management and not depend on alliances with third parties. This development is called the SAP Advanced Planner and Optimizer (APO).

SAP–APO provides a suite of specially tailored optimization routines for each one of the blocks in the cube shown in Figure 19.1. A dual modeling approach allows for aggregate planning (on the strategic and tactical levels) via LP solvers as well as detailed planning (on the operational level) via scheduling algorithms. SAP–APO allows for medium term planning as well as detailed scheduling of all

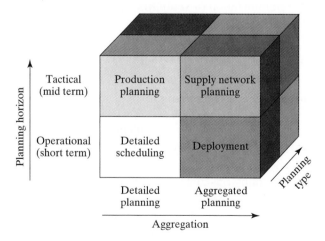

Figure 19.1   SAP-AOP classification of planning problems.

the activities in the supply chain. The timing of all the activities for each one of the individual resources in the supply chain (transportation units, production machinery) can be determined with a high level of precision. Several schedulers can work simultaneously on the same plan, each one focusing on a different part of the supply chain.

The medium term planning procedures take a certain level of detail into account while considering the degree of uncertainty in the planning data (due to the unexpected changes that occur before the plan becomes reality: new orders, order cancellations, forecast errors, market changes, etc.).

The medium term planning problem is solved via a linear programming (LP) relaxation. This LP relaxation is solved with the CPLEX LP solver of the ILOG library. To deal with integrality constraints, SAP–APO has a discretization heuristic that can take into account the actual meaning of each one of the integer decision variables. After solving the LP relaxation, the variables are stepwise discretized using in each step again an LP relaxation. The discretization process is done gradually: Lot sizes for later time buckets are discretized later. The planning problem may include linear as well as integer constraints. Linear constraints may be necessary because of due date constraints, maximum delay constraints, storage capacity constraints, and so on. The integer constraints may be necessary because of minimal lot sizes, full truck capacities, and so on. Such optimization problems are modeled as Mixed Integer Programs (MIPs).

The detailed scheduling problem is modeled in its most generic way as a so-called Multi-Mode Resource Constrained Project Scheduling Problem with minimum and maximum time lags. Maximum time constraints such as deadlines or shelf life (expiration dates), storage capacities, sequence-dependent setup times, precedence constraints, processing interruptions due to breakdowns, and objectives (such as the minimization of setup times, setup costs, and due date delays) can all be included.

SAP–APO provides a generic framework that integrates a suite of scheduling algorithms that includes the following three basic approaches (see Figure 19.2):

  (i) Constraint Based Programming,
 (ii) Genetic Algorithms, and
(iii) Repair Algorithms.

The Constraint Based Programming (CBP) scheduler is the most generic scheduler in the SAP–APO library. It is based on the ILOG libraries that use dynamic constraint propagation in each step.

The Genetic Algorithms (GA) are based on the evolutionary approach. The genetic representation contains the schedule information used by a fast scheduler

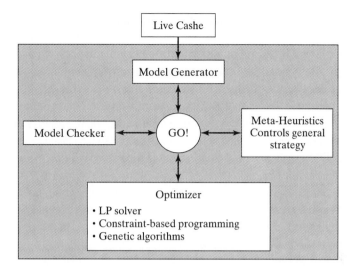

Figure 19.2   SAP-APO optimizer architecture.

for generating new solutions. Because its scheduler uses no dynamic constraint propagation and only limited backtracking, this approach has limitations on the use of maximal time constraints.

The Repair Algorithms (RA) can solve a resource relaxation problem in polynomial time. The resource capacity constraints are enforced by adding precedence constraints that deal dynamically with capacity violations and then continue with the resource relaxation.

The performance of each type of algorithm depends on the setting as well as the instance under consideration. The user of the system may select the most appropriate algorithm after some experimental analysis.

In its generic framework, SAP–APO provides the option to combine the algorithms described earlier using at the same time one or more decomposition techniques. The decomposition techniques enable the user to partition a problem instance according to

(i) workcenter or machine,

(ii) time,

(iii) product type or job, and

(iv) job priority.

The decomposition techniques enable a user to also scale the neighborhood up or down (i.e., the user can adjust the decomposition width). SAP–APO provides a feature that allows fine-tuning of the decomposition width.

19.2 IBM'S INDEPENDENT AGENTS ARCHITECTURE

Production scheduling and distribution of paper and paper products is an extremely complex task that must take into account numerous objectives and constraints. The complexity of the problem is compounded by the interactions between the schedules of consecutive stages in the production process. Often a good schedule of the jobs at one stage is a terrible schedule at the next stage. This makes the overall optimization problem hard.

IBM developed a cooperative multi-objective decision-support system for the paper industry. The system generates multiple scheduling alternatives using several types of algorithms that are based on linear and integer programming, network flow methods, and heuristics. To generate multiple solutions and facilitate the interaction with the scheduler, IBM's system is implemented using the agent-based Asynchronous Team (A-Team) architecture, in which multiple scheduling methods cooperate in the generation of a shared population of schedules (see Figures 19.3 and 19.4).

There are three types of agents in an A-Team:

(i) constructors,
(ii) improvers, and
(iii) destroyers.

Constructors have as input a description of the problem. Based on this description, they generate new schedules. Improvers attempt to improve on the schedules in the current population by modifying and combining existing schedules. Destroyers attempt to keep the population size in check by removing bad

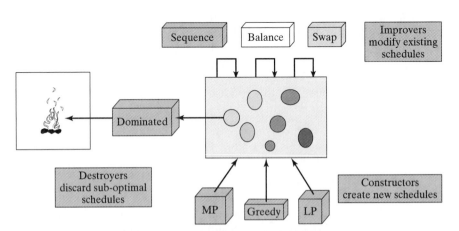

Figure 19.3   The essential features of an IBM A-Team.

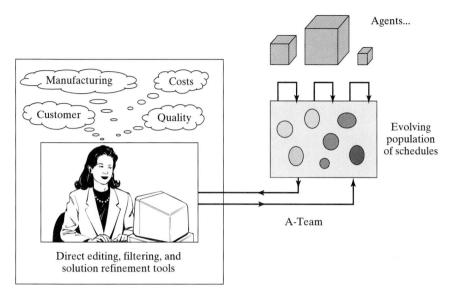

Figure 19.4   Interactive optimization with IBM's A-Team.

schedules from the population. A human scheduler can interact with the A-Team by assuming the role of an agent; the scheduler can add new schedules, modify and improve existing schedules, or remove existing schedules.

One of the more important algorithms in the system considers a set of jobs in an environment with a number of nonidentical machines in parallel. The jobs have due dates and tardiness penalties and are subject to sequence-dependent setup times. The problem setting is similar to the $Rm \mid r_j, s_{jk} \mid \sum w_j T_j$ model. For this particular problem, IBM's algorithmic framework provides constructor agents as well as improver agents.

There are basically two types of constructor agents for this problem—namely:

 (i) Agents that allocate jobs to machines and do an initial sequencing.
 (ii) Agents that do a thorough sequencing of all the jobs on a given machine.

The algorithmic framework used for the job-machine allocation and initial sequencing is called allocate-and-sequence. The objective is to optimize the allocation of the jobs to the machines based on due dates, machine load balance, and transportation costs considerations (the machines may be at various different locations).

The main ideas behind the allocate-and-sequence heuristic include a sorting of all the jobs according to certain criteria (e.g., due date, processing time, tardiness penalty). The job at the top of the list of remaining jobs is allocated to the machine with the lowest index $I_1(i, j, k)$ of the machine-job (i, j) pair. The index $I_1(i, j, k)$ of job j on machine i is computed by considering several combinations of job and machine properties (assuming that job j is processed on machine i immediately after job k). Some examples of indexes used in actual implementations are:

(i) The processing time of job j on machine i plus the setup time required between jobs k and j.

(ii) The completion time of job j.

(iii) The weighted tardiness of job j.

(iv) The absolute difference between the due date of job j and its completion time on machine i.

The allocate-and-sequence heuristic can be summarized as follows.

Algorithm 19.2.1. (Allocate and Sequence Heuristic)

Step 1. *Rank the jobs according to the given sort criteria.*

Step 2. *Select the next job on the list.*
Compute the machine-job index I, (i, j, k) for all machines that can process this job.
Assign the job to the machine with the lowest machine-job index.

The goal of the second type of constructor agent is to sequence the jobs on a given machine based on their due dates and then group jobs of the same kind in batches (in the case of paper machine scheduling, this is analogous to the aggregation of orders for items of the same grade). The algorithmic framework used for sequencing jobs on a given machine is called single-dispatch. Assume the set of jobs allocated to machine i is already known (as an output of allocate-and-sequence). Call this set U_i.

An index $I_2(j, k)$ for a job j in U_i can be computed by considering several combinations of job properties. Some of the index computations use the *slack*, which is defined as the difference between the due date of the job and the earliest possible completion time. Other examples of this index used in implementations of the single-dispatch framework include the processing time of job j on machine i, the setup time required between jobs k and j, the tardiness of job j, and a weighted combination of any of these measures with the setup time.

The single-dispatch heuristic can be summarized as follows.

Algorithm 19.2.2. (Single-Dispatch Heuristic)

Step 1. *Let the set U_i initially be equal to all jobs allocated to machine i.*

Step 2. *For each job in U_i, compute the index $I_2(j, k)$ that corresponds to the scheduling of job j as the next one on machine i.*

Step 3. *Schedule the job with the smallest index as the next job and remove the job from U_i. If U_i is not empty, go to Step 2.*

Improver agents consider the schedules in the current population and attempt to improve on them in several different dimensions. For example, an improver agent may either move a single job to improve its tardiness or aggregate batches of jobs to decrease the number of batches and reduce total setup times.

A job can be moved to a new position on the same machine or it can be moved to a different machine. The goal of any exchange may be to improve a single objective, such as the total weighted tardiness, or a combination of objectives, such as machine load balance and total weighted tardiness. Each agent selectively picks a solution to work on based on certain criteria. For example, a tardiness improver agent picks a schedule from the population that needs an improvement with regard to tardiness. A load balancing agent picks a schedule of which the load is unevenly balanced and attempts to rebalance, and so on.

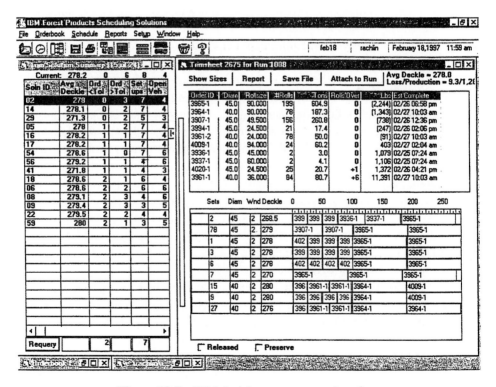

Figure 19.5   IBM decision support user interface.

This scheduling system has been implemented in many facilities in the paper industry. Figure 19.5 shows a user interface of such an implementation.

19.3 i2's TRADEMATRIX PRODUCTION SCHEDULER

This section describes the i2 TradeMatrix Production Scheduler, which is a commercial scheduling system developed by i2 Technologies. Production Scheduler is one of several scheduling solutions in the Scheduler suite of products offered by i2 to meet the scheduling needs of various different manufacturing industries. The Scheduler solution suite includes Sequencer and Master Scheduler for automotive and industrial companies, Mill Scheduler for the metals industries, and Paper Mill Manager for the paper and board industries. All of these products are part of a complete and integrated supply chain management solution. These products are designed to capture the specific needs and complexities of the different manufacturing environments and use patented Genetic Algorithm (GA) technology to generate optimized manufacturing schedules. This technology has been licensed for nearly 200 implementations and enables representation of a wide range of production situations easily adapting to new and changing constraints. This section describes the various elements of the Production Scheduler model, the key functionalities and system requirements of Production Scheduler, the functionalities of the user interfaces, and some implementation examples.

Production Scheduler is a configurable scheduling system that can produce detailed schedules for multistage manufacturing processes. The patented technology combines a fast constraint calculation engine and an optimization algorithm to schedule all manufacturing stages simultaneously. Production Scheduler has the capability to model either discrete (batch) or continuous processes, with multiple resources at each stage. The core product is Java-based, with an easily customizable, proprietary optimal scheduler language defining much of its functionality. The manufacturing model data is maintained in a database—typically MS Access or Oracle. The product is configured to run in client–server mode, with a Windows NT client and either a Windows NT or Unix operating system for the server. Demand data may be entered directly into the Production Scheduler database or imported from an ERP or customer order entry system.

Production Scheduler is a tool for detailed scheduling that is appropriate for a variety of manufacturing industries, including consumer goods, textile, apparel, footwear, automotive stamping plants, gear and transmission shops, pharmaceuticals, and chemicals. A given manufacturing process can be modeled in such a way that the various business rules or constraints that specify how the process is run are captured. Typical constraints include capacity limits, setup times and costs, due date requirements, Work-in-Process (WIP) minimization, labor restrictions, and material availability. Production Scheduler generates optimal schedules that balance all of these different constraints while generating the best feasible solution.

A model is basically a data representation of the rules that govern the given manufacturing process. This representation includes:

(i) the products and component parts that the factory produces (items and item families);

(ii) the physical process of production (BOMs, routings, operations, and manufacturing resources); and

(iii) production preferences, policies, and requirements (constraints).

Given the basic building blocks of the model, Production Scheduler generates production schedules that satisfy the demand (in the form of work orders) subject to the operating constraints specified. The basic Production Scheduler model contains the following elements:

Items. An item is a part, component, subassembly, or finished good that is produced or consumed by the factory. In a Production Scheduler model, an item can be a finished part, intermediate part, or raw material. The Bill of Material (BOM) is a list of intermediate items that are needed to produce a given item.

Operations. An operation is a production activity that is carried out by one or more resources (e.g., a mixing tank and the labor resource required in the mixing process). It is one step in a route that produces an item. An operation may be partitioned into time segments that all share a common resource (e.g., the mixing tank is the common resource used throughout the mixing process, but the labor is used just initially to fill the tank).

Resources. A resource can be a machine, tool, labor pool or storage container needed for processing an item during an operation.

Routes. A route is a sequence of one or more operations that produce one or more items.

Capacity Profiles. A capacity profile defines the capacity of a resource over time.

Calendars. A calendar defines the days and times a resource is not available for the processing of an item. It specifies which regions of the defined capacity calendar are unavailable for processing. If this unavailability calendar is not defined, then the resource is assumed to be available all day every day over the entire scheduling horizon.

Figure 19.6 depicts the relationships between these modeling elements. To generate a good schedule, Production Scheduler uses a set of scheduling criteria and con-

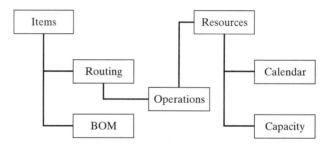

Figure 19.6   Relationship among modeling elements in TradeMatrix Production Scheduler.

straints that enforce preferences and policies and impose physical limitations on the operations of the factory. Production Scheduler maintains a suite of common constraints to handle requirements such as:

 (i) tardiness objectives,
 (ii) minimization of WIP,
 (iii) enforcement of routing constraints,
 (iv) selection of resources for given operations, and
 (v) setup times and resource costs due to changeovers.

Production Scheduler has the flexibility to capture process limitations and constraints across a range of diverse industries from automotive to, for example, pharmaceutical. This may include upper bounds on the number of product changeovers on a set of resources within a given time period or constraints on the amount of time a certain item may be held in a storage vessel before it begins to spoil. The prior list is only a small sample of the large library of common constraints maintained within Production Scheduler, enabling easy implementation of many detailed and complex manufacturing constraints.

Constraints may be categorized as either *hard* or *soft*. Hard constraints represent rules that, if violated, could result in a production problem. Soft constraints, when violated, result in a feasible schedule that is less than ideal.

Given a set of constraints with associated penalties, Production Scheduler uses a genetic algorithm (GA) to generate schedules. For a given model, Production Scheduler determines the best scheduling strategy by determining trade-offs among soft constraints. Therefore, while Production Scheduler never violates a hard constraint when it schedules a task, it will invariably violate one or more soft constraints. However, the optimizer always generates a schedule that minimizes the total number of soft constraint violations. Assigning violation penalties to constraints is an important aspect of the fine-tuning of a given model and gives an indication of the relative priorities of the various conflicting rules in a typical manufacturing environment.

In addition to the basic set of constraints, Production Scheduler is configured for incorporating customized constraints easily to model the unique characteristics of any particular implementation.

Production Scheduler has two User Interfaces (UIs). The first is the Modelling UI, through which most of the model can be specified (resources, resource calendars, capacity profiles, items, operations, routes, and constraints). While defining modeling elements, Production Scheduler provides the flexibility to customize the names of the resources and the constraints, as well as the violation messages displayed when constraints are violated. Once a satisfactory model has been built and demand has been imported, all interactions with the model occur through the Scheduling UI (Figure 19.7).

The Scheduling UI consists of three grids (displayed along the left hand side), three display panes (displayed along the bottom), and the schedule board (Gantt chart). This part of the UI is quite frequently used in practice and is highly inter-

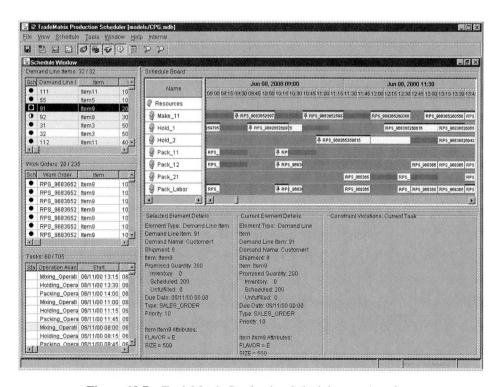

Figure 19.7   TradeMatrix Production Scheduler user interface.

active. It contains a lot of useful information and tools designed to make the job of coming up with a good schedule easy and intuitive.

The three grids display the following three categories of information:

Demand line items. A list of all the current demand, along with promised quantities, current inventory levels, due dates, and priorities.

Work orders. Production Scheduler aggregates demand based on the model-defined manufacturing quantity. Each demand line item may be broken down into a number of work orders or be aggregated with other demand line items (depending on the relative size of the demand quantity compared with the manufacturing quantity).

Tasks. Each work order has a number of tasks that must be scheduled to satisfy the demand. Each individual task represents an operation in the routing that produces the final item.

Each one of the grids can be manipulated (through sorting and filtering mechanisms) to help the user organize and display information as necessary.

The three display panes also provide an abundance of information: The first display pane can provide selected details concerning any element; when one clicks on a demand line item, work order, or task, this pane displays information about that element. The second display pane provides details concerning the current element; this pane provides details regarding the object (demand line item, work order, task, etc.) on which the cursor is positioned. The third display pane shows constraint violations associated with the manual or automatic scheduling of a given task. The violation message consists of the penalty points associated with the constraint, violation information, constraint type, and specific constraint name.

The Schedule Board (or Gantt Chart) is the standard two-dimensional grid with time on the X-axis and resources on the Y-axis. The time increments are defined by the model. For example, the time line may be divided into 15-minute intervals (i.e., 4 time slots per hour). The Schedule Board can be configured to zoom out to various coarser granularities (e.g., 30 minutes, 1 hour, etc.). Tasks are assigned to slots that correspond to resources and time intervals.

The Schedule Board can be configured to display both a frozen history of scheduled tasks that have been released to the shop floor and the current scheduling horizon. This enables the user to view previously scheduled tasks as they relate to the current schedule. In addition, Production Scheduler supports defining an end buffer in which tasks that start within the scheduling horizon may be completed. The Schedule Board may be in either one of two modes: The Manual Scheduling may be either on or off. The current mode is indicated in the bottom right corner of the Schedule Window. When the Manual Scheduling is off, the user has a view of the current schedule with certain tasks having violation colors. So, for example, if two tasks of differently flavored items are scheduled one after another in a mixing tank, the second task appears in yellow if there is a changeover penalty associated with a change in flavor in a mixing tank. When Manual Scheduling is turned on, by selecting a task in the tasks pane or in the Schedule Board, the user is looking at a What If? scenario associated with the manual scheduling of the selected task. That is, the user sees the violations on the Schedule Board associated with placing the task on certain resources over the scheduling horizon. This is accomplished by a unique dynamic constraint checking environment, which provides instantaneous feedback when the Schedule Board is in the interactive manual scheduling mode. The color red indicates that a hard constraint violation occurs if the task is moved into an appropriate position. Yellow indicates a soft violation, and white indicates no violations. Again, as the user moves the cursor over the highlighted area in the Gantt chart, the Constraint Violations pane provides feedback with regard to the associated violations (based on the customized messages created in the modeling UI).

Any schedule generated by the system can be manipulated manually. The What If? mode provides the necessary guidance and feedback with regard to the consequences of scheduling a particular task on a particular resource at a particular time. The Scheduling UI provides simple drag-and-drop capabilities to move tasks that still have to be scheduled from the Task Grid to the Schedule Board, or to move tasks from resource to resource, or from one time slot to another on the Schedule Board. Production Scheduler also has the ability to pin (or freeze) certain tasks (within a time range or on a specific resource). This gives the end user the flexibility to reschedule a set of tasks while keeping the pinned tasks in their positions.

The Scheduling UI also provides a number of performance measures and metrics with respect to the schedule (number of tasks and work orders scheduled vs. number not yet scheduled; a summary of hard and soft constraint violations; resource utilization information; number of late work orders, etc.). The Scheduling UI provides an item production grid that displays the demand versus the produced quantities. The UI also provides time-varying charts that display resource capacity and item inventory over the scheduling horizon (see the Figure 19.8).

The remaining part of this section describes two implementations of the TradeMatrix Production Scheduler.

Example 19.3.1 (An Implementation at Ford Motor Co.)

The TradeMatrix Production Scheduler has been installed at the Woodhaven Stamping Plant at Ford Motor Co. This stamping plant has over 175 production lines with three basic stages of operation: blanking, die pressing, and assembly. The final products are fenders, roofs, and doors used in the final assembly of Ford cars and trucks. The sheet metal (coils) are converted into blanks in the blanking or press stage (for direct coil feed lines). The press lines shape the blanks into a form that is ready for assembly. These parts are either sent to the assembly line or other assembly plants.

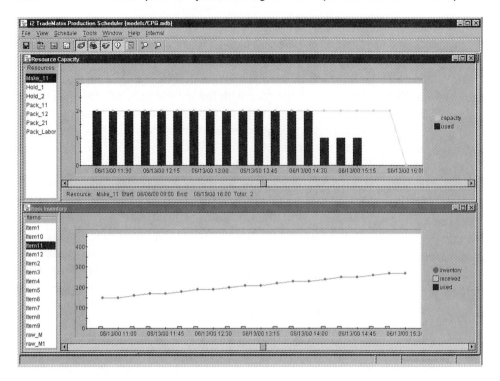

Figure 19.8   TradeMatrix Production Scheduler capacity buckets.

The goals of the scheduling implementation are:

(i) better utilization of plant resources (both labor and machine) to reduce direct overtime and increase throughput,

(ii) reduction of premium freight charges,

(iii) improved customer service,

(iv) schedules that minimize the number of die setups, and

(v) reduction of end item inventory.

The Production Scheduler at the Woodhaven Stamping Plant is used to do cycle planning as well as detailed scheduling. The cycle planning lays out a month-long production plan for each line with the goal of maintaining minimal WIP and utilizing the staffing levels optimally. It does this by considering due dates, die sets, labor requirements, storage rack availability, and so forth. The end result of the planning is a series of manufacturing codes denoting frequency of production of each part and the run days. This information is uploaded to the Ford MRP system on a monthly basis. The detailed scheduling lays out a detailed week-long schedule for all three stages at a finer time granularity, considering the same constraints outlined earlier, and a few additional constraints related to the finer time granularity. The schedule is uploaded daily into the Ford MRP system.

Since i2's inception at Ford Woodhaven, the plant has seen a significant reduction in average daily WIP inventories. It has achieved further savings through increased throughput, reduced overhead from unplanned overtime, and a tighter control on inventories and cycle times.

Example 19.3.2 (An Implementation in a Gear Manufacturing Facility)

The TradeMatrix Production Scheduler has also been installed in a gear manufacturing facility of one of the world's leading heavy equipment manufacturers, with plans for implementation in other facilities of the company.

This gear manufacturing facility is based on cellular manufacturing. A typical route consists of seven primary operations: heat treatment, gear cell, heat treatment, grinding, heat treatment, washing, and checkout. This route is a generic one as some parts could skip some of these operations and others can go through additional operations. The goal of the Production Scheduler implementation was to schedule the gear shop, minimizing the number of changeovers and taking into consideration runtimes, demand priorities, and the flexibility of alternate resource choices. The implementation went live in less than 6 months, including both the initial knowledge acquisition sessions and complete end-to-end testing of the solution. Since implementing Production Scheduler, the company has realized an increased productivity and improved its service levels.

Increased productivity was achieved as follows. Prior to implementing Production Scheduler, the plant required four to five foremen to schedule their shop floor area. Now a dedicated scheduler uses Production Scheduler and generates detailed schedules, freeing up resources for other productive uses in the plant.

The improvement in the service levels were a result of an increased visibility. Prior to implementing Production Scheduler, each foreman only had access to information about his or her shop floor and not upstream and downstream operations. With Production Scheduler, the dedicated scheduler has visibility across different areas of the plant, resulting in more efficient utilization of resources and, as a result, improved customer service levels.

19.4 AN IMPLEMENTATION OF CYBERTEC'S CYBERPLAN

Cybertec is the largest planning and scheduling software house in Italy. It is based in Trieste. Its main product is a software system called Cyberplan, which is a suite of six software modules—namely,

 (i) Supply Chain Design,
 (ii) Capacity Requirements Planning,
 (iii) Material Requirements Planning,

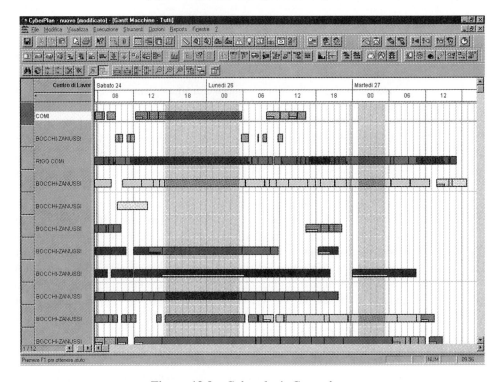

Figure 19.9   Cyberplan's Gantt chart.

(iv) Master Production Scheduling,

(v) Finite Capacity Planning, and

(vi) Finite Capacity Scheduling.

Many different combinations of the various modules have been implemented in numerous companies. All modules have elaborate user interfaces—not only the typical interfaces such as Gantt charts (see Figure 19.9), but also the less common interfaces such as capacity buckets (see Figure 17.5). The optimization techniques used include constraint programming, local search (genetic algorithms), and mathematical programming procedures.

The remaining part of this section describes an implementation of Cybertec's Finite Capacity Planning and Finite Capacity Scheduling Modules.

Example 19.4.1 (An Implementation in a Yogurt Production Facility)

The Finite Capacity Planning and Scheduling modules have been implemented in a facility that produces jars of yogurt. The plant consists of three basic areas—namely, the preparation area (in which the so-called *white masses* are prepared), the production area (in which the yogurt is produced in bulk), and the packaging area (in which the yogurt is packaged in jars).

The incoming milk (light and fat) are stored in the preparation area. From this storage, the milk goes to the concentration area where it is condensed. The condensed milk enters the production area, which consists of two separate flexible flow lines that are identical. The production process consists of four stages: mixing, pasteurization, fermentation, and cooling (see Figure 19.10). The mixing stage in each one of the two flow lines has two mixers. The pasteurization stage in each one of the lines has a single pasteurizer. The fermentation stage of each line has 10 tanks. Each fermentation stage is followed by a cooling center.

Both lines feed into a buffer area. There are two buffer areas, and each area has four buffers. The yogurt may remain in a buffer for a maximum of 6 hours.

The two buffer areas feed into the packaging line. The input of the packaging line consists of the white mass, the fruit and other raw material. The output consists of the packages, each of which contains a number of yogurt jars. There are almost

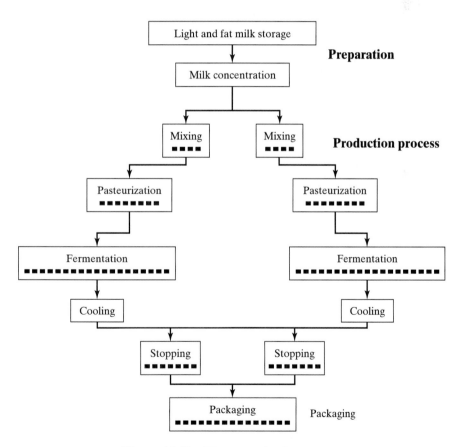

Figure 19.10   Yogurt production process.

100 different finished products. The differences in packaging are due to the number of jars in a package, the weight of the jar, the flavor, and so on.

The production plan is established as follows. Sales forecasts and orders arrive at the plant on a weekly basis. This information is incorporated in the Master Production Schedule. Planning and scheduling is done on a daily basis. The daily input into the production scheduling module includes the following information:

   (i) machines availability,
  (ii) milk availability, and
 (iii) personnel availability.

There are several objectives that the scheduling system takes into consideration. The primary objective is to meet the due dates of the orders. An important secondary objective is the minimization of the number of different pieces of equipment used since this helps control energy costs. The schedules are subject to several classes of constraints, an important one being the environmental constraints. The daily output of the scheduling system includes, besides the schedules, the data that are needed to maintain quality control.

The scheduling procedure is a backward procedure (see Figure 19.11). Based on the orders and their delivery dates (the finished products requirements), the net material requirements are determined. The basic input to the scheduling module is

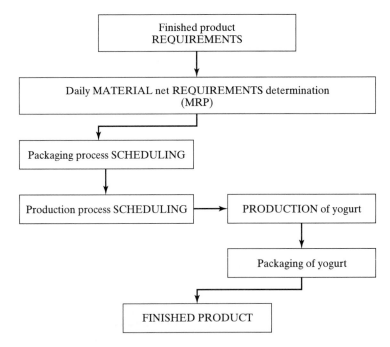

Figure 19.11   Scheduling process of yogurt facility.

the delivery data and the material availability; the output is a packaging schedule. The packaging scheduling problem is basically a parallel machine scheduling problem. There are certain heuristic rules that have to be followed. Each packaging line has to follow a preferred sequence: natural yogurt, white puree, dark puree, white pieces, and dark pieces. The reason for following such a sequence on a packaging line is obvious: This way the number of jars rejected because of quality considerations is minimized. This problem is comparable to the $Pm \mid r_j, s_{jk} \mid \sum w_j T_j$ problem considered previously.

The solution of the packaging scheduling problem is an input to the production process scheduling problem. The production process is a form of continuous production (in contrast to the packaging, which is a form of discrete production). In the production process, the quantities are measured in weights or volumes rather than in units. The process has a large number of constraints and a limited amount of data (the number of different types of products to be produced is significantly smaller than the number of different items to be produced on the packaging line). The scheduling of the production process is done via a constraint guided heuristic search procedure.

The system is used on a daily basis in the following manner:

Phase 1. Every day before scheduling the production, the machine availabilities are entered into the program.

Phase 2. The packaging program function is selected; after the packaging schedule has been determined, the final result is displayed on the screen.

Phase 3. Before calling the procedure for scheduling the production, the settings of the pasteurizer are verified and set (i.e., the number of washing machines, duration of the washing, minimum time between washing, and maximum pasteurization time).

Phase 4. After setting the parameters of the pasteurizer, the production scheduling procedure is called. The resulting schedules for the pasteurizers and fermenters are displayed on the screen separately.

Before the installation of the Cybertec system, schedules were generated manually after determining the daily requirements of finished products, taking into consideration the scheduled arrivals of milk, the amount of milk on hand, and the personnel present. The resulting orders, still without starting and finishing times, were assigned to the packaging lines taking into account the product types. For each line, a sequence was generated manually, subject to production sequence constraints. This resulted then in a packaging sequence with corresponding starting times and completion times. This information was then entered into a scheduler that generated a Gantt chart and determined the total requirements of white mass. The quantities of white mass were divided into different lots, and a manual schedule of the production process mixes was done. This manual scheduling took into account the daily recipes, the fermenter and pasteurizer capacities, and the constraints on the time and on the buffers. The recipes could change every day because of the different protein contents of the milk. When the fermentation of the mixtures on the fermentation lines had to be scheduled, constraints with regard to milk preservation had to be taken into account.

Manual scheduling was a complicated process. It took at least 2 hours a day and was error-prone. Constraints were at times violated, resulting in quality problems, reduced plant capacity utilization, and environmental problems.

After the installation of the Cybertec system, the situation improved considerably. Schedules are generated in less than 2 or 3 minutes, and no human errors are made. The pasteurizers and fermenters are used at capacity, and the utilization of the plant capacity and the energy consumption are optimized.

19.5 SYNQUEST'S VIRTUAL PRODUCTION ENGINE

The SynQuest system provides an integrated approach to production management by taking strategic, tactical, and operational aspects into consideration. The Syn-Quest approach considers production management in the context of the total logistics network of an enterprise or an extended enterprise that is managed in a collaborative fashion. The network may include suppliers at several tiers, production facilities, logistical resources, intermediaries and marketing channels, as well as customers.

In all its decision support modules, SynQuest applies mathematical optimization techniques to every problem to find a solution with maximum total benefit. An optimal solution may be defined as a solution that simultaneously meets all customer requirements, satisfies all constraints, and optimizes the enterprise's objective. The total benefit may include, for example, operational profit, economic profit, or market share. SynQuest models include all the relevant variables explicitly. SynQuest has established a certain position in the production management software market by using mathematical optimization throughout its suite of models.

SynQuest conceptualizes management according to three interrelated dimensions:

  (i) management purposes,
 (ii) management processes, and
(iii) constituent processes.

With regard to management purposes, SynQuest makes a distinction among the strategic, tactical, and operational aspects. Strategic management deals with long-term decisions concerning resource deployment in the entire logistics network; tactical management deals with medium-term decisions concerning resource allocation in the logistics network; and operational management deals with short-term decisions concerning resource utilization in the individual facilities.

With regard to management processes, SynQuest follows Deming's model that assumes four management processes: planning, execution, control, and learning.

Constituent processes are processes at the strategic, tactical, and operational levels that are managed through planning, execution, control, and learning. In the context of logistics and production management, SynQuest defines four constituent processes: procurement, transformation, delivery, and support.

The remaining part of this section focuses on the salient aspects of the Syn-Quest integrated approach to management processes (i.e., planning, execution, control, and learning) in terms of its strategic, tactical, and operational modules, the interfaces between modules, and the interfaces with other systems.

The planning process takes place at three interrelated levels: strategic, tactical, and operational. From a strategic point of view, production planning deals with the optimum number, location, and mission of production facilities determined through the optimization of the entire logistics network, from suppliers' suppliers, to customers' customers. Production facilities are located simultaneously with the selection of suppliers, location of logistics facilities, selection of intermediaries and channels, determination of product flows and inventory levels, and assignment of customers to shipping facilities.

From a tactical point of view, production planning deals with the optimum allocation of demand to existing production—and logistics—facilities across multiple periods. Production facilities are optimally loaded, period by period, over the entire planning horizon. The load reflects tactical conditions such as seasonality, promotions, and planned price changes. It prescribes the optimal accumulation or reduction of inventories, number of shifts, numbers of temporary workers to be added or reduced, bills of materials to be used, and other tactical parameters for each production facility and process by period.

From an operational point of view, production planning deals with the optimum utilization of production assets, including people, facilities, equipment, and materials in real time. Each process, machine, or operation, within each production facility, is already optimally loaded with production orders that have been determined by the tactical planning. These are then transformed into detailed production sequences that specify for each product the number of production batches, their sizes, the machines at which they will be produced, and the start and end times of work at each machine.

The execution process takes place at each level by translating the optimal plans into specific actions that have to be executed by specific individuals at given times. These are displayed at appropriate points throughout the production facilities.

The control process is provided through informational feedback of actual performance throughout the supply chain. This information is compared to the original plans to determine whether there are deviations from the plans that are outside preestablished control limits. In strategic and tactical control, these comparisons take place at periodic intervals. In operational control, the comparisons take place continuously in real time. Thus, in operational production management, SynQuest offers a completely synchronized system that is continuously re-planning, executing, and controlling all the activities of each production facility as needed. The objective of the synchronization system is to optimize performance, not merely to plan.

The learning process occurs when the SynQuest modules are used to run analyses of actual performance against planning standards. This learning process ensures a continuous improvement of production performance.

The interfaces between the modules and with other systems are elaborate. The interfacing among the strategic, tactical, and operational modules takes place through two mechanisms: restriction propagation and data consolidation (see Figure 19.12).

In restriction propagation, the strategic solution establishes the optimum logistics network for the individual or extended enterprise, including the determination of the production facilities required, their locations, missions, and ownership (e.g., company owned, outsourcing). Those aspects of the optimum strategy become restrictions for tactical planning: In the medium term, only the facilities optimized strategically can be used. The optimum tactical plans, in turn, establish restrictions for the operational, real time re-planning of production operations at each facility (e.g., inventory accumulation or reduction, number of work shifts by process).

Data consolidation completes the interfacing of the management models. It provides operational data that are appropriately aggregated to support tactical

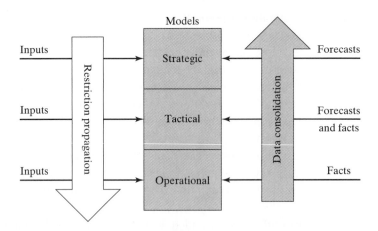

Figure 19.12   SynQuest's planning system interfacing.

models, and tactical planning data that are in turn further aggregated to support strategic planning.

The SynQuest modules also have to interface with other systems, such as transaction processing and legacy systems, to ensure their efficient integration in each enterprise. These interfaces take place through standard interfaces with popular systems and through custom interfaces with special systems. Current developments at SynQuest are directed toward using the Internet as the interface-enabling tool among modules and systems.

The Virtual Production Engine and its Algorithms are responsible for the optimization process. Strategic and tactical optimization are done using Mixed Integer Programming as the main optimizer. This enables the explicit modeling of, for example, fixed costs, economies and diseconomies of scale and logical restrictions.

Through restriction propagation, SynQuest systems restrict its operational models optimally, thus limiting optimally their solution space. This is an important consideration because it allows the use of solvers that provide near optimal solutions in very short computation times while not using mathematical optimization techniques. Operational problem characteristics are changing almost continuously: For example, order quantities and deadlines are revised, materials supplies arrive late or are rejected for quality reasons, equipment breaks down, and so on. Under these very dynamic conditions, true operational optimization is not critical because the solutions calculated are seldom fully implemented. Typically they must be recalculated well before their full implementation. Since restriction propagation ensures near optimal solutions, computation speed becomes the most important practical consideration for this module.

SynQuest's current operational production scheduling module uses Genetic Algorithms (GA) as its preferred solvers. This approach enables the calculation of near optimal production schedules in very short times. It also has the benefit of offering users up to three objectives to guide the solver: minimize cost, minimize delays, and minimize makespan. The user can choose any combination of these three objectives and assign them weights. In this fashion, the production scheduling module can consider simultaneously many practical aspects by satisficing up to three objectives instead of optimizing a single one.

As described in Chapter 14, GA are procedures that represent sequences of activities as chromosomes of a population characterized by their fitness. A chromosome's fitness is defined by the corresponding value of the problem's objective function. GA arrive at a solution by successive approximations, in which every iteration, known as a generation, produces an improved objective function. Every generation consists of chromosomes from the previous generation plus children or new schedules resulting from the previous generation. The children are generated by reproduction and mutation of chromosomes from prior generations, known as parents.

The general production scheduling problem is characterized by multiple products to be produced on multiple machines. Under these conditions, a chromosome is formed by subchromosomes, each one containing the work sequence for a specific machine. In each iteration, many different schedules are generated and transferred to the succeeding iteration as a set of schedules. A new schedule is then built through the combination of subchromosomes from different generations.

SynQuest also uses solvers based on other local search techniques, such as tabu-search and simulated annealing. While less general than GA, those techniques can be tuned to produce very good schedules, comparable to those obtained with GA, much faster.

The main advantages of GA in solving production scheduling problems can be summarized as follows: First, they constitute a very general approach that can model efficiently a wide variety of scheduling problems in a relatively simple manner. Second, they typically can find near optimal solutions for very complex problems. When a restriction propagation approach provides optimal restrictions for a GA, the solutions obtained are very close to estimated theoretical optima, and the computation times are substantially reduced. Third, GA are not hampered by one of the most difficult problems found in mathematical optimization: Many optimization techniques assume that variables are continuous functions. However, in most scheduling problems, discontinuities are the norm. Since GA do not assume continuity, they are especially useful for solving discrete manufacturing problems in Just-in-Time and lean manufacturing environments, where discrete quantities are required.

Example 19.5.1 (An Implementation at Herman Miller, Inc.)

> SynQuest's Virtual Production Engine has been installed in numerous facilities. One example is Miller SQA, which is a capability within the furniture giant Herman Miller, Inc. The company receives an average of 300 customer orders a day. Unlike the typical sequential approach to manufacturing, where a product moves from department to department until it is finished, Miller SQA has adopted a cellular approach. Each of 19 cells specializes in a particular product line and oversees its assembly from start to finish. To slash lead times and operate effectively in its Just-in-Time order assembly environment, Miller SQA needed to synchronize all cell activities in real time. SynQuest software enables the company to do just that. Its Virtual Production Engine combines advanced planning, execution, and control in real time, with a finite capacity scheduler, to dynamically calculate the best schedule for all orders through each routed operation. The software does so by prioritizing each order according to the production time it requires and the time it must leave the plant. After implementation of the SynQuest software, Miller SQA has not only seen its revenues climb more than 35 percent per year in the last 5 years, but it has established itself as a leader in customer satisfaction by consistently providing unparalleled ontime delivery that is higher than 99 percent.

19.6 THE LEKIN SYSTEM FOR RESEARCH AND TEACHING

The LEKIN system contains a number of scheduling algorithms and heuristics. It is designed to allow the user to link and test his or her own heuristics and compare their performances with the heuristics and algorithms that are embedded in the system. The system can handle a number of machine environments:

  (i) single machine,
 (ii) parallel machines,
(iii) flow shop,
 (iv) flexible flow shop,
  (v) job shop, and
 (vi) flexible job shop.

Furthermore, it is capable of dealing with sequence-dependent setup times in all the environments listed. The system can handle up to 18 jobs, up to 10 workcenters or workstations, and up to 4 machines at each workstation.

The educational version of the LEKIN system is a teaching tool for job shop scheduling. It can be downloaded from the site

$$http://www.stern.nyu.edu/{\sim}mpinedo$$

(after clicking on LEKIN). The system has been designed for use in either a Windows 98 or a Windows NT environment.

Installation on a Windows 98 PC is straightforward. The user has to click on the *start* button and open *a:setup.exe* in the run menu.

Installation on a network server in a Windows NT environment may require some (minor) system adjustments. The program will attempt to write in the directory of the network server (which is usually read only). The program can be installed in one of the following two ways. The system administrator can create a public directory on the network server where the program can write, or a user can create a new directory on a local drive and write a link routine that connects the new directory to the network server.

When LEKIN is run for the first time a Welcome page appears. If the user does not want to see this page again in the future, he has to uncheck the check box. Closing the welcome page makes the main menu appear. The main menu can also be accessed during a scheduling session by clicking on *start over* under *file*.

The main menu allows the user to select the machine environment in which he is interested. If the user selects a machine environment, he has to enter all the necessary machine data and job data manually. However, the user also has the option of opening an existing file in this window. An existing data file contains data with regard to one of the machine environments and a set of jobs. The user can open such an existing file, make changes in the file, and work with the modified file. At the end of the session, the user may save the modified file under a new name.

If the user wants to enter a data set that is completely new, he must first select a machine environment and then a dialog box appears, where he has to enter the most basic information (i.e., the number of workstations and the number of jobs to be scheduled). After the user has done this, a second dialog box appears and he has to enter the more detailed workstation information (i.e., the number of machines at the workstation, their availability, and the details needed to determine the setup times on each machine; if there are setup times). In the third dialog box, the user has to enter the detailed information with regard to the jobs (i.e., release dates, due dates, priorities or weights, routing, and processing times of the various operations). If the jobs require sequence-dependent setup times, then the machine settings that are required for the processing have to be entered. The structure of the setup times is similar to the one described in Example 10.3.2. However, in the LEKIN system, every job has just a single parameter in contrast to the three parameters in Example 10.3.2.

After all the data have been entered, four windows appear simultaneously:

  (i) the machine park window,
 (ii) the job pool window,
(iii) the sequence window, and
 (iv) the Gantt chart window

(see Figure 19.13).

The machine park window displays all the information regarding the workstations and machines. This information is organized in the format of a tree. This window first shows a list of all the workstations. If the user clicks on a workstation, the individual machines of that workstation appear.

The job pool window contains the starting time, completion time, and more information with regard to each job. The information with regard to the jobs is also organized in the form of a tree. First, the jobs are listed. If the user clicks on a specific job, then immediately a list of the various operations that belong to that job appear.

The sequence window contains the lists of jobs in the order in which they are processed on each one of the various machines. The presentation here also has a tree structure. First, all the machines are listed. If the user clicks on a machine, then all the operations that are processed on that machine appear in the sequence in which they are processed. This window is equivalent to the dispatch list interface described in Chapter 17. At the bottom of this sequence window, there is a summary of the various performance measures of the current schedule.

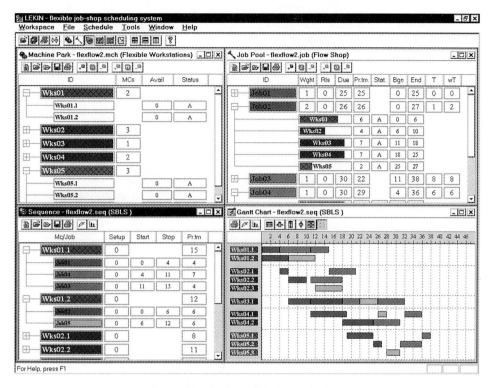

Figure 19.13   LEKIN's four windows.

The Gantt chart window contains a conventional Gantt chart. This Gantt chart window enables the user to do a number of things. For example, the user can click on an operation and a window pops up displaying the detailed information with regard to the corresponding job (see Figure 19.14). The Gantt chart window also has a button that activates a window where the user can see the current values of all the objectives.

The windows described earlier can be displayed simultaneously on the screen in a number of ways (e.g., in a quadrant style [see Figure 19.13], tiled horizontally, or tiled vertically). Besides these four windows, there are two other windows, which are described in more detail later. These two windows are the log book window and the objective chart window. The user can print out the windows separately or all together by selecting the print option in the appropriate window.

The data set of a particular scheduling problem can be modified in a number of ways. First, information with regard to the workstations can be modified in the machine park window. When the user double clicks on the workstation, the relevant information appears. Machine information can be accessed in a similar manner.

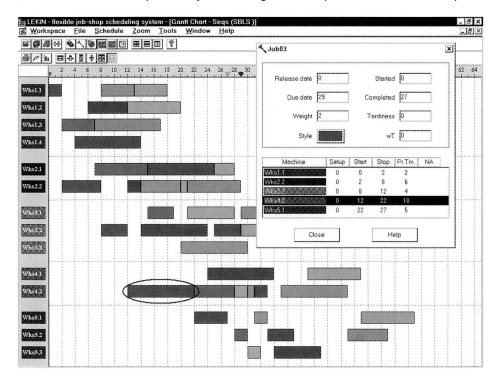

Figure 19.14   LEKIN's Gantt chart window.

Jobs can be added, modified, or deleted in the job list window. Double clicks on the job display all the relevant information.

After the user has entered the data set, all the information is displayed in the machine park window and the job pool window. However, the sequence and Gantt chart windows remain empty. If the user in the beginning had opened an existing file, then the sequence and Gantt chart windows may display information pertaining to a sequence that had been generated during an earlier session.

The user can select a schedule from any window. Typically the user would do so either from the sequence window or from the Gantt chart window by clicking on schedule and selecting a heuristic or algorithm from the drop-down menu. A schedule is then generated and displayed in both the sequence and Gantt chart windows. The schedule generated and displayed in the Gantt chart is a so-called *semi-active schedule*. A semi-active schedule is characterized by the fact that the start time of any operation of any given job on any given machine is determined by either the completion of the preceding operation of the same job on a different machine or the completion of an operation of a different job on the same machine.

The system contains a number of algorithms for several of the machine environments and objective functions. These algorithms include

 (i) dispatching rules,
 (ii) heuristics of the shifting bottleneck type,
 (iii) local search techniques, and
 (iv) a heuristic for the flexible flow shop with the total weighted tardiness as objective (SB-LS).

The dispatching rules include EDD and WSPT. The way these dispatching rules are applied in a single machine environment and a parallel machine environment is standard. However, they can also be applied in the more complicated machine environments such as the flexible flow shop and the flexible job shop. They are then applied as follows: Each time a machine is freed, the system checks which jobs have to go on that machine next. The system then uses the following data for the priority rules: The due date of a candidate job is then the due date at which the job has to leave the system. The processing time that is plugged in the WSPT rule is the sum of the processing times of all the remaining operations of that job.

The system also has a general purpose routine of the shifting bottleneck type that can be applied to each one of the machine environments and every objective function. Since this routine is quite generic and designed for many different machine environments and objective functions, it cannot compete against a shifting bottleneck heuristic that is tailored to a specific machine environment and a specific objective function.

The system also contains a neighborhood search routine that is applicable to the flow shop and job shop (but not to the flexible flow shop or flexible job shop) with either the makespan or the total weighted tardiness as objective. If the user selects the shifting bottleneck or the local search option, then he must also select the objective he wants to minimize. When the user selects the local search option and the objective, a window appears in which the user has to enter the number of seconds he wants the local search to run.

The system also has a specialized routine for the flexible flow shop with the total weighted tardiness as objective; this routine is a combination of a shifting bottleneck routine and a local search (SB–LS). This routine tends to yield schedules that are reasonably good.

If the user wants to construct a schedule manually, he can do so in one of two ways. First, he can modify an existing schedule in the Gantt chart window as much as he wants by clicking, dragging, and dropping operations. After such modifications, the resulting schedule is, again, semi-active. However, the user can also construct this way a schedule that is *not* semi-active. To do this, he has to activate the Gantt chart, hold down the shift button, and move the operation to the desired position. When the user releases the shift button, the operation remains fixed.

A second way of creating a schedule manually is the following. After clicking on schedule, the user must select *manual entry*. The user then has to enter for each machine a job sequence. The jobs in a sequence are separated from one another by a ; .

Whenever the user generates a schedule for a particular data set, the schedule is stored in a log book. The system automatically assigns an appropriate name to every schedule generated. If the user wants to compare the different schedules, he has to click on *logbook*. The user can change the name of each schedule and give each schedule a different name for future reference. The system can store and retrieve a number of different schedules (see Figure 19.15).

The schedules stored in the log book can be compared with one another by clicking on the *performance measures* button. The user may then select one or more objectives. If the user selects a single objective, a bar chart appears that compares

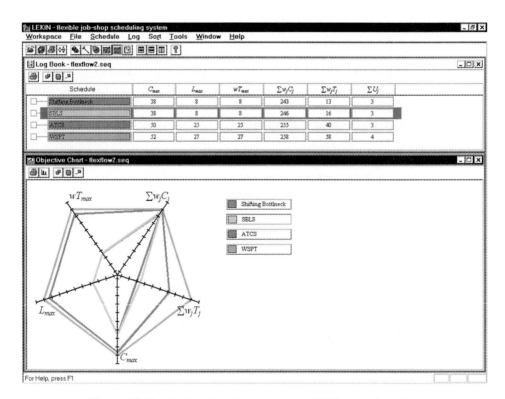

Figure 19.15   Logbook and comparison of different schedules.

the schedules stored in the log book with respect to the objective selected. If the user wants to compare the schedules with regard to two objectives, an $x - y$ coordinate system appears and each schedule is represented by a dot. If the user selects three or more objectives, then a multi-dimensional graph appears that displays the performance measures of the schedules stored in the log book.

Some users may want to incorporate the concept of setup times. If there are setup times, then the relevant data have to be entered together with all the other job and machine data at the very beginning of a session. (However, if at the beginning of a session an existing file is opened, then such a file may already contain setup data.)

The structure of the setup times is as described in Example 16.4.1. Each operation has a single parameter or attribute, which is represented by a letter (e.g., A, B, etc.). This parameter represents the machine setting required for processing that operation. When the user enters the data for each machine, he has to fill in a setup time matrix for that machine. The setup time matrix for a machine specifies the time that it takes to change that machine from one setting to another—that is, from B to E (see Figure 19.16). The setup time matrixes of all the machines at any given

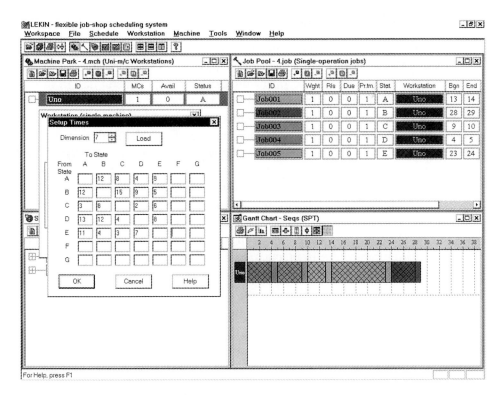

Figure 19.16   Setup time matrix window.

workstation have to be the same (the machines at a workstation are assumed to be identical). This setup time structure does not allow the user to implement *arbitrary* setup time matrixes.

A more advanced user can also link his own algorithms to the system. This feature allows the developer of a new algorithm to test his algorithm using the interactive Gantt chart features of the system. The process of making such an external program recognizable by the system consists of two steps—namely, the preparation of the code (programming, compiling, and debugging) and the linking of the code to the system.

The linking of an external program is done by clicking on *Tools* and selecting *Options*. A window appears with a button for a New Folder and a button for a New Item. Clicking on *New Folder* creates a new submenu. Clicking on *New Item* creates a placeholder for a new algorithm. After the user has clicked on *New Item*, he has to enter all the data with respect to the New Item. Under *Executable* he has to enter the full path to the executable file of the program. The development of the code can be done in any environment under Win3.2.

19.7 DISCUSSION

In this chapter, an outline is given of the general architectural designs of the IBM and SAP systems. A comparison of the IBM and SAP–APO systems makes it clear that there may already be major differences in the general design of scheduling systems. The detailed description of the i2 system highlights the importance of the role of the user interfaces in a scheduling system. The description of the Cybertec system focuses on an actual implementation. From the setting of this implementation, it is clear that an implementation can have many peculiarities, making it very hard to install a system just of the shelf. Customization seems to be always necessary. The overall description of the SynQuest system shows how a detailed scheduling system can be integrated in a system that combines long-term planning, medium-term planning and detailed scheduling.

The various systems described in this chapter clearly show how popular Genetic Algorithms are in practice. SAP–APO, i2, Cybertec, as well as SynQuest consider Genetic Algorithms an important element in their libraries of algorithms.

COMMENTS AND REFERENCES

Braun (2000) gives an overall description of the SAP–APO system. Akkiraju, Keskinocak, Murthy, and Wu (1998a, 1998b) describe IBM's A-Team architecture and an application of this architecture in the paper industry. Bell (2000) gives a detailed description of i2's TradeMatrix Production Scheduler. Marchesi, Rusconi, and Tiozzo (1999) describe the implementation of the Cybertec system in the yogurt factory. A detailed description of The LEKIN system is discussed in Feldman and Pinedo (1998).

20

What Lies Ahead?

This chapter describes a number of research and development topics that are likely to receive some attention in the near future. A distinction is made between theoretical research, applied research, and developments in system design and systems integration.

The first section focuses on avenues for theoretical research. It describes the types of models that may become of interest as well as the types of results to be expected. The second section discusses research areas that are more applied (i.e., more experimental and more empirical). This section discusses some specific types of problems that may be investigated as well as the results that can be expected. The third section discusses systems development and integration issues. It analyzes the functional links among the scheduling algorithms, the system components, and the user.

There are many other research avenues that are not discussed. This chapter is not meant to be exhaustive, but rather aims at giving a view of some of the possibilities.

502

20.1 THEORETICAL RESEARCH

In the future, theoretical research may well focus on theory and models that have not been covered in Parts I and II of this book. This section discusses

(i) the theoretical underpinnings of scheduling,

(ii) new scheduling formats and assumptions, and

(iii) theoretical properties of specific models.

Theoretical Underpinnings of Scheduling

The theoretical underpinnings will always receive a certain amount of research attention.

One theoretical field in deterministic scheduling is the area of polyhedral combinatorics and cutting planes. This research area has already generated exact solution methods of the branch and cut type (see Appendix A) and also approximation algorithms with performance guarantees for some basic scheduling models. It is likely that this research will be extended in the future to more general scheduling models. Another form of branch and bound—namely branch and price, has recently been applied successfully to several parallel machine scheduling problems.

Another research direction in deterministic scheduling is the area of Polynomial Time Approximation Schemes (PTAS) for NP-hard problems. This area has received a significant amount of attention in the recent past. It is likely that this area will receive more attention in the future and that schemes will be developed for models that are more complicated than those considered so far. However, it is not clear when these developments will start contributing to the effectiveness of heuristics used in practice. Other types of approximation methods will also be investigated. One very promising class of approximation algorithms is based on Linear Programming relaxations.

New Scheduling Formats and Assumptions

The most typical scheduling format and the basic assumptions that underlie most models discussed in Part I are the following: In a given machine environment, there are n jobs, and all the information concerning the n jobs is available at time 0. A specific objective function has to be minimized and an optimal (or at least a very good) schedule has to be generated from scratch. Interesting new research areas concern models based on scheduling assumptions that are totally different from the assumptions made in Part I of this book.

One avenue of research is the area of online scheduling. Online scheduling is important since it is in a very basic way different from the conventional (off-line) deterministic models, which assume that all information is known a priori (i.e., before any decision is made). In online scheduling, decisions have to be made based only on information regarding the jobs that already have been released and not on information regarding jobs still to be released in the future. The relationship between online scheduling and stochastic scheduling may also merit some attention. This seems to be a relatively open research area.

The concept of rescheduling may become more important in the future. This concept has been briefly touched upon in Chapter 18. However, a formal theoretical framework for the analysis of rescheduling problems has not yet been established. A rescheduling problem may have multiple objectives: the objective of the original problem (e.g., the total weighted tardiness) and the minimization of the difference between the new schedule (after rescheduling) and the old schedule (before rescheduling). It may be necessary to have formal definitions or functions that measure the difference or the similarity between two schedules for the same job set or between two schedules for two slightly different job sets (e.g., one job set having all the jobs of a second set plus one additional job—a rush job).

The rescheduling process may also have to deal with frozen jobs (i.e., jobs that have been assigned earlier to certain time slots and that may not be moved). Sometimes the constraints on the frozen jobs do allow the scheduler to make some minor adjustments in the timing of the processing of these jobs. A frozen job may be started slightly earlier or slightly later. That is, there may be a time range in which a frozen job can be processed (there may be a limited amount of slack). Scheduling around frozen jobs is somewhat similar to dealing with machine breakdowns or a preventive maintenance schedule. However, there may be a difference because of the fact that data regarding frozen jobs tend to be fixed, whereas data concerning machine breakdowns tend to be stochastic.

The concept of robustness is related to rescheduling and deserves more attention in the future as well. Chapter 18 gives a relatively short treatment of this topic. It presents a number of robustness measures as well as several practical rules for constructing robust schedules. However, very little theoretical research has been done with regard to these rules. At this point, it is not clear how useful these rules are in the various scheduling environments.

Theoretical Properties of Specific Models

One research area concerns models that combine deterministic features with stochastic features. For example, consider a model with jobs that have deterministic processing times and due dates and machines that are subject to a random breakdown process. Such a model may be quite realistic in many environments. Only very special cases are tractable. For example, a single machine with up times that are i.i.d. exponential and downtimes that all have the same mean can be analyzed. The WSPT rule then minimizes the total expected weighted completion time. However, it may be of interest to study more complicated machine environments in which the machines are subject to more general forms of breakdown processes.

Since such models tend to be more complicated than the more classical models described in Parts I and II of this book, the types of results to be expected may be more general or structural. These structural results may, for example, consist of proofs for dominance criteria or proofs for monotonicity results.

20.2 APPLIED RESEARCH

Applied research may go a little bit deeper into some of the topics covered in Part III of this book. The applied topics described in this section concern

  (i) the performance analysis of heuristics,

  (ii) robustness and reactive scheduling, and

  (iii) new, more applied, models.

Performance Analysis of Heuristics

This is a very important area of empirical and experimental research. Currently, many job shop problems can only be dealt with on a small scale. For example, it is still very hard to find an optimal solution for an instance of $Jm \mid\mid \sum w_j T_j$ with only 10 machines and 30 jobs. If there are multiple objectives and a parametric analysis has to be done, then the problem becomes even harder. Many different types of heuristics are available, but it is not clear how effective they are in dealing with large-scale scheduling problems (e.g., job shop scheduling problems with two or three objectives and with, say, 50 machines and 1,000 jobs). The heuristics for these large-scale job shops may require continuous improvement and fine tuning in the future.

Heuristics can be compared to one another with respect to various criteria. In practice, three criteria play an important role. First, the quality of the solution obtained; second, the amount of computer time needed to reach a good solution; third, the time required to develop and implement the heuristic. Comparisons of heuristics done in academic environments typically take only the first two criteria into account. However, in an industrial environment, the third criterion is of critical importance. In industry, it is important that the time it takes to adapt a heuristic to a given problem is short. This is one of the reasons that in practice local search heuristics are often more popular than very sophisticated decomposition techniques such as the shifting bottleneck heuristic.

The performance of any heuristic depends, of course, on the structure of the scheduling problem (e.g., the type of routing constraints in job shops). The performance may even depend on the particular data set. Often when one deals with an NP-hard problem, it turns out that most instances can be solved to optimality in a reasonably short time; but certain instances are very hard to solve and require an enormous amount of computing time before reaching optimality. It is of interest to find out what makes those instances hard to solve. Empirical studies indicate that a heuristic often performs quite well when the data of an instance are generated at random, whereas that same heuristic performs poorly when applied to another instance of that same problem with data that come from a real-life situation. (It may be that industrial data have certain dependencies and correlations that make an instance hard to solve.) It would be useful to establish rules that indicate the type of algorithm that is most suitable for the type of instance under consideration.

To characterize an instance of a problem properly, one must have various suitable descriptive factors, such as the due date tightness factor τ defined in Chapter 14. Besides having such factors, one also has to assess proper weights for each one of the factors. It would be useful to know which type of algorithm is the most suitable for a given instance when the following is known: First, the size of the instance (the scale); second, the values of characteristic factors; third, the computer time available.

An important class of heuristic methods are the local search procedures. The last two decades have seen an enormous amount of work on applications and implementations of local search procedures. This research has yielded interesting results with regard to neighborhood structures. However, most of this research has focused on nonpreemptive scheduling problems. Preemptive scheduling problems have received little attention from researchers specializing in local search procedures. One reason is that it tends to be more difficult to design an effective neighborhood structure for a preemptive environment than for a nonpreemptive environment. It may be of interest to focus attention first on problems that allow preemptions only at certain well-defined points in time (e.g., when new jobs are released).

It is likely that in the future there will be a demand for industrial strength heuristics that are applicable to standard scheduling problems common in industry. Consider, for example, the problem $Qm \mid r_j, s_{jk} \mid \theta_1 \sum w_j T_j + \theta_2 C_{\max}$. This scheduling problem is typical in many process industries. The two objectives are common: One objective focuses on the due dates and the other tries to balance the loads over the machines and minimize setup times. In the future, heuristics may be developed that are problem-specific and that can be easily linked to a variety of scheduling systems (as easily as Linear Programming packages can be linked to many decision support systems).

Robustness and Reactive Scheduling

A completely different line of empirical research involves robustness and rescheduling. As stated in the previous section, the concepts of robustness and rescheduling may lead to interesting theoretical research. However, they may lead to even more interesting empirical and experimental research. New measures for robustness have to be developed. The definition of these measures may depend on the machine environment. Rescheduling procedures may be based on some very specific general purpose procedures that may have similarities with the procedures described in Chapters 14 and 15.

New and More Applied Models

More practical models combine machine scheduling aspects with other aspects, such as personnel scheduling, inventory control, or maintenance scheduling. For example, in the airline industry, airlines have to schedule their planes and crews in a coherent way. An extensive amount of research has been done on pure personnel scheduling (independent of machine scheduling), but little research has been done on models that combine personnel scheduling with machine scheduling. Some theoretical research has been done in other areas related to these types of problems—namely, resource constrained scheduling (i.e., a limited number of personnel may be equivalent to a constraining resource). However, research in resource constrained scheduling has typically focused on complexity analysis and worst case analysis of heuristics. It may be of interest in the future to study more specific models that combine machine scheduling with personnel scheduling.

20.3 SYSTEMS DEVELOPMENT AND INTEGRATION

In the future, systems development and integration may focus a bit more on some of the topics covered in Part III of this book. In this section, the topics discussed are

(**i**) distributed scheduling,
(**ii**) user interfaces and interactive optimization,
(**iii**) scheduling description languages, and
(**iv**) integration with other supply chain management modules.

Distributed Scheduling

Dealing with large-scale scheduling problems may lead to implementations of distributed scheduling. Many industrial problems are so large that they cannot be solved on just a single workstation. The computational effort has to be divided over a number of workstations or computers that may even reside at different locations. With certain procedures, the computational effort can be divided up rather easily, whereas with other procedures it may not be that easy. For example, when a problem is solved via branch and bound, it may be relatively easy to decompose the branching tree and partition the computational work involved. At periodic intervals, the different workstations still have to compare their progress and share information (e.g., compare their best solutions found so far). If a problem is solved via time-based decomposition, then distributed scheduling may also be applicable (as long as the schedules of the different periods are somewhat independent of one another). With the latest Internet technologies, distributed scheduling may soon become even more important.

User Interfaces and Interactive Optimization

The development of user interfaces and interactive optimization may face interesting hurdles. The designs of the user interfaces have to be such that interactive optimization can be accomplished easily and effectively. A scheduler must maintain at all times a good overview of the schedule, even when the schedule contains over 1,000 jobs. The abilities of the user interface to zoom in and out are important. To allow for interactive optimization, the user interface must have provisions for clicking, dragging and dropping operations, freezing operations, dealing with cascading and propagation effects, and rescheduling. After the user makes some manual changes in the system, the system may reschedule automatically to maintain feasibility (without any user input). The (internal) algorithms that are used to maintain schedule feasibility may be relatively simple; they may only postpone some operations. However, internal algorithms may also be more involved and may perform some internal reoptimization (automatically). Yet the reoptimization process may also be managed by the user; he may want to specify the appropriate objective functions for the reoptimization process. Reoptimization algorithms may be very different from optimization algorithms that operate from scratch. The main reason why reoptimizing is often harder than optimizing from scratch is because an algorithm that reoptimizes has to deal with boundary conditions and constraints that are dictated by the original schedule. The integration of rescheduling algorithms in a user interface that enables the user to optimize interactively is not an easy problem.

Scheduling Description Languages

Composition and integration of procedures have led to the development of so-called *scheduling description languages*. A scheduling description language is a high-level language that enables a scheduler to write the code for a complex algorithm with only a limited number of concise statements or commands. Each statement in a description language involves the application of a relatively powerful procedure. For example, a statement may give an instruction to apply a tabu-search procedure on a given set of jobs in a given machine environment. The input to such a statement consists of the set of jobs, the machine environment, the processing restrictions and constraints, the length of the tabu-list, an initial schedule, and the maximum number of iterations. The output consists of the best schedule obtained with the procedure. Other statements may be used to set up various different procedures in parallel or concatenate two different procedures. Scheduling description languages are not yet widely used. The existing languages are still somewhat cumbersome and need streamlining. However, it is likely that that there will be some improvement in the future.

Integration with Other Supply Chain Management Modules

Many companies started out developing scheduling software for the manufacturing industry. However, they realized that, to compete in the marketplace, they must offer software that deals with all aspects of supply chain management. The types of modules that are required in supply chain optimization include, besides planning and scheduling, forecasting, demand management, inventory control, and so on. Scheduling problems in supply chain management have to take forecasts, inventories, and routings into consideration. These integrated scheduling problems are considerably harder than the more elementary problems studied in the research literature. The structure and organization of the software must be well designed and modular.

COMMENTS AND REFERENCES

Some research has already been done on polyhedral combinatorics of scheduling problems. Queyranne and Wang (1991) analyze the polyhedra of scheduling problems with precedence constraints, and Queyranne (1993) studies the structure of another simple scheduling polyhedron. Queyranne and Schulz (1994) present a general overview of polyhedral approaches to machine scheduling. Chen, Potts, and Woeginger (1998) discuss approximation algorithms. Schuurman and Woeginger (1999) present 10 open problems with regard to Polynomial Time Approximation Schemes. Sgall gives a survey of online scheduling and mentions some open problems.

For very good overviews of heuristic design as well as performance analysis of heuristics, see Morton and Pentico (1993) and Ovacik and Uzsoy (1997).

An enormous amount of research and development is going on in user interfaces and interactive decision making in general. An annual conference on Intelligent User Interfaces focuses on this topic. This annual conference typically does include papers concerning scheduling systems; see, for example, Kerpedjiev and Roth (2000).

Some research groups have already started developing scheduling description languages; see, for example, Smith and Becker (1997).

McKay, Pinedo, and Webster (2001) present a comprehensive practice-focused agenda for scheduling research.

Appendixes

A

Mathematical Programming: Formulations and Applications

This appendix gives an overview of the types of problems that can be formulated as mathematical programs.

The applications discussed all concern scheduling problems. To understand these examples, the reader should be familiar with the notation introduced in Chapter 2.

This appendix is aimed at people who are already familiar with elementary Operations Research techniques. It makes an attempt to put various notions and problem definitions in perspective. Relatively little is said about the standard solution techniques for dealing with such problems.

A.1 LINEAR PROGRAMMING FORMULATIONS

The most basic mathematical program is the *Linear Program (LP)*. The LP refers to an optimization problem in which the objective and constraints are linear in the variables to be determined. It can be formulated as follows:

$$\text{minimize } c_1 x_1 + c_2 x_2 + \cdots + c_n x_n$$

subject to

$$a_{11} x_1 + a_{12} x_2 + \cdots + a_{1n} x_n \leq b_1$$
$$a_{21} x_1 + a_{22} x_2 + \cdots + a_{2n} x_n \leq b_2$$
$$\vdots$$
$$a_{m1} x_1 + a_{m2} x_2 + \cdots + a_{mn} x_n \leq b_m$$
$$x_j \geq 0 \qquad \text{for } j = 1, \ldots, n.$$

The objective is the minimization of costs. The $c_1, \ldots, c_n$ vector is usually referred to as the cost vector. The variables $x_1, \ldots, x_n$ have to be determined so that the objective function $c_1 x_1 + \cdots + c_n x_n$ is minimized. The column vector $a_{1j}, \ldots, a_{mj}$ is referred to as activity vector j. The value of the variable x_j refers to the level at which this activity j is utilized. The $b_1, \ldots, b_m$ is usually referred to as the resources vector. The fact that in linear programming n denotes the number of activities has nothing to do with the fact that in scheduling theory n refers to the number of jobs; that m denotes the number of resources in linear programming also has nothing to do with the fact that m refers to the number of machines in scheduling theory. The above representation is usually given in the following matrix form:

$$\text{minimize } \bar{c}\bar{x}$$

subject to

$$\mathbf{A}\bar{x} \leq \bar{b}$$
$$\bar{x} \geq 0.$$

There are several algorithms or classes of algorithms for dealing with an LP. The two most important ones are

(i) the simplex methods, and
(ii) the interior point methods.

Although simplex methods work very well in practice, it is not known if there is any version of it that solves the LP problem in polynomial time. The best known example of an interior point method is *Karmarkar's Algorithm*, which is known to solve the LP problem in polynomial time. There are many texts that cover these subjects in depth.

A special case of the linear program is the so-called *transportation* problem. In the transportation problem, the matrix $\mathbf{A}$ takes a special form. The matrix has mn columns and $m + n$ rows and takes the form

$$\mathbf{A} = \begin{bmatrix} \bar{1} & 0 & \cdots & 0 \\ 0 & \bar{1} & \cdots & 0 \\ \vdots & \vdots & \ddots & \vdots \\ 0 & 0 & \cdots & \bar{1} \\ \mathbf{I} & \mathbf{I} & \cdots & \mathbf{I} \end{bmatrix},$$

where the $\bar{1}$ is a row vector with n 1s and the $\mathbf{I}$ is an $n \times n$ identity matrix. All but two entries in each column (activity) of this $\mathbf{A}$ matrix are zero; the two nonzero entries are equal to 1. This matrix is associated with the following problem. Consider a situation in which items have to be shipped from m sources to n destinations. A column (activity) in the $\mathbf{A}$ matrix represents a route from a given source to a given destination. The cost associated with this column (activity) is the cost of transporting one item from the given source to the given destination. The first m entries in the $b_1, \ldots, b_{m+n}$ vector represent the supplies at the m sources, whereas the last n entries of the $b_1, \ldots, b_{m+n}$ vector represent the demands at the n destinations. Usually it is assumed that the sum of the demands equals the sum of the supplies, and the problem is to transport all the items from the sources to the demand points and minimize the total cost incurred. (When the sum of the supplies is less than the sum of the demands there is no feasible solution. When the sum of the supplies is larger than the sum of the demands, an artificial destination can be created where the surplus is sent at zero cost.)

The matrix $\mathbf{A}$ of the transportation problem is an example of a matrix with the so-called *total unimodularity property*. A matrix has the total unimodularity property if the determinant of every square submatrix within the matrix has value $-1, 0$, or 1. It can be easily verified that this is the case with the matrix of the transportation problem. This total unimodularity property has an important consequence: If the values of the supplies and demands are all integers, then there is an optimal solution $x_1, \ldots, x_n$ that is a vector of integers, and the simplex method will find such a solution.

The transportation problem is important in scheduling theory for a number of reasons. First, there are many scheduling problems that can be formulated as transportation problems. Second, transportation problems are often used to obtain bounds in branch and bound procedures applied to NP-hard problems (see Section 3.5).

In the following example, a scheduling problem is described that can be formulated as a transportation problem.

Example A.1.1 (A Transportation Problem)

Consider $Qm \mid p_j = 1 \mid \sum h_j(C_j)$. The speed of machine i is v_i. The variable x_{ijk} is 1 if job j is scheduled as the kth job on machine i and 0 otherwise. So the variable x_{ijk} is associated with an activity. The cost of operating this activity at unit level is

$$c_{ijk} = h_j(C_j) = h_j(k/v_i).$$

Assume that there are a total of $n \times m$ positions (a maximum of n jobs can be assigned to any one machine). Clearly, not all positions will be filled. The n jobs are equivalent to the n sources in the transportation problem and the $n \times m$ positions are the destinations.

The problem can be formulated easily as an LP.

$$\text{minimize} \sum_{i=1}^{m} \sum_{j=1}^{n} \sum_{k=1}^{n} c_{ijk} x_{ijk}$$

subject to

$$\sum_i \sum_k x_{ijk} = 1 \qquad \text{for } j = 1, \dots, n$$

$$\sum_j x_{ijk} \leq 1 \qquad \text{for } i = 1, \dots, m, k = 1, \dots, n$$

$$x_{ijk} \geq 0 \qquad \text{for } i = 1, \dots, m, j = 1, \dots, n, k = 1, \dots, n.$$

The first set of constraints ensures that job j is assigned to one and only one position. The second set of constraints ensures that each position i, k has at most one job assigned to it. Actually from the LP formulation, it is not immediately clear that the optimal values of the variables x_{ijk} have to be either 0 or 1. From the constraints, it may appear at first sight that an optimal solution of the LP formulation may result in x_{ijk} values between 0 and 1. Because of the total unimodularity property, the constraints do not specifically have to require that the variables be either 0 or 1.

An important special case of the transportation problem is the *weighted bipartite matching* problem. This problem can be described as follows. Let $G = (N_1, N_2, A)$ be an undirected bipartite graph. This implies that there are two sets of nodes N_1 and N_2 with arcs connecting nodes from N_1 with nodes from N_2. There are m nodes in N_1 and n nodes in N_2. The set A denotes a set of undirected arcs. The arc $(j, k) \in A$, that connects node $j \in N_1$ with node $k \in N_2$, has a weight w_{jk}. The objective is to find a matching for which the sum of the weights of the arcs is minimum. Let the variable x_{jk} correspond to arc (j, k). The variable x_{jk} equals 1 if the arc (j, k) is selected for the matching and 0 otherwise. The relationship with the transportation problem is clear. Without loss of generality, it may be assumed that $m > n$ (if this is not the case, then sets N_1 and N_2 can be interchanged). The nodes in N_1 then correspond to the sources while the nodes in N_2 correspond to the destinations. At each source, there is exactly one item and at each destination there is a demand for exactly one item as well. The cost of transporting one item from one source to one destination is equal to the weight of the matching. The problem can be formulated as the following LP.

$$\text{minimize} \sum_{j=1}^{m} \sum_{k=1}^{n} w_{jk} x_{jk}$$

subject to

$$\sum_{k=1}^{n} x_{jk} \leq 1 \qquad \text{for } j = 1, \dots, m,$$

$$\sum_{j=1}^{m} x_{jk} \leq 1 \qquad \text{for } k = 1, \dots, n,$$

$$x_{jk} \geq 0 \qquad \text{for } j = 1, \dots, m, k = 1, \dots, n.$$

Again, it is not necessary to explicitly require integrality for the x_{jk} variables. The internal structure of the problem is such that the solution of the linear program is integral. The weighted bipartite matching problem is also important from the point of view of scheduling.

Example A.1.2 (A Weighted Bipartite Matching Problem)

Consider $Rm \mid\mid \sum C_j$. Position $(i, 1)$ now refers to the position of the last job scheduled on machine i; position $(i, 2)$ refers to the position of the job immediately before the last on machine i. Position (i, k) refers to that job on machine i that still has $k - 1$ jobs following it. So in contrast to Example A.1.1, the count of job positions starts at the end and not at the beginning. The variable x_{ijk} is 1 if job j is the kth last job on machine i and 0 otherwise. One set of nodes consists of the n jobs, while the second set of nodes consists of the $n \times m$ positions. The arc that connects job j with position (i, k) has a weight kp_{ij}.

A special case of the weighted bipartite matching problem is the *assignment* problem. A weighted bipartite matching problem is referred to as an assignment problem when $n = m$ (the number of sources is equal to the number of destinations). The assignment problem is also important in scheduling theory. Deterministic as well as stochastic single machine problems with the n jobs having identically distributed processing times can be formulated as assignment problems.

Example A.1.3 (An Assignment Problem)

Consider a special case of the problem discussed in Example A.1.1—namely, $1 \mid p_j = 1 \mid \sum h_j(C_j)$. There are n jobs and n positions, and the assignment of job j to position k has cost $h_j(k)$ associated with it.

A.2 INTEGER PROGRAMMING FORMULATIONS

An *Integer Program (IP)* is basically a linear program with the additional require-ment that the variables $x_1, \ldots, x_n$ have to be integers. If only a subset of the vari-ables are required to be integer and the remaining ones are allowed to be real, the problem is referred to as a *Mixed Integer Program (MIP)*. In contrast to the LP, an efficient (polynomial time) algorithm for the IP or MIP does *not* exist (see Appendix C).

Many scheduling problems can be formulated as integer programs. In what follows, two examples of integer programming formulations are given. The first example describes an integer programming formulation for $1 \mid\mid \sum w_j C_j$. Even though the $1 \mid\mid \sum w_j C_j$ problem is quite easy and can be solved by a simple priority rule, the problem still serves as an illustrative and useful example. This formulation is a generic one and can be used for scheduling problems with multiple machines as well.

Example A.2.1 (An Integer Programming Formulation with Time-Indexed Variables)

To formulate $1 \mid\mid \sum w_j C_j$ as an integer program, let the the integer variable x_{jt} be 1 if job j starts at the integer time t and 0 otherwise. Let $l = C_{\max} - 1$.

$$\text{minimize} \sum_{j=1}^{n} \sum_{t=0}^{l} w_j(t + p_j)x_{jt}$$

subject to

$$\sum_{t=0}^{l} x_{jt} = 1 \qquad \text{for } j = 1, \ldots, n,$$

$$\sum_{j=1}^{n} \sum_{s=\max(t-p_j,0)}^{t-1} x_{js} = 1 \qquad \text{for } t = 0, \ldots, l,$$

$$x_{jt} \in \{0, 1\} \qquad \text{for } j = 1, \ldots, n, t = 0, \ldots, l.$$

The first set of constraints ensures that only one job can start on the machine at any point in time. The second set of constraints ensures that only one job can be processed at any point in time while the last set contains the integrality constraints on the variables. The major disadvantage of this formulation is the number of variables required. There are $nC_{\max}$ variables x_{jt}.

Often there is more than one integer programming formulation of the same problem. In the next example, a different integer programming formulation is given for $1 \mid\mid \sum w_j C_j$. This second formulation can also be applied to $1 \mid prec \mid \sum w_j C_j$.

Example A.2.2 (An Integer Programming Formulation with Sequencing Variables)

Consider $1 \mid prec \mid \sum w_j C_j$. Let x_{jk} denote a 0–1 decision variable that assumes the value 1 if job j precedes job k in the sequence and 0 otherwise. The values x_{jj} have to be 0 for all j. The completion time of job j is then equal to $\sum_{k=1}^{n} p_k x_{kj} + p_j$. The integer programming formulation of the problem without precedence constraints thus becomes

$$\text{minimize} \sum_{j=1}^{n} \sum_{k=1}^{n} w_j p_k x_{kj} + \sum_{j=1}^{n} w_j p_j$$

subject to

$$x_{kj} + x_{jk} = 1 \qquad \text{for } j, k = 1, \ldots, n, \, j \neq k,$$
$$x_{kj} + x_{lk} + x_{jl} \geq 1 \qquad \text{for } j, k, l = 1, \ldots, n, \, j \neq k, \, j \neq l, k \neq l,$$
$$x_{jk} \in \{0, 1\} \qquad \text{for } j, k = 1, \ldots, n,$$
$$x_{jj} = 0 \qquad \text{for } j = 1, \ldots, n.$$

The third set of constraints can be replaced by a combination of (i) a set of linear constraints that require all x_j to be nonnegative, (ii) a set of linear constraints requiring all x_j to be less than or equal to 1, and (iii) a set of constraints requiring all x_j to be integer. Constraints requiring certain precedences between the jobs can be easily added by specifying the corresponding x_{jk} values.

There are several ways to deal with Integer Programs. The best known approaches are:

(i) cutting plane (polyhedral) techniques, and

(ii) branch and bound techniques.

The first class of techniques focuses on the linear program relaxation of the integer program. The techniques aim at generating *additional* linear constraints that have to be satisfied for the variables to be integer. These additional inequalities constrain the feasible set more than the original set of linear inequalities without cutting off integer solutions. Solving the LP relaxation of the IP with the additional inequalities then yields a different solution, which may be integer. If the solution is integer, the procedure stops as the solution obtained is an optimal solution for the original IP. If the variables are not integer, more inequalities are generated. As this method is not used in the main body of the book, it is not further discussed here.

The second approach, branch and bound, is basically a sophisticated way of doing complete enumeration that can be applied to many combinatorial problems. The branching refers to a partitioning of the solution space. Each part of the solution space is then considered separately. The bounding refers to the development of lower bounds for parts of the solution space. If a lower bound on the objectives in one part of the solution space is larger than a solution already obtained in a different part of the solution space, the corresponding part of the former solution space can be disregarded.

Branch and bound can easily be applied to integer programs. Suppose that one solves the LP relaxation of an IP (i.e., solves the IP without the integrality constraints). If the solution of the LP relaxation happens to be integer—say $\bar{x}^0$—then this solution is optimal for the original integer program as well. If $\bar{x}^0$ is not integer, then the value of the optimal solution of the LP relaxation, $\bar{c}\bar{x}^0$, still serves as a lower bound for the value of the optimal solution for the original integer program.

If one of the variables in $\bar{x}^0$ is not integer—say $x_j = r$—then the branch and bound procedure proceeds as follows. The integer programming problem is split into two subproblems by adding two mutually exclusive and exhaustive constraints. In one subproblem, say *Problem (1)*, the original integer program is modified by adding the additional constraint

$$x_j \leq \lfloor r \rfloor,$$

where $\lfloor r \rfloor$ denotes the largest integer smaller than r, whereas in the other subproblem, say *Problem (2)*, the original integer program is modified by adding the additional constraint

$$x_j \geq \lceil r \rceil,$$

where $\lceil r \rceil$ denotes the smallest integer larger than r. It is clear that the optimal solution of the original integer program has to lie in the feasible region of one of these two subproblems.

The branch and bound procedure now considers the LP relaxation of one of the subproblems—say *Problem (1)*—and solves it. If the solution is integer, then this branch of the tree does not have to be explored further as this solution is the optimal solution of the original integer programming version of *Problem (1)*. If the solution is not integer, *Problem (1)* has to be split into two subproblems—say *Problem (1.1)* and *Problem (1.2)*—through the addition of mutually exclusive and exhaustive constraints.

Proceeding in this manner, a tree is constructed. From every node that corresponds to a noninteger solution, a branching occurs to two other nodes, and so on.

The procedure stops when all nodes of the tree correspond to problems of which the linear program relaxations have either an integer solution or a fractional solution that is higher than a feasible integer solution found elsewhere. The node with the best solution is the optimal solution of the original integer program.

An enormous amount of research and experimentation has been done on branch and bound techniques. For example, the bounding technique described earlier based on LP relaxations is relatively simple. There are other bounding techniques that generate lower bounds that are substantially better (higher) than the LP relaxation bounds and better bounds typically cut down the overall computation time substantially. One of the most popular bounding techniques is referred to as *Lagrangian relaxation*. This strategy, instead of dropping the integrality constraints, relaxes some of the main constraints. However, the relaxed constraints are not totally dropped. Instead, they are dualized or weighted in the objective function with suitable Lagrange multipliers to discourage violations.

Two ways of applying branch and bound have proved to be very useful:

 (i) branch and cut, and
 (ii) branch and price (also known as column generation).

Branch and cut combines branch and bound with cutting plane techniques. Branch and cut uses in each subproblem of the branching tree a cutting plane algorithm to generate a lower bound. That is, a cutting plane algorithm is applied to the problem formulation that includes the additional constraints introduced at that node.

Branch and price, also referred to as *column generation*, is often used to solve integer programs that have a huge number of variables (columns). A branch and price algorithm always works with a restricted problem in a sense that only a subset of the variables is taken into account; the variables outside the subset are fixed at 0. From the theory of Linear Programming, it is known that after solving this restricted problem to optimality, each variable that is included has a non-negative so-called *reduced cost*. If each variable that is not included in the restricted problem also has a non-negative reduced cost, then an optimal solution for the original problem is found. However, if there are variables with a negative reduced cost, then one or more of these variables should be included in the restricted problem. The main idea behind column generation is that the occurrence of variables with negative reduced cost is not verified by enumerating all variables, but rather by solving an optimization problem. This optmization problem is called the *pricing problem* and is defined as the problem of finding the variable with minimum reduced cost. To apply column generation effectively, it is important to find a good method for solving the pricing problem. A branch and bound algorithm in which the lower bounds are computed by solving LP relaxations through column generation is called a branch and price algorithm.

Column generation has been applied successfully to various parallel machine scheduling problems.

A.3 DISJUNCTIVE PROGRAMMING FORMULATIONS

There is a large class of mathematical programs in which the constraints can be divided into a set of *conjunctive* constraints and one or more sets of *disjunctive* constraints. A set of constraints is called *conjunctive* if each one of the constraints has to be satisfied. A set of constraints is called *disjunctive* if at least one of the constraints has to be satisfied but not necessarily all.

In the standard linear program, all constraints are conjunctive. The mixed integer program described in Example A.2.1 in essence contains pairs of disjunctive constraints. The fact that the integer variable x_{jk} has to be either 0 or 1 can be enforced by a pair of disjunctive linear constraints: either $x_{jk} = 0$ or $x_{jk} = 1$. This implies that the problem $1 \mid prec \mid \sum w_j C_j$ can be formulated as a *disjunctive program* as well.

Example A.3.1 (A Disjunctive Programming Formulation)

Before formulating $1 \mid prec \mid \sum w_j C_j$ as a disjunctive program, it is of interest to represent the problem by a disjunctive graph model. Let N denote the set of nodes that correspond to the n jobs. Between any pair of nodes (jobs) j and k in this graph, exactly one of the following three conditions has to hold:

(i) job j precedes job k,
(ii) job k precedes job j, and
(iii) jobs j and k are independent with respect to one another.

The set of directed arcs A represent the precedence relationships between the jobs. These arcs are the so-called conjunctive arcs. Let set I contain all the pairs of jobs that are independent of one another. Each pair of jobs $(j, k) \in I$ are now connected with one another by two arcs going in opposite directions. These arcs are referred to as disjunctive arcs. The problem is to select from each pair of disjunctive arcs between two independent jobs j and k one arc that indicates which one of the two jobs goes first. The selection of disjunctive arcs has to be such that these arcs, together with the conjunctive arcs, do not contain a cycle. The selected disjunctive arcs, together with the conjunctive arcs, determine a schedule for the n jobs.

Let the variable x_j in the disjunctive program formulation denote the completion time of job j. The set A denotes the set of precedence constraints $j \to k$ that require job j to be processed before job k.

$$\text{minimize} \sum_{j=1}^{n} w_j x_j$$

subject to

$$x_k - x_j \geq p_k \qquad\qquad \text{for all } j \rightarrow k \in A,$$

$$x_j \geq p_j \qquad\qquad \text{for } j = 1, \dots, n,$$

$$x_k - x_j \geq p_k \text{ or } x_j - x_k \geq p_j \qquad \text{for all } (j, k) \in I.$$

The first and second set of constraints are sets of conjunctive constraints. The third set is a set of disjunctive constraints.

The same techniques that are applicable to integer programs are also applicable to disjunctive programs. The application of branch and bound to a disjunctive program is straightforward. The LP relaxation of the disjunctive program has to be solved (i.e., the LP obtained after deleting the set of disjunctive constraints). If the optimal solution of the LP by chance satisfies all disjunctive constraints, then the solution is optimal for the disjunctive program as well. However, if one of the disjunctive constraints is violated, say the constraint

$$(x_k - x_j) \geq p_k \text{ or } (x_j - x_k) \geq p_j,$$

then two additional LPs are generated. One has the additional constraint $(x_k - x_j) \geq p_k$ and the other has the additional constraint $(x_j - x_k) \geq p_j$. The procedure is in all other respects similar to the branch and bound procedure for integer programming.

COMMENTS AND REFERENCES

Many books have been written on linear programming, integer programming, and combinatorial optimization. Examples of some relatively recent ones are Papadimitriou and Steiglitz (1982), Parker and Rardin (1988), Nemhauser and Wolsey (1988), Schrijver (1998), and Wolsey (1998).

Blazewicz, Dror, and Weglarz (1991) give an overview of mathematical programming formulations for machine scheduling. The thesis by Van de Velde (1991) contains many examples (and references) of integer programming formulations for scheduling problems. Dauzère-Pérès and Sevaux (1998) present an interesting comparison of four different integer programming formulations for $1 \mid r_j \mid \sum U_j$.

The first example of branch and bound combined with Lagrangian relaxation applied to scheduling is due to Fisher (1976); he presents a solution method for $1 \mid\mid \sum w_j T_j$. Dauzère-Pérès and Sevaux (1999) and Baptiste, Peridy, and Pinson (2000) apply Lagrangian relaxation to $1 \mid r_j \mid \sum w_j U_j$. Barnhart, Johnson, Nemhauser, Savelsbergh, and Vance (1998) provide an excellent general overview of branch and price (column generation), and Van den Akker, Hoogeveen, and Van de Velde (1999) as well as Chen and Powell (1999) apply this technique specifically to scheduling.

B

Deterministic and Stochastic Dynamic Programming

Dynamic programming is one of the more widely used techniques for dealing with combinatorial optimization problems. Dynamic Programming can be applied to problems that are solvable in polynomial time, as well as problems that cannot be solved in polynomial time (see Appendix C). It has proved to be very useful for stochastic problems as well.

B.1 DETERMINISTIC DYNAMIC PROGRAMMING

Dynamic programming is basically a complete enumeration scheme that attempts, via a divide and conquer approach, to minimize the amount of computation to be done. The approach solves a series of subproblems until it finds the solution of the original problem. It determines the optimal solution for each subproblem and its contribution to the objective function. At each iteration, it determines the optimal solution for a subproblem, which is larger than all previously solved subproblems. It finds a solution for the current subproblem by utilizing all the information obtained earlier in the solutions of the previous subproblems.

Dynamic programming is characterized by three types of equations:

(i) initial conditions,
(ii) a recursive relation, and
(iii) an optimal value function.

In scheduling a choice can be made between forward dynamic programming and backward dynamic programming. The following example illustrates the use of forward dynamic programming.

Example B.1.1 (A Forward Dynamic Programming Formulation)

Consider $1 \parallel \sum h_j(C_j)$. This problem is a very important problem in scheduling theory as it comprises many of the objective functions studied in Part I of the book. The problem is, for example, a generalization of $1 \parallel \sum w_j T_j$ and is therefore NP-hard in the strong sense. Let J denote a subset of the n jobs and assume the set J is processed first. Let

$$V(J) = \sum_{j \in J} h_j(C_j)$$

provided the set of jobs J is processed first. The dynamic progamming formulation of the problem is based on the following initial conditions, recursive relation, and optimal value function.

Initial Conditions:

$$V(\{j\}) = h_j(p_j), \qquad j = 1, \ldots, n$$

Recursive Relation:

$$V(J) = \min_{j \in J} \left(V(J - \{j\}) + h_j \left(\sum_{k \in J} p_k \right) \right)$$

Optimal Value Function:

$$V(\{1, \ldots, n\}).$$

The idea behind this dynamic programming procedure is relatively straightforward. At each iteration, the optimal sequence for a subset of the jobs (say a subset J, which contains l jobs) is determined, assuming this subset goes first. This is done for *every* subset of size l. There are $n!/(l!(n-l)!)$ subsets. For each subset the contribution of the l scheduled jobs to the objective function is computed. Through the recursive relation, this is expanded to every subset that contains $l + 1$ jobs. Each one of the $l + 1$ jobs is considered as a candidate to go last. When using the recursive relation, the actual sequence of the l jobs of the smaller subset does not have to be taken into consideration; only the contribution of the l jobs to the objective has to be known. After the value $V(\{1, \ldots, n\})$ has been determined, the optimal sequence is obtained through a simple backtracking procedure.

The computational complexity of this problem can be determined as follows. The value of $V(J)$ has to be determined for all subsets that contain l jobs. There are $n!/l!(n-l)!$ subsets. So the total number of evaluations that have to be done are

$$\sum_{l=1}^{n} \frac{n!}{l!(n-l)!} = O(2^n).$$

Example B.1.2 (An Application of Forward Dynamic Programming)

Consider the problem described in the previous example with the following jobs.

jobs	1	2	3
p_j	4	3	6
$h_j(C_j)$	$C_1 + C_1^2$	$3 + C_1^3$	$8C_3$

So $V(\{1\}) = 20$, $V(\{2\}) = 30$ and $V(\{3\}) = 48$. The second iteration of the procedure considers all sets containing two jobs. Applying the recursive relation yields

$$V(\{1, 2\}) = \min\left(V(\{1\}) + h_2(p_1 + p_2), V(\{2\}) + h_1(p_2 + p_1)\right)$$

$$= \min(20 + 346, 30 + 56) = 86.$$

So if jobs 1 and 2 precede job 3, then job 2 has to go first and job 1 has to go second. In the same way, it can be determined that $V(\{1, 3\}) = 100$ with job 1 going first and job 3 going second and that $V(\{2, 3\}) = 102$ with job 2 going first and job 3 going second. The last iteration of the procedure considers set $\{1, 2, 3\}$.

$$V(\{1, 2, 3\}) = \min\left(V(\{1, 2\}) + h_3(p_1 + p_2 + p_3), V(\{2, 3\})\right.$$

$$\left. + h_1(p_1 + p_2 + p_3), V(\{1, 3\}) + h_2(p_1 + p_2 + p_3)\right).$$

So

$$V(\{1, 2, 3\}) = \min\left(86 + 104, 102 + 182, 100 + 2200\right) = 190.$$

It follows that jobs 1 and 2 have to go first and job 3 last. The optimal sequence is 2, 1, 3 with objective value 190.

In the following example, the same problem is handled through the backward dynamic programming procedure. In scheduling problems, the backward version typically can be used only for problems with a makespan that is schedule independent (e.g., single machine problems without sequence dependent setups, multiple machine problems with jobs that have identical processing times).

The use of backward dynamic programming is nevertheless important as it is somewhat similar to the dynamic programming procedure discussed in the next section for stochastic scheduling problems.

Example B.1.3 (A Backward Dynamic Programming Formulation)

Consider again $1 \mid\mid \sum h_j(C_j)$. It is clear that the makespan $C_{\max}$ is schedule independent and that the last job is completed at $C_{\max}$, which is equal to the sum of the n processing times.

Again, J denotes a subset of the n jobs, and it is assumed that J is processed first. Let J^C denote the complement of J. So set J^C is processed last. Let $V(J)$ denote the minimum contribution of the set J^C to the objective function. In other words, $V(J)$ represent the minimum additional cost to complete all *remaining* jobs after all jobs in set J have already been completed.

The backward dynamic programming procedure is now characterized by the following initial conditions, recursive relation, and optimal value function.

Initial Conditions:

$$V(\{1, \ldots, j-1, j+1, \ldots, n\}) = h_j(C_{\max}) \qquad j = 1, \ldots, n$$

Recursive Relation:

$$V(J) = \min_{j \in J^C} \left(V(J \cup \{j\}) + h_j \left(\sum_{k \in J \cup \{j\}} p_k \right) \right)$$

Optimal Value Function:

$$V(\emptyset).$$

Again, the procedure is relatively straightforward. At each iteration, the optimal sequence for a subset of the n jobs—say a subset J^C of size l—is determined, assuming this subset goes *last*. This is done for every subset of size l. Through the recursive relation, this is expanded for every subset of size $l + 1$. The optimal sequence is obtained when the subset comprises all jobs. Note that, as in Example B.1.1, subset J goes first. However, in Example B.1.1, set J denotes the set of jobs already scheduled while in this example set J denotes the set of jobs still to be scheduled.

Example B.1.4 (An Application of Backward Dynamic Programming)

Consider the same instance as in Example B.1.2. The makespan $C_{\max}$ is 13. So

$$V(\{1, 3\}) = h_2(C_{\max}) = 2200$$

$$V(\{1, 2\}) = h_3(C_{max}) = 104$$
$$V(\{2, 3\}) = h_1(C_{max}) = 182.$$

The second iteration of the procedure results in the following recursive relations.

$$V(\{1\}) = \min(V(\{1, 2\}) + h_2(p_1 + p_2), V(\{1, 3\}) + h_3(p_1 + p_3))$$
$$= \min(104 + 346, 2200 + 80) = 450.$$

In the same way, $V(\{2\})$ and $V(\{3\})$ can be determined: $V(\{2\}) = 160$ and $V(\{3\}) = 914$. The last iteration results in the recursive relation

$$V(\emptyset) = \min(V(\{1\}) + h_1(p_1), V(\{2\}) + h_2(p_2), V(\{3\}) + h_3(p_3))$$
$$= \min(450 + 20, 160 + 30, 914 + 48) = 190.$$

Of course, dynamic programming can also be used for problems that are polynomial time solvable. Examples of such dynamic programming algorithms are the $O(n^2)$ procedure for $1 \mid prec \mid h_{max}$ and the pseudopolynomial time $O(n^4 \sum p_j)$ procedure for $1 \mid\mid \sum T_j$ described in Chapter 3.

Dynamic programming concepts can also be used to prove the optimality of certain rules (e.g., LRPT for $P \mid prmp \mid C_{max}$).

The proofs are then done through a combination of induction and contradiction. The induction argument assumes that the priority rule is optimal for $k - 1$ jobs. To show that the rule is optimal for k jobs, a contradiction argument is used. Assume that at time 0 an action is taken that is not prescribed by the priority rule. At the first job completion there is one less job (i.e., $k - 1$ jobs), and the scheduler has to revert back to the priority rule because of the induction hypothesis.

It has to be shown now that starting out at time 0 following the priority rule results in a lower objective than not acting according to the priority rule at time 0 and switching over to the priority rule at the first job completion. This proof technique is usually applied in a preemptive setting. The proof of optimality of the LRPT rule for $Pm \mid prmp \mid C_{max}$ in Section 5.2 is an example of this technique.

B.2 STOCHASTIC DYNAMIC PROGRAMMING

Dynamic programming is often used in stochastic sequential decision processes, especially when the random variables are exponentially distributed. This class of decision processes is usually referred to as *Markovian Decision Processes (MDPs)*. An MDP can be characterized, in the same way as a deterministic dynamic program, by

  (i) initial conditions,
 (ii) a recursive relation, and
(iii) an optimal value function.

The setup of an MDP formulation of a scheduling problem is fairly similar to the setup of a backward dynamic program as described in Example B.1.3.

Example B.2.1 (An MDP Formulation of a Stochastic Scheduling Problem)

Consider the following stochastic counterpart of $Pm \mid prmp \mid C_{\max}$ with m machines in parallel and n jobs. The processing time of job j is exponentially distributed with rate λ_j. Consider a particular time t. Let J denote the set of jobs already completed and let J^C the set of jobs still in the process. Let $V(J)$ denote the expected value of the remaining completion time under the optimal schedule when the set of jobs J already have been completed. In this respect, notation $V(J)$ is somewhat similar to notation used in Example B.1.3. The following initial conditions, recursive relation, and optimal value function characterize this Markov Decision Process.

Initial Conditions:

$$V(\{1, \dots, j-1, j+1, \dots, n\}) = \frac{1}{\lambda_j}$$

Recursive Relation:

$$V(J) = \min_{j,k \in J^C} \left(\frac{1}{\lambda_j + \lambda_k} + \frac{\lambda_j}{\lambda_j + \lambda_k} V(J \cup \{j\}) + \frac{\lambda_k}{\lambda_j + \lambda_k} V(J \cup \{k\}) \right)$$

Optimal Value Function:

$$V(\emptyset).$$

The initial conditions are clear. If only job j remains to be completed, then the expected time until all jobs have completed their processing is, because of the memoryless property, $1/\lambda_j$. The recursive relation can be explained as follows. Suppose two or more jobs remain to be completed. If jobs j and k are selected for processing, then the expected remaining time until all jobs have completed their processing can be computed by conditioning on which one of the two jobs finishes first with its processing. The first completion occurs after an expected time $1/(\lambda_j + \lambda_k)$. With probability $\lambda_j/(\lambda_j + \lambda_k)$, it is job j that is completed first; the expected remaining time needed to complete remaining jobs is then $V(J \cup \{j\})$. With probability $\lambda_k/(\lambda_j + \lambda_k)$, it is job k that is completed first; the expected time needed to complete then all remaining jobs is $V(J \cup \{k\})$.

Dynamic programming is also used for stochastic models as a basis to verify the optimality of certain priority rules. The proofs are then done through a combination of induction and contradiction, very much as they are done for the deterministic models.

COMMENTS AND REFERENCES

Many books have been written on deterministic as well as stochastic dynamic programming. See, for example, Denardo (1982), Ross (1983b), and Bertsekas (1987).

C

Complexity Theory

Complexity theory is based on a mathematical framework developed by logicians and computer scientists. This theory was developed to study the intrinsic difficulty of algorithmic problems and has proved very useful for combinatorial optimization. This appendix presents a brief overview of this theory and its ramifications for scheduling.

The applications discussed again concern scheduling problems. To understand these examples, the reader should be familiar with the notation introduced in Chapter 2.

C.1 PRELIMINARIES

In complexity theory, the term *problem* refers to the generic description of a problem. For example, the scheduling problem $Pm \mid\mid C_{\max}$ is a problem as is the linear programming problem described in the first section of Appendix A. The term *instance* refers to a problem with a given set of numerical data. For example, the setting with two machines, five jobs with processing times 2, 3, 5, 5, 8, and the makespan objective is an instance of the $Pm \mid\mid C_{\max}$ problem.

With each instance, there is a size associated. The size of an instance refers to the length of the data string necessary to specify the instance. It is also referred to as the length of the encoding. For example, consider instances of the problem $Pm \parallel C_{max}$ and assume that a convention is made to encode every instance in a certain way. The number of machines is specified, followed by the number of jobs and then the processing times of each one of n jobs. The different pieces of data may be separated from one another by commas, which perform the functions of separators. The instance of the $Pm \parallel C_{max}$ problem described earlier is then encoded as

$$2, 5, 2, 3, 5, 5, 8.$$

One could say that the size of the instance under this encoding scheme is 7 (not counting the separators). Of course, the length of the encoding heavily depends on the encoding conventions. In the prior example, the processing times were encoded using the decimal system. If all the data were presented in binary form, the length of the encoding changes. The example then becomes

$$10, 101, 10, 11, 101, 101, 1000.$$

The length of the encoding is clearly larger. Another form of encoding is the unary encoding. Under this encoding, the integer number n is represented by n ones. According to this encoding, the prior example becomes

$$11, 11111, 11, 111, 11111, 11111, 11111111.$$

Clearly, the size of an instance under unary encoding is larger than that under binary encoding. The size of an instance depends not only on the number of jobs, but also on the length of the processing times of the jobs. One instance is larger than another instance with an equal number of jobs if all the processing times in the first instance are larger than all the processing times in the second instance.

However, in practice, the size of an instance is often simply characterized by the number of jobs (n). Although this may appear at first sight somewhat crude, it is sufficiently accurate for making distinctions between the complexities of different problems. In this book, the size of an instance is measured by the number of jobs.

The efficiency of an algorithm for a given problem is measured by the maximum (worst-case) number of computational steps needed to obtain an optimal solution as a function of the size of the instance. This, in turn, requires a definition of a computational step. To define a computational step, a standard model of computing is used—the *Turing* machine. Any standard text on computational complexity contains the assumptions of the Turing machine; however, these assumptions are beyond the scope of this appendix. In practice, however, a computational step in an algorithm is either a comparison, a multiplication, or any data manipulation step concerning one job. The efficiency of an algorithm is then measured by the maximum number of computational steps needed to obtain an optimal solution (as a function of the size of the instance; i.e., the number of jobs). The number of computational steps may often be just the maximum number of iterations the algorithm has to go through. Even this number of iterations is typically approximated.

For example, if careful analysis of an algorithm establishes that the maximum number of iterations needed to obtain an optimal solution is $1500 + 100n^2 + 5n^3$, then only the term that, as a function of n, increases the fastest is of importance. This algorithm is then referred to as an $O(n^3)$ algorithm. In spite of the fact that for small n the first two terms have a larger impact on the number of iterations, these first two terms are not of interest for large-scale problems. For large numbers of n, the third term has the largest impact on the maximum number of iterations required. Even the coefficient of this term (the 5) is not that important. An $O(n^3)$ algorithm is usually referred to as a polynomial time algorithm; the number of iterations is polynomial in the size (n) of the problem. Such a polynomial time algorithm is in contrast to an algorithm that is either $O(4^n)$ or $O(n!)$. The number of iterations in an $O(4^n)$ or an $O(n!)$ algorithm is, in the worst case, exponential in the size of the problem.

Some of the easiest scheduling problems can be solved through a simple priority rule (e.g., WSPT, EDD, LPT, etc.). To determine the optimal order then requires a simple sorting of the jobs based on one or two parameters. In the following example, it is shown that a simple sort can be done in $O(n \log(n))$ time.

Example C.1.1 (MERGESORT)

The input of the algorithm is a sequence of numbers $x_1, \ldots, x_n$ and the desired output is the sequence $y_1, \ldots, y_n$, a permutation of the input, satisfying $y_1 \leq y_2 \leq \cdots \leq y_n$. One procedure for this problem is in the literature referred to as MERGESORT. This method takes two sorted sequences S_1 and S_2 of equal size as its input and produces a single sequence S containing all elements of S_1 and S_2 in sorted order. This algorithm repeatedly selects the larger of the largest elements remaining in S_1 and S_2 and then deletes the element selected. This recursive procedure requires for n elements the following number of comparisons:

$$T(1) = 0$$

$$T(n) = 2T\left(\frac{n}{2}\right) + (n - 1).$$

It can be easily shown that if n is a power of 2, the total number of comparisons is $O(n \log(n))$. Examples of problems that can be solved this way in polynomial time are $1 \ || \ \sum w_j C_j$ and $1 \ || \ L_{max}$.

Example C.1.2 (Complexity of the Assignment Problem)

The assignment problem described in the first part of Appendix A can be solved in polynomial time; actually in $O(n^3)$.

Actually, any LP can be solved in polynomial time. However, in contrast to the linear program, there exists no polynomial time algorithm for the integer program.

C.2 POLYNOMIAL TIME SOLUTIONS VERSUS NP-HARDNESS

In complexity theory, a distinction is made between *optimization* problems and *decision* problems. The question raised in a decision problem requires either a yes or no answer. These decision problems are therefore often also referred to as yes–no problems. With every optimization problem, one can associate a decision problem. For example, in the problem $Fm \parallel C_{\max}$ the makespan has to be minimized. In the associated decision problem, the question is raised whether there exists a schedule with a makespan less than a given value z. It is clear that the optimization problem and the related decision problem are strongly connected. Actually, if there exists a polynomial time algorithm for the optimization problem, then there exists a polynomial time algorithm for the decision problem and vice versa. A fundamental concept in complexity theory is the concept of *problem reduction.* Very often it occurs that one combinatorial problem is a special case of another problem, equivalent to another problem, or more general than another problem. Often an algorithm that works well for one combinatorial problem works as well for another after only minor modifications. Specifically, it is said that problem P *reduces* to problem P′ if for any instance of P an equivalent instance of P′ can be constructed. In complexity theory, usually a more stringent notion is used. Problem P *polynomially* reduces to problem P′ if a polynomial time algorithm for P′ implies a polynomial time algorithm for P. Polynomial reducibility of P to P′ is denoted by $P \propto P'$. If it is known that if there does not exist a polynomial time algorithm for problem P, then there does not exist a polynomial time algorithm for problem P′ either.

Some formal definitions of problem classes are now in order.

Definition C.2.1. (Class $\mathcal{P}$)   *The class $\mathcal{P}$ contains all decision problems for which there exists a Turing machine algorithm that leads to the right yes–no answer in a number of steps bounded by a polynomial in the length of the encoding.*

The definition of the class $\mathcal{P}$ is based on the time it takes a Turing machine to *solve* a decision problem. There exists a larger class of problems that is based on the time it takes a Turing machine to *verify* whether a *given* solution of a decision problem is correct or not. This solution may be in the form of a clue (e.g., for a scheduling problem a clue may be a sequence or a schedule).

Definition C.2.2. (Class $\mathcal{NP}$)   *The class $\mathcal{NP}$ contains all decision problems for which the correct answer, given a proper clue, can be verified by a Turing machine in a number of steps bounded by a polynomial in the length of the encoding.*

For example, if the decision problem associated with the $F3 \parallel C_{\max}$ problem is considered, then a proper clue could be an actual schedule or permutation of jobs that results in a certain makespan, which is less than the given fixed value z. To verify whether the correct answer is yes, the algorithm takes the given permutation and computes the makespan under this permutation to show that it is less than the given constant z. Verifying that a sequence satisfies such a condition is, of course, simpler than *finding* a sequence that satisfies such a condition. The class $\mathcal{P}$ is, clearly, a subclass of the class $\mathcal{NP}$.

One of the most important open issues in mathematical logic and combinatorial optimization is the question of whether $\mathcal{P} = \mathcal{NP}$. If $\mathcal{P}$ were equal to $\mathcal{NP}$, then there would exist polynomial time algorithms for a very large class of problems for which up to now no polynomial time algorithm has been found.

Definition C.2.3. (NP-Hardness)   *A problem P, either a decision problem or an optimization problem, is called NP-hard if the entire class of $\mathcal{NP}$ problems polynomially reduces to P.*

Actually, not all problems within the NP-hard class are equally difficult. Some problems are more difficult than others. For example, it may be that a problem can be solved in polynomial time as a function of the size in unary encoding, whereas it cannot be solved in polynomial time as a function of the size of the problem in binary encoding. For other problems, there may not exist polynomial time algorithms under either unary or binary encoding. The first class of problems are not as hard as the second class of problems. The problems in this first class are usually referred to as *NP-hard in the ordinary sense* or simply *NP-hard*. The algorithms for this class of problems are called *pseudopolynomial*. The second class of problems are usually referred to as *strongly NP-hard*.

A great variety of decision and optimization problems have been shown to be either NP-hard or strongly NP-hard. Some of the more important decision problems in this class are listed next.

The most important NP-hard decision problem is a problem in Boolean logic, the so-called SATISFIABILITY problem. In this problem, there are n Boolean variables $x_1, \ldots, x_n$. These variables may assume either the value 0 (false) or the value 1 (true). A so-called *clause* is a function of a subset of these variables. A variable x_j may appear in a clause as x_j or as its *negation* $\rceil x_j$. If $x_j = 0$, then its negation $\rceil x_j = 1$ and vice versa. The clause $(x_2 + \rceil x_5 + x_1)$ is 1 (true) if at least one of the elements in the clause gives rise to a true value (i.e., $x_2 = 1$ and/or $x_5 = 0$ and/or $x_1 = 1$). An expression can consist of a number of clauses. For the entire expression to be true, there must exist an assignment of 0s and 1s to the n variables which makes *each* clause true. More formally, the SATISFIABILITY problem is defined as follows.

Definition C.2.4. (SATISFIABILITY)   *Given a set of variables and a collection of clauses defined over the variables, is there an assignment of values to the variables for which each one of the clauses is true?*

Example C.2.5 (SATISFIABILITY)

Consider the expression

$$(x_1+\rceil x_4 + x_3+\rceil x_2)(\rceil x_1+\rceil x_2 + x_4+\rceil x_3)(\rceil x_2+\rceil x_3 + x_1 + x_5)(\rceil x_5+\rceil x_1 + x_4 + x_2).$$

It can be easily verified that the assignment $x_1 = 0, x_2 = 0, x_3 = 0, x_4 = 0,$ and $x_5 = 0$ gives a truth assignment to each one of the four clauses.

The following four NP-hard problems are very important from the scheduling point of view.

Definition C.2.6. (PARTITION)   *Given positive integers $a_1, \ldots, a_t$ and $b = (1/2) \sum_{j=1}^{t} a_j$, do there exist two disjoint subsets S_1 and S_2 such that*

$$\sum_{j \in S_i} a_j = b$$

for $i = 1, 2$?

Definition C.2.7. (3-PARTITION)   *Given positive integers $a_1, \ldots, a_{3t}, b$ with*

$$\frac{b}{4} < a_j < \frac{b}{2}, \qquad j = 1, \ldots, 3t,$$

and

$$\sum_{j=1}^{3t} a_j = tb,$$

do there exist t pairwise disjoint three element subsets $S_i \subset \{1, \ldots, 3t\}$ such that

$$\sum_{j \in S_i} a_j = b$$

for $i = 1, \ldots, t$?

Definition C.2.8. (HAMILTONIAN CIRCUIT)   *For a graph $G = (N, A)$ with node set N and arc set A, does there exist a circuit (or tour) that connects all nodes in N exactly once?*

Definition C.2.9. (CLIQUE) *For a graph $G = (N, A)$ with node set N and arc set A, does there exist a clique of size c? That is, does there exist a set $N^* \subset N$, consisting of c nodes, such that for each distinct pair of nodes $u, v \in N^*$, the arc $\{u, v\}$ is an element of A?*

Problems of which the complexity is established through a reduction from PAR-TITION typically allow for pseudopolynomial time algorithms and are therefore NP-hard in the ordinary sense.

C.3 EXAMPLES

This section contains a number of examples illustrating several simple problem reductions.

Example C.3.1 (The Knapsack Problem)

Consider the so-called *knapsack* problem, which is equivalent to the scheduling problem $1 \mid d_j = d \mid \sum w_j U_j$. It is clear why this problem is usually referred to as the knapsack problem. The value d refers to the size of the knapsack and the jobs are the items that have to be put into the knapsack. The size of item j is p_j and the benefit obtained by putting item j into the knapsack is w_j. It can be shown that PARTITION reduces to the knapsack problem by taking

$$n = t,$$

$$p_j = a_j,$$

$$w_j = a_j,$$

$$d = \frac{1}{2} \sum_{j=1}^{t} a_j = b,$$

$$z = \frac{1}{2} \sum_{j=1}^{t} a_j = b.$$

It can be verified that there exists a schedule with an objective value less than or equal to $1/2 \sum_{j=1}^{n} w_j$ if and only if there exists a solution for the PARTITION problem.

Example C.3.2 (Minimizing Makespan on Parallel Machines)

Consider $P2 \parallel C_{\max}$. It can be shown that PARTITION reduces to this problem by taking

$$n = t,$$

$$p_j = a_j,$$

$$z = \frac{1}{2} \sum_{j=1}^{t} a_j = b.$$

It is trivial to verify that there exists a schedule with an objective value less than or equal to $1/2 \sum_{j=1}^{n} p_j$ if and only if there exists a solution for the PARTITION problem.

The following example illustrates how $P2 \parallel C_{\max}$ can be solved in pseudopolynomial time via a simple dynamic programming algorithm.

Example C.3.3 (Application of a Pseudopolynomial Time Algorithm)

Consider $P2 \parallel C_{\max}$ with five jobs.

jobs	1	2	3	4	5
p_j	7	8	2	4	1

The question is: Does there exist a partition with a makespan equal to 11 (half of the sum of the processing times)? The following dynamic program (see Appendix B) results in an optimal partition. Let the indicator variable $I(j, z)$ be 1 if there is a subset of jobs $\{1, 2, \ldots, j-1, j\}$ for which the sum of the processing times is exactly z and 0 otherwise. The values of $I(j, z)$ have to be determined for $j = 1, \ldots, 5$ and $z = 0, \ldots, 11$. The procedure basically fills in 0-1 entries in the $I(j, z)$ matrix row by row.

z	0	1	2	3	4	5	6	7	8	9	10	11
$j = 1$	1	0	0	0	0	0	0	1	0	0	0	0
$j = 2$	1	0	0	0	0	0	0	1	1	0	0	0
$j = 3$	1	0	1	0	0	0	0	1	1	1	1	0
$j = 4$	1	0	1	0	1	0	1	1	1	1	1	1
$j = 5$	1	1	1	1	1	1	1	1	1	1	1	1

From the table entries, it follows that there is a partition of the jobs that results in a makespan of 11. From simple backtracking, it follows that jobs 4 and 1 have a total processing time of 11. Clearly, this table entry algorithm is polynomial in $n \sum p_j/2$. If the instance is encoded according to the *unary* format, the length of the encoding is $O(\sum p_j)$. The algorithm is polynomial in the size of the problem. However, if the instance is encoded according to the *binary* format, the length of the encoding is $O(\log(\sum p_j))$ and the algorithm, being of $O(n \sum p_j)$, is *not* bounded by any polynomial function of $O(\log(\sum p_j))$. The algorithm is therefore pseudopolynomial.

NP-hard problems of which the complexity is established via a reduction from SATISFIABILITY, 3-PARTITION, HAMILTONIAN CIRCUIT or CLIQUE are strongly NP-hard.

Example C.3.4 (Minimizing Makespan in a Job Shop)

Consider $J2 \mid recrc, prmp \mid C_{max}$. It can be shown that 3-PARTITION reduces to $J2 \mid recrc, prmp \mid C_{max}$. The reduction to the scheduling problem is based on the following transformation. The number of jobs, n, is chosen to be equal to $3t + 1$. Let

$$p_{1j} = p_{2j} = a_j \qquad j = 1, \dots, 3t.$$

Each one of these $3t$ jobs has to be processed first on machine 1 and then on machine 2. These $3t$ jobs do *not* recirculate. The last job, job $3t + 1$, has to start its processing on machine 2 and then has to alternate between machines 1 and 2. It has to be processed this way t times on machine 2 and t times on machine 1, and each one of these $2t$ processing times is equal to b. For a schedule to have a makespan

$$C_{max} = 2tb,$$

this last job has to be scheduled without interruption. The remaining time slots can be filled without idle times by jobs $1, \dots, 3t$ if and only if 3-PARTITION has a solution.

Example C.3.5 (Sequence-Dependent Setup Times)

Consider the Traveling Salesman Problem or, in scheduling terms, the $1 \mid s_{jk} \mid C_{max}$ problem. That HAMILTONIAN CIRCUIT can be reduced to $1 \mid s_{jk} \mid C_{max}$ can be shown as follows. Let each node in the HAMILTONIAN CIRCUIT correspond to a city in the Traveling Salesman Problem. Let the distance between two cities equal to 1 if there exists an arc between the two corresponding nodes in the HAMILTONIAN CIRCUIT. Let the distance between two cities equal to 2 if there does *not* exist an arc between the two corresponding nodes. The bound on the objective is equal to the number of nodes in the HAMILTONIAN CIRCUIT. It is easy to see that the two problems are equivalent.

Example C.3.6 (Scheduling with Precedence Constraints)

Consider $1 \mid p_j = 1, prec \mid \sum U_j$. That CLIQUE can be reduced to $1 \mid p_j = 1, prec \mid \sum U_j$ can be shown as follows. Given a graph $G = (N, A)$ and an integer c, assume there are l nodes (i.e., $N = \{1, \dots, l\}$). Let a denote the number of arcs (i.e., $a = \mid A \mid$) and let

$$\ell = \frac{c(c - 1)}{2}.$$

So a clique of size c has exactly ℓ arcs.

The reduction involves two types of jobs: a node-job j correspond to node $j \in N$ and an arc-job (j, k) corresponds to arc $\{j, k\} \in A$. Each arc-job (j, k) is subject to the constraint that it must follow the two corresponding node-jobs j and k.

Consider the following instance of $1 \mid p_j = 1, prec \mid \sum U_j$. Let the number of jobs be $n = l + a$. All node-jobs have the same due date $l + a$ (i.e., due date $d_j = l + a$ for $j = 1, \dots, l$). All arc-jobs (j, k) have the same due date $c + \ell$ (i.e., due date $d_{(j,k)} = c + \ell$ for all $\{j, k\} \in A$). The precedence constraints are such that $j \rightarrow (j, k)$ and $k \rightarrow (j, k)$ for all $\{j, k\} \in A$. Let $z = a - \ell$.

A schedule with at least $l + \ell$ early jobs and at most z late jobs exists if and only if CLIQUE has a solution. If G has a clique of size c on a subset N^*, the corresponding node-jobs $j \in N^*$ have to be scheduled first. The ℓ corresponding arc-jobs can then be completed before their due dates by time $c + \ell$. The remaining $a - \ell$ arc-jobs are late and $\sum U_j = a - \ell = z$.

If G has no clique of size c, then at most $\ell - 1$ arc jobs can be on time and the treshold cannot be met. So the two problems are equivalent.

COMMENTS AND REFERENCES

The classic on computational complexity is Garey and Johnson (1979). A number of books have chapters on this topic; see, for example, Papadimitriou and Steiglitz (1982), Parker and Rardin (1988), Papadimitriou (1994), Schrijver (1998), and Wolsey (1998).

D

Complexity Classification of Deterministic Scheduling Problems

In scheduling theory, it is often of interest to determine the borderline between polynomial time problems and NP-hard problems. To determine the exact boundaries, it is necessary to find the hardest or most general problems that still can be solved in polynomial time. These problems are characterized by the fact that any generalization (e.g., the inclusion of precedence constraints) results in NP-hardness either in the ordinary sense or strongly. In the same vein it is of interest to determine the simplest or least general problems that are NP-hard either in the ordinary sense or strongly. Making such a strongly NP-hard problem easier in any respect (e.g., setting all w_j equal to 1) results in a problem that is either solvable in polynomial time or NP-hard in the ordinary sense. In addition, it is also of interest to determine the most general problems that are NP-hard in the ordinary sense, but not strongly NP-hard.

A significant amount of research has focused on these boundaries. However, the computational complexity of a number of scheduling problems has not yet been determined, and the borderlines are therefore still somewhat fuzzy.

In the following problem classification it is assumed throughout that the number of machines (m) is fixed. If an algorithm is said to be polynomial, then the algorithm is polynomial in the number of jobs n, but not necessarily polynomial in the number of machines m. If a problem is said to be NP-hard, either in the ordinary sense or strongly, then the assumption is made that the number of machines is fixed.

Table D.1 presents a sample of fairly general problems that are solvable in polynomial time. The table is organized according to machine environments. Some of the problems in this table are, however, not the *most* general problems solvable in polynomial time (e.g., $1 \mid\mid \sum U_j$ is a special case of the proportionate flow shop problem $Fm \mid p_{ij} = p_j \mid \sum U_j$). Also, the fact that $Fm \mid p_{ij} = p_j \mid \sum U_j$ can be solved in polynomial time implies that $Fm \mid p_{ij} = p_j \mid L_{\max}$ can be solved in polynomial time as well.

Table D.2 presents a number of problems that are NP-hard in the ordinary sense. This table contains some of the simplest as well as some of the most general problems that fall in this class. The complexity of these problems is determined through a reduction from PARTITION. However, not for every one of these problems a pseudopolynomial algorithm is known. For the problems followed by a $(*)$, there exists a pseudopolynomial time algorithm. For example, as no pseudopolynomial time algorithm is known for $O2 \mid prmp \mid \sum C_j$, this problem may still turn out to be strongly NP-hard even though there is a reduction from PARTITION.

TABLE D.1 POLYNOMIAL TIME SOLVABLE PROBLEMS

Single machine	Parallel machines	Shops
$1 \mid r_j, p_j = 1, prec \mid \sum C_j$	$P2 \mid p_j = 1, prec \mid L_{\max}$	$O2 \mid\mid C_{\max}$
$1 \mid r_j, prmp \mid \sum C_j$	$P2 \mid p_j = 1, prec \mid \sum C_j$	
$1 \mid tree \mid \sum w_j C_j$		$Om \mid r_j, prmp \mid L_{\max}$
	$Pm \mid p_j = 1, tree \mid C_{\max}$	
$1 \mid prec \mid L_{\max}$	$Pm \mid prmp, tree \mid C_{\max}$	$F2 \mid block \mid C_{\max}$
$1 \mid r_j, prmp, prec \mid L_{\max}$	$Pm \mid p_j = 1, outtree \mid \sum C_j$	$F2 \mid nwt \mid C_{\max}$
	$Pm \mid p_j = 1, intree \mid L_{\max}$	
$1 \mid\mid \sum U_j$	$Pm \mid prmp, intree \mid L_{\max}$	$Fm \mid p_{ij} = p_j \mid \sum C_j$
$1 \mid r_j, prmp \mid \sum U_j$		$Fm \mid p_{ij} = p_j \mid L_{\max}$
$1 \mid r_j, p_j = 1 \mid \sum w_j U_j$		$Fm \mid p_{ij} = p_j \mid \sum U_j$
	$Q2 \mid prmp, prec \mid C_{\max}$	
	$Q2 \mid r_j, prmp, prec \mid L_{\max}$	$J2 \mid\mid C_{\max}$
$1 \mid r_j, p_j = 1 \mid \sum w_j T_j$		
	$Qm \mid r_j, p_j = 1 \mid C_{\max}$	
	$Qm \mid r_j, p_j = 1 \mid \sum C_j$	
	$Qm \mid prmp \mid \sum C_j$	
	$Qm \mid p_j = 1 \mid \sum w_j C_j$	
	$Qm \mid p_j = 1 \mid L_{\max}$	
	$Qm \mid prmp \mid \sum U_j$	
	$Qm \mid p_j = 1 \mid \sum w_j U_j$	
	$Qm \mid p_j = 1 \mid \sum w_j T_j$	
	$Rm \mid\mid \sum C_j$	
	$Rm \mid r_j, prmp \mid L_{\max}$	

TABLE D.2   NP-HARD PROBLEMS IN THE ORDINARY SENSE

Single machine	Parallel machines	Shops
$1 \mid\mid \sum w_j U_j$ (*) $1 \mid r_j, prmp \mid \sum w_j U_j$ (*) $1 \mid\mid \sum T_j$ (*)	$P2 \mid\mid C_{\max}$ (*) $P2 \mid r_j, prmp \mid \sum C_j$ $P2 \mid\mid \sum w_j C_j$ (*) $P2 \mid r_j, prmp \mid \sum U_j$ $Pm \mid prmp \mid \sum w_j C_j$ $Qm \mid\mid \sum w_j C_j$ (*) $Rm \mid r_j \mid C_{\max}$ (*) $Rm \mid\mid \sum w_j U_j$ (*) $Rm \mid prmp \mid \sum w_j U_j$	$O2 \mid prmp \mid \sum C_j$ $O3 \mid\mid C_{\max}$ $O3 \mid prmp \mid \sum w_j U_j$

Table D.3 contains problems that are strongly NP-hard. The problems tend to be the simplest problems that are strongly NP-hard. However, in this table, there are also some exceptions. For example, the fact that $1 \mid r_j \mid L_{\max}$ is strongly NP-hard implies that $1 \mid r_j \mid \sum U_j$ and $1 \mid r_j \mid \sum T_j$ are strongly NP-hard as well.

TABLE D.3   STRONGLY NP-HARD PROBLEMS

Single machine	Parallel machines	Shops
$1 \mid s_{jk} \mid C_{\max}$ $1 \mid r_j \mid \sum C_j$ $1 \mid prec \mid \sum C_j$ $1 \mid r_j, prmp, tree \mid \sum C_j$ $1 \mid r_j, prmp \mid \sum w_j C_j$ $1 \mid r_j, p_j = 1, tree \mid \sum w_j C_j$ $1 \mid p_j = 1, prec \mid \sum w_j C_j$ $1 \mid r_j \mid L_{\max}$ $1 \mid r_j \mid \sum U_j$ $1 \mid p_j = 1, chains \mid \sum U_j$ $1 \mid r_j \mid \sum T_j$ $1 \mid p_j = 1, chains \mid \sum T_j$ $1 \mid\mid \sum w_j T_j$	$P2 \mid chains \mid C_{\max}$ $P2 \mid chains \mid \sum C_j$ $P2 \mid prmp, chains \mid \sum C_j$ $P2 \mid p_j = 1, tree \mid \sum w_j C_j$ $R2 \mid prmp, chains \mid C_{\max}$	$F2 \mid r_j \mid C_{\max}$ $F2 \mid r_j, prmp \mid C_{\max}$ $F2 \mid\mid \sum C_j$ $F2 \mid prmp \mid \sum C_j$ $F2 \mid\mid L_{\max}$ $F2 \mid prmp \mid L_{\max}$ $F3 \mid\mid C_{\max}$ $F3 \mid prmp \mid C_{\max}$ $F3 \mid nwt \mid C_{\max}$ $O2 \mid r_j \mid C_{\max}$ $O2 \mid\mid \sum C_j$ $O2 \mid prmp \mid \sum w_j C_j$ $O2 \mid\mid L_{\max}$ $O3 \mid prmp \mid \sum C_j$ $J2 \mid recrc \mid C_{\max}$ $J3 \mid p_{ij} = 1, recrc \mid C_{\max}$

These tables have to be used in conjunction with the figures presented in Chapter 2. When attempting to determine the status of a problem that does not appear in any one of the three tables, it is necessary to search for related problems that are either easier or harder to determine the complexity status of the given problem.

COMMENTS AND REFERENCES

The complexity classification of scheduling problems has its base in the work of Lenstra, and Rinnooy Kan (1979), and Lageweg, Lawler, Lenstra and Rinnooy Kan (1981, 1982). Brucker (1998) presents a very thorough and up-to-date complexity classification of scheduling problems, and Timkovsky (2000) discusses reducibility among scheduling problems.

A significant amount of research attention has focused on the complexity statuses of scheduling problems that are close to the boundaries; see, for example, Du and Leung (1989, 1990, 1993a, 1993b), Du, Leung, and Wong (1992), Du, Leung, and Young (1990, 1991), Leung and Young (1990), Timkovski (1998), and Baptiste (1999).

E

Overview of Stochastic Scheduling Problems

No framework or classification scheme has ever been introduced for stochastic scheduling problems. It is more difficult to develop such a scheme for stochastic scheduling problems than for deterministic scheduling problems. To characterize a stochastic scheduling problem, more information is required. For example,the distributions of the processing times have to be specified as well as the distributions of the due dates (which may be different). It has to be specified whether the processing times of the n jobs are independent or correlated (e.g., equal to the same random variable) and also which class of policies is considered. For these reasons, no framework has been introduced in this book either.

Table E.1 outlines a number of scheduling problems of which stochastic versions are tractable. This list refers to most of the problems discussed in Part II of the book. In the distribution column, the distribution of the processing times is specified. If the entry in this column specifies a form of stochastic dominance, then the n processing times are arbitrarily distributed and ordered according to the form of stochastic dominance specified.The due dates in this table are considered fixed (deterministic).

However, the list in Table E.1 is far from complete. For example, it is mentioned that the stochastic counterpart of $Pm \mid\mid \sum C_j$ leads to the SEPT rule when the processing times are exponentially distributed. However, as stated in Chap-

TABLE E.1   TRACTABLE STOCHASTIC SCHEDULING PROBLEMS

Deterministic Counterpart	Distributions	Optimal policy	Section
$1 \| \| \sum w_j C_j$	arbitrary	WSEPT	10.1
$1 \| r_j, prmp \| \sum w_j C_j$	exponential	WSEPT (preemptive)	10.4
$1 \| \| \sum w_j (1 - e^{-rC_j})$	arbitrary	DWSEPT	10.1
$1 \| prmp \| \sum w_j (1 - e^{-rC_j})$	arbitrary	Gittins Index	10.2
$1 \| \| L_{\max}$	arbitrary	EDD	10.1
$1 \| d_j = d \| \sum w_j U_j$	exponential	WSEPT	10.4
$1 \| d_j = d \| \sum w_j T_j$	exponential	WSEPT	10.4
$Pm \| \| C_{\max}$	exponential	LEPT	12.1, 12.2
$Pm \| prmp \| C_{\max}$	exponential	LEPT	12.1, 12.2
$Pm \| \| \sum C_j$	exponential	SEPT	12.2
$Pm \| prmp \| \sum C_j$	exponential	SEPT	12.2
$P2 \| p_j = 1, intree \| C_{\max}$	exponential	CP	12.2
$P2 \| p_j = 1, intree \| \sum C_j$	exponential	CP	12.2
$F2 \| \| C_{\max}$	exponential	$(\lambda_j - \mu_j) \downarrow$	13.1
$Fm \| p_{ij} = p_j \| C_{\max}$	$\geq_{as}$	SEPT-LEPT	13.1
$Fm \| p_{ij} = p_j \| \sum C_j$	$\geq_{as}$	SEPT	13.1
$F2 \| block \| C_{\max}$	arbitrary	TSP	13.2
$F2 \| p_{ij} = p_j, block \| C_{\max}$	$\geq_{st}$	SEPT-LEPT	13.2
$F2 \| p_{ij} = p_j, block \| C_{\max}$	$\geq_{sv}$	LV-SV	13.2
$Fm \| p_{ij} = p_j, block \| C_{\max}$	$\geq_{as}$	SEPT-LEPT	13.2
$Fm \| p_{ij} = p_j, block \| \sum C_j$	$\geq_{as}$	SEPT	13.2
$J2 \| \| C_{\max}$	exponential	Theorem 13.3.1	13.3
$O2 \| p_{ij} = p_j \| C_{\max}$	exponential	Theorem 13.4.1	13.4
$O2 \| p_{ij} = 1, prmp \| \sum C_j$	exponential	SERPT	13.4

ter 12, a much more general result holds: If the processing times (from distributions $F_1, \ldots, F_n$) are independent, then SEPT is optimal provided the n distributions can be ordered stochastically.

Comparing Table E.1 with the tables in Appendix D reveals that there are a number of stochastic scheduling problems that are tractable while their deterministic counterparts are NP-hard. The four NP-hard deterministic problems are:

(i) $1 \mid r_j, prmp \mid \sum w_j C_j$,

(ii) $1 \mid d_j = d \mid \sum w_j U_j$,

(iii) $1 \mid d_j = d \mid \sum w_j T_j$, and

(iv) $Pm \mid\mid C_{\max}$.

The first problem allows for a nice solution when the processing times are exponential and the release dates are arbitrarily distributed. The optimal policy is then the preemptive WSEPT rule. When the processing time distributions are anything but exponential, it appears that the preemptive WSEPT rule is not necessarily optimal. The stochastic counterparts of the second and third problems also lead to the WSEPT rule when the processing time distributions are exponential and the jobs have a common due date that is arbitrarily distributed. Also, if the processing times are anything but exponential, the optimal rule is not necessarily WSEPT.

The stochastic counterparts of $Pm \parallel C_{\max}$ are slightly different. When the processing times are exponential, the LEPT rule minimizes the expected makespan in all classes of policies. However, this holds for other distributions also. If the processing times are DCR (e.g., hyperexponentially distributed) and satisfy a fairly strong form of stochastic dominance, the LEPT rule remains optimal. Note that when preemptions are allowed and the processing times are DCR, the nonpreemptive LEPT rule remains optimal. Note also that if the n processing times have the same mean and are hyperexponentially distributed, as in Example 12.1.7, then the LV rule minimizes the expected makespan.

There are many other problems of which the stochastic versions exhibit very strong similarities to their deterministic counterparts. Examples of such problems are

(i) $1 \mid r_j, prmp \mid L_{\max}$,
(ii) $1 \mid prec \mid h_{\max}$,
(iii) $F2 \parallel C_{\max}$, and
(iv) $J2 \parallel C_{\max}$.

It can be shown that the preemptive EDD rule is optimal for the deterministic problem $1 \mid r_j, prmp \mid L_{\max}$ and that it remains optimal when the processing times are random variables that are arbitrarily distributed. In Chapter 10, it is shown that the algorithm for the stochastic counterpart of $1 \mid prec \mid h_{\max}$ is very similar to the algorithm for the deterministic version. The same can be said with regard to $F2 \parallel C_{\max}$ and $J2 \parallel C_{\max}$ when the processing times are exponential.

Of course, there are also problems of which the deterministic version is easy and the version with exponential processing times is hard. Examples of such problems are

(i) $Pm \mid p_j = 1, tree \mid C_{\max}$, and
(ii) $O2 \parallel C_{\max}$.

For the deterministic problem $Pm \mid p_j = 1, tree \mid C_{\max}$, the CP rule is optimal. For the version of the same problem with all processing times i.i.d. exponential, the optimal policy is not known and may depend on the form of the tree.

For the $O2 \parallel C_{\max}$ problem, the LAPT rule is optimal; when the processing times are exponential, the problem appears to be very hard.

COMMENTS AND REFERENCES

For an early overview of stochastic scheduling on parallel machines, see Weiss (1982). For a discussion of tractable stochastic scheduling problems of which the deterministic counterparts are NP-hard, see Pinedo (1983). For a more recent and comprehensive overview of stochastic scheduling, see Righter (1994).

F

Selected Scheduling Systems

Over the last two decades, hundreds of scheduling systems have been developed. These developments have taken place in industry and academia in various countries. An up-to-date list or annotated bibliography of all systems most likely does not exist. However, several survey papers have been written describing a number of systems.

In this appendix, a distinction is made among commercial generic systems, industrial systems that are application-specific, and academic prototypes (research systems). The considerations in the design and the development of the various classes of systems are usually quite different. In this appendix systems are *not* categorized according to the approach used for generating schedules (i.e., whether it is knowledge-based or based on algorithms since most knowledge-based systems have an algorithmic component as well).

Commercial generic systems are designed for implementation in a wide variety of settings with only minor customization. The software houses that develop generic systems are usually not associated with a single company. However, they may focus on a specific industry. Examples of such systems are given in Table F.1.

Application-specific systems are designed for either a single installation or a single type of installation. The algorithms embedded in these systems are typically quite elaborate. A number of the application-specific systems in industry have been

TABLE F.1   COMMERCIAL GENERIC SYSTEMS

System	Company	Website
Quick Response Engine	Computer Associates	www.interbiz.cai.com
Cyberplan	Cybertec	www.cybertec.it
SKEP	DynaSys Group	www.adaptasolutions.com
BaanSCS Scheduler	Invensys plc	www.baan.com
Quick Response	Isera Group	www.isera.com
TradeMatrix		
* Production Scheduler*	i2 Technologies, Inc.	www.i2.com
NetWORKS Scheduling	Manugistics Group, Inc.	www.manugistics.com
APO	SAP AG	www.sap.com
Virtual Production Engine	SynQuest Inc.	www.synquest.com

developed in collaboration with an academic partner. Examples of application-specific systems are listed in Table F.2.

Academic prototypes are usually developed for research and teaching. The programmers are typically graduate students who work part time on the system. The design of these systems is often completely different from the design of commercial systems. Some attempts have been made to commercialize academic prototypes. Examples of academic prototypes are presented in Table F.3.

TABLE F.2   APPLICATION-SPECIFIC SYSTEMS

System	Company	Reference
BPSS	International Paper	Adler et al. (1993)
GATES	Trans World Airways	Brazile and Swigger (1988)
Jobplan	Siemens	Kanet and Sridharan (1990)
LMS	IBM	Sullivan and Fordyce (1990)
MacMerl	Pittsburgh Plate & Glass	Hsu et al. (1993)
SAIGA	Aeroports de Paris	Ilog (1999)

TABLE F.3   ACADEMIC PROTOTYPES

System	Institution	Reference
Lekin	New York University	Feldman and Pinedo (1998)
OPIS	Carnegie-Mellon University	Smith (1994)
TOSCA	University of Edinburgh	Beck (1993)
TTA	Universidad Catolica de Chile	Nussbaum and Parra (1993)

COMMENTS AND REFERENCES

Several reviews and survey papers have been written on scheduling systems. See Steffen (1986), Adelsberger and Kanet (1991), Smith (1992), Arguello (1994), and Yen and Pinedo (1994).

References

J.O. Achugbue and F.Y. Chin (1982) "Scheduling the Open Shop to Minimize Mean Flow Time," *SIAM Journal of Computing,* Vol. 11, pp. 709–720.

J. Adams, E. Balas, and D. Zawack (1988) "The Shifting Bottleneck Procedure for Job Shop Scheduling," *Management Science,* Vol. 34, pp. 391–401.

H.H. Adelsberger and J.J. Kanet (1991) "The LEITSTAND—A New Tool for Computer-Integrated-Manufacturing," *Production and Inventory Management Journal,* Vol. 32, pp. 43–48.

L. Adler, N.M. Fraiman, E. Kobacker, M.L. Pinedo, J.C. Plotnicoff, and T.-P. Wu (1993) "BPSS: A Scheduling System for the Packaging Industry," *Operations Research,* Vol. 41, pp. 641–648.

A.K. Agrawala, E.G. Coffman, Jr., M.R. Garey, and S.K. Tripathi (1984) "A Stochastic Optimization Algorithm Minimizing Exponential Flow Times on Uniform Processors," *IEEE Transactions on Computers,* C–33, pp. 351–356.

R. Akkiraju, P. Keskinocak, S. Murthy, and F. Wu (1998a) "A New Decision Support System for Paper Manufacturing," in *Proceedings of the Sixth International Workshop on Project Management and Scheduling (1998),* pp. 147–150, Bogazici University Printing Office, Istanbul, Turkey.

R. Akkiraju, P. Keskinocak, S. Murthy, and F. Wu (1998b) "Multi-Machine Scheduling: An Agent Based Approach," in *Proceedings of the Fifteenth National Conference on Artificial Intelligence and Tenth Innovative Applications of Artificial Intelligence Conference (AAAI98, IAAI98),* pp. 1013–1019, AAAI Press/The MIT Press, Cambridge, Massachusetts.

M.S. Akturk and E. Gorgulu (1999) "Match-Up Scheduling under a Machine Breakdown," *European Journal of Operational Research,* Vol. 112, pp. 81–97.

N. Alon, Y. Azar, G.J. Woeginger, and T. Yadid (1998) "Approximation Schemes for Scheduling on Parallel Machines," *Journal of Scheduling,* Vol. 1, pp. 55–66.

D. Applegate and W. Cook (1991) "A Computational Study of the Job-Shop Scheduling Problem," *ORSA Journal on Computing,* Vol. 3, pp. 149–156.

M. Arguello (1994) "Review of Scheduling Software," Technology Transfer 93091822A–XFR, SEMATECH, Austin, Texas.

H. Atabakhsh (1991) "A Survey of Constraint Based Scheduling Systems Using an Artificial Intelligence Approach," *Artificial Intelligence in Engineering*, Vol. 6, No. 2, pp. 58–73.

H. Aytug, S. Bhattacharyya, G.J. Koehler, and J.L. Snowdon (1994) "A Review of Machine Learning in Scheduling," *IEEE Transactions on Engineering Management,* Vol. 41, pp. 165–171.

P.C. Bagga (1970) "n-Job, 2-Machine Sequencing Problem with Stochastic Service," *Opsearch,* Vol. 7, pp. 184–197.

K.R. Baker (1974) *Introduction to Sequencing and Scheduling*, John Wiley, New York.

K.R. Baker (1975) "A Comparative Survey of Flowshop Algorithms," *Operations Research,* Vol. 23, pp. 62–73.

K.R. Baker (1995) *Elements of Sequencing and Scheduling,* K. Baker, Amos Tuck School of Business Administration, Dartmouth College, Hanover, New Hampshire.

K.R. Baker and G.D. Scudder (1990) "Sequencing with Earliness and Tardiness Penalties: A Review," *Operations Research,* Vol. 38, pp. 22–36.

E. Balas, J.K. Lenstra, and A. Vazacopoulos (1995) "The One-Machine Scheduling Problem with Delayed Precedence Constraints and Its Use in Job Shop Scheduling," *Management Science,* Vol. 41, pp. 94–109.

P. Baptiste (1999) "Polynomial Time Algorithms for Minimizing the Weighted Number of Late Jobs on a Single Machine when Processing Times are Equal," *Journal of Scheduling,* Vol. 2, pp. 245–252.

P. Baptiste, C. Le Pape, and W. Nuijten (1995) "Constraint-Based Optimization and Approximation for Job-Shop Scheduling," in *Proceedings of the AAAI-SIGMAN Workshop on Intelligent Manufacturing Systems IJCAI-95,* Montreal, Canada.

P. Baptiste, L. Peridy, and E. Pinson (2000) "A Branch and Bound to Minimize the Number of Late Jobs on a Single Machine with Release Time Constraints," Research Report, Université de Technologie de Compiègne, Compiègne, France.

J.R. Barker and G.B. McMahon (1985) "Scheduling the General Job-Shop," *Management Science,* Vol. 31, pp. 594–598.

R.E. Barlow and F. Proschan (1975) *Statistical Theory of Reliability and Life Testing: Probability Models,* Holt, Rinehart and Winston, Inc., New York.

C. Barnhart, E.L. Johnson, G.L. Nemhauser, M.W.P. Savelsbergh and P.H. Vance (1998) "Branch and Price: Column Generation for Solving Huge Integer Programs," *Operations Research,* Vol. 46, pp. 316–329.

K.M. Baumgartner and B.W. Wah (1991) "Computer Scheduling Algorithms: Past, Present and Future," *Information Sciences,* Vol. 57–58, pp. 319–345.

J. Bean (1994) "Genetics and Random Keys for Sequencing and Optimization," *ORSA Journal of Computing,* Vol. 6, pp. 154–160.

J. Bean, J. Birge, J. Mittenthal and C. Noon (1991) "Matchup Scheduling with Multiple Resources, Release Dates and Disruptions," *Operations Research,* Vol. 39, pp. 470–483.

H. Beck (1993) "The Management of Job Shop Scheduling Constraints in TOSCA," in *Intelligent Dynamic Scheduling for Manufacturing Systems,* L. Interrante (ed.), Proceedings of a Workshop Sponsored by the National Science Foundation, the University of Alabama in Huntsville and Carnegie Mellon University, held at Cocoa Beach, January 1993.

M. Bell (2000) "i2's Tradematrix Production Scheduler Architecture," Technical Report, i2 Technologies, Dallas, Texas.

E. Bensana, G. Bell, and D. Dubois (1988) "OPAL: A Multi-Knowledge-Based System for Industrial Job Shop Scheduling," *International Journal of Production Research,* Vol. 26, pp. 795–819.

D.P. Bertsekas (1987) *Dynamic Programming: Deterministic and Stochastic Models*, Prentice-Hall, New Jersey.

K. Bhaskaran and M. Pinedo (1992) "Dispatching," Chapter 83 in *Handbook of Industrial Engineering,* G. Salvendy (ed.), pp. 2184–2198, John Wiley, New York.

L. Bianco, S. Ricciardelli, G. Rinaldi, and A. Sassano (1988) "Scheduling Tasks with Sequence-Dependent Processing Times," *Naval Research Logistics Quarterly,* Vol. 35, pp. 177–184.

J. Birge, J.B.G. Frenk, J. Mittenthal, and A.H.G. Rinnooy Kan (1990) "Single Machine Scheduling Subject to Stochastic Breakdowns," *Naval Research Logistics Quarterly,* Vol. 37, pp. 661–677.

G.R. Bitran and D. Tirupati (1988) "Planning and Scheduling for Epitaxial Wafer Production," *Operations Research,* Vol. 36, pp. 34–49.

J. Blazewicz, W. Cellary, R. Slowinski, and J. Weglarz (1986) *Scheduling under Resource Constraints—Deterministic Models, Annals of Operations Research,* Vol. 7, Baltzer, Basel.

J. Blazewicz, M. Dror, and J. Weglarz (1991) "Mathematical Programming Formulations for Machine Scheduling: A Survey," *European Journal of Operational Research,* Vol. 51, pp. 283–300.

J. Blazewicz, K. Ecker, G. Schmidt, and J. Weglarz (1993) *Scheduling in Computer and Manufacturing Systems*, Springer Verlag, Berlin.

J. Blazewicz, K. Ecker, E. Pesch, G. Schmidt, and J. Weglarz (1996) *Scheduling Computer and Manufacturing Processes*, Springer Verlag, Berlin.

G. Booch (1994) *Object-Oriented Analysis and Design with Applications,* 2nd ed., Benjamin/Cummings Scientific, Menlo Park, California.

O.J. Boxma and F.G. Forst (1986) "Minimizing the Expected Weighted Number of Tardy Jobs in Stochastic Flow Shops," *Operations Research Letters*, Vol. 5, pp. 119–126.

H. Braun (2000) "Optimizing the Supply Chain—Challenges and Opportunities," *SAP Inside*, SAP AG, Walldorf, Germany.

R.P. Brazile and K.M. Swigger (1988) "GATES: An Airline Assignment and Tracking System," *IEEE Expert,* Vol. 3, pp. 33–39.

R.P. Brazile and K.M. Swigger (1991) "Generalized Heuristics for the Gate Assignment Problem," *Control and Computers,* Vol. 19, pp. 27–32.

A. Brown and Z.A. Lomnicki (1966) "Some Applications of the Branch and Bound Algorithm to the Machine Sequencing Problem," *Operational Research Quarterly,* Vol. 17, pp. 173–186.

D.E. Brown and W.T. Scherer (eds.) (1995) *Intelligent Scheduling Systems,* Kluwer Academic Publishers, Boston.

M. Brown and H. Solomon (1973) "Optimal Issuing Policies under Stochastic Field Lives," *Journal of Applied Probability,* Vol. 10, pp. 761–768.

S. Browne and U. Yechiali (1990) "Scheduling Deteriorating Jobs on a Single Processor," *Operations Research,* Vol. 38, pp. 495–498.

P. Brucker (1995) *Scheduling Algorithms* (1st ed.), Springer Verlag, Berlin.

P. Brucker (1998) *Scheduling Algorithms* (2nd ed.), Springer Verlag, Berlin.

P. Brucker, B. Jurisch, and M. Jurisch (1993) "Open Shop Problems with Unit Time Operations," *Zeitschrift für Operations Research,* Vol. 37, pp. 59–73.

P. Brucker, B. Jurisch, and B. Sievers (1994) "A Branch and Bound Algorithm for the Job Shop Problem," *Discrete Applied Mathematics,* Vol. 49, pp. 107–127.

P. Brucker, B. Jurisch, and A. Krämer (1994) "The Job Shop Problem and Immediate Selection," *Annals of Operations Research,* Vol. 50, pp. 73–114.

J. Bruno, P. Downey, and G. Frederickson (1981) "Sequencing Tasks with Exponential Service Times to Minimize the Expected Flow Time or Makespan," *Journal of the Association of Computing Machinery,* Vol. 28, pp. 100–113.

J. Bruno and T. Gonzalez (1976) "Scheduling Independent Tasks with Release Dates and Due Dates on Parallel Machines," *Technical Report 213,* Computer Science Department, Pennsylvania State University.

L. Burns and C.F. Daganzo (1987) "Assembly Line Job Sequencing Principles," *International Journal of Production Research,* Vol. 25, pp. 71–99.

G. Buxey (1989) "Production Scheduling: Practice and Theory," *European Journal of Operational Research,* Vol. 39, pp. 17–31.

C. Buyukkoc, P. Varaiya, and J. Walrand (1985) "The $c\mu$ Rule Revisited," *Advances in Applied Probability,* Vol. 17, pp. 237–238.

H.G. Campbell, R.A. Dudek, and M.L. Smith (1970) "A Heuristic Algorithm for the n Job m Machine Sequencing Problem," *Management Science,* Vol. 16, pp. B630–B637.

J. Carlier (1982) "The One-Machine Sequencing Problem," *European Journal of Operational Research,* Vol. 11, pp. 42–47.

J. Carlier and E. Pinson (1989) "An Algorithm for Solving the Job Shop Problem," *Management Science,* Vol. 35, pp. 164–176.

D.C. Carroll (1965) *Heuristic Sequencing of Single and Multiple Component Jobs,* Ph.D. Thesis, Sloan School of Management, M.I.T., Cambridge, Massachusetts.

S. Chand, R. Traub, and R. Uzsoy (1996) "Single Machine Scheduling with Dynamic Arrivals: Decomposition Results and an Improved Algorithm," *Naval Research Logistics,* Vol. 31, pp. 709–719.

S. Chand, R. Traub, and R. Uzsoy (1997) "Rolling Horizon Procedures for the Single Machine Deterministic Total Completion Time Scheduling Problem with Release Dates," *Annals of Operations Research,* Vol. 70, pp. 115–125.

K.M. Chandy and P.F. Reynolds (1975) "Scheduling Partially Ordered Tasks with Probabilistic Execution Times," in *Proceedings of the fifth Symposium on Operating Systems Principles, Operating Systems Review,* Vol. 9, pp. 169–177.

C.-S. Chang, R. Nelson, and M. Pinedo (1992) "Scheduling Two Classes of Exponential Jobs on Parallel Processors: Structural Results and Worst Case Analysis," *Advances in Applied Probability,* Vol. 23, pp. 925–944.

C.-S. Chang, X.L. Chao, M. Pinedo, and R.R. Weber (1992) "On the Optimality of *LEPT* and $c\mu$ Rules for Machines in Parallel," *Journal of Applied Probability,* Vol. 29, pp. 667–681.

C.-S. Chang and D.D. Yao (1993) "Rearrangement, Majorization and Stochastic Scheduling," *Mathematics of Operations Research,* Vol. 18, pp. 658–684.

X. Chao and M. Pinedo (1992) "A Parametric Adjustment Method for Dispatching," Technical Report, Department of Industrial Engineering and Operations Research, Columbia University, New York.

C. Chekuri, R. Motwani, B. Natarajan, and C. Stein (1997) "Approximation Techniques for Average Completion Time Scheduling," in *Proceedings of the Annual ACM-SIAM Symposium on Discrete Algorithms (SODA),* pp. 609–617.

B. Chen, C.N. Potts, and G.J. Woeginger (1998) "A Review of Machine Scheduling: Complexity, Algorithms, and Approximability," in *Handbook of Combinatorial Optimization,* D.-Z. Du and P. Pardalos (eds.), pp. 21–169, Kluwer Academic Press, Boston.

C.-L. Chen and R.L. Bulfin (1993) "Complexity of Single Machine, Multi-Criteria Scheduling Problems," *European Journal of Operational Research,* Vol. 70, pp. 115–125.

C.-L. Chen and R.L. Bulfin (1994) "Scheduling a Single Machine to Minimize Two Criteria: Maximum Tardiness and Number of Tardy Jobs," *IIE Transactions,* Vol. 26, pp. 76–84.

N.-F. Chen and C.L. Liu (1975) "On a Class of Scheduling Algorithms for Multiprocessors Computing Systems," in *Parallel Processing* (Lecture Notes in Computer Science 24), T.-Y. Feng (ed.), pp. 1–16, Springer, Berlin.

Y.R. Chen and M.N. Katehakis (1986) "Linear Programming for Finite State Multi-Armed Bandit Problems," *Mathematics of Operations Research,* Vol. 11, pp. 180–183.

Z.-L. Chen and W.B. Powell (1999) "Solving Parallel Machine Scheduling Problems by Column Generation," *INFORMS Journal of Computing,* Vol. 11, pp. 78–94.

C.-C. Cheng and S.F. Smith (1997) "Applying Constraint Satisfaction Techniques to Job Shop Scheduling," *Annals of Operations Research,* Vol. 70, pp. 327–357.

T.C.E. Cheng and M.C. Gupta (1989) "Survey of Scheduling Research Involving Due Date Determination Decisions," *European Journal of Operational Research,* Vol. 38, pp. 156–166.

Y. Cho and S. Sahni (1981) "Preemptive Scheduling of Independent Jobs with Release and Due Times on Open, Flow and Job Shops," *Operations Research,* Vol. 29, pp. 511–522.

P. Chrétienne, E.G. Coffman, Jr., J.K. Lenstra, and Z. Liu (eds.) (1995) *Scheduling Theory and Applications,* John Wiley, New York.

A. Cobham (1954) "Priority Assignment in Waiting Line Problems," *Operations Research,* Vol. 2, pp. 70–76.

E.G. Coffman, Jr. (ed.) (1976) *Computer and Job Shop Scheduling Theory,* John Wiley, New York.

E.G. Coffman, Jr., L. Flatto, M.R. Garey, and R.R. Weber (1987) "Minimizing Expected Makespans on Uniform Processor Systems," *Advances in Applied Probability,* Vol. 19, pp. 177–201.

E.G. Coffman, Jr., M.R. Garey, and D.S. Johnson (1978) "An Application of Bin-Packing to Multiprocessor Scheduling," *SIAM Journal of Computing,* Vol. 7, pp. 1–17.

A. Collinot, C. LePape, and G. Pinoteau (1988) "SONIA: A Knowledge-Based Scheduling System," *Artificial Intelligence in Engineering,* Vol. 2, pp. 86–94.

R.W. Conway (1965a) "Priority Dispatching and Work-in-Process Inventory in a Job Shop," *Journal of Industrial Engineering,* Vol. 16, pp. 123–130.

R.W. Conway (1965b) "Priority Dispatching and Job Lateness in a Job Shop," *Journal of Industrial Engineering,* Vol. 16, pp. 228–237.

R.W. Conway, W.L. Maxwell, and L.W. Miller (1967) *Theory of Scheduling,* Addison-Wesley, Reading, Massachusetts.

D.R. Cox and W.L. Smith (1961) *Queues,* John Wiley, New York.

T.B. Crabill and W.L. Maxwell (1969) "Single Machine Sequencing with Random Processing Times and Random Due Dates," *Naval Research Logistics Quarterly,* Vol. 16, pp. 549–554.

A.A. Cunningham and S.K. Dutta (1973) "Scheduling Jobs with Exponentially Distributed Processing Times on Two Machines of a Flow Shop," *Naval Research Logistics Quarterly,* Vol. 20, pp. 69–81.

D.G. Dannenbring (1977) "An Evaluation of Flowshop Sequencing Heuristics," *Management Science,* Vol. 23, pp. 1174–1182.

S. Dauzère-Pérès and J.-B. Lasserre (1993) "A Modified Shifting Bottleneck Procedure for Job Shop Scheduling," *International Journal of Production Research,* Vol. 31, pp. 923–932.

S. Dauzère-Pérès and J.-B. Lasserre (1994) *An Integrated Approach in Production Planning and Scheduling,* Lecture Notes in Economics and Mathematical Systems, Vol. 411, Springer Verlag, Berlin.

S. Dauzère-Pérès and M. Sevaux (1998) "An Efficient Formulation for Minimizing the Number of Late Jobs in Single Machine Scheduling," Research Report 98/9/AUTO, Dept. of Automatic Control and Production Engineering, Ecole des Mines de Nantes, France.

S. Dauzère-Pérès and M. Sevaux (1999) "Using Lagrangean Relaxation to Minimize the (Weighted) Number of Late Jobs on a Single Machine," Research Report 99/8/AUTO, Dept. of Automatic Control and Production Engineering, Ecole des Mines de Nantes, France.

E. Davis and J.M. Jaffe (1981) "Algorithms for Scheduling Tasks on Unrelated Processors," *Journal of the Association of Computing Machinery,* Vol. 28, pp. 721–736.

M. Dell'Amico and M. Trubian (1991) "Applying Tabu-Search to the Job Shop Scheduling Problem," *Annals of Operations Research,* Vol. 41, pp. 231–252.

F. Della Croce, R. Tadei, and G. Volta (1992) "A Genetic Algorithm for the Job Shop Problem," *Computers and Operations Research,* Vol. 22, pp. 15–24.

M.A.H. Dempster, J.K. Lenstra, and A.H.G. Rinnooy Kan (eds.) (1982) *Deterministic and Stochastic Scheduling,* Reidel, Dordrecht.

E.V. Denardo (1982) *Dynamic Programming: Models and Applications,* Prentice-Hall, New Jersey.

C. Derman, G. Lieberman, and S.M. Ross (1978) "A Renewal Decision Problem," *Management Science,* Vol. 24, pp. 554–561.

G. Dobson (1984) "Scheduling Independent Tasks on Uniform Processors," *SIAM Journal of Computing,* Vol. 13, pp. 705–716.

D.-Z. Du and P. Pardalos (eds.) (1998) *Handbook of Combinatorial Optimization,* Kluwer Academic Press, Boston.

J. Du and J.Y.-T. Leung (1989) "Scheduling Tree-Structured Tasks on Two Processors to Minimize Schedule Length," *SIAM Journal of Discrete Mathematics,* Vol. 2, pp. 176–196.

J. Du and J.Y.-T. Leung (1990) "Minimizing Total Tardiness on One Machine is NP-Hard," *Mathematics of Operations Research,* Vol. 15, pp. 483–495.

J. Du and J.Y.-T. Leung (1993a) "Minimizing Mean Flow Time in Two-Machine Open Shops and Flow Shops," *Journal of Algorithms,* Vol. 14, pp. 24–44.

J. Du and J.Y.-T. Leung (1993b) "Minimizing Mean Flow Time with Release Times and Deadline Constraints," *Journal of Algorithms,* Vol. 14, pp. 45–68.

J. Du, J.Y.-T. Leung, and C.S. Wong (1992) "Minimizing the Number of Late Jobs with Release Time Constraint," *Journal of Combinatorial Mathematics and Combinatorial Computing,* Vol. 11, pp. 97–107.

J. Du, J.Y.-T. Leung, and G.H. Young (1990) "Minimizing Mean Flow Time with Release Time Constraint," *Theoretical Computer Science,* Vol. 75, pp. 347–355.

J. Du, J.Y.-T. Leung, and G.H. Young (1991) "Scheduling Chain-Structured Tasks to Minimize Makespan and Mean Flow Time," *Information and Computation,* Vol. 92, pp. 219–236.

B. Eck and M. Pinedo (1988) "On the Minimization of the Flow Time in Flexible Flow Shops," Technical Report, Department of Industrial Engineering and Operations Research, Columbia University, New York.

B. Eck and M. Pinedo (1993) "On the Minimization of the Makespan Subject to Flow Time Optimality," *Operations Research,* Vol. 41, pp. 797–800.

A. Elkamel and A. Mohindra (1999) "A Rolling Horizon Heuristic for Reactive Scheduling of Batch Process Operations," *Engineering Optimization,* Vol. 31, pp. 763–792.

S.E. Elmaghraby and S.H. Park (1974) "Scheduling Jobs on a Number of Identical Machines," *AIIE Transactions,* Vol. 6, pp. 1–12.

H. Emmons (1969) "One-Machine Sequencing to Minimize Certain Functions of Job Tardiness," *Operations Research*, Vol. 17, pp. 701–715.

H. Emmons (1975) "A Note on a Scheduling Problem with Dual Criteria," *Naval Research Logistics Quarterly,* Vol. 22, pp. 615–616.

H. Emmons and M. Pinedo (1990) "Scheduling Stochastic Jobs with Due Dates on Parallel Machines," *European Journal of Operational Research,* Vol. 47, pp. 49–55.

A. Federgruen and H. Groenevelt (1986) "Preemptive Scheduling of Uniform Machines by Ordinary Network Flow Techniques," *Management Science,* Vol. 32, pp. 341–349.

A. Feldman (1999) *Scheduling Algorithms and Systems,* Ph.D. Thesis, Department of Industrial Engineering and Operations Research, Columbia University, New York.

A. Feldman and M. Pinedo (1998) "The Design and Implementation of an Educational Scheduling System," Technical Report, Department of Operations Management, Stern School of Business, New York University, New York.

M.L. Fisher (1976) "A Dual Algorithm for the One-Machine Scheduling Problem," *Mathematical Programming,* Vol. 11, pp. 229–251.

M.L. Fisher (1981) "The Lagrangean Relaxation Method for Solving Integer Programming Problems," *Management Science,* Vol. 27, pp. 1–18.

R. Fleischer and M. Wahl (2000) "Online Scheduling Revisited," *Journal of Scheduling,* Vol. 3, pp. 343–355.

R.D. Foley and S. Suresh (1984a) "Minimizing the Expected Flow Time in Stochastic Flow Shops," *IIE Transactions,* Vol. 16, pp. 391–395.

R.D. Foley and S. Suresh (1984b) "Stochastically Minimizing the Makespan in Flow Shops," *Naval Research Logistics Quarterly,* Vol. 31, pp. 551–557.

R.D. Foley and S. Suresh (1986) "Scheduling n Nonoverlapping Jobs and Two Stochastic Jobs in a Flow Shop," *Naval Research Logistics Quarterly,* Vol. 33, pp. 123–128.

F.G. Forst (1984) "A Review of the Static Stochastic Job Sequencing Literature," *Opsearch,* Vol. 21, pp. 127–144.

M.S. Fox (1987) *Constraint Directed Search: A Case Study of Job-Shop Scheduling*, Morgan Kaufmann Publishers, San Mateo, California.

M.S. Fox and S.F. Smith (1984) "ISIS—A Knowledge-Based System for Factory Scheduling," *Expert Systems,* Vol. 1, pp. 25–49.

N.M. Fraiman, M.L. Pinedo, and P.-C. Yen (1993) "On the Architecture of a Prototype Scheduling System," in *Proceedings of the 1993 NSF Design and Manufacturing Systems Conference,* Vol. 1, pp. 835–838, Society of Manufacturing Engineers, Dearborn, Michigan.

S. French (1982) *Sequencing and Scheduling: An Introduction to the Mathematics of the Job Shop*, Horwood, Chichester.

J.B.G. Frenk (1991) "A General Framework for Stochastic One Machine Scheduling Problems with Zero Release Times and no Partial Ordering," *Probability in the Engineering and Informational Sciences,* Vol. 5, pp. 297–315.

D.K. Friesen (1984a) "Tighter Bounds for the Multifit Processor Scheduling Algorithm," *SIAM Journal of Computing,* Vol. 13, pp. 170–181.

D.K. Friesen (1984b) "Tighter Bounds for LPT Scheduling on Uniform Processors," *SIAM Journal of Computing,* Vol. 16, pp. 554–560.

D.K. Friesen and M.A. Langston (1983) "Bounds for Multifit Scheduling on Uniform Processors," *SIAM Journal of Computing,* Vol. 12, pp. 60–70.

E. Frostig (1988) "A Stochastic Scheduling Problem with Intree Precedence Constraints," *Operations Research,* Vol. 36, pp. 937–943.

E. Frostig and I. Adiri (1985) "Three-Machine Flow Shop Stochastic Scheduling to Minimize Distribution of Schedule Length," *Naval Research Logistics Quarterly,* Vol. 32, pp. 179–183.

G. Galambos and G.J. Woeginger (1995) "Minimizing the Weighted Number of Late Jobs in UET Open Shops," *ZOR-Mathematical Methods of Operations Research,* Vol. 41, pp. 109–114.

J. Gaylord (1987) *Factory Information Systems,* Marcel Dekker, New York.

M.R. Garey and D.S. Johnson (1979) *Computers and Intractability—A Guide to the Theory of NP-Completeness,* W.H. Freeman and Company, San Francisco.

M.R. Garey, D.S. Johnson, and R. Sethi (1976) "The Complexity of Flowshop and Jobshop Scheduling," *Mathematics of Operations Research,* Vol. 1, pp. 117–129.

M.R. Garey, D.S. Johnson, B.B. Simons, and R.E. Tarjan (1981) "Scheduling Unit-Time Tasks with Arbitrary Release Times and Deadlines," *SIAM Journal of Computing,* Vol. 10, pp. 256–269.

L. Gelders and P.R. Kleindorfer (1974) "Coordinating Aggregate and Detailed Scheduling Decisions in the One-Machine Job Shop: Part I. Theory," *Operations Research*, Vol. 22, pp. 46–60.

L. Gelders and P.R. Kleindorfer (1975) "Coordinating Aggregate and Detailed Scheduling Decisions in the One-Machine Job Shop: Part II. Computation and Structure," *Operations Research*, Vol. 23, pp. 312–324.

G.V. Gens and E.V. Levner (1981) "Fast Approximation Algorithm for Job Sequencing with Deadlines," *Discrete Applied Mathematics,* Vol. 3, pp. 313–318.

B. Giffler and G.L. Thompson (1960) "Algorithms for Solving Production Scheduling Problems," *Operations Research,* Vol. 8, pp. 487–503.

P.C. Gilmore and R.E. Gomory (1964) "Sequencing a One-State Variable Machine: A Solvable Case of the Travelling Salesman Problem," *Operations Research*, Vol. 12, pp. 655–679.

J.C. Gittins (1979) "Bandit Processes and Dynamic Allocation Indices," *Journal of the Royal Statistical Society Series B,* Vol. 14, pp. 148–177.

J.C. Gittins (1981) "Multiserver Scheduling of Jobs with Increasing Completion Rates," *Journal of Applied Probability,* Vol. 18, pp. 321–324.

K.D. Glazebrook (1981a) "On Nonpreemptive Strategies for Stochastic Scheduling Problems in Continuous Time," *International Journal of System Sciences,* Vol. 12, pp. 771–782.

K.D. Glazebrook (1981b) "On Nonpreemptive Strategies in Stochastic Scheduling," *Naval Research Logistics Quarterly,* Vol. 28, pp. 289–300.

K.D. Glazebrook (1982) "On the Evaluation of of Fixed Permutations as Strategies in Stochastic Scheduling," *Stochastic Processes and Applications,* Vol. 13, pp. 171–187.

K.D. Glazebrook (1984) "Scheduling Stochastic Jobs on a Single Machine Subject to Breakdowns," *Naval Research Logistics Quarterly,* Vol. 31, pp. 251–264.

K.D. Glazebrook (1987) "Evaluating the Effects of Machine Breakdowns in Stochastic Scheduling Problems," *Naval Research Logistics,* Vol. 34, pp. 319–335.

F. Glover (1990) "Tabu Search: A Tutorial," *Interfaces,* Vol. 20, Issue 4, pp. 74–94.

T. Gonzalez (1979) "A Note on Open Shop Preemptive Schedules," *IEEE Transactions on Computers,* Vol. C–28, pp. 782–786.

T. Gonzalez and S. Sahni (1976) "Open Shop Scheduling to Minimize Finish Time," *Journal of the Association of Computing Machinery,* Vol. 23, pp. 665–679.

T. Gonzalez and S. Sahni (1978a) "Preemptive Scheduling of Uniform Processor Systems," *Journal of the Association of Computing Machinery,* Vol. 25, pp. 92–101.

T. Gonzalez and S. Sahni (1978b) "Flowshop and Jobshop Schedules: Complexity and Approximation," *Operations Research,* Vol. 26, pp. 36–52.

S.K. Goyal and C. Sriskandarajah (1988) "No-Wait Shop Scheduling: Computational Complexity and Approximate Algorithms," *Opsearch,* Vol. 25, pp. 220–244.

R.L. Graham (1966) "Bounds for Certain Multiprocessing Anomalies," *Bell System Technical Journal,* Vol. 45, pp. 1563–1581.

R.L. Graham (1969) "Bounds on Multiprocessing Timing Anomalies," *SIAM Journal of Applied Mathematics,* Vol. 17, pp. 263–269.

S.C. Graves (1981) "A Review of Production Scheduling," *Operations Research*, Vol. 29, pp. 646–676.

S.C. Graves, H. C. Meal, D. Stefek, and A. H. Zeghmi (1983) "Scheduling of Reentrant Flow Shops," *Journal of Operations Management,* Vol. 3, pp. 197–207.

J.N.D. Gupta (1972) "Heuristic Algorithms for the Multistage Flow Shop Problem," *AIIE Transactions,* Vol. 4, pp. 11–18.

N.G. Hall and M.E. Posner (1991) "Earliness-Tardiness Scheduling Problems, I: Weighted Deviation of Completion Times about a Common Due Date," *Operations Research,* Vol. 39, pp. 836–846.

N.G. Hall, W. Kubiak, and S.P. Sethi (1991) "Earliness-Tardiness Scheduling Problems, II: Weighted Deviation of Completion Times about a Common Due Date," *Operations Research,* Vol. 39, pp. 847–856.

N.G. Hall and C. Sriskandarajah (1996) "A Survey of Machine Scheduling Problems with Blocking and No-Wait in Process," *Operations Research,* Vol. 44, pp. 421–439.

J.M. Harrison (1975a) "A Priority Queue with Discounted Linear Costs," *Operations Research,* Vol. 23, pp. 260–269.

J.M. Harrison (1975b) "Dynamic Scheduling of a Multiclass Queue: Discount Optimality," *Operations Research,* Vol. 23, pp. 270–282.

J. Herrmann, C.-Y. Lee, and J. Snowdon (1993) "A Classification of Static Scheduling Problems," in *Complexity in Numerical Optimization*, P.M. Pardalos (ed.), World Scientific, pp. 203–253.

D.P. Heyman and M.J. Sobel (1982) *Stochastic Models in Operations Research, Volume I (Stochastic Processes and Operating Characteristics),* McGraw-Hill, New York.

D.S. Hochbaum and D.B. Shmoys (1987) "Using Dual Approximation Algorithms for Scheduling Problems: Theoretical and Practical Results," *Journal of the ACM*, Vol. 34, pp. 144–162.

T.J. Hodgson (1977) "A Note on Single Machine Sequencing with Random Processing Times," *Management Science,* Vol. 23, pp. 1144–1146.

T.J. Hodgson and G.W. McDonald (1981a) "Interactive Scheduling of a Generalized Flow Shop, Part I: Success Through Evolutionary Development," *Interfaces,* Vol. 11, No. 2, pp. 42–47.

T.J. Hodgson and G.W. McDonald (1981b) "Interactive Scheduling of a Generalized Flow Shop, Part II: Development of Computer Programs and Files," *Interfaces,* Vol. 11, No. 3, pp. 83–88.

T.J. Hodgson and G.W. McDonald (1981c) "Interactive Scheduling of a Generalized Flow Shop, Part III: Quantifying User Objectives to Create Better Schedules," *Interfaces,* Vol. 11, No. 4, pp. 35–41.

D.J. Hoitomt, P.B. Luh, and K.R. Pattipati (1993) "A Practical Approach to Job Shop Scheduling Problems," *IEEE Transactions on Robotics and Automation,* Vol. 9, pp. 1–13.

J.A. Hoogeveen and S.L. Van de Velde (1995) "Minimizing Total Completion Time and Maximum Cost Simultaneously Is Solvable in Polynomial Time," *Operations Research Letters,* Vol. 17, pp. 205–208.

E. Horowitz and S. Sahni (1976) "Exact and Approximate Algorithms for Scheduling Nonidentical Processors," *Journal of the Association of Computing Machinery,* Vol. 23, pp. 317–327.

E.C. Horvath, S. Lam, and R. Sethi (1977) "A Level Algorithm for Preemptive Scheduling," *Journal of the Association of Computing Machinery,* Vol. 24, pp. 32–43.

W.-L. Hsu, M. Prietula, G. Thompson, and P.S. Ow (1993) "A Mixed-Initiative Scheduling Workbench: Integrating AI, OR and HCI," *Decision Support Systems,* Vol. 9, pp. 245–257.

T.C. Hu (1961) "Parallel Sequencing and Assembly Line Problems," *Operations Research,* Vol. 9, pp. 841–848.

O.H. Ibarra and C.E. Kim (1978) "Approximation Algorithms for Certain Scheduling Problems," *Mathematics of Operations Research,* Vol. 3, pp. 179–204.

Ilog (1997) "Reducing Congestion at Paris Airports—SAIGA," Ilog ADP Success Story, Ilog, Paris, France.

L. Interrante (ed.) (1993) *Intelligent Dynamic Scheduling for Manufacturing Systems,* Proceedings of a Workshop Sponsored by National Science Foundation, the University of Alabama in Huntsville and Carnegie Mellon University, held at Cocoa Beach, January 1993.

J.R. Jackson (1955) "Scheduling a Production Line to Minimize Maximum Tardiness," Research Report 43, Management Science Research Project, University of California, Los Angeles.

J.R. Jackson (1956) "An Extension of Johnson's Results on Job Lot Scheduling," *Naval Research Logistics Quarterly,* Vol. 3, pp. 201–203.

S.M. Johnson (1954) "Optimal Two and Three-Stage Production Schedules with Setup Times Included" *Naval Research Logistics Quarterly,* Vol. 1, pp. 61–67.

T. Kämpke (1987a) "On the Optimality of Static Priority Policies in Stochastic Scheduling on Parallel Machines," *Journal of Applied Probability,* Vol. 24, pp. 430–448.

T. Kämpke (1987b) "Necessary Optimality Conditions for Priority Policies in Stochastic Weighted Flow Time Scheduling Problems," *Advances in Applied Probability,* Vol. 19, pp. 749–750.

T. Kämpke (1989) "Optimal Scheduling of Jobs with Exponential Service Times on Identical Parallel Processors," *Operations Research,* Vol. 37, pp. 126–133.

J.J. Kanet and H.H. Adelsberger (1987) "Expert Systems in Production Scheduling," *European Journal of Operational Research,* Vol. 29, pp. 51–59.

J.J. Kanet and V. Sridharan (1990) "The Electronic Leitstand: A New Tool for Shop Scheduling," *Manufacturing Review,* Vol. 3, pp. 161–170.

R.M. Karp (1972) "Reducibility among Combinatorial Problems," in *Complexity of Computer Computations,* R.E. Miller and J.W. Thatcher (eds.), Plenum Press, New York.

K.R. Karwan and J.R. Sweigart (eds.) (1989) *Proceedings of the Third International Conference on Expert Systems and the Leading Edge in Production and Operations Management,* Conference held on Hilton Head Island, South Carolina, 1989, sponsored by Management Science Department, College of Business Administration, University of South Carolina.

M.N. Katehakis and A.F. Veinott, Jr. (1987) "The Multi-Armed Bandit Problem: Decomposition and Computation," *Mathematics of Operations Research,* Vol. 12, pp. 262–268.

T. Kawaguchi and S. Kyan (1986) "Worst Case Bound of an LRF Schedule for the Mean Weighted Flow Time Problem," *SIAM Journal of Computing,* Vol. 15, pp. 1119–1129.

K.G. Kempf (1989) "Intelligent Interfaces for Computer Integrated Manufacturing," in *Proceedings of the Third International Conference on Expert Systems and the Leading Edge in Production and Operations Management,* K.R. Karwan and J.R. Sweigert (eds.), pp. 269–280, Management Science Department, College of Business Administration, University of South Carolina.

K.G. Kempf (1994) "Intelligently Scheduling Semiconductor Wafer Fabrication," in *Intelligent Scheduling,* M. Zweben and M. Fox (eds.), Morgan and Kaufmann, San Mateo, California.

S. Kerpedjiev and S.F. Roth (2000) "Mapping Communicative Goals into Conceptual Tasks to Generate Graphics in Discourse," in *Proceedings of Intelligent User Interfaces (IUI '00),* January 2000.

M. Kijima, N. Makimoto, and H. Shirakawa (1990) "Stochastic Minimization of the Makespan in Flow Shops with Identical Machines and Buffers of Arbitrary Size," *Operations Research,* Vol. 38, pp. 924–928.

Y.-D. Kim and C.A. Yano (1994) "Minimizing Mean Tardiness and Earliness in Single-Machine Scheduling Problems with Unequal Due Dates," *Naval Research Logistics,* Vol. 41, pp. 913–933.

S. Kirkpatrick, C.D. Gelatt, and M. P. Vecchi (1983) "Optimization by Simulated Annealing," *Science,* Vol. 220, pp. 671–680.

H. Kise, T. Ibaraki, and H. Mine (1978) "A Solvable Case of the One-Machine Scheduling Problem with Ready and Due Times," *Operations Research,* Vol. 26, pp. 121–126.

L. Kleinrock (1976) *Queueing Systems, Vol. II: Computer Applications,* John Wiley, New York.

S.A. Kravchenko (2000) "On the Complexity of Minimizing the Number of Late Jobs in Unit Time Open Shops," *Discrete Applied Mathematics,* Vol. 100, pp. 127–132.

S. Kreipl (2000) "A Large Step Random Walk for Minimizing Total Weighted Tardiness in a Job Shop," *Journal of Scheduling,* Vol. 3, pp. 125–138.

P. Ku and S.-C. Niu (1986) "On Johnson's Two-Machine Flow Shop with Random Processing Times," *Operations Research,* Vol. 34, pp. 130–136.

M. Kunde (1976) "Beste Schranken Beim LP-Scheduling," *Bericht 7603, Institut für Informatik and und Praktische Mathematik,* Universität Kiel.

A. Kusiak and M. Chen (1988) "Expert Systems for Planning and Scheduling Manufacturing Systems," *European Journal of Operational Research,* Vol. 34, pp. 113–130.

E. Kutanoglu and S.D. Wu (1999) "On Combinatorial Auction and Lagrangean Relaxation for Distributed Resource Scheduling," *IIE Transactions,* Vol. 31, pp. 813–826.

J. Labetoulle, E.L. Lawler, J.K. Lenstra, and A.H.G. Rinnooy Kan (1984) "Preemptive Scheduling of Uniform Machines subject to Release Dates" in *Progress in Combinatorial Optimization,* W.R. Pulleyblank (ed.), pp. 245–261, Academic Press, New York.

B.J. Lageweg, E.L. Lawler, J.K. Lenstra, and A.H.G. Rinnooy Kan (1981) "Computer-Aided Complexity Classification of Deterministic Scheduling Problems," Technical Report BW 138/81, the Mathematical Centre, Amsterdam, the Netherlands.

B.J. Lageweg, E.L. Lawler, J.K. Lenstra, and A.H.G. Rinnooy Kan (1982) "Computer-Aided Complexity Classification of Combinatorial Problems," *Communications of the ACM,* Vol. 25, pp. 817–822.

E.L. Lawler (1973) "Optimal Sequencing of a Single Machine subject to Precedence Constraints," Management Science, Vol. 19, pp. 544–546.

E.L. Lawler (1977) "A 'Pseudopolynomial' Time Algorithm for Sequencing Jobs to Minimize Total Tardiness," *Annals of Discrete Mathematics,* Vol. 1, pp. 331–342.

E.L. Lawler (1978) "Sequencing Jobs to Minimize Total Weighted Completion Time Subject to Precedence Constraints," *Annals of Discrete Mathematics,* Vol. 2, pp. 75–90.

E.L. Lawler (1982) "A Fully Polynomial Approximation Scheme for the Total Tardiness Problem," *Operations Research Letters,* Vol. 1, pp. 207–208.

E.L. Lawler and J. Labetoulle (1978) "On Preemptive Scheduling of Unrelated Parallel Processors by Linear Programming," *Journal of the Association of Computing Machinery,* Vol. 25, pp. 612–619.

E.L. Lawler, J.K. Lenstra, and A.H.G. Rinnooy Kan (1981) "Minimizing Maximum Lateness in a Two-Machine Open Shop," *Mathematics of Operations Research,* Vol. 6, pp. 153–158; Erratum Vol. 7, pp. 635.

E.L. Lawler, J.K. Lenstra, and A.H.G. Rinnooy Kan (1982) "Recent Developments in Deterministic Sequencing and Scheduling: A Survey," in *Deterministic and Stochastic Scheduling*, Dempster, Lenstra and Rinnooy Kan (eds.), pp. 35–74, Reidel, Dordrecht.

E.L. Lawler, J.K. Lenstra, A.H.G. Rinnooy Kan, and D. Shmoys (1993) "Sequencing and Scheduling: Algorithms and Complexity," in *Handbooks in Operations Research and Management Science, Vol. 4: Logistics of Production and Inventory,* S. S. Graves, A. H. G. Rinnooy Kan, and P. Zipkin, (eds.), pp. 445–522, North-Holland, New York.

E.L. Lawler and C.U. Martel (1989) "Preemptive Scheduling of Two Uniform Machines to Minimize the Number of Late Jobs," *Operations Research,* Vol. 37, pp. 314–318.

G. Lawton (1992) "Genetic Algorithms for Schedule Optimization," *AI Expert*, May Issue, pp. 23–27.

C.-Y. Lee and L. Lei (eds.) (1997) "Scheduling: Theory and Applications," *Annals of Operations Research,* Vol. 70, Baltzer, Basel.

C.-Y. Lee and C.S. Lin (1991) "Stochastic Flow Shops with Lateness-Related Performance Measures," *Probability in the Engineering and Informational Sciences*, Vol. 5, pp. 245–254.

C.-Y. Lee, L.A. Martin-Vega, R. Uzsoy, and J. Hinchman (1993) "Implementation of a Decision Support System for Scheduling Semiconductor Test Operations," *Journal of Electronics Manufacturing,* Vol. 3, pp. 121–131.

C.-Y. Lee and J.D. Massey (1988) "Multi-Processor Scheduling: Combining *LPT* and *MULTIFIT*," *Discrete Applied Mathematics*, Vol. 20, pp. 233–242.

C.-Y. Lee, R. Uzsoy, and L.A. Martin-Vega (1992) "Efficient Algorithms for Scheduling Semiconductor Burn-In Operations," *Operations Research*, Vol. 40, pp. 764–995.

Y.H. Lee, K. Bhaskaran, and M.L. Pinedo (1997) "A Heuristic to Minimize the Total Weighted Tardiness with Sequence Dependent Setups," *IIE Transactions,* Vol. 29, pp. 45–52.

P. Lefrancois, M.H. Jobin, and B. Montreuil "An Object-Oriented Knowledge Representation in Real-Time Scheduling," in *New Directions for Operations Research in Manufacturing,* G. Fandel, Th. Gulledge, and A. Jones (eds.), Springer Verlag, pp. 262–279.

J.K. Lenstra (1977) "Sequencing by Enumerative Methods," Mathematical Centre Tracts 69, Centre for Mathematics and Computer Science, Amsterdam.

J.K. Lenstra and A.H.G. Rinnooy Kan (1978) "Computational Complexity of Scheduling under Precedence Constraints," *Operations Research,* Vol. 26, pp. 22–35.

J.K. Lenstra and A.H.G. Rinnooy Kan (1979) "Computational Complexity of Discrete Optimization Problems," *Annals of Discrete Mathematics*, Vol. 4, pp. 121–140.

J.K. Lenstra, A.H.G. Rinnooy Kan, and P. Brucker (1977) "Complexity of machine scheduling problems," *Annals of Discrete Mathematics*, Vol. 1, pp. 343–362.

V.J. Leon and S.D. Wu (1994) "Robustness Measures and Robust Scheduling for Job Shops," *IIE Transactions,* Vol. 26, pp. 32–43.

V.J. Leon, S.D. Wu, and R. Storer (1994) "A Game Theoretic Approach for Job Shops in the Presence of Random Disruptions," *International Journal of Production Research,* Vol. 32, pp. 1451–1476.

J.Y.-T. Leung and G.H. Young (1990) "Minimizing Total Tardiness on a Single Machine with Precedence Constraint," *ORSA Journal on Computing,* Vol. 2, pp. 346–352.

E.M. Levner (1969) "Optimal Planning of Parts Machining on a Number of Machines," *Automation and Remote Control,* Vol. 12, pp. 1972–1981.

C.Y. Liu and R.L. Bulfin (1985) "On the Complexity of Preemptive Open-Shop Scheduling Problems," *Operations Research Letters,* Vol. 4, pp. 71–74.

Z.A. Lomnicki (1965) "A Branch and Bound Algorithm for the Exact Solution of the Three-Machine Scheduling Problem," *Operational Research Quarterly,* Vol. 16, pp. 89–100.

P.B. Luh and D.J. Hoitomt (1993) "Scheduling of Manufacturing Systems Using the Lagrangean Relaxation Technique," *IEEE Transactions on Automatic Control, Special Issue: Meeting the Challenge of Computer Science in the Industrial Applications of Control,* Vol. 38, pp. 1066–1080.

P.B. Luh, D.J. Hoitomt, E. Max, and K.R. Pattipati (1990) "Schedule Generation and Reconfiguration for Parallel Machines," *IEEE Transactions on Robotics and Automation,* Vol. 6, pp. 687–696.

M. Marchesi, E. Rusconi, and F. Tiozzo (1999) "Yogurt Process Production: A Flow Shop Scheduling Solution, the Process Industry," Cybertec Technical Report, Cybertec, Trieste, Italy.

A.W. Marshall and I. Olkin (1979) *Inequalities: Theory of Majorization and Its Applications*, Academic Press, New York.

C.U. Martel (1982) "Preemptive Scheduling with Release Times, Deadlines and Due Times," *Journal of the Association of Computing Machinery,* Vol. 29, pp. 812–829.

J. Martin (1993) *Principles of Object-Oriented Analysis and Design,* Prentice-Hall, Englewood Cliffs, New Jersey.

H. Matsuo (1990) "Cyclic Sequencing Problems in the Two-machine Permutation Flow Shop: Complexity, Worst-Case and Average Case Analysis," *Naval Research Logistics,* Vol. 37, pp. 679–694.

H. Matsuo, C.J. Suh, and R.S. Sullivan (1988) "A Controlled Search Simulated Annealing Method for the General Job Shop Scheduling Problem," Working Paper 03-44-88, Graduate School of Business, University of Texas, Austin.

S.T. McCormick and M.L. Pinedo (1995) "Scheduling n Independent Jobs on m Uniform Machines with Both Flow Time and Makespan Objectives: A Parametric Analysis," *ORSA Journal of Computing,* Vol. 7, pp. 63–77.

S.T. McCormick, M.L. Pinedo, S. Shenker, and B. Wolf (1989) "Sequencing in an Assembly Line with Blocking to Minimize Cycle Time," *Operations Research,* Vol. 37, pp. 925–936.

S.T. McCormick, M.L. Pinedo, S. Shenker, and B. Wolf (1990) "Transient Behavior in a Flexible Assembly System," *The International Journal of Flexible Manufacturing Systems,* Vol. 3, pp. 27–44.

K. McKay, M. Pinedo, and S. Webster (2001) "A Practice-Focused Agenda for Production Scheduling Research," *Production and Operations Management,* Vol. 10.

K. McKay, F. Safayeni, and J. Buzacott (1988) "Job Shop Scheduling Theory: What is Relevant?" *Interfaces*, Vol. 18, No. 4, pp. 84–90.

G.B. McMahon and M. Florian (1975) "On Scheduling with Ready Times and Due Dates to Minimize Maximum Lateness," *Operations Research,* Vol. 23, pp. 475–482.

R. McNaughton (1959) "Scheduling with Deadlines and Loss Functions," *Management Science,* Vol. 6, pp. 1–12.

S.V. Mehta and R. Uzsoy (1999) "Predictable Scheduling of a Single Machine Subject to Breakdowns," *International Journal of Computer Integrated Manufacturing,* Vol. 12, pp. 15–38.

R.H. Möhring and F.J. Radermacher (1985a) "Generalized Results on the Polynomiality of Certain Weighted Sum Scheduling Problems," *Methods of Operations Research,* Vol. 49, pp. 405–417.

R.H. Möhring and F.J. Radermacher (1985b) "An Introduction to Stochastic Scheduling Problems," in *Contributions to Operations Research,* K. Neumann and D. Pallaschke (eds.), Lecture Notes in Economics and Mathematical Systems 240, Springer Verlag, Berlin, pp. 72–130.

R.H. Möhring, F.J. Radermacher, and G. Weiss (1984) "Stochastic Scheduling Problems, I: General Strategies," *Zeitschrift für Operations Research,* Vol. 28, pp. 193–260.

R.H. Möhring, F.J. Radermacher, and G. Weiss (1985) "Stochastic Scheduling Problems, II: Set Strategies," *Zeitschrift für Operations Research,* Vol. 29, pp. 65–104.

C.L. Monma and A.H.G. Rinnooy Kan (1983) "A Concise Survey of Efficiently Solvable Special Cases of the Permutation Flow-Shop Problem," *RAIRO Recherche Operationelle,* Vol. 17, pp. 105–119.

C.L. Monma and J.B. Sidney (1979) "Sequencing with Series-Parallel Precedence Constraints," *Mathematics of Operations Research,* Vol. 4, pp. 215–224.

C.L. Monma and J.B. Sidney (1987) "Optimal Sequencing via Modular Decomposition: Characterizations of Sequencing Functions," *Mathematics of Operations Research,* Vol. 12, pp. 22–31.

J.M. Moore (1968) "An *n* Job, One Machine Sequencing Algorithm for Minimizing the Number of Late Jobs," *Management Science,* Vol. 15, pp. 102–109.

T.E. Morton and D. Pentico (1993) *Heuristic Scheduling Systems*, John Wiley, New York.

E.J. Muth (1979) "The Reversibility Property of a Production Line," *Management Science,* Vol. 25, pp. 152–158.

J.F. Muth and G.L. Thompson (eds.) (1963) *Industrial Scheduling*, Prentice-Hall, New Jersey.

P. Nain, P. Tsoucas, and J. Walrand (1989) "Interchange Arguments in Stochastic Scheduling," *Journal of Applied Probability,* Vol. 27, pp. 815–826.

R.T. Nelson, R.K. Sarin, and R.L. Daniels (1986) "Scheduling with Multiple Performance Measures: The One Machine Case," *Management Science,* Vol. 32, pp. 464–479.

G.L. Nemhauser and L.A. Wolsey (1988) *Integer and Combinatorial Optimization*, John Wiley, New York.

S.J. Noronha and V.V.S. Sarma (1991) "Knowledge-Based Approaches for Scheduling Problems: A Survey," *IEEE Transactions on Knowledge and Data Engineering,* Vol. 3, pp. 160–171.

E. Nowicki and C. Smutnicki (1996) "A Fast Taboo Search Algorithm for the Job Shop Problem," *Management Science,* Vol. 42, pp. 797–813.

E. Nowicki and S. Zdrzalka (1986) "A Note on Minimizing Maximum Lateness in a One-Machine Sequencing Problem with Release Dates," *European Journal of Operational Research,* Vol. 23, pp. 266–267.

W.P.M. Nuijten (1994) *Time and Resource Constrained Scheduling: A Constraint Satisfaction Approach*, Ph.D. Thesis, Eindhoven University of Technology, Eindhoven, the Netherlands.

W.P.M. Nuijten and E.H.L. Aarts (1996) "A Computational Study of Constraint Satisfaction for Multiple Capacitated Job Shop Scheduling," *European Journal of Operational Research,* Vol. 90, pp. 269–284.

M. Nussbaum and E.A. Parra (1993) "A Production Scheduling System," *ORSA Journal on Computing,* Vol. 5, pp. 168–181.

M.D. Oliff (ed.) (1988) *Expert Systems and Intelligent Manufacturing, Proceedings of the Second International Conference on Expert Systems and the Leading Edge in Production Planning and Control,* held May 1988 in Charleston, South Carolina, Elsevier, New York.

I.M. Ovacik and R. Uzsoy (1997) *Decomposition Methods for Complex Factory Scheduling Problems,* Kluwer Academic Publishers, Boston.

P.S. Ow (1985) "Focused Scheduling in Proportionate Flowshops," *Management Science,* Vol. 31, pp. 852–869.

P.S. Ow and T.E. Morton (1988) "Filtered Beam Search in Scheduling," *International Journal of Production Research,* Vol. 26, pp. 297–307.

P.S. Ow and T.E. Morton (1989) "The Single Machine Early/Tardy Problem," *Management Science,* Vol. 35, pp. 177–191.

P.S. Ow, S.F. Smith, and R. Howie (1988) "A Cooperative Scheduling System," in *Expert Systems and Intelligent Manufacturing,* M. D. Oliff (ed.), Elsevier, Amsterdam, pp. 43–56.

D.S. Palmer (1965) "Sequencing Jobs Through a Multi-Stage Process in the Minimum Total Time— A Quick Method of Obtaining a Near Optimum," *Operational Research Quarterly,* Vol. 16, pp. 101–107.

S.S. Panwalkar and W. Iskander (1977) "A Survey of Scheduling Rules," *Operations Research,* Vol. 25, pp. 45–61.

S.S. Panwalkar, M.L. Smith, and A. Seidmann (1982) "Common Due Date Assignment to Minimize Total Penalty for the One–Machine Scheduling Problem," *Operations Research,* Vol. 30, pp. 391–399.

C.H. Papadimitriou (1994) *Computational Complexity,* Addison-Wesley, Reading, Massachusetts.

C.H. Papadimitriou and P.C. Kannelakis (1980) "Flowshop Scheduling with Limited Temporary Storage," *Journal of the Association of Computing Machinery,* Vol. 27, pp. 533–549.

C.H. Papadimitriou and K. Steiglitz (1982) *Combinatorial Optimization: Algorithms and Complexity,* Prentice-Hall, New Jersey.

R.G. Parker (1995) *Deterministic Scheduling Theory,* Chapman & Hall, London.

R.G. Parker and R.L. Rardin (1988) *Discrete Optimization,* Academic Press, San Diego.

S. Park, N. Raman, and M.J. Shaw (1997) "Adaptive Scheduling in Dynamic Flexible Manufacturing Systems: A Dynamic Rule Selection Approach," *IEEE Transactions on Robotics and Automation,* Vol. 13, pp. 486–502.

E. Pesch (1994) *Learning in Automated Manufacturing—A Local Search Approach,* Physica-Verlag (A Springer-Verlag Company), Heidelberg, Germany.

J.R. Pimentel (1990) *Communication Networks for Manufacturing,* Prentice-Hall, New Jersey.

M. Pinedo (1981a) "Minimizing Makespan with Bimodal Processing Time Distributions," *Management Science,* Vol. 27, pp. 582–586.

M. Pinedo (1981b) "A Note on the Two-Machine Job Shop with Exponential Processing Times," *Naval Research Logistics Quarterly,* Vol. 28, pp. 693–696.

M. Pinedo (1982) "Minimizing the Expected Makespan in Stochastic Flow Shops," *Operations Research,* Vol. 30, pp. 148–162.

M. Pinedo (1983) "Stochastic Scheduling with Release Dates and Due Dates," *Operations Research,* Vol. 31, pp. 559–572.

M. Pinedo (1984) "A Note on the Flow Time and the Number of Tardy Jobs in Stochastic Open Shops," *European Journal of Operational Research,* Vol. 18, pp. 81–85.

M. Pinedo (1985) "A Note on Stochastic Shop Models in Which Jobs have the Same Processing Requirements on Each Machine," *Management Science,* Vol. 31, pp. 840–845.

M. Pinedo and X. Chao (1999) *Operations Scheduling with Applications in Manufacturing and Services,* Irwin/McGraw-Hill, Burr Ridge, Illinois.

M. Pinedo and E. Rammouz (1988) "A Note on Stochastic Scheduling on a Single Machine Subject to Breakdown and Repair," *Probability in the Engineering and Informatiional Sciences,* Vol. 2, pp.41–49.

M. Pinedo and S.M. Ross (1980) "Scheduling Jobs under Nonhomogeneous Poisson Shocks," *Management Science,* Vol. 26, pp. 1250–1258.

M. Pinedo and S.M. Ross (1982) "Minimizing Expected Makespan in Stochastic Open Shops," *Advances in Applied Probability,* Vol. 14, pp. 898–911.

M. Pinedo, R. Samroengraja, and P.C. Yen (1994) "Design Issues with Regard to Scheduling Systems in Manufacturing," in *Control and Dynamic Systems,* C. Leondes (ed.), Vol. 60, pp. 203–238, Academic Press, San Diego, California.

M. Pinedo and Z. Schechner (1985) "Inequalities and Bounds for the Scheduling of Stochastic Jobs on Parallel Machines," *Journal of Applied Probability,* Vol. 22, pp. 739–744.

M. Pinedo and M. Singer (1999) "A Shifting Bottleneck Heuristic for Minimizing the Total Weighted Tardiness in a Job Shop," *Naval Research Logistics,* Vol. 46, pp. 1–12.

M. Pinedo and R.R. Weber (1984) "Inequalities and Bounds in Stochastic Shop Scheduling," *SIAM Journal of Applied Mathematics,* Vol. 44, pp. 869–879.

M. Pinedo and G. Weiss (1979) "Scheduling of Stochastic Tasks on Two Parallel Processors," *Naval Research Logistics Quarterly,* Vol. 26, pp. 527–535.

M. Pinedo and G. Weiss (1984) "Scheduling Jobs with Exponentially Distributed Processing Times and Intree Precedence Constraints on Two Parallel Machines," *Operations Research,* Vol. 33, pp. 1381–1388.

M. Pinedo and G. Weiss (1987) "The 'Largest Variance First' Policy in some Stochastic Scheduling Problems," *Operations Research,* Vol. 35, pp. 884–891.

M. Pinedo and S.-H. Wie (1986) "Inequalities for Stochastic Flow Shops and Job Shops," *Applied Stochastic Models and Data Analysis,* Vol. 2, pp. 61–69.

M. Pinedo, B. Wolf, and S.T. McCormick (1986) "Sequencing in a Flexible Assembly Line with Blocking to Minimize Cycle Time," in *Proceedings of the Second ORSA/TIMS Conference on Flexible Manufacturing Systems,* K. Stecke and R. Suri (eds.), Elsevier, Amsterdam, pp. 499–508.

M. Pinedo and B. P.-C. Yen (1997) "On the Design and Development of Object-Oriented Scheduling Systems," *Annals of Operations Research,* Vol. 70, C.-Y. Lee and L. Lei (eds.), pp. 359–378.

E. Pinson (1995) "The Job Shop Scheduling Problem: A Concise Survey and Some Recent Developments," in *Scheduling Theory and Applications,* P. Chrétienne, E.G. Coffman, Jr., J.K. Lenstra, and Z. Liu (eds.), John Wiley, New York, pp. 177–293.

M.E. Posner (1985) "Minimizing Weighted Completion Times with Deadlines," *Operations Research,* Vol. 33, pp. 562–574.

C.N. Potts (1980) "Analysis of a Heuristic for One Machine Sequencing with Release Dates and Delivery Times," *Operations Research,* Vol. 28, pp. 1436–1441.

C.N. Potts and L.N. van Wassenhove (1982) "A Decomposition Algorithm for the Single Machine Total Tardiness Problem," *Operations Research Letters,* Vol. 1, pp. 177–181.

C.N. Potts and L.N. van Wassenhove (1983) "An Algorithm for Single Machine Sequencing with Deadlines to Minimize Total Weighted Completion Time," *European Journal of Operational Research,* Vol. 12, pp. 379–387.

C.N. Potts and L.N. van Wassenhove (1985) "A Branch and Bound Algorithm for the Total Weighted Tardiness Problem," *Operations Research,* Vol. 33, pp. 363–377.

C.N. Potts and L.N. van Wassenhove (1987) "Dynamic Programming and Decomposition Approaches for the Single Machine Total Tardiness Problem," *European Journal of Operational Research,* Vol. 32, pp. 405–414.

C.N. Potts and L.N. van Wassenhove (1988) "Algorithms for Scheduling a Single Machine to Minimize the Weighted Number of Late Jobs," *Management Science,* Vol. 34, pp. 843–858.

M. Queyranne (1993) "Structure of a Simple Scheduling Polyhedron," *Mathematical Programming,* Vol. 58, pp. 263–286.

M. Queyranne and A.S. Schulz (1994) "Polyhedral Approaches to Machine Scheduling," Preprint No. 408/1994, Fachbereich 3 Mathematik, Technische Universität, Berlin.

M. Queyranne and Y. Wang (1991) "Single Machine Scheduling Polyhedra with Precedence Constraints," *Mathematics of Operations Research,* Vol. 16, pp. 1–20.

R.M.V. Rachamadugu (1987) "A Note on the Weighted Tardiness Problem," *Operations Research,* Vol. 35, pp. 450–452.

M. Raghavachari (1988) "Scheduling Problems with Non-Regular Penalty Functions: A Review," *Opsearch,* Vol. 25, pp. 144–164.

S.S. Reddi and C.V. Ramamoorthy (1972) "On the Flowshop Sequencing Problem with No Wait in Process," *Operational Research Quarterly,* Vol. 23, pp. 323–330.

J. Rickel (1988) "Issues in the Design of Scheduling Systems," in *Expert Systems and Intelligent Manufacturing,* M.D. Oliff (ed.), pp. 70–89, Elsevier, New York.

R. Righter (1988) "Job Scheduling to Minimize Expected Weighted Flow Time on Uniform Processors," *System and Control Letters,* Vol. 10, pp. 211–216.

R. Righter (1992) "Loading and Sequencing on Parallel Machines," *Probability in the Engineering and Informational Sciences,* Vol. 6, pp. 193–201.

R. Righter (1994) "Stochastic Scheduling," Chapter 13 in *Stochastic Orders*, M. Shaked and G. Shanthikumar (eds.), Academic Press, San Diego.

R. Righter and S. Xu (1991a) "Scheduling Jobs on Heterogeneous Processors," *Annals of Operations Research,* Vol. 28, pp. 587–602.

R. Righter and S. Xu (1991b) "Scheduling Jobs on Nonidentical IFR Processors to Minimize General Cost Functions," *Advances in Applied Probability,* Vol. 23, pp. 909–924.

A.H.G. Rinnooy Kan (1976) *Machine Scheduling Problems: Classification, Complexity and Computations,* Nijhoff, The Hague.

H. Röck (1984) "The Three-Machine No-Wait Flow Shop Problem Is NP-Complete," *Journal of the Association of Computing Machinery,* Vol. 31, pp. 336–345.

F.A. Rodammer and K.P. White (1988) "A Recent Survey of Production Scheduling," *IEEE Transactions on Systems, Man and Cybernetics*, Vol. 18, pp. 841–851.

S.M. Ross (1983a) *Stochastic Processes,* John Wiley, New York.

S.M. Ross (1983b) *Introduction to Stochastic Dynamic Programming,* Academic Press, New York.

S.M. Ross (1981) *Introduction to Probability Models (2nd ed.),* Academic Press, New York.

M.H. Rothkopf (1966a) "Scheduling Independent Tasks on Parallel Processors," *Management Science,* Vol. 12, pp. 437–447.

M.H. Rothkopf (1966b) "Scheduling with Random Service Times," *Management Science,* Vol. 12, pp. 707–713.

M.H. Rothkopf and S.A. Smith (1984) "There are no Undiscovered Priority Index Sequencing Rules for Minimizing Total Delay Costs," *Operations Research,* Vol. 32, pp. 451–456.

R. Roundy (1992) "Cyclic Schedules for Job Shops with Identical Jobs" *Mathematics of Operations Research,* Vol. 17, pp. 842–865.

R. Roundy, W. Maxwell, Y. Herer, S. Tayur, and A. Getzler (1991) "A Price-Directed Approach to Real-Time Scheduling of Production Operations," *IIE Transactions,* Vol. 23, pp. 449–462.

B. Roy and B. Sussmann (1964) "Les Problemes d'Ordonnancement avec Constraintes Disjonctives," Note DS No. 9 bis, SEMA, Montrouge.

I. Sabuncuoglu and M. Bayiz (2000) "Analysis of Reactive Scheduling Problems in a Job Shop Environment," *European Journal of Operational Research,* Vol. 126, pp. 567–586.

I. Sabuncuoglu and A. Toptal (1999) "Distributed Scheduling, I: A Review of Concepts and Applications," Technical Paper No: IEOR 9910, Department of Industrial Engineering, Bilkent University, Ankara, Turkey.

I. Sabuncuoglu and A. Toptal (1999) "Distributed Scheduling, II: Bidding Algorithms and Performance Evaluations," Technical Paper No: IEOR 9911, Department of Industrial Engineering, Bilkent University, Ankara, Turkey.

S. Sahni (1976) "Algorithms for Scheduling Independent Tasks," *Journal of the Association of Computing Machinery,* Vol. 23, pp. 116–127.

S. Sahni and Y. Cho (1979a) "Complexity of Scheduling Jobs with No Wait in Process," *Mathematics of Operations Research,* Vol. 4, pp. 448–457.

S. Sahni and Y. Cho (1979b) "Scheduling Independent Tasks with Due Times on a Uniform Processor System," *Journal of the Association of Computing Machinery,* Vol. 27, pp. 550–563.

T. Sandholm (1993) "An Implementation of the Contract Net Protocol Based on Marginal Cost Calculations," in *Proceedings of the 11th National Conference on Artificial Intelligence (AAAI-93),* pp. 256–262.

S.C. Sarin, S. Ahn, and A.B. Bishop (1988) "An Improved Branching Scheme for the Branch and Bound Procedure of Scheduling n Jobs on m Machines to Minimize Total Weighted Flow Time," *International Journal of Production Research,* Vol. 26, pp. 1183–1191.

S.C. Sarin, G. Steiner, and E. Erel (1990) "Sequencing Jobs on a Single Machine with a Common Due Date and Stochastic Processing Times," *European Journal of Operational Research,* Vol. 51, pp. 188–198.

J. Sauer (1993) "Dynamic Scheduling Knowledge for Meta-Scheduling," in *Proceedings of the Sixth International Conference on Industrial Engineering Applications of Artificial Intelligence and Expert Systems (IEA/AIE 93),* Edinburgh, Scotland.

A.-W. Scheer (1988) *CIM—Computer Steered Industry,* Springer Verlag, New York.

L. Schrage (1968) "A Proof of the Optimality of the Shortest Remaining Processing Time Discipline," *Operations Research,* Vol. 16, pp. 687–690.

A. Schrijver (1998) *Theory of Linear and Integer Programming,* John Wiley, New York.

P. Schuurman and G. Woeginger (1999) "Polynomial Time Approximation Algorithms for Machine Scheduling: Ten Open Problems," *Journal of Scheduling,* Vol. 2, pp. 203–214.

R. Sethi (1977) On the Complexity of Mean Flow Time Scheduling," *Mathematics of Operations Research,* Vol. 2, pp. 320–330.

S.V. Sevastianov and G. Woeginger (1998) "Makespan Minimization in Open Shops: A Polynomial Time Approximation Scheme," *Mathematical Programming,* Vol. 82, pp. 191–198.

J. Sgall (1998) "On-line Scheduling," Chapter 9 in *Online Algorithms: The State of the Art,* A. Fiat and G. Woeginger (eds.), Lecture Notes in Computer Science, Vol. 1442, Springer Verlag, Berlin.

N. Shakhlevich, H. Hoogeveen, and M. Pinedo (1998) "Minimizing Total Weighted Completion Time in a Proportionate Flow Shop," *Journal of Scheduling,* Vol. 1, pp. 157–168.

G. Shanthikumar and D. Yao (1991) "Bivariate Characterization of Some Stochastic Order Relations," *Advances in Applied Probability,* Vol. 23, pp. 642–659.

M.J. Shaw (1987) "A Distributed Scheduling Method for Computing Integrated Manufacturing: The Use of Local Area Networks in Cellular Systems," *International Journal of Production Research,* Vol. 25, pp. 1285–1303.

M.J. Shaw (1988a) "Dynamic Scheduling in Cellular Manufacturing Systems; A Framework for Networked Decision Making," *Journal of Manufacturing Systems,* Vol. 7, pp. 83–94.

M.J. Shaw (1988b) "Knowledge-Based Scheduling in Flexible Manufacturing Systems: An Integration of Pattern-Directed Inference and Heuristic Search," *International Journal of Production Research,* Vol. 6, pp. 821–844.

M.J. Shaw (1989) "FMS Scheduling as Cooperative Problem Solving," *Annals of Operations Research,* Vol. 17, pp. 323–346.

M.J. Shaw, S. Park, and N. Raman (1992) "Intelligent Scheduling with Machine Learning Capabilities: The Induction of Scheduling Knowledge," *IIE Transactions on Design and Manufacturing,* Vol. 24, pp. 156–168.

M.J. Shaw and A.B. Whinston (1989) "An Artificial Intelligence Approach to the Scheduling of Flexible Manufacturing Systems," *IIE Transactions,* Vol. 21, pp. 170–183.

D.B. Shmoys, J. Wein, and D.P. Williamson (1995) "Scheduling Parallel Machines On-line," *SIAM Journal of Computing,* Vol. 24, pp. 1313–1331.

J.B. Sidney (1977) "Optimal Single Machine Scheduling with Earliness and Tardiness Penalties," *Operations Research,* Vol. 25, pp. 62–69.

J.B. Sidney and G. Steiner (1986) "Optimal Sequencing by Modular Decomposition: Polynomial Algorithms," *Operations Research,* Vol. 34, pp. 606–612.

B. Simons (1983) Multiprocessor Scheduling of Unit-Time Jobs with Arbitrary Release Times and Deadlines," *SIAM Journal of Computing,* Vol. 12, pp. 294–299.

M. Singer and M. Pinedo (1998) "A Computational Study of Branch and Bound Techniques for Minimizing the Total Weighted Tardiness in Job Shops," *IIE Transactions Scheduling and Logistics,* Vol. 30, pp. 109–118.

M.L. Smith, S.S. Panwalkar, and R.A. Dudek (1975) "Flow Shop Sequencing with Ordered Processing Time Matrices," *Management Science,* Vol. 21, pp. 544–549.

M.L. Smith, S.S. Panwalkar, and R.A. Dudek (1976) "Flow Shop Sequencing Problem with Ordered Processing Time Matrices: A General Case," *Naval Research Logistics Quarterly,* Vol. 23, pp. 481–486.

S.F. Smith (1992) "Knowledge-Based Production Management: Approaches, Results and Prospects," *Production Planning and Control*, Vol. 3, pp. 350–380.

S.F. Smith (1994) "OPIS: A Methodology and Architecture for Reactive Scheduling," in *Intelligent Scheduling,* M. Zweben and M. Fox (eds.), Morgan and Kaufmann, San Mateo, California.

S.F. Smith and M.A. Becker (1997) "An Ontology for Constructing Scheduling Systems," in Working Notes from 1997 AAAI Spring Symposium on Ontological Engineering, Stanford University, Stanford, California.

S.F. Smith, M.S. Fox, and P.S. Ow (1986) "Constructing and Maintaining Detailed Production Plans: Investigations into the Development of Knowledge-Based Factory Scheduling Systems," *AI Magazine,* Vol. 7, pp. 45–61.

S.F. Smith and O. Lassila (1994) "Configurable Systems for Reactive Production Management," in *Knowledge-Based Reactive Scheduling (B-15),* E. Szelke and R.M. Kerr (eds.), Elsevier Science, North Holland, Amsterdam.

S.F. Smith, N. Muscettola, D.C. Matthys, P.S. Ow, and J. Y. Potvin (1990) "OPIS: An Opportunistic Factory Scheduling System," *Proceedings of the Third International Conference on Industrial and Expert Systems (IEA/AIE 90),* Charleston, South Carolina.

W.E. Smith (1956) "Various Optimizers for Single Stage Production," *Naval Research Logistics Quarterly*, Vol. 3, pp. 59–66.

J.J. Solberg (1989) "Production Planning and Scheduling in CIM," *Information Processing,* Vol. 89, pp. 919–925.

C. Sriskandarajah and S.P. Sethi (1989) Scheduling Algorithms for Flexible Flow Shops: Worst and Average Case Performance," *European Journal of Operational Research,* Vol. 43, pp. 143–160.

M.S. Steffen (1986) "A Survey of Artificial Intelligence-Based Scheduling Systems," in *Proceedings of Fall 1986 Industrial Engineering Conference,* Institute of Industrial Engineers, pp. 395–405.

R.H. Storer, S.D. Wu, and R. Vaccari (1992) "New Search Spaces for Sequencing Problems with Application to Job Shop Scheduling," *Management Science,* Vol. 38, pp. 1495–1509.

D.R. Sule (1996) *Industrial Scheduling,* PWS Publishing Company, Boston.

G. Sullivan and K. Fordyce (1990) "IBM Burlington's Logistics Management System," *Interfaces,* Vol. 20, pp. 43–64.

S. Suresh, R.D. Foley, and S.E. Dickey (1985) "On Pinedo's Conjecture for Scheduling in a Stochastic Flow Shop," *Operations Research,* Vol. 33, pp. 1146–1153.

W. Szwarc (1971) "Elimination Methods in the $m \times n$ Sequencing Problem," *Naval Research Logistics Quarterly,* Vol. 18, pp. 295–305.

W. Szwarc (1973) "Optimal Elimination Methods in the $m \times n$ Flow Shop Scheduling Problem," *Operations Research,* Vol. 21, pp. 1250–1259.

W. Szwarc (1978) "Dominance Conditions for the Three-Machine Flow-Shop Problem," *Operations Research,* Vol. 26, pp. 203–206.

W. Szwarc (1998) "Decomposition in Single-Machine Scheduling," *Annals of Operations Research,* Vol. 83, pp. 271–287.

W. Szwarc and S.K. Mukhopadhyay (1995) "Optimal Timing Schedules in Earliness-Tardiness Single Machine Sequencing," *Naval Research Logistics,* Vol. 42, pp. 1109–1114.

E. Taillard (1990) "Some Efficient Heuristic Methods for the Flow Shop Sequencing Problem," *European Journal of Operational Research,* Vol. 47, pp. 65–74.

P.P. Talwar (1967) "A Note on Sequencing Problems with Uncertain Job Times," *Journal of the Operations Research Society of Japan,* Vol. 9, pp. 93–97.

C.S. Tang (1990) "Scheduling Batches on Parallel Machines with Major and Minor Setups," *European Journal of Operational Research,* Vol. 46, pp. 28–37.

T. Tautenhahn and G.J. Woeginger (1997) "Minimizing the Total Completion Time in a Unit Time Open Shop with Release Times," *Operations Research Letters,* Vol. 20, pp. 207–212.

V.G. Timkovsky (1998) "Is a Unit-Time Job Shop not Easier than Identical Parallel Machines?," *Discrete Applied Mathematics,* Vol. 85, pp. 149–162.

V.G. Timkovsky (2000) "Reducibility among Scheduling Classes," Technical Report, Star Data Systems Inc., Toronto, Canada.

R. Uzsoy (1993) "Decomposition Methods for Scheduling Complex Dynamic Job Shops," in *Proceedings of the 1993 NSF Design and Manufacturing Systems Conference,* Vol. 2, pp. 1253–1258, Society of Manufacturing Engineers, Dearborn, Michigan.

R. Uzsoy, C.-Y. Lee, and L.A. Martin-Vega (1992a) "Scheduling Semiconductor Test Operations: Minimizing Maximum Lateness and Number of Tardy Jobs on a Single Machine," *Naval Research Logistics,* Vol. 39, pp. 369–388.

R. Uzsoy, C.-Y. Lee, and L.A. Martin-Vega (1992b) "A Review of Production Planning and Scheduling Models in the Semiconductor Industry, Part I: System Characteristics, Performance Evaluation and Production Planning," *IIE Transactions,* Vol. 24, pp. 47–61.

G. Vairaktarakis and S. Sahni (1995) "Dual Criteria Preemptive Open-Shop Problems with Minimum Makespan," *Naval Research Logistics,* Vol. 42, pp. 103–122.

J.M. Van den Akker, J.A. Hoogeveen, and S.L. Van de Velde (1999) "Parallel Machine Scheduling by Column Generation," *Operations Research,* Vol. 47, pp. 862–872.

L. Van der Heyden (1981) "Scheduling Jobs with Exponential Processing and Arrival Times on Identical Processors so as to Minimize the Expected Makespan," *Mathematics of Operations Research,* Vol. 6, pp. 305–312.

S.L. Van de Velde (1991) *Machine Scheduling and Lagrangean Relaxation,* Ph.D. thesis, Eindhoven University of Technology, Eindhoven, the Netherlands.

H. Van Dyke Parunak (1991) "Characterizing the Manufacturing Scheduling Problem," *Journal of Manufacturing Systems,* Vol. 10, pp. 241–259.

P.J.M. Van Laarhoven, E.H.L. Aarts, and J.K. Lenstra (1992) "Job Shop Scheduling by Simulated Annealing," *Operations Research,* Vol. 40, pp. 113–125.

L.N. Van Wassenhove and F. Gelders (1980) "Solving a Bicriterion Scheduling Problem," *European Journal of Operational Research,* Vol. 4, pp. 42–48.

A. Vepsalainen and T.E. Morton (1987) "Priority Rules and Lead Time Estimation for Job Shop Scheduling with Weighted Tardiness Costs," *Management Science,* Vol. 33, pp. 1036–1047.

H.M. Wagner (1959) "An Integer Programming Model for Machine Scheduling," *Naval Research Logistics Quarterly,* Vol. 6, pp. 131–140.

R.R. Weber (1982a) "Scheduling Jobs with Stochastic Processing Requirements on Parallel Machines to Minimize Makespan or Flow Time," *Journal of Applied Probability,* Vol. 19, pp. 167–182.

R.R. Weber (1982b) "Scheduling Stochastic Jobs on Parallel Machines to Minimize Makespan or Flow Time," *Applied Probability—Computer Science: The Interface,* R. Disney and T. Ott (eds.), pp. 327–337, Birkhauser, Boston.

R.R. Weber (1992) "On the Gittins Index for Multi-Armed Bandits," *Annals of Applied Probability,* Vol. 2, pp. 1024–1033.

R.R. Weber, P. Varaiya, and J. Walrand (1986) "Scheduling Jobs with Stochastically Ordered Processing Times on Parallel Machines to Minimize Expected Flow Time," *Journal of Applied Probability,* Vol. 23, pp. 841–847.

S. Webster (2000) "Frameworks for Adaptable Scheduling Algorithms," *Journal of Scheduling,* Vol. 3, pp. 21–50.

L.M. Wein (1988) "Scheduling Semi-Conductor Wafer Fabrication," *IEEE Transactions on Semiconductor Manufacturing,* Vol. 1, pp. 115–129.

L.M. Wein and P.B. Chevelier (1992) "A Broader View of the Job-Shop Scheduling Problem," *Management Science,* Vol. 38, pp. 1018–1033.

G. Weiss (1982) "Multiserver Stochastic Scheduling," in *Deterministic and Stochastic Scheduling,* Dempster, Lenstra, and Rinnooy Kan (eds.), D. Reidel, Dordrecht, pp. 157–179.

G. Weiss (1990) "Approximation Results in Parallel Machines Stochastic Scheduling," *Annals of Operations Research,* Vol. 26, pp. 195–242.

G. Weiss and M.L. Pinedo (1980) "Scheduling Tasks with Exponential Processing Times to Minimize Various Cost Functions," *Journal of Applied Probability,* Vol. 17, pp. 187–202.

M.P. Wellman, W.E. Walsh, P. Wurman, and J.K. MacKie-Mason (2001) "Auction Protocols for Decentralized Scheduling," *Games and Economic Behavior*, Vol. 35, pp. 271–303.

P. Whittle (1980) "Multi-Armed Bandits and the Gittins Index," *Journal of the Royal Statistical Society Series B,* Vol. 42, pp. 143–149.

P. Whittle (1981) "Arm Acquiring Bandits," *Annals of Probability,* Vol. 9, pp. 284–292.

M. Widmer and A. Hertz (1989) "A New Heuristic Method for the Flow Shop Sequencing Heuristic," *European Journal of Operational Research,* Vol. 41, 186–193.

V.C.S. Wiers (1997) *Human Computer Interaction in Production Scheduling: Analysis and Design of Decision Support Systems in Production Scheduling Tasks,* Ph.D. Thesis, Eindhoven University of Technology, Eindhoven, the Netherlands.

D.A. Wismer (1972) "Solution of Flowshop Scheduling Problem with No Intermediate Queues," *Operations Research,* Vol. 20, pp. 689–697.

R.J. Wittrock (1985) "Scheduling Algorithms for Flexible Flow Lines," *IBM Journal of Research and Development,* Vol. 29, pp. 401–412.

R.J. Wittrock (1988) "An Adaptable Scheduling Algorithm for Flexible Flow Lines," *Operations Research,* Vol. 36, pp. 445–453.

R.J. Wittrock (1990) "Scheduling Parallel Machines with Major and Minor Setup Times," *International Journal of Flexible Manufacturing Systems,* Vol. 2, pp. 329–341.

I.W. Woerner and E. Biefeld (1993) "HYPERTEXT-BASED Design of a User Interface for Scheduling," in *Proceedings of the AIAA Computing in Aerospace 9,* San Diego, California.

R.W. Wolff (1970) "Work-Conserving Priorities," *Journal of Applied Probability,* Vol. 7, pp. 327–337.

R.W. Wolff (1989) *Stochastic Modeling and the Theory of Queues,* Prentice-Hall, Englewood Cliffs, New Jersey.

L.A. Wolsey (1998) *Integer Programming,* John Wiley, New York.

S.D. Wu, E.S. Byeon, and R.H. Storer (1999) "A Graph-Theoretic Decomposition of Job Shop Scheduling Problems to Achieve Schedule Robustness," *Operations Research,* Vol. 47, pp. 113–124.

S.D. Wu, R.H. Storer, and P.C. Chang (1991) "A Rescheduling Procedure for Manufacturing Systems under Random Disruptions," in *New Directions for Operations Research in Manufacturing,* T. Gulledge and A. Jones (eds.), Springer Verlag, Berlin.

S.H. Xu (1991a) "Minimizing Expected Makespans of Multi-Priority Classes of Jobs on Uniform Processors," *Operations Research Letters,* Vol. 10, pp. 273–280.

S.H. Xu (1991b) "Stochastically Minimizing Total Delay of Jobs Subject to Random Deadlines," *Probability in the Engineering and Informational Sciences,* Vol. 5, pp. 333–348.

S.H. Xu, P.B. Mirchandani, S.P. Kumar, and R.R. Weber (1990) "Stochastic Dispatching of Multi-Priority Jobs to Heterogenous Processors," *Journal of Applied Probability,* Vol. 28, pp. 852–861.

Y. Yang, S. Kreipl, and M. Pinedo (2000) "Heuristics for Minimizing Total Weighted Tardiness in Flexible Flow Shops," *Journal of Scheduling,* Vol. 3, pp. 89–108.

C.A. Yano and A. Bolat (1989) "Survey, Development and Applications of Algorithms for Sequencing Paced Assembly Lines," *Journal of Manufacturing and Operations Management,* Vol. 2, pp. 172–198.

P.C. Yen (1995) *On the Architecture of an Object-Oriented Scheduling System,* Ph.D. thesis, Department of Industrial Engineering and Operations Research, Columbia University, New York.

P.C. Yen (1997) "Interactive Scheduling Agents on the Internet," in *Proceedings of the Hawaii International Conference on System Science (HICSS-30),* Hawaii.

P.C. Yen and M.L. Pinedo (1994) "Scheduling Systems: A Survey," Technical Report, Department of Industrial Engineering and Operations Research, Columbia University, New York.

Y. Yih (1990) "Trace-driven Knowledge Acquisition (TDKA) for Rule-Based Real Time Scheduling Systems," *Journal of Intelligent Manufacturing,* Vol. 1, pp. 217–230.

E. Yourdon (1994) "Object-Oriented Design: An Integrated Approach," Prentice-Hall, Englewood Cliffs, New Jersey.

M. Zweben and M. Fox (eds.) (1994) *Intelligent Scheduling,* Morgan and Kaufmann, San Mateo, California.

Name Index

Aarts, E.H.L., 358, 390
Achugbue, J.O., 208
Adams, J., 184
Adelsberger, H.H., 442, 553
Adiri, I., 331
Adler, L., 418
Agrawala, A.K., 307
Ahn, S., 127
Akkiraju, R., 467, 501
Akturk, M.S., 418
Alon, N.,92
Applegate, D., 184
Arguello, M., 553
Atabakhsh, H., 9, 443
Aytug, H., 466
Azar, Y., 92

Bagga, P.C., 331
Baker, K.R., 8, 9, 32, 92, 155
Balas, E., 184, 185
Baptiste, P., 390, 523, 546
Barker, J.R., 184
Barlow, R.E., 229
Barnhart, C., 523
Baumgartner, K.M., 418
Bayiz, M., 466

Bean, J., 358, 418
Becker, M.A., 510
Bell, G., 501
Bertsekas, D., 531
Bhaskaran, K., 358
Bhattacharyya, S., 466
Bianco, L., 92
Biefeld, E., 443
Birge, J., 257, 418
Bishop, A.B., 127
Bitran, G., 418
Blazewicz, J., 8, 9, 523
Bolat, A., 418
Booch, G., 467
Boxma, O.J., 332
Braun, H., 501
Brazile, R.P., 418
Brown, A., 184,
Brown, D.E., 9,
Brown, M., 229, 257
Browne, S., 257
Brucker, P., 9, 61, 184, 208, 546
Bruno, J., 128, 306
Bulfin, R.L., 92, 208
Burns, L., 418
Buxey, G., 418

Subject Index